THE ENCHANTMENT OF URANIA

25 CENTURIES OF EXPLORATION OF THE SKY

THE ENCHANTMENT OF URANIA

25 CENTURIES OF EXPLORATION OF THE SKY

Massimo Capaccioli

University of Naples Federico II, Italy

World Scientific

NEW JERSEY · LONDON · SINGAPORE · BEIJING · SHANGHAI · HONG KONG · TAIPEI · CHENNAI · TOKYO

Published by

World Scientific Publishing Co. Pte. Ltd.

5 Toh Tuck Link, Singapore 596224

USA office: 27 Warren Street, Suite 401-402, Hackensack, NJ 07601

UK office: 57 Shelton Street, Covent Garden, London WC2H 9HE

British Library Cataloguing-in-Publication Data
A catalogue record for this book is available from the British Library.

THE ENCHANTMENT OF URANIA
25 Centuries of Exploration of the Sky

ISBN 978-981-12-4777-4 (hardcover)
ISBN 978-981-12-4927-3 (paperback)
ISBN 978-981-12-4778-1 (ebook for institutions)
ISBN 978-981-12-4779-8 (ebook for individuals)

For any available supplementary material, please visit
https://www.worldscientific.com/worldscibooks/10.1142/12572#t=suppl

Desk Editor: Rhaimie Wahap

Typeset by Stallion Press
Email: enquiries@stallionpress.com

Printed in Singapore

To the memory of my mother Lilian,
in whose eyes I first saw the sky

Contents

Preface

The best prophet of the future is the past.
George Byron, *Journal*

Science is not only a disciple of reason,
but, also, one of romance and passion.
Stephen Hawking,
interview with PARADE Magazine

Let me begin with an *excusatio non petita*, an unsolicited justification. I am not a historian, not even of astronomy alone. I might have tried to become one, as some of my friends know, but in the end, many years ago now, I chose something else, preferring Urania to Clio. More than a systematic study, this book is the result of readings and personal reflections, mediated by my experience as a scientist and teacher, and by direct knowledge of some facts and figures from the last half century. I am sure that my more experienced colleagues in one or another of the many fields covered in this book, not to mention true historians, will detect here and there some gaps, imperfections, inaccuracies, oversights, and even mistakes. I will come to terms with it if, as I hope, the overall sense of the story I was trying to build does not suffer too much.

Indeed, the purpose of this book is to trace, against the background of human history, the long and often tortuous path that has led us to our present knowledge of the cosmos and its phenomena: a pilgrimage of consciousness and awareness, a metaphor for the growth of human society, consumed by a thousand moral and material pitfalls, enlightened by genius and sacrifice, and humbled by fear, selfishness, and superstition. Every great civilization that has flourished on Earth has developed its own astronomy and cosmology; and the various streams, fed by the same impulses but governed by different methods of investigation, have not always crossed and flowed into a single river. Thus, a fresco that claims to be complete should narrate and compare all these lines of knowledge, relating them to each other and to the intertwining of history. But this would require knowledge and skills that I do not possess. For this reason, I have limited myself to wandering through the science of the Western world. Blossoming from the roots of the classical world, lovingly protected by the Arabs, and miraculously rediscovered in Europe after the luminous Dark Ages, it has undoubtedly produced extraordinary results, as witnessed by the amazing technological applications of our time; and the so-called Galilean method has established itself as a rule in every corner of the globe. Since I cannot pretend that this is the only or even the best way, I have taken it as a postulate.

The cost of this program is an unavoidable slalom between people, places, successes and failures, epochs of human history, and technological breakthroughs; a going back and forth that is consequent to maintaining time as the sole coordinate of the narrative, while at the same time preserving a minimum of logical internal coherence in the exposition of topics. Was I successful in my intent? In any case, I tried.

Finally, I would like to ask for the understanding of those who may find some of the explanations of concepts and more or less lexical and technical terms rather elementary. I felt it was more comfortable to skip the reading of what is known than to try to find the meaning of what is not known, although I am aware that today's tools can make the search relatively easy. I have also left many of the Latin locutions already present in the Italian edition of this book, putting the English translation alongside to encourage those who are not too familiar with a language now in disuse.

The book was originally written in Italian and published in 2020 by Carocci Editore, Rome. The English translation of the Italian text and of some quotations from languages other than English is my responsibility; but I must acknowledge some outside help in cleaning up my Italianisms, removing typos, and homogenizing the style. I also took the opportunity to make a few updates to the Italian text, correcting some mistakes discovered during the many re-readings. No surprise: learning in this long story is a never-ending process.

Naples, May 30, 2023

Chapter 1

Introduction:
The first time

> The natural desire of good men is knowledge
> Leonardo da Vinci, *Codex Atlanticus*

> So too, as a child, I believed that
> there was nothing beyond Vesuvius,
> since I couldn't see anything beyond it
> Giordano Bruno,
> *De Immenso et Innumerabilis*

We do not know precisely how, when, and where Galileo Galilei first learned of the invention of the "Dutch cylinder".[1] In the absence of documentary evidence, we can only piece together the clues and try to figure it out. It probably happened in Venice at the early summer of 1609, within the walls of the convent annexed to the Gothic church of Santa Fosca, in the popular and populous Cannaregio district, the northernmost of the six historic *sestieri* of the city of canals. The place, which was not far from the ghetto where the *"Concilio dei Pregadi"*[2] had relegated Jews about a century earlier,[3] was easy to reach by boat for those arriving at the Lagoon city from the mainland. The then forty-five-year-old professor of the Padua

---

[1] The literature on Galilei is boundless. See, for example, L. Geymonat, *Galileo Galilei*, Einaudi, Turin 1997 (in Italian); A. Naess, *Galileo Galilei: When the World Stood Still*, Springer, New York 2010; S. Drake, *Galileo: Pioneer Scientist*, University of Toronto Press, Toronto 1990; P. Greco, *Galileo, l'artista toscano*, Springer, Milan 2014 (in Italian). Specifically on the "Dutch cylinder": E. Rosen, *Galileo and the Telescope*, in "The Scientific Monthly", 72, 3, 1951, pp. 180–182.

[2] The Venetian Senate, main deliberative and legislative body of *La Serenessima* Republic.

[3] D. Calabi, *Venice and its jews. 500 years since the founding of the ghetto*, Officina Libraria, Rome 2017; R. Calimani, *Storia del ghetto di Venezia*, Rusconi, Milan 1985 (in Italian).

1

Studium (Latin names of universities in the Middle Ages) went there from time to time to meet his friend Paolo Sarpi, who had been severely disabled after a vicious attack. Two years earlier, the friar of the mendicant order of the *Servorum Beatae Mariae Virginis* (Servants of Mary) had been seriously wounded in the face by the dagger of some assassins in retaliation for his critical attitude towards the Roman Curia and for having taken the side of Venice in the dispute with Pope Paul V Borghese for civil liberties.[4] The ambush of the peaceful man of God brought a bloody end to a hitherto bloodless tug-of-war between the *Serenissima* (Most Serene) Republic and the temporal power of the Church, punctuated by the sudden excommunication of Doge Leonardo Donà and the whole Senate and by an interdict beautifully ignored by Venetians.[5]

Friendship with Sarpi[6] was hence a potentially dangerous connection even within the boundaries of a state that was still proud of its independence, but less and less capable of keeping the pontiff's claws at bay. Galilei could not ignore the risks that Giordano Bruno's sentence to the stake, just nine years earlier, had made crystal clear. The former Dominican had sought in vain for protection under the wings of the lion of St. Mark. Reported by the same person who had invited him to Venice and was hosting him in his own house with the purpose of carping the secrets of mnemonics (the art of memory), Bruno was handed over to the emissaries of Inquisition and translated to Rome before the Holy Office.[7] There, he was tried for a crime of opinion, sentenced to death, and burned alive at dawn of February 17, 1600, with his mouth gagged.[8] It was a very sad

---

[4]Id., *Storia della Repubblica di Venezia*, Mondadori, Milan 2019 (in Italian); J.J. Norwich, *A history of Venice*, Vintage Books, New York 1982.

[5]It is said that, one day, Cardinal Camillo Borghese, scion of a noble and wealthy Sienese family, complained to the Venetian Leonardo Donà about the arrogance of his fellow citizens toward the papacy, and concluded by saying that, if he were pope, he would excommunicate the Venetians. "And I, if I were doge, would laugh at the excommunication", was Donà's response. The Borghese became Paul V and the Venetian fellow became doge.

[6]E. Reeves, *Kingdoms of Heaven: Galileo and Sarpi on the Celestial*, in "Representations", 105, 1, 2009, pp. 61–84.

[7]Created in 1542 by Pope Paulus III Farnese as a key instrument of Counter Reformation to fight the heresies, the "Supreme Sacred Congregation of the Roman and Universal Inquisition" had the mission «of maintaining and defending the integrity of the faith and of examining and proscribing errors and false doctrines».

[8]On December 21, 1599, at the request of Cardinal Robert Bellarmine, Bruno was visited in prison to verify if he was willing to abjure some of his propositions declared to be heretical. A record of the Holy Inquisition reports his answer: «*Dixit quod non debet nec vult rescipiscere, et non habet quid recsipiscat nec habet materiam rescipiscendi, et nescit super quod debet rescipisci*» (He said that he neither wants nor should repent,

day for the whole of humanity, but also a clear, unambiguous signal to dissidents.

With little concern for the impending dangers, Galilei had continued to cultivate a long-standing relationship with *"Fra Paolo"*, whom he admired for his profound doctrine and unparalleled freedom of thought. This somewhat too casual attitude, together with the incautious expression of some scientific opinions and the astrological activities which he practiced with a venality imposed on him by heavy family burdens, had already attracted the interest of the officers of the Inquisition, the not-so-secret police of the Pope, as when he assigned the place of the new star that appeared in 1604 *in pede Serpentarii* (in the foot of the Serpent Handler), then named "Kepler nova".[9] Noting the absence of diurnal parallax,[10] Galilei had concluded that, whatever it was, this inconstant celestial flame should lie beyond the orbit of the Moon. This observation[11] undermined the idea of perfection

---

and has nothing to repent, nor has matters of repentance, and does not know what he should repent of). A month later, on January 20, 1600, Pope Clement VIII Aldobrandini ordered the sentence of condemnation, which was read to the Nolan monk on the same day he was led to the stake. «*Maiori forsan cum timore sententiam in me fertis quam ego accipiam*» (Perhaps you tremble more when pronouncing this sentence against me than I do when listening to it) was his brave reaction. In the "Public Notices" to the people of Rome of February 19, 1600, we read: «On Thursday [February 17] was burned alive in Campo di Fiore that friar of St. Dominic, of Nola, a pertinacious heretic, with his tongue "in yoke" [immobilized by a hook], for the very ugly words he spoke, not wanting to listen to either the comforters or the others». G. Maifreda, *Io dirò la verità. Il processo di Giordano Bruno*, Laterza, Rome-Bari 2020 (in Italian).

[9] F.R. Stephenson, D.H. Clark, *Historical Supernovas*, in "Scientific American", 234, 6, 1976, pp. 100–108.

[10] The term parallax, from the Greek παράλλαχις (change of direction), indicates the angular displacement of a target with respect to a background (possibly to infinity) as the point of view changes. To visualize the phenomenon, just look at a nearby object and alternately close one of your eyes; the object will be seen jumping against of the horizon, with a greater amplitude the closer it is. The parallax angle, typically very small in astronomy, is what an assigned base subtends when viewed perpendicularly. Astronomers call "diurnal" the parallax referred to the average radius of the Earth (6,373 km), "annual" the one related to the radius of the Earth's orbit, that is, to the Astronomical Unit (1 AU = 149.6 million km).

[11] «I have postponed this publication and still will postpone it for a few days, since writing what has been the main objective of my lectures, that is, to prove that the new star lies far beyond the lunar orbit, is in itself very easy [...] It is a subject I had to deal with because of the young students and the multitude of people who wanted to understand the geometrical demonstrations, despite the fact that they were trite and trite, and also trivial exercises in astronomy. In spite of the risk of censorship, I decided to state my thoughts [...] and I believe I reached a conclusion free of contradiction. Therefore I had to go at an extremely slow pace». Letter from Galilei to Onofrio Castelli, Padua, January 13, 1605. Quoted in Italian by A. De Angelis, *Galileo and the Supernova of 1604*, Castelvecchi, Rome 2022.

commonly attributed to the sidereal world and therefore could only displease the Aristotelians, and hence the Church.

Friendship with Sarpi certainly entailed some risk, but also some significant benefits, as the elderly friar was usefully contiguous to governmental power of the Republic. He was also kept constantly informed of what happened around the world through a web of correspondents scattered throughout Europe, with whom he exchanged all kinds of news: a sort of low-tech (and low-speed) social network. It was in this way that Sarpi had come to know of a magic tube, a «*vetro prospettico*» (perspective glass) which made things seem three times closer than they really were.[12] There were rumors that Hans Lippershey, an obscure German glasses-maker who worked in Middelburg, a city in South Holland renowned for its glassworks, had come up with this devilry. Actually, the craftsman had even presented an application to the *Staten-Generaal* of the Hague for a patent.

Not surprising! The idea of entrusting to the State the protection of intellectual property has much older origins.[13] There are traces of it as early as classical Greece. In the modern age, the institute of patent reappeared in 1421, when the Municipality of Florence granted the architect Filippo Brunelleschi a three-year right on a machine to handle the marble used for the great dome of the Cathedral of *Santa Maria del Fiore*. Activated occasionally in different corners of Europe and finally regulated in Venice in the 16th century, then as now, patents are only meaningful and effective for inventions that lent themselves to protection. On the basis of this argument, Lippershey's claim was rejected by the Dutch government. The object for which the license was requested appeared to be of interest, but, he was told, too easy to copy. It would have been enough to look at it closely to quickly understand how to replicate it. Thus the unfortunate craftsman had been dismissed with a meager gratuity. In any case, the news of the invention caused quite a stir — as well as sparking a legal fight among a handful of Dutchmen who claimed paternity of the invention — and reached Sarpi in Venice through a dispatch from the apostolic nuncio to Flanders; the ambassadors of the sovereigns, both secular and religious, also had their own internal networks, for obvious commercial and intelligence reasons.

---

[12]P. Moore, *Eyes on the Universe: The Story of the Telescope*, Springer, London 1997.
[13]P.O. Long, *"Intellectual Property", and the Origin of Patents: Notes toward a Conceptual History,* in "Technology and Culture", 32, 4, Special Issue: *Patents and Invention*, 1991, pp. 846–884.

Intrigued by this news, Sarpi asked for enlightenment from an acquaintance in France, Giacomo Badoer, a nobleman of Venetian origin who had been a student of Galilei in Padua some ten years earlier. Badoer quickly confirmed the rumor and reported the presence in Paris of some specimens of the "cannon-spectacles" for sale in a store on the banks of the Seine: small and coarse instruments, little better than toys.[14] Galilei learned of Badoer's reply during a visit to his friend the friar. Realizing the practical importance of this amazing application of the properties of «*refrazione*» (refraction), he decided on the spot to reap the benefits. There was no time to lose, especially since — he would later write — a Flemish merchant was already roaming the territory of *La Serenissima* seeking a meeting with the doge with the purpose of placing his instrument.

We can imagine the haste with which Galilei left the convent of Santa Fosca to hurry to the small boat that would take him back to Fusina, on the mainland: a flat-bottomed "*sandolino*" suitable for the shallow waters of the Lagoon and the canals between sandbanks, but wobbly in the wind-curled sea. By then the Tuscan was moving well in Venice, among the splendors of the palaces and the countless churches, the smell of salt air and the cries of a picturesque and multi-ethnic people, the colors of the sky hovering between the smoky walls of the *calli*, and the narrow passages where lovers for hire could lurk, but so could flying daggers and the cold eyes of the secret police. Galilei returned there whenever he could. On his first visit, in September 1592, he had stayed in Venice for almost a month, the guest of another Tuscan fellow. At the time, he was a brilliant and penniless young man looking for a job, eager to be appreciated by the Lagoon high society for his talent in the liberal arts of the *trivium* and the *quadrivium*: grammar, rhetoric, and dialectics, arithmetic, geometry, and astronomy, which he had studied in Pisa and cultivated in Florence, and music, which his father Vincenzo had taught him. In the years Galileo spent at the University of Padua, he had collected many cultured and influential friends, and through them had met interesting characters passing through Venice. He was also granted access to the off-limit warehouses of the Arsenal, where he enjoyed spending his days observing the work of naval engineers.

Who knows what he was thinking about as he returned to Padua along the *Riviera del Brenta*, with the river banks full of «palaces and

---

[14]E. Reeves, *Galileo's Glassworks: The Telescope and the Mirror*, Harvard University Press, Cambridge (MA) 2008.

the delightful mansions of the nobles, and more opulent citizens»[15] and then on the Piovego canal up to the Portello pier, not far from his house. Perhaps he was foretasting the success he had vainly pursued during his 17 years of service in a Studium where he was poorly paid and forced to teach Aristotelian physics in spite of a claimed *"patavina libertas"*.[16] And perhaps his thoughts went for a moment to his father, a talented musician whose untimely death had forced him, as the firstborn of numerous off-spring, to take care of the family, albeit unwillingly. Vincenzo, who knew well the practical difficulties of an intellectual's life spent rowing against the tide, would have liked to make his son a physician as that ancestor from whom the surname Galilei came from,[17] and for this reason he had sent him to study in Pisa, also to remove him from the influence of his first educators, the Benedictines. The mother, Giulia Venturi degli Ammannati, born into an artisan family with an illustrious past, had instead worried, and continued to worry even then[18] that her son was adequately fearing God. Galileo had disappointed them both, leaving the university without graduating and leading, in Florence, and then in Padua, a life that Giulia, an authoritarian and bigoted woman, considered almost dissolute. Neither parent had been able to grasp that heroic fury burning in their son's chest.[19]

---

[15]V.M. Coronelli, *Viaggio dall'Italia all'Inghilterra*, Gio. Battista Tramontino, Venice, 1697 (in Italian); cited in http://tesi.cab.unipd.it/57146/1/Irene_Reffo_2017.pdf.

[16]The motto of Padua University, founded in 1222, is *"Universa Universis Patavina Libertas"*, meaning "all and for everyone the freedom at Padua".

[17]The Florentine Galileo Bonaiuti (c. 1370–c. 1450), a well-known doctor. With him the family lost the original surname Bonaiuti to acquire that of de Galileis, which might appear to be of Jewish origin. But Galilei himself excludes this hypothesis, as we read in his *Contro il portar la toga* (vv. 148–153):

> *Ma ch'io sia per voler portar la toga,*
> *Come s'io fussi qualche Fariseo,*
> *O qualche scriba o archisinagoga,*
> *Non lo pensar; ch'io non son mica Ebreo,*
> *Se bene e' pare al nome e al casato*
> *Ch'io sia disceso da qualche Giudeo.*

(And pray don't think I'll ever don a gown / As if I were a Pharisaic professor: / I couldn't be convinced, not for a golden crown. / Not Pharisee nor Jew I count a predecessor: / In spite of both my name and my descendence / I'm not a Jew, nor will be my successor); translation by G.F. Bignami, *Against the donning of the gown*, Moon-Books, London 2005.

[18]While Vincenzo died in 1591, Giulia survived up to 1620.

[19]C.S. Slichter, *Galileo*, in "American Scientist", 31, 2, 1943, pp. 168–176.

Back home, that same evening Galilei put all his genius into figuring out the principles of the «*occhiali in canna*» (glasses in barrels) and producing a working sample. He was not entirely unfamiliar with optics and knew the ancient problem of bringing distant things closer through more or less prodigious mirrors. Shortly after his enrollment at the University of Padua, he had personally met Giovanni Battista Della Porta,[20] a Neapolitan of multifaceted talent who, in a page of his monumental *Magiae naturalis sive de miraculis rerum naturalium* (Natural magic or about the miracles of natural things), had hinted at intriguing combinations of lenses capable of magnifying and sharpening objects both far away and very close, but never really playing with them. At a practical level, Galilei could take advantage of the excellent glass blanks[21] produced by the Murano foundries and, to turn them into "*lenticchie*" (lentils), of the technological capabilities provided by his workshop of scientific instruments. In fact, in order to supplement his meager teacher's salary with some extra income, about ten years earlier he had established, in the basement of his house, a laboratory for the production of precision equipments and innovative machinery that he designed and then built with the help of a skilled assistant.[22] The desire to live well and to help his brethren required more money than his university was willing to pay him for his services.

In a single night — as he would later write in the first lines of an account suggestively titled *Sidereus Nuncius* (Starry Messenger) — he managed to solve the problem. By combining simple eyeglass lenses and perhaps following an approach we would now call reverse engineering, he succeeded in fabricating an instrument whose performance was comparable to those of the Flemish. It was only a beginning, but a very promising one and not too difficult if, as we believe,[23] he had been even able to get his hands on a specimen of a Dutch telescope. Understandably satisfied, he hastened to communicate the news of his success to Sarpi, asking him to do his best to prevent the foreign merchant, who was trying to place his instrument, from

---

[20] S. Kodera, *Giambattista della Porta*, in *The Stanford Encyclopedia of Philosophy*, E.N. Zalta (ed.), Stanford University, Stanfort (CA) 2015.

[21] A blank is a piece of raw material suitable for machining into an optical element of which it usually already has the shape.

[22] R.J. Seeger, *Men of Physics: Galileo Galilei, his Life and his Works*, Pergamon Press, Oxford 1966.

[23] Z. Shore, *A Sense of the Enemy: The High-Stakes History of Reading Your Rival's Mind*, Oxford University Press, Oxford 2014.

reaching the doge's ears before him.[24] He still needed time to perfect his prototype and make it better than those of his competitors. The recommendation worked well, as Sarpi was highly regarded within the Venetian government, and also widely recognized for his skills in mathematics and natural philosophy. With the intent to please him, the hapless Dutchman was put on hold in the doge's antechamber until he took the hint and decided to leave with his tail between his legs.

While freed from the pressure of competition, Galilei wasted no time and, in just one month, prepared a new model of what would now be known as the telescope. The instrument produced an image of much higher quality than the Flemish one, with almost three times the magnification. It took the form of a tin cylinder wrapped in a rough sailcloth and just under a meter long, connecting a flat-convex lens the size of a «*scudo d'argento*» (silver shield) to a plane-concave eyepiece, both grounded and polished in his workshop using the finest Murano glass.[25] It was a high-tech late Renaissance jewel made in a record time, more by intuition than by in-depth knowledge of the principles of optics, even if later, at the height of his fame, Galilei tried to credit a different, much more convenient self-image by flaunting an insight that perhaps he did not possess.[26]

The summer of 1609 was winding down, but the improvements of the telescope were far from completed. However, any further delay could prove fatal. Galilei knew that he had to strike while the iron was hot. Thanks to Sarpi's good offices, shortly after mid-August he was able to present his instrument to a delegation of senators led by the doge himself, Leonardo Donà, a serious and rigorous man, not easy to charm.[27] As ambassador to Spain, he had succeeded in convincing Emperor Charles V to arm the fleet that in 1571 had routed the Ottomans at Lepanto, thereby relieving the pressure of Islamic influences on Europe; and as the head of *La Serenissima*

---

[24] E. Rosen, *Did Galileo Claim He Invented the Telescope?*, in "Proceedings of the American Philosophical Society", 98, 5, 1954, pp. 304–312.

[25] V. Ronchi, *Galileo e il suo telescopio*, Boringhieri, Turin 1964 (in Italian).

[26] From a letter of Galilei to Leonardo Donato, dated Padua, August 24, 1609 (*Dal Carteggio e dai Documenti, Pagine di Vita di Galileo per cura di Isidoro Del Lungo ed Antonio Favero*, G.C. Sansoni, Firenze 1915, p. 66): «*un nuovo artifizio di un occhiale cavato dalle più recondite speculazioni di prospettiva, il quale conduce gl'oggetti visibili così vicini all'occhio, et così grandi et distinti gli rappresenta, che quello che è distante, v.g., nove miglia, ci apparisce come se fusse lontano un miglio solo*» (a new artifice of a spectacle derived from the most recondite speculations of perspective, which brings visible objects so near to the eye, and represents them so large and distinct, that what is distant, e.g., nine miles, appears to us as if it were only one mile away).

[27] Although a layman, Donà had taken a vow of chastity so that he could devote himself fully to public affairs.

he had opposed the tough Pope Paul V. Since the surprise effect had now worn off due to word of mouth about the ultramontane news, in order to impress the notable audience and his employers, Galilei decided to adopt a marketing strategy focused on the characteristics and performance of the product rather than on its novelty. It was not merely a sophisticated toy such as the devices one could find in Holland and France, but rather a tool capable of amazing applications. Above all, he insisted on the tactical advantage that early sighting of enemies at sea could offer the galleys of *La Serenissima* (something comparable, *mutatis mutandis,*[28] to radar in World War II). The argument, presented with the skill of a son of the Tuscan Renaissance, made inroads into the hearts of men deeply linked to the sea, but it was not enough to overcome the shrewdness that had made them the most savvy merchants in the world. As a good professor would, Galilei prepared a practical demonstration of the telescope from the top of the bell tower soaring above San Marco square. He pointed the instrument at the Malamocco inlet and proved that it was able to spot vessels in transit long before his prudent and parsimonious judges had even been able to see them with their own eyes.[29] He did not achieve the resounding success he had hoped for, but he fared much better than Lippershey did. For the

---

[28] Medieval Latin saying meaning "with the necessary changes".

[29] Antonio Priuli, Procurator of the San Marco Bell Tower and future doge, so described his direct experience (*Dal Carteggio e dai Documenti, Pagine di Vita di Galileo per cura di Isidoro Del Lungo ed Antonio Favero*, cit., p. 65):

«*21 Agosto (1609). Andai io* [...] *con l'Ecc.te Gallileo* [sic!] *a veder le meraviglie et effetti singolari del cannon di detto Gallileo, che era di banda* [...] *con due veri, un* [...] *cavo, l'altro no, per parte; con il quale, posto a un ochio e serando l'altro, ciasched'uno di noi vide distintamente, oltre Liza Fusina e Marghera, anco Chioza, Treviso et sino Conegliano* [∼50 km], *et il campaniel et cubbe [cupole] con la facciata della chiesa de Santa Giustina de Padoa: si discernivano quelli che entravano et uscivano di chiesa di San Giacomo di Muran; si vedevano le persone a montar et dismontar le gondola al traghetto alla Colonna nel principio del Rio de' Verieri, con molti altri particolari nella laguna et nella città veramente amirabili. E poi da lui presentato in Collegio li 24 del medesimo, moltiplicando la vista con quello 9 volte più.*»

(August 21 (1609). I went [...] with Ex.nt Gallileo [sic!] [...] to see the marvels and the singular effects of the cannon of the said Gallileo, which was made of tin [...] with two glasses on each side, one [...] hollow, the other not; with which, placed at one eye and closing the other, each of us saw distinctly, besides Liza Fusina and Marghera, also Chioggia, Treviso and as far as Conegliano [∼50 km], and the bell tower and the domes with the facade of the Church of Santa Giustina in Padua: one discerned those who entered and left the church of San Giacomo in Murano; one saw people getting on and off the gondola at the Colonna ferry at the beginning of the Rio de' Verieri, with many other details in the lagoon and the city that are truly admirable. And then presented by him in the Senate on 24 of the same month, multiplying the sight with that 9 times more).

transfer of the patent to the Republic, the professor-inventor was rewarded with the life-long reconfirmation of the position of *"lettore di matematiche"* (mathematics lecturer) at the University of Padua and the promise of a pay rise.[30]

But Galilei aspired to something more.[31] Moved by necessity, by the anger he felt at the ungrateful stinginess of the doge's government, and above all by an inexhaustible curiosity, in the following months he continued to experiment with new solutions to further improve the instrument. He studied and rebuilt, tried and tried again. By early autumn, he had refined his lens-manufacturing technique to the point he made a telescope with an aperture of nearly four centimeters and a field of 15 arc minutes,[32] as the apparent radius of the Moon. «Sparing no labor or expense, I progressed so far that I constructed for myself an instrument so excellent that things seen through it appear about a thousand times larger and more than thirty times closer than when observed with the natural faculty only».[33] Finally, after familiarizing himself with his own creation through an astonishing series of laboratory experiments,[34] on a splendid November evening he pointed his telescope towards the sky from a window of his house in Borgo dei Vignali, not far from the Bo Palace where he held his lectures.

From that moment on, nothing was the same in the path of knowledge, in man's consciousness and approach to faith, as well as in the life of Galilei himself, who from a middle-aged teacher with no particular prospects became the very icon of science: a late but bursting flowering, as it had been for Julius Caesar.[35] In 55 clear and sleepless nights, with growing excitement he gathered evidence of the falsity of Aristotelian postulates on the centrality of the Earth and the duality between the terrestrial world, corruptible and imperfect, and a celestial one, freed from time and

---

[30] G. Donà Dalle Rose, *Antipapa veneziano. Vita del doge Leonardo Donà (1536–1612)*, Giunti, Florence 2019 (in Italian).

[31] R.S. Westfall, *Science and Patronage: Galileo and the Telescope*, in "Isis", 76, 1, 1985, pp. 11–30.

[32] The angle of a degree, which is the ninetieth part of the right angle, is in turn subdivided into 60 arc minutes, and each of them in 60 arc seconds.

[33] G. Galilei, *Sidereus Nuncius*, Baglioni, Venice 1610. For an English translation, see: http://people.reed.edu/~wieting/mathematics537/SideriusNuncius.pdf.

[34] H.I. Brown, *Galileo on the Telescope and the Eye*, in "Journal of the History of Ideas", 46, 4, 1985, pp. 487–501.

[35] Julius Caesar was already 51 years old when he crossed the Rubicon river with his legions (no magistrate invested with military offices could cross this border of Roman territory in arms without the authorization of the Senate), starting his dazzling adventure.

its ravages. His discovery of the imperfections of the lunar surface and of the «four wandering stars around Jupiter» strengthened the evidence in favor of heliocentric cosmology, re-proposed seventy years earlier by Nicolaus Copernicus, thereby reassuring those who, while inclined to accept the transfer of the very center of the celestial motions to the Sun, could not understand the fact that the operation did not include the Moon too. He opened a window to the unknown universe of stars and to that Milky Way which, for centuries, had made «very wise men to wonder», as Dante had said in Canto XIV of *Paradise* (vv. 97–99). With the bold account of his discoveries, and while the crackle of the flames of Giordano Bruno's stake still resounded grimly within Catholic countries, he set science on the road to cultural independence from faith.

It was an unparalleled booty, whose significance for knowledge and for man's very identity was immediately perceived by philosophers and theologians. The first to realize this was Galilei himself, who, as a good manager of himself, hastened to recount and comment on his discoveries in the *Sidereus Nuncius*: a dry booklet written in Latin, the *lingua franca* of the learned, to ensure its widest diffusion. It was published in record time to avoid being overtaken at the last minute, and was dedicated to the Grand Duke of Tuscany Cosimo II de' Medici for *captatio benevolentiae* (winning of goodwill). Kepler, the foolish German genius, happily exclaimed: «*Vicisti Galilee*» (Galileo, you have won);[36] and from his harsh Neapolitan prison the rebellious Dominican philosopher Tommaso Campanella wrote with indomitable courage: «You purged the eyes of men, and you show a new sky and a new Earth in the Moon. [...] After your Nuncius, Galileo, all knowledge must be renewed».[37] For their part, the Jesuits were not distracted by the curiosity aroused by the new technology for observing the sky, which they quickly mastered anyway. Ever prudent and very practical, they equipped themselves to neutralize the potentially disruptive impact of the Galilean discoveries on the hold of the Aristotelian-Ptolemaic theories that the Church of Rome had made its own at the time of Thomas Aquinas. The Reformation of Luther and Calvin had already deeply undermined the

---

[36]Cf. L. Fermi, G. Bernardini, *Galileo and the Scientific Revolution*, Dover, New York 2003 (1st ed. 1962). Alluding to the war *contra Aristotelem* that at that point was far more important to him than to the Paduan professor, Kepler had taken up a phrase attributed by Christian tradition to the Emperor Julian the Apostate, who in 363 A.D., mortally wounded, with these words acknowledged the victory of Christ, the Galilean.

[37]T. Campanella, *Apologia per Galileo*, P. Ponzio (ed.), Bompiani, Milan 2001 (in Italian).

authority and temporal power of the pope. It was crucial to avoid any further uncontrollable shocks, and above all, to prevent the secularization of thought. It was for this reason that the professors of the Roman College,[38] who were among the first to recognize the veracity of the telescopic observations,[39] could not make an exception for Galilei, although he had always been on excellent terms with the Society of Jesus to the point of daring to lament (privately) the expulsion of Order from Venice in 1606.

The matter of *Sidereus Nuncius* was so hot[40] that even the ambassador of King James I of England to *La Serenissima* had his say on the Galilei case. On the same day the book appeared, commenting on the sending of a copy to his sovereign, Sir Henry Wotton declared that it was the strangest news that King James had ever received from anywhere in the world and that «the author runneth a fortune to be either exceedingly famous or exceedingly ridiculous»;[41] and a year later the English poet John Donne summarized in some celebrated verses the astonishment felt by a humanity made orphan of ancient certainties and at last set out to confront his own ignorance of natural phenomena, in heaven as on Earth, without the constraints of a superstitious and irrational faith: «And new philosophy calls all in doubt, / The element of fire is quite put out, / The Sun is lost, and th'earth, and no man's wit / Can well direct him where to look for it».[42]

The *Sidereus Nuncius*, 60 pages in-quarter with drawings by the author's hand, printed in 550 copies, was a sensational best-seller for those times, and was literally snapped up — so much so that in the same year a pirated edition was reproduced in Frankfurt. These priceless copies reached every corner of Europe, and the Old Continent was rocked by yet another formidable revolution after the one unleashed by Copernicus seventy years earlier with the heliocentric model of the world.[43] A paradigm reversal that

---

[38] The *Collegium Romanum* (Roman College) was established in 1551 by Ignatius of Loyola just eleven years after he founded the Society of Jesus (Jesuits), to cover the entire educational spectrum, from elementary to university studies.

[39] G.V. Coyne, *The Jesuits and Galileo: Fidelity to Tradition and the Adventure of Discovery*, in "Forum Italicum. A Journal of Italian Studies", 4, 1, 2015, pp. 3–23.

[40] O. Gingerich, *Galileo, the Impact of the Telescope, and the Birth of Modern Astronomy*, in "Proceedings of the American Philosophical Society", 155, 2, 2011, pp. 134–141.

[41] *Henry Watton to the Earl of Salusbury*, March 13, 1610, quoted in M. Nicolson, *The New Astronomy and the English Literary Imagination*, in "Studies in Philology", 32, 1995, p. 440.

[42] For the full poem see: https://www.poetryfoundation.org/poems/44092/an-anatomy-of-the-world.

[43] A. Koyre, *Galileo and the Scientific Revolution of the Seventeenth Century*, in "The Philosophical Review", 52, 4, 1943, pp. 333–348.

this time did not stem from an idea, but from the ingenious use of an optical instrument.

Yet, as we have seen, Galilei was by no means the inventor of the telescope, as the spyglass was later christened by the Greek Giovanni Demisiani, mathematician to Cardinal Ferdinando Gonzaga, in a toast for the Pisan scientist's association with the Lincean Academy.[44] They were probably not even those Flemings to whom Galilei himself referred when he recounted the genesis of his technological feat in the *Sidereus Nuncius*: «About ten months ago a rumor came to our ears that a spyglass had been made by a certain Dutchman by means of which visible objects, although far removed from the eye of the observer, were distinctly perceived as though they were nearby. Some accounts were spread abroad about this truly wonderful effect, to which some gave credence while others denied them».

Perhaps the credit for the idea should be given to the multifaceted Neapolitan Giovan Battista Della Porta, author of the first known drawing of a telescope,[45] or to the Florentine nobleman Raffaele Gualterotti, minor poet at the Medici court, amateur astronomer and correspondent of Galilei, to whom he contested the primacy of the observation of the sky; or to Leonard Digges, father of the mathematician and astronomer Thomas.[46] We could even scrape the bottom of the barrel of history and dredge up a certain Juan Roget, a Catalan craftsman pulled out of the memoirs of a contemporary Milanese essayist.[47] He would have been the one who manufactured that first telescope which, having arrived by some way in Holland, would have inspired the Middelburg spectacles. More intriguing

---

[44] Founded in Rome in 1603 by the aristocratic scientist Federico Cesi, the academy was named after the lynx, whose sharp vision symbolizes the observational prowess required by science.

[45] «The invention of the spyglasses in that tube was my invention, and Galileo, lecturer of Padua, set it up, with which he found 4 other planets in the sky, and a number of thousands of fixed stars, and in the milky way as many not yet seen, and great things in the globe of the Moon, which fill the world with amazement». In G. Gabrieli, *Il Carteggio linceo*, Rome 1996, pp. 148–149. But Kepler did not trust him at all: «After I began to work on my "Optics", the emperor [Rudolf II] questioned me quite frequently about Della Porta's aforementioned devices. I must confess that I disparaged them most vigorously, and no wonder, for he obviously mixes up the incredible with the probable»; *Kepler's Conversation with Galileo's Sidereal Messenger*, in "The Sources of Science", http://digitalcollections.library.cmu.edu/awweb/awarchive?type=file&item=393654.

[46] F.R. Johnson, *The Influence of Thomas Digges on the Progress of Modern Astronomy in Sixteenth-Century England*, in "Osiris", 1, 1936, pp. 390–410.

[47] *Hieronymi Sirturi, Telescopium, sive Ars perficiendi novum illud Galilaei visorium instrumentum ad sydera*, Paul Jacobi, Francofurti 1618.

is the position of Girolamo Fracastoro, physician, astronomer, and writer from Verona, who in 1538 had already discussed the combinations of lenses capable of magnifying things,[48] and of Leonardo da Vinci who, even a hundred years before the *Sidereus Nuncius*, wrote in his notes: «*Per vedere la natura delli pianeti apri il tetto e mostra alla basa un sol pianeta; el moto refresso da tal basa dirà la complessione del predetto pianeta*» (In order to observe the nature of the planets, open the roof and bring the image of a single planet onto the "*basa*"; the image of the planet reflected by the "*basa*" will show the structure of the surface of the planet).[49] It reads like the operating instructions of a modern astronomical observation station: open the dome, point the mirror («*basa*»), and observe the reflected target («*refresso*») on the focus of the instrument to grasp its details («*complessione*»). On one page of the Arundel codex, there is even a drawing of what is presumed to be a telescope: a sort of bombard that can be tilted in elevation. But this whole suggestive affair was probably just another of Leonardo's numerous and extraordinary intuitions. Had he really made an instrument of such quality as to enable him to observe the Moon and the planets, he would certainly have written more about it and would have accompanied his notes with a wealth of drawings; and his genius would have conceived other applications of the instrument, for example to the art of war or to sea travels. However, a doubt remains, supported by the studies of Leonardo's codices by great modern opticians and by a seductive annotation by the artist himself: «made glasses to see the Moon big».[50]

Leaving aside the sociological aspects of the pathetic quest to grab the primacy — a communication fair to which some cultures seem more prone than others — the uncertain origin of the invention shows that by the end of the 16th century the time was fully ripe for the birth of the telescope. Many people, almost everywhere in Europe, had already guessed what should be done to enlarge and sharpen the images of distant objects, though without being able to do it well, convincingly, and functionally for human activities such as travels, wars, and science. The last step was essentially technological, and the solution floated in the air. A quality leap from the optical

---

[48]V. Ilardi, *Renaissance Vision from Spectacles to Telescopes*, American Philosophical Society, Philadelphia (PA) 2007, p. 208.

[49]Leonardo da Vinci, *Codex Arundel 263*, f. 279v.

[50]Leonardo da Vinci, *Codex Atlantic*, f. 190r a. Cf. also A. Bettini, *2008: Il cinquecentesimo anniversario del cannocchiale?*, Padua Academy of Sciences, Letters, and Arts, Padua 2008 (in Italian).

elements used by the many pioneers and also the Flemish for their rudimentary terrestrial telescopes was needed. Galilei took this last step in his workshop. If he was not the inventor of the telescope or *perspicillum*,[51] as he had christened it in the front page of the *Starry Messenger*, then he made it capable of observing things on Earth and heaven, characterized it, and used it masterfully.

On closer inspection, Galilei was not even the first man to make an astronomical use of the telescope. The Englishman Thomas Harriot, also known for his geographical survey of Virginia during the exploration of the New World with Sir Walter Raleigh, at the end of July 1609 was already peering at the Moon with an instrument with six magnifications. He obtained a rough drawing of the satellite's visible face and the right to enter the Guinness Book of Records. Almost at the same time, the German Adam Elsheimer, a painter active in Rome, burst onto the scene by looking at the sky through a telescope perhaps a few months before Galilei.[52] He saw the Milky Way dotted with stars and so he represented them in a small oil on copper of religious subject now on display at the *Alte Pinakothek* in Munich. But before Galilei's feat, telescopic observations of the heavens had, if anything, aroused curiosity, amusement, and amazement, but had not contributed to the real progress of knowledge.

The Pisan scientist deserves the greatest credit for being able to see — and it was already not an easy thing, as anyone who has looked at the sky with a small instrument knows well, especially given the tiny pupil of the Galilean *perspicillum* — and then to understand what the telescope revealed to him, and to have the courage of promoting and defending his discoveries. None of this would have happened, however, if Galilei had not built an exceptional device for those times, developing also techniques of use and methods of interpretation that placed him in a position of absolute advantage over any other observer. Before him, the exploration of the heavens could only make use of the human eye, which at best was assisted by an instrument to point the source and a protractor to estimate angles. After his work, the advancement of observational machines would acquire a leading role in marking the course of astronomy, making subsequent important

---

[51] The term *perspicillum* derives from the Latin *perspicere*, to look deeply, to examine. Cf. the beautiful digression (in Italian) at: https://tuespetrus.wordpress.com/2009/02/26/galileo-il-perspicillum-e-il-messaggio-del-cielo/.

[52] A. Ottani Cavina, *On the Theme of Landscape. II: Elsheimer and Galileo*, in "The Burlington Magazine", 118, 1976, pp. 139–144.

scientific milestones often coincide with those of technological revolutions: conceptually new or just more powerful instruments, which, with an ever-increasing pace, have in turn triggered authentic paradigm changes in the knowledge of astronomical bodies and phenomena.

This book also speaks of them, of their inventors and their brilliant users, without pretensions to completeness and with the original sin of having to come to terms with the tastes and cognition of the writer. The intent is not to produce another handbook of the history of astronomy, of which there is no need. It is rather to give a view of our accumulation of the scientific know-how about the cosmos that uses technological leaps as reference points in history, where the term "leap" must be understood in the sense given to it by the American historian of science Thomas Kuhn, as a real discontinuity. This also offers us an opportunity to describe the wonders of today, contextualizing them in the process of shaping the world's knowledge, and to hint at the projects for tomorrow.

Before we get into the subject, it seems useful to underline how Galilei's astronomical discoveries are the prime example of a fruitful alliance between technological progress, scientific maturity, and courage: a triad that still guides most of the developments in the study of the sky and its phenomena. Any newly conceived or simply more powerful tool, associated with someone's ingenuity and good luck — «I prefer a lucky general to a good one», Napoleon liked to say — is almost always at the basis of small and big revolutions in astronomy. The main reason lies in the peculiar nature of this ancient science, an experimental discipline whose laboratory existed long before mankind, with all possible experiments in progress or documented by their consequences, and with an endless number of targets which are spread in the immensity of space-time.

Innovative and/or more powerful tools are essential for finding new things in the sky, curiosity and genius serve to recognize them, but the apparatus of knowledge needed for their interpretation may not yet be available at the time of observations. In short, astronomy lends itself to serendipitous[53] discoveries and, more than any of the other sciences, is exposed

---

[53]This neologism, now widely used not only in the Anglo-Saxon world, characterizes those discoveries in which luck is combined with the ability to seize an opportunity. It was first used in the 18th century by the Englishman Horace Walpole, a son of the better-known politician Robert Walpole, inspired by the tale of *The Three Princes of Serendippo* by the Italian Cristoforo Armeno (mid-16th century) in which, not by chance but with sagacity, the story of a lost camel was reconstructed, based on the clever collection of a few clues.

to the risk of forced interpretations to account for new, unexpected, and incomprehensible appearances, as we will see several times in the rest of this book. Astronomers not infrequently find themselves in a position comparable to that of an intelligent ape entering the spectacular laboratories of the Gran Sasso National Institute for Nuclear Physics. The animal, stimulated by what it sees, tries to explain these appearances with the tools at its disposal. Thus, in most cases, it oversimplifies, gets confused, and makes mistakes, but in the meantime it learns and grows in knowledge and self-awareness.

Since, as Newton wrote taking up a phrase of the Platonic philosopher of the 12th century, Bernard of Chartres, "to look far we must be like dwarfs on the shoulders of giants",[54] our story must necessarily begin from afar, from those people who built the cult and the culture of the sky well before the age of the telescope. They are our first shoulders.

---

[54] It seems, however, that with this suggestive sentence Newton intended to mock the short stature of his scientific opponent, Robert Hooke (see p. 158). If this were true, one would like to react by taking up in turn the response to a similar allusion made to the Prince of Taillerand by Napoleon: «It is a pity that so great a man should be so ill-mannered».

Chapter 2

With the naked eye: .
Astronomy and cosmology
of the ancients

Since the dawn of reason, the heavens have captured the imagination of mankind. This interest stemmed not only from the inimitable beauty of the firmament and the undeniable usefulness of celestial bodies for measuring time and navigating oceans and deserts. In fact, because of its apparent perfection, the sky was seen as a kind of open window to the supernatural world. Among the consequences of this attitude is the birth, for divinatory purposes, of astronomy, the most ancient philosophy of nature we have and the only science with its own muse, Urania, and lay enthusiasts, the amateur astronomers. It is astonishing to think that all this came from a misunderstanding, a misinterpretation of the phenomena! The Sun and the black vault of the night sky, dotted with stars and the planets projecting their orbits onto it, gave the terrestrial observer the impression of being at the center of a perfect mechanical system of rotating spheres. By treating this erroneous perception as true, the ancients attempted to explain it away. In doing so, they ended up elevating themselves to the rank of creatures favored by a god-architect, a demiurge[1]: someone who had brought order out of primordial chaos and started the clock of time. It was believed that this god-architect had made the world as a cradle for his own children, for whom he had reserved, among other things, a better life after death. One might say — to paraphrase the Aristotelian Don Ferrante from Alessandro Manzoni's novel *I Promessi Sposi* (The Betrothed)[2] — that the fault of this misunderstanding lies entirely with the stars and, we must add, with the fate that has made us see them.

The ability to see the night sky is not actually a necessary trait for a species to adapt and thrive in its environment. Human eyes, for example, are not primarily designed to see light sources in the starry night sky. They are tuned to the needs of terrestrial diurnal predators to detect the slightest changes in the landscape in order to be effective hunters and avoid becoming prey. However, despite the fact that we share the planet with many species of animals that are unable to perceive the faint light of the stars and sometimes even the intense rays of the Sun, we find that our eyes are indeed good enough to look up into a clear night sky and see the objects that allow us to communicate with the universe. Is this ability an accident, or does it hide a greater purpose? For those who believe in chance rather

---

[1] The word "demiurge" originally referred to the Greek public servant, but in Plato's time was also understood to refer to the divine, philosophical, and mythological figure who created and maintains the physical universe. Myth is a primitive form of science not about the method it uses or the answers it gives but about the questions it asks.

[2] Here is an *excerptum* of this wonderful page: «From the commencement of the pestilence, Don Ferrante was one of the most resolute in denying its existence, not indeed like the multitude, with cries of rage, but with arguments which none could accuse of want of concatenation. *"In rerum natura"*, said he, "there are but two kinds of things, substances and accidents; and if I prove that the contagion can neither be one nor the other of these I shall have proved that it does not exist; that it is a chimera" [...] "Here is the true reason", said he, "and even those who maintain other fancies are obliged to acknowledge it. Let them deny, if they can, that there is a fatal conjunction of Jupiter and Saturn. And when has it been said that influences propagate? And would these gentlemen deny the existence of influences? Will they say there are no planets? Or will they say that they keep up above, doing nothing, as so many pins in a pincushion? But that which I cannot understand from these doctors is that they confess we are under so malign a conjunction, and then they tell us, don't touch this, don't touch that, and you will be safe! As if, in avoiding the material contact of terrestrial bodies, we could prevent the virtual effect of celestial bodies. And all this work to burn a few rags. Poor people! Will you burn Jupiter? Will you burn Saturn?"
His fretus, that is to say, on these grounds, he took no precautions against the pestilence; he caught it, and died, like Metastasio's hero, complaining of the stars.» (https://en.wikisource.org/wiki/The_Betrothed_(Manzoni)/Chapter_36).

than necessity,[3] it seems to be a rather grand conspiracy of events involving our visual apparatus and the atmosphere surrounding our planet.

We live at the bottom of a vast and turbulent ocean of air, an atmosphere that could have been opaque to electromagnetic radiation, just as it is on Venus, the blue planet, and on Titan, Earth's young twin orbiting Saturn. Both celestial bodies are shrouded in perennial clouds thick enough to hide not only the starry sky but even the sunlight from the potential and unfortunate inhabitants of these gloomy worlds, if there were any. We earthlings have it much better. Although clouds sometimes obscure the Sun's light, down here where we live and vegetate, as Giordano Bruno would say,[4] the divine Helios always returns to shine in the clear sky. It is a fact that we have naturally come to be able to see the Sun and to live thanks to the effect of its light. But not all of the radiation that this yellow star sends towards us manages to reach the ground. The atmosphere that surrounds our planet acts as a providential filter that, in addition to low-frequency radio waves, only lets through radiation in a range of wavelengths whose extremes we perceive as violet and red, respectively. Fortunately for us, ultraviolet (UV) and the even more deadly high-frequency X and γ rays are blocked (at least until humans succeed in completely destroying their habitat) by atmospheric oxygen, which greedily absorbs them, while a derivative of the water molecule strips the incoming solar light of harmless infrared (IR) radiation.

What remains between the two Herculean columns of UV and IR is a mere sliver, no more than a tiny keyhole, through which the light carrying all the colors of the rainbow passes. Now, the natural order of things dictates that the energy emitted by most common stars is mainly concentrated in this "keyhole", in the interval from violet to red. And so we come full circle. The human eye has learned to use the radiation that reaches the ground during the day in a way that is functional for the needs of day hunters; and this light has the same characteristics as that which makes the majority of stars shine in the night sky! But this favorable coincidence is neither a rule nor a law of nature. No one yet knows if there are other forms of life than ours in the Milky Way. However, it is possible to imagine environmental conditions sufficiently different from those on Earth that,

---

[3]Cf. J. Monod, *Chance and Necessity: An Essay on the Natural Philosophy of Modern Biology*, Knopf, New York 1971.

[4]G. Bruno, *De infinito, universo e mondi*, dialogue V.

even without *a priori* restricting the existence and flourishing of an intelligent species, would deprive it of the possibility of direct visual contact with its sky. The firmament would thus remain hidden to the senses of these hypothetical and unfortunate creatures. Then no metaphysical speculation could be stimulated until a suitable means of perceiving the starry sky was invented, at a late stage in the evolution of this alien species and after the development of conscious scientific thought. Is this an improbable way? Not at all! Less than a century ago, we ourselves discovered the existence of invisible radio and X-ray sources, thanks to a better understanding of the physical world and new observational tools, including the ability to look at the universe from space beyond the atmosphere (see chapters 18 and 19). It is the metaphor of the lobster that lives happily at the bottom of the sea without ever seeing the stars, thus ignoring the existence of that sky the luckiest coyote unknowingly howls at on full moon nights. But if this tasty crustacean could develop sufficient intelligence and technology, sooner or later it too would discover the sky and be forced to ask difficult and uncomfortable questions, just as we did a few millennia ago. Better than ending up alive in boiling water, one might say to downplay.

Since ancient times, the sky has been an object of intense human interest, not only observed but also actively measured. This fact is attested by numerous monuments of antiquity; for example, the pyramids of Giza, which were built at the northeastern edge of the Sahara more than 3,000 years before the birth of Christ, or Stonehenge, a ring of megaliths erected between 2,800 and 1,500 B.C. in southern England.[5] These imposing and evocative constructions are symbols of timeless eternity. They are oriented to indicate celestial events and the positions of stars and landmarks of particular astronomical importance: for example, the point on the horizon where the Sun rises at the summer solstice, when the day is the longest of the year, or the direction to the star that (*pro tempore*)[6] is the fulcrum of the diurnal rotation of the sky (today the Polaris, α Canis Minoris).

The heavenly bodies that our enterprising ancestors knew, observed, and worshipped were the same ones that we can admire with the naked eye today: first of all, the Sun, the main source of light and heat, the undisputed

---

[5]G. Magli, *Archaeoastronomy: Introduction to the Science of Stars and Stones*, Springer, Cham 2016.

[6]Due to the precession of the equinoxes, i.e. the conical motion of the Earth's axis of rotation, the role of the North Star changes with a period of about twenty-six thousand years; see also p. 140.

lord of the day and arbiter of human life. It appears on the horizon, accompanied by the «rosy-fingered» dawn, and, like a charioteer who has spurred his horses into a gallop, to stay with the Homeric metaphors, it begins to rise higher and higher in the sky.[7] Halfway through the ride, having reached its maximum height above the horizon, the Sun begins to descend until it sets, giving way to darkness. In the ancient world, darkness was perceived as a real entity. This is evidenced by a beautiful passage from the Biblical *Genesis* (1, 3–5): «And God said: "Let there be light"; and there was light. And God saw the light, and it was good; and God divided the light from the darkness, and called the light "day", and the darkness "night"». The latter, ruled by the dim lights of the stars and the fickle Moon, reigns until the fiery Sun, born to new life as in Egyptian mythology, extinguishes the nocturnal sources by resuming its journey across the sky in a periodic path[8] which admits no pause, and which inspires wonder and awe in the creatures to whom fate gives instead a single round.

The Sun has a second important cycle in addition to the daily one. We discovered this by noticing that, over the course of a year, the maximum height reached each day by the star is not always the same. A rod stuck in a flat ground casts a shorter shadow in the middle of the day[9] (sign of a higher Sun) when the season is hot and nature shows all its life and vigor. Conversely, the shadow appears longer (and the Sun lower) when it is cold and nature is asleep. A course that is completed within one calendar year and is perfectly synchronized with the cycle of the seasons. The two extremes between which it takes place are called solstices (from *"sol"*, Sun in Latin, and *"sistere"*, to stand still) because then the ascending or descending motion of our star on the meridian line is reversed, and therefore it "stops" its vertical ascent/descent. Since the different meridian heights are associated with a longer or shorter presence of the Sun above the horizon,

---

[7]In Greek mythology, the Sun — the Titan Helios, son of Hyperion and Theia and brother of Selene, the Moon — rises at dawn from the great river Ocean in his chariot drawn by four fiery-nosed horses and travels across the sky from east to west, then plunges into the waters surrounding the Earth.

[8]In the *Carmen Saeculare*, the Latin poet Horace describes the Sun as follows: «*Alme Sol, curru nitido diem qui / promis et celas aliusque et idem / nasceris*» (Kindly Sun, you who bring forth and hide the day in your shining chariot, born both altered and the same).

[9]The shadow cast by the rod varies greatly throughout the day; very long at dawn, when the Sun is low on the horizon, it gradually shortens until it reaches a minimum and then begins to lengthen again until sunset. The epoch of the minimum is called "solar mid-day". When this happens, the shadow points in the north-south direction.

the solstices also mark the epochs of maximum and minimum day length relative to night. When the Sun appears halfway between the two solstices, day and night are equally long. This happens twice a year, at the equinoxes (from Latin *"aequus"*, equal, and *"nox"*, night), respectively rising (spring in the northern hemisphere) or falling (autumn), depending on whether the star crosses one of them as it rises or falls over the equator.[10]

The monotonous carousel of these phenomena results from the combination of the two major motions in which the Earth is involved: a diurnal rotation about its polar axis and a revolution around the Sun, which takes place in about 365.25 average solar days at a respectable average speed of 107,225 kilometers per hour.[11] The first motion determines the alternation of day and night that presides over our biorhythms. The second motion would not have much effect on us were it not for the considerable tilt of the Earth's axis on the plane of revolution. This "obliquity", as wide as about a quarter of a right angle (23°.5), embodies the real reason for the different heights of the Sun at noon during the course of a year. The phenomenon is purely geometric and nevertheless governs the seasonal changes because it affects the level of insolation. In fact, while the flow of solar energy does not change over time — except for the modest effects related to the orbital eccentricity,[12] the solar cycle, and the much more relevant vagaries of the weather — the surface over which it is distributed increases as the height of the source falls on the horizon, as demonstrated by the daily trends of the average temperature.

Unaware of the causes, but armed with infinite patience, our ancestors discovered and classified the regularities of the Sun's motion, and at some point began to wonder if and how it was possible to predict its future movements and anticipate its behavior and trajectory in the sky. In doing so, they unwittingly laid down the rudimentary foundations of a predictive

---

[10]For an in-depth analysis of a wide range of astronomical topics, it is possible to consult the large and very articulate work of A. Franknoi, D. Morrison, S.C. Wolff, *Astronomy*, 2016, Kindle Edition; https://openstax.org/details/books/astronomy.

[11]Note that, after firing the third-stage Saturn V engine again to head toward the Moon, the initial speed of the command module Columbia of the Apollo 11 mission was "only" 39,000 km/h.

[12]The small flattening of the Earth's orbit produces a secondary effect on the climate. The small differences (less than 3.5%) in the planet's distance from the Sun modulate the solar flux during the year with an amplitude of about 10%. In the northern hemisphere, the Sun reaches its minimum distance in December. Therefore, the phenomenon moderates our winter while it enhances the austral summer.

activity called science (from the Latin "*scire*", to know). Needless to say, the Moon, the stars, and the other bodies that roam the velvet of the night sky also played their part in inspiring man's ambitious plan to seize the secrets of the heavens. It was a monumental saga with heroes and numerous soldiers, with many battles won and some sensational defeats. Without this effort, we would not have today, for example, the theory of relativity and quantum mechanics, nor the countless technological "prostheses" that help us to live and that have grown out of this knowledge.

What do we see on a clear night? The brightest source is the Moon which, with its 29.5-day cycles, is like a slow analog clock available to everyone. Even today, there are peoples and nations that base their calendars and their religious lives on the phases of the Moon. And there are the stars, numerous, colorful, and of varying brightness, which move in much the same way as the Sun. They rise to a certain height above the horizon, then descend until they set, all moving together with such a precision that they seem to be firmly anchored to the surface of a uniformly rotating sphere, capable of maintaining the rhythm of time without being affected by it. Only those celestial objects that are closest to the visible celestial pole never set,[13] but of course they also cannot be viewed during the day. The stars surrounding the opposite pole, on the other hand, never cross the horizon unless the observer changes his latitude to make them visible.

On the carpet of the night flames, flowing unceasingly with constancy and regularity, wander some luminous bodies, the planets (from the Greek verb πλανάομαι, to go astray). We have inherited the custom of calling them by the Latin names of the Greek Olympian deities: Mercury, Venus, Mars, Jupiter, and Saturn. One does not even have to grasp their peculiar pacing to realize their diversity. Planetary images appear more colorful and less pointlike and flickering than ordinary stars, making them easy to recognize and trace. Mercury is an exception, but only because, in its tight revolution around the Sun, it shows up when the elongation[14] is at its maximum, either in the ashen light of twilight or when the night is tinted with the colors of the dawn. This is why the Greeks named it after Apollo, the morning star, and Hermes, the evening star, although they knew it was always the same object. Even the blue Venus, the divine Ishtar of the Babylonians, moves hand in hand with Helios, while keeping a more modest

---

[13]The amplitude of the effect depends on the latitude of the observer.

[14]The angle that two celestial bodies form with respect to the terrestrial observer.

and demure distance. Sometimes she anticipates the Sun, appearing before dawn, and sometimes she follows the star, showing up after sunset. Because of this "feminine shyness", first noticed by Pythagoras in the fifth century B.C., the blue planet has the double name of Hesperus and Phosphorus (evening and morning star, respectively).

Red Mars, bright Jupiter, and slow Saturn move in the same direction among the night stars. But sometimes they seem to repent, and, like a traveler who realizes that he has left behind something he longs for, they retrace their steps for a while, reversing their direction of travel, or retrograding,[15] as astronomers say. Strange behaviors such as these are difficult to understand at first, unless they are seen as the result of some divine whim. It took enormous patience and a long time to understand that the planets follow regular celestial paths like the Sun and the Moon, each on its own trajectory and in its own time. A more accurate topography of the sky was needed to achieve this awareness.

For many people, especially in this day and age, the night sky is a magnificent but confusing array of lights that can seem impossible to navigate. With a little practice, however, it becomes quite easy. The trick to memorizing the star map is the same one we use to find our way home after moving to a new neighborhood. We need to pick some reference points that are easy to remember, and perhaps draw on some personal details from our lives to create a mental association: signs that we share with our friends when we want them to join us for dinner. "Turn right at the red house" or "Go to the Crown Tavern and make a left". On the celestial vault, the primary references are given by groups of bright stars, arbitrarily identified by different cultures and usually called by names chosen on the basis of religious impulses or celebratory intentions. A toponomy that is fairly easy to memorize thanks to a certain similarity between the morphology of the asterism and its name. With a little practice, it makes navigating among the stars fairly practical.[16]

Let us consider some examples. In winter, a short parade of three bright stars stands out in the northern sky. According to the Greeks, this string

---

[15]Sometimes the direct apparent motion (counterclockwise to the boreal observer) of a planet is reversed as it moves along the zodiacal belt. The cause of this temporary "retrogradation" is the same as that of parallax: a progressive change in perspective due to the motion of the terrestrial observer in its heliocentric orbit.

[16]A. Aveni, *Star Stories: Constellations and People*, Yale University Press, New Haven (CT) 2019.

marked the belt of the giant Orion.[17] From there it is easy to find the
red Betelgeuse and the blue Rigel, the shoulder and foot of the handsome hunter. This classic myth helps us remember the rest.[18] Orion is
accompanied by his faithful dogs, two constellations in which shine the
bright Sirius, a star sacred to the Egyptians, and the white Procyon. The
giant faces the long horns of the raging bull, another neighboring constellation called Taurus, where the red Aldebaran, which in Arabic means
"the one who hunts the Pleiades", represents the bloody eye of the fierce
animal. In this way, the map of the night is mastered and it becomes a
tool useful for orientation when the Sun is asleep and indispensable for
following the trajectories of wandering bodies. Curiously, the same asterism that the Greeks interpreted as the stylized image of a young giant
reminded the Sumerians of the shape of a sheep, with all due respect
to the mythological hero. But that's okay. Constellations are just conven-
tions, established in different ways by different peoples, always with the
same practical purposes. Of another use of them, much more abstract,
intriguing, and fraught with unpredictable consequences, we will speak later
(p. 30).

Shortly after the First World War, in the years when men's minds
were still very much in the grip of the great idealistic dreams of a global
democracy — before they were replaced by the much more concrete and

---

[17] Here is an example to appreciate the importance of knowing the mythology of the
ancient Greeks and Romans to better read the starry sky. According to Greek mythology,
Orion was the son of Poseidon and of the daughter of the king of Crete. Young and
handsome, he was blinded by Princess Merope, whom he had tried to rape. He regained
his sight thanks to Eos, the goddess of dawn, whom Orion gratefully took as his wife. A
heavenly-eyed hunter, he went out at night in search of prey, accompanied by his faithful
dog Sirius. During one of these expeditions, he was spotted by Artemis, the goddess of
hunting, who, despite her vow of chastity, lost her head for him. Orion politely recoiled,
claiming to be a faithful husband. Artemis got over it but, when she discovered that
Orion was philandering with the Pleiades, the beautiful seven daughters of Atlas and
Pleione, the goddess, blinded by rage, sent the scorpion, one of her faithful servants, to
avenge her. The predatory arachnid crept into the hunters' hut at night. When Orion
and his dog fell asleep, the scorpion's poisoned sting killed first Orion and then Sirius,
who had tried to defend his master. The Latin myth tells a slightly different story, but
both versions agree that after death, Orion and even the Scorpion were placed in the
sky where the constellations of the same name are found. Sirius instead became the
brightest star of Canis Majoris.

[18] For an overview of the classic myths, see, for example, R. Watherfield, *The Greek
Myths. Stories of the Greek Gods and Heroes Vividly Retold*, Quercus Publishing,
London 2011.

monopolizing plutocracy that rules the planet today by bartering work for wagers — the entire celestial vault, even that part of the southern hemisphere that Western civilization had neglected because of poor attendance, was divided up by the International Astronomical Union.[19] This kind of United Nations of astronomers definitively fixed the number of constellations at 88 and defined their mutual boundaries. More than half of them bear names and symbolic meanings established in Classical and Hellenistic Greece, although the probable origins of some are much more remote. We know, for example, that Ursa existed by this name before the end of the last Ice Age. It was used by peoples on both sides of the Bering Strait, although they had ceased to communicate with each other more than 12,000 years ago with the melting of the ice bridge between Asia and America.[20] The constellations of the southern hemisphere, on the other hand, have more secular names, such as Fornax and Circinus, associated mainly with advances in science and technology (Latin words for furnace and compass). Signs of the victorious revolt of the Enlightenment against the elitist conservatism of the classicists, now followed by the vindication of emerging and endangered cultures. Thus the new trans-Neptunian planets, Pluto-sized bodies recently discovered in large numbers at the periphery of the Solar System,[21] have been given unlikely names, taken from the legends of the peoples of the cold north or the Pacific islands, such as Makemake (creator god of the indigenous people of Easter Island), Sedna (goddess of Inuit mythology), and Quaoar (deity of the Tongva, indigenous people of California). An example of the positive face of globalization, or perhaps a demonstration of political correctness.

Of particular interest are the 12 constellations that the Sun passes through in its apparent annual motion. Together they form the so-called "zodiacal belt". They are recalled in their proper order in the verses of the late Latin poet Ausonius Decimus Magnus: «*Aries, Taurus, Gemini, Cancer, Leo, Virgo, / Libra, Scorpius, Arquitenens, Capricornus, et urnam / qui tenet, et Pisces*».[22] Obviously, it is not the Sun that goes around in

---

[19]W. Adams, *The History of the International Astronomical Union*, in "Publications of the Astronomical Society of the Pacific", 64, 385, 1945, pp. 5–12.

[20]E.W. Hetherington, N.S. Hetherington, *Astronomy and Culture*, Greenwood Pub Group-ABC-CLIO, Santa Barbara (CA) 2009, p. 32.

[21]Cf. D. Prialnik, M.A. Barucci, L.A. Young, *The Trans-Neptunian Solar System*, Elsevier, Amsterdam 2020.

[22]*D. Magni Ausonii Opuscula*, p. 282.

a circle among the stars, but the small Earth that revolves about it in a year,[23] according to the Newtonian rule that the more massive a body is, relative to its context, the less it will move. But the observer standing on the surface of the planet, feeling himself very still in relation to the sidereal sphere, perceives the phenomenon in a different way. As the months pass, he will notice that the Sun projects itself at different points of a circle ideally drawn at the center of the zodiacal constellations, which astronomers call the "ecliptic".[24]

The Sun therefore behaves like the hand of a majestic celestial clock whose cycle is one solar year and whose dial is the zodiac: a dazzling and rather bizarre clock hand, because it makes visible the part of the dial opposite to that in which it projects (the night sky "turns on" when the Sun sets). There are other clocks in the sky that are easier to read and to some extent more accurate, such as the Earth's diurnal rotation, which determines the alternation of day and night, and the phases of the Moon. But the annual motion of the Sun is the most intriguing because it seems to be in tune with the events that cyclically regulate life on Earth and the destiny of human beings. In fact, all the manifestations and consequences of the seasons, such as animal migrations, great river floods that fertilize the fields, and in Egypt, for example, were the sources of life for the Nile farmers,[25] epidemics of fever, climatic and thermal excursions, and many other rhythms of animal and plants are synchronized with the annual motions of the Sun; and they correspond to the cyclical appearance of characteristic asterisms in the night sky.

In short, heaven seems to mark the changes without explicitly sharing them. This quality has associated it with a God — however people have understood this word — who rules time without being a victim of it. Echoing Plato, Augustine of Hippo writes in his *Confessions* that God was

---

[23]The correct form is that both bodies revolve around the center of gravity of the Solar System. But since the Sun is by far the most massive object, the center of gravity of the whole system falls always very close to that of the star.

[24]The name comes from eclipse (from the Greek verb "to leave" and also "to disappear") because the alignment of the Earth, Moon, and Sun occurs on an ecliptic plane, which is the cause of the spectacular astronomical phenomenon.

[25]R.A. Parker, *Ancient Egyptian Astronomy*, in F.R. Hodson (ed.), *The Place of Astronomy in the Ancient World*, monographic issue of "Philosophical Transactions of the Royal Society of London. Series A. Mathematical and Physical Sciences", 276, 1257, 1974, pp. 51–65.

not and will not be: he is «the eternal present».[26] The human beings, on the other hand, have a past and a future centered in the present, trapped between two extremes, birth and death, α and ω (the first and the last letters of the Greek alphabet). Prisoner of a world where everything is changing and where fate is looming, frightened by a nature much stronger than himself and seemingly irrational if not ruthless, man nevertheless discovers that he is at the center of a divinely uncorrupted celestial music box that cycles his existence, and he understands this peculiar position as a recognition of his role, a promotion to the rank of beloved son of the Creator. It is a short step from there to the idea that Father writes his messages in the sky through combinations of wandering stars.

These are the deep motivations behind judicial astrology. The art of interpreting divine judgments written in the stars[27] was seen as an almost sacred activity, to be entrusted to the wise and in the service of the powerful. It required an ever more precise knowledge of the positions of the wandering stars and the construction of predictive tools capable of updating planetary coordinates at different epochs for divinatory purposes. Astronomy, the oldest of the natural sciences, was ennobled to such a subordinate role and liberated from its humble role as a mere aid to farmers and travelers. For millennia, by means of angular measurements and complicated geometric constructions, men have sought to build a kinematic model of an Earth-centered cosmos enclosed within the narrow confines of the sphere of fixed stars, capable of imitating as accurately as possible the image provided by observations. It was a titanic undertaking because the first astronomers, priests and philosophers, lacked all the necessary tools.

The devices available at the time for observing and measuring angles and time were ingenious but inadequate for the task they were intended

---

[26]Augustine, *Le Confessioni*, XI, 11, 13. See also Plato, *The Timaeus*: «For there were no days and nights and months and years before the heaven was created, but when he constructed the heaven he created them also. They are all parts of time, and the past and future are created species of time, which we unconsciously but wrongly transfer to the eternal essence; for we say that he "was", he "is", he "will be", but the truth is that "is" alone is properly attributed to him, and that "was" and "will be" only to be spoken of becoming in time, for they are motions, but that which is immovably the same cannot become older or younger by time, nor ever did or has become, or hereafter will be, older or younger, nor is subject at all to any of those states which affect moving and sensible things and of which generation is the cause. These are the forms of time, which imitates eternity and revolves according to a law of number».

[27]See P. Whitfield, *Astrology: A History*, British Library Publishing Division, London 2004.

to perform:[28] gnomons, rulers, plumb lines, dioptres, armillas, dials, astro-labes, hourglasses, along with the eye — a sophisticated and inexpensive telescope with its own detector. The eye was considered the noblest of the senses and was worshipped as a god by the Egyptians, but it has limited astronomical capabilities. In fact, our visual organ has an aperture thou-sands of times smaller than modern large reflectors, a linear resolution 500 times worse than that of the Hubble Space Telescope, a sensitivity that varies with environmental conditions and is not easily quantifiable, and, above all, a very limited capacity for signal integration. The human visual system refreshes the image about 30 times per second to allow for the rapid reactions that are essential for offense and defense. This is equivalent to pro-ducing a movie in the brain whose frames have very short exposure times. Modern astronomical detectors, on the contrary, accumulate the signal on a single image for intervals up to hundreds of thousands of times longer, with obvious advantages for detecting weaker or more distant sources.[29] From this comparison, the eye comes out soundly defeated, like a pierced trouser pocket as compared to a piggy bank.

But with boundless perseverance, a lot of time on their hands, no union guarantees for workers, and a great deal of passion, the ancient astronomers made some truly fundamental discoveries. They noticed, for example, that solar and lunar eclipses follow each other in a sequence called the "Saros cycle", which repeats identically every 19 years or so. It was the Greek Meton who, in the 5th century B.C., discovered the rule that gave great power over men to those who knew how to use it — but perhaps he sim-ply appropriated the work of the Fertile Crescent astronomers. Before 129 B.C., while the Roman legions were busy giving the known world a single language and a single law (and a single subject to whom to pay taxes), Hipparchus of Nicaea, comparing his catalog of star positions with those made by the Babylonians a few hundred years earlier, noticed that the equinoxes moved forward each year by about 20 minutes. This is a subtle phenomenon, difficult to detect, yet extraordinarily important, because it determines the Zodiac's shift in relation to the seasons. It is as if the face of the clock that marks the year, instead of standing still, undergoes a very

---

[28] J. Evans, *The History and Practice of Ancient Astronomy*, Oxford University Press, Oxford 1998.

[29] The flux (energy received by a unit area) is proportional to the intrinsic luminosity of the source (energy produced per unit time) and inversely proportional to the square of its distance.

slow but constant drift in the opposite direction to that of the solar hand. The cause, as Isaac Newton would demonstrate in the late 17th century in the third volume of *Philosophiae Naturalis Principia Mathematica* (see chapter 7), is a very slow conical motion by which the Earth's axis of rotation responds to the twisting forces exerted by the Moon and Sun on the planet's equatorial bulge. Forces that would straighten the axis and instead have the effect of turning the Earth into a spinning top with a period of 25,772 years. The difference between our planet and the childhood toy is that the tin top tilts more and more due to friction slowing its axial rotation until it collapses. The Earth, on the contrary, being almost a free rotator,[30] has continued its dance undaunted for billions of years.

But the most important goal of ancient astronomy was to construct a kinematic model of the world that could reproduce observations, despite ignorance of the mechanisms driving these phenomena. Such a model would be able to withstand the wear and tear of time for two millennia, longer than any other known scientific theory. Historians attribute the first robust formulation to Eudoxus of Cnidus, a Greek born in 408 B.C. in the hotbed of genius that was the Anatolian peninsula.[31] We know that he was a student of Plato and of the great mathematicians of Magna Graecia. He used his talent, in the wake of what he had heard from the masters, to devise a sort of cosmic automaton, composed of concentric and rotating spheres, with which he made the planets revolve around a common center, the Earth. His concern was not to know why things happened in the heavens the way observations showed, but rather to build a mental model capable of reproducing them as faithfully as possible, in the Platonic belief that perfection belongs only to ideas. In any case, he did it better than Plato's construction, which was rather elementary in that it was based on pure circular motions having the Earth as their center, one for each planet plus one for the fixed stars.

In the scheme of Eudoxus,[32] each wandering star took part in the motion of four spheres, hinged together, apparently transparent and therefore "crystalline" if not purely geometrical. The first was responsible for the diurnal motion, the second for the monthly motion. The two were kept separate in

---

[30] Actually, the Earth is not a rigid body; thus any accelerated motion implies dissipation. For example, its daily rotation is systematically slowed by the tides.

[31] H. Baker, *Eudoxus of Cnidus: A Proto-Classical Life*, in "The Sewanee Review", 81, 2, 1973, pp. 237–281.

[32] D.R. Dicks, *Early Greek Astronomy to Aristotle*, Thames & Hudson, London 1970.

order to reproduce the inclination of the planets' orbits on the ecliptic. The last two spheres, tilted and rotating in opposite directions to each other, were used to create the phenomenon of retrogradation by the geometric artifice of superimposing, in a synodic period,[33] a path in the form of a figure of eight on the direct motion. This allowed the planets of Eudoxus, among other things, to reverse their trajectories among the stars and retrace their steps for a while, just as it appeared.[34] By calculating three spheres for the motions of both the Sun and the Moon and only one for the fixed stars — since they all rotated rigidly around the Earth as a single system — the total number of spheres already reached $4 \times 5 + 2 \times 3 + 1 = 27$, without taking into account the connecting spheres between the different groups, essential to give kinematic coherence to the model.

Although highly complicated, this geometric tool made it possible to predict with consistent accuracy the positions assumed over time by the various planets — proving the divine nature of mathematics (in this case geometry) advocated by Pythagoras.[35] However, it failed miserably to represent the variations in brightness that were observed in the wandering stars, which we now know are due to the different distances from the observer (mainly due to the Earth's orbital motion). Not too bad for a student of Plato, who placed more importance on ideas than on phenomena.[36] Also a fruit of ideology was the assertion that the immobile Earth

---

[33] The synodic period is the time it takes for a circumsolar object, observed from the Earth, to return to the same position in the sky with respect to the Sun.

[34] G.V. Schiaparelli, *Scritti sulla storia della astronomia antica*, Mimesis, Rome, 1998, pp. 42–66 (in Italian).

[35] F.M. Cornford, *Mysticism and Science in the Pythagorean Tradition*, in "The Classical Quarterly", 16, 3–4, 1922, pp. 137–150.

[36] P. Duhem, *To save the phenomena, an essay on the idea of physical theory from Plato to Galileo*, University of Chicago Press, Chicago 1969. Immanuel Kant expressed his disagreement to Plato's ideology with these words in the introduction to *The Critique of Pure Reason*: «The light dove cleaving in free flight the thin air, whose resistance it feels, might imagine that her movements would be far more free and rapid in airless space. Just in the same way did Plato, abandoning the world of sense because of the narrow limits it sets to the understanding, venture upon the wings of ideas beyond it, into the void space of pure intellect. He did not reflect that he made no real progress by all his efforts; for he was met with no resistance which might serve him for a support, as it were, whereon to rest, and on which he might apply his powers, in order to let the intellect acquire momentum for its progress. It is, indeed, the common fate of human reason in speculation, to finish the imposing edifice of thought as rapidly as possible, and then for the first time to begin to examine whether the foundation is a solid one or no».

must be at the center of all motion; a prejudice based on erroneous physical considerations and the fideistic belief in the centrality of man. Anyone who had attempted to challenge this dogma by advocating, for example, a heliocentric cosmology would have risked much;[37] even an accusation of impiety, as is said to have happened to Aristarchus of Samos, although this is probably just a legend popularized by the 19th century Italian poet Giacomo Leopardi in his *History of Astronomy*.

Eudoxus was a mathematician and a theorist.[38] Aristotle, born in 384 B.C. and a student of Plato too, was also a physicist (to the extent that he could be at that time). With him the geometric model of homocentric spheres[39] became a cosmology. Apart from further complications introduced to make it more consistent with observations, the philosopher of Stagira[40] characterized it from the point of view of ingredients and behavior. The first thing to be resolved was the contrast between the seemingly eternal cyclicity of the heavens and the transience of things on Earth. Aristotle was convinced that in the sublunar world, that is, in the innermost of the celestial spheres, the four elements listed as fundamental by Empedocles of Akragas in the 5th century B.C. — earth, water, air, and fire — were those that competed for the stage.[41] Each pursued its own "natural place",[42] in a fair where changes occur in a linear time and in only one direction. In the celestial world, the region between the spheres of the Moon and the fixed stars, time had instead to be cyclic and thus powerless to induce permanent changes in the only substance at play, ether (from the Greek "pure air") or quintessence (fifth element in addition to the four of Empedocles). Physics also had to follow different rules in the two Aristotelian worlds. In the sublunar world, natural motion had to be rectilinear, though not uniform, and always supported by the action of a force (a law that would be falsified by

---

[37] W.H.H. Stahl, *The Greek Heliocentric Theory and Its Abandonment*, in "Transactions and Proceedings of the American Philological Association", 76, 1945, pp. 321–332.

[38] O. Neugebauer, *A History of Ancient Mathematical Astronomy*, Springer-Verlag, New York 1975.

[39] H.J. Easterling, *Homocentric Spheres in 'De Caelo'*, in "Phronesis", 6, 2, 1961, pp. 138–153.

[40] Known for being the birthplace of Aristotle, Stagira is an ancient Greek city on the east coast of the Chalkidiki peninsula, now in the Greek province of Central Macedonia.

[41] P. Kingsley, *Empedocles and His Interpreters: The Four-Element Doxography*, in "Phronesis", 39, 3, 1994, pp. 235–254.

[42] Natural places are the result of experience. Earth, which is the heaviest, goes in search of the center of the universe (i.e. the Earth). Water, which is lighter, finds peace in a shell surrounding the earth, which is in turn surrounded by the shell of air. Above it all, but below the sphere of the Moon is the place of fire.

Galileo's and Newton's principle of inertia); in the heavens, instead, natural motion had to be circular.

The issue of the circle, a perfect plane figure and therefore worthy of the empyrean, was destined to become an obsession for ancient cosmologists. Nearly five centuries after Aristotle, another great astronomer of Greek culture, the Alexandrian Claudius Ptolemy, would construct his geocentric model of the world by cleverly manipulating the circular motions to reproduce with remarkable fidelity the orbits of the Sun, the Moon, and the planets, which instead are not circular. It is worth spending a few words on these brilliant theoretical artifices in order to touch upon one of the greatest problems of science of all time: the preservation of a preconceived notion at all costs, for ethical, aesthetic, and ideological reasons, or simply to satisfy tradition or convenience, even when the evidence is to the contrary. A recent and glaring example of this psychological conditioning is the model of a static universe to which Albert Einstein was led by his adherence to Spinozian pantheism. The slogan «*Deus sive natura*» (God, that is, nature) implied global immobility. The German genius would later recognize this prejudicial worldview as "the greatest blunder" of his life, as we will say in chapter 14.

In the second century A.D., when Ptolemy speculated on his celestial constructions, data on the positions and motions of the stars were more accurate and abundant than in the time of Eudoxus and Aristotle.[43] A new model of the world was needed that would more accurately reproduce the different velocities exhibited by the Sun in its annual motion on the celestial sphere, somewhat faster in winter than in summer, which we now know is mainly due to the slight elongation of the Earth's orbit. The gross errors of the earlier models regarding the retrograde motions of the planets should also be corrected. Hipparchus's invention[44] of combining two uniform circular motions was no longer sufficient to explain the phenomena. The astronomer of Nicaea had imagined that each wandering star was drawn by the rotation of a first circle, the epicycle, whose center slid over a second circle, the deferent, centered on the Earth. It is interesting to note that, for a suitable choice of velocities and directions, the combination of epicycle and deferent produces an orbit in the shape of an ellipse, but with

---

[43] O. Neugebauer, *The History of Ancient Astronomy: Problems and Methods*, in "Publications of the Astronomical Society of the Pacific", 58, 341, 1946, pp. 104–142.

[44] Actually, the first proposal was made by Apollonius of Perga, who lived in the 3rd century B.C. and is better known for his work on conic sections.

the Earth in the center.[45] In conclusion, Hipparchus had come close to a geocentric version of what would later become the Keplerian solution for planetary orbits. However, he had been unable to recognize it because he was clouded by a preconceived ideology.

Ignorant of the causes but determined to keep the dogma of uniform circular motions intact, Ptolemy[46] recklessly made some clever but bizarre corrections to the epicyclic motions that greatly complicated the model. For example, he assumed that the solar deferent did not have its center exactly on the Earth, while for the capricious planets he adopted the solution of the equant, a point other than the center of the orbit (i.e. the center of the deferent). It was placed symmetrically with respect to the position of the Earth, from which the center of the epicycle would appear to move at constant angular velocity, which is not the case. Although an extravagant and vaguely hypocritical solution, it nonetheless worked quite well in reproducing appearances.[47] Ptolemy expounded it, with a rich set of theorems, in the *Mathēmatikē Syntaxis* (Mathematical Syntax), the monumental treatise on cosmology that was to shape science in Europe and the Islamic world for the next fifteen centuries. The Arabs, adapting to their language the Greek superlative for "the greatest", renamed it *al-majistī*, that is, *Almagest*,[48] a title that became popular in the Latin translation of the book made in the 12th century in Toledo, Spain, by the Italian Gerardo da Cremona (the same one who made the first complete translation from Arabic of *The Compendious Book on Calculation by Completion and Balancing* by the Persian polymath Muhammad ibn Musa al-Khwarizmi, from which the terms "algorithm" and "algebra" are derived).

As for the sidereal sphere, inspired by Hipparchus[49] who had qualitatively introduced three classes of brightness, Ptolemy placed the stars in a discrete scale of magnitudes consisting of six steps. In the first there were the brightest stars and in the sixth and last were those barely visible to the

---

[45]This happens when the angular velocities of the epicycle and deferent are equal and opposite.

[46]J. North, *Cosmos: An Illustrated History of Astronomy and Cosmology*, The University of Chicago Press, Chicago (IL) 2008, pp. 67–133.

[47]It is worth noting that a Ptolemaic model with epicycle and deferent counter-rotating at the same angular velocity is geometrically and kinematically very similar to the Keplerian solution; see the next chapter.

[48]G.J. Toomer, *Ptolemy's Almagest*, Princeton University Press, Princeton 1998; O. Pedersen, *A Survey of the Almagest: With Annotation and New Commentary by Alexander Jones*, Springer, New York 2010.

[49]M.T. Riley, *Ptolemy's Use of His Predecessors' Data*, in "Transactions of the American Philological Association (1974–2014)", 125, 1995, pp. 221–250.

naked eye. It is believed that the Alexandrian astronomer used the changing brightness of the sky background at sunset as a touchstone for his photometric estimates.[50] The procedure should be something like this. First, he ideally divided into six equal parts the time interval between the appearance of the brightest stars (of the first magnitude) and the moment when the faintest stars (of the sixth magnitude) became visible. He then placed on this sequence the time at which, due to the progressive darkening of the sky, the object whose magnitude he wished to measure showed up. In practice, he had transformed a temporal sequence into a scale of brightness,[51] without knowing what we would now call the transform function. In short, he could not say what the brightness ratio was between two adjacent stars on his magnitude scale, nor whether this ratio remained the same along the entire scale. He was thus able to produce a catalog, which appeared around 140 A.D. in the seventh and eighth volumes of the *Almagest*, in which he listed, in addition to the ecliptic positions, the magnitudes of 1,022 stars arranged in 48 constellations.

Ptolemy's approach gained a considerable following. Recovered in the Middle Ages by the Arabs,[52] it was the starting point of an important field of study. At the end of the first millennium, the Ptolemaic scale would have been refined by the Persian Abd al-Rahman al-Sufi with the introduction of intermediate states between two successive classes of magnitudes through qualitative expressions such as "brighter than" or "fainter than". This implies that the uncertainties of the measurements were within half a magnitude. Al-Sufi's estimates persisted for hundreds of years, until in 1437 Ulugh Begh, the astronomer of Samarkand, grandson of the Turkish-Mongol conqueror Tamerlane, recovered them in their entirety in his catalog, where, instead, thanks to the results of more precise instrumentation, he was able to present a revision of the positions measured by Ptolemy. The work of the Alexandrian astronomer reached the West after the end of the millennium, mediated by the Arabs of the Iberian Peninsula, but also translated directly from Greek into Latin in Norman Sicily, and formed the basis of the studies of Tycho Brahe, one of the greatest naked-eye observers of the heavens, in the 16th century.

---

[50]R. Miles, *A Light History of Photometry: From Hipparchus to the Hubble Space Telescope*, in "Journal of the British Astronomical Association", 117, 4, 2007, pp. 172–186.

[51]Magnitude is the brightness of a source expressed on a logarithmic scale to adapt it to the response of the human eye, which is a relative instrument. In fact, the visual sensation depends on the relative variation of the stellar flux, $\Delta F/F$.

[52]O. Gingerich, *Islamic Astronomy*, in "Scientific American", 254, 4, 1986, pp. 74–83.

Overall, Ptolemy's model satisfied many of the observational requirements, but it was difficult to use. Moreover, it completely neglected the causes, which did not seem to be of much interest to its author. His main goal was accuracy in astrology,[53] a pseudoscience with rules rigorously derived from a few and entirely arbitrary principles, and associated with a precise knowledge of the changing configurations of the heavens. On this subject he wrote the famous *Tetrabiblos* (*Four Books*). In this thousand-year-old bestseller, he railed, among other things, against the charlatans who confused the seriousness of judicial astrology, the art of reading destiny in the stars, with the magic and occultism of which modern horoscopes are the most vulgar tail. Who knows what this honest scholar would say today if he were confronted with the havoc wreaked on common sense by the media.

Claudius Ptolemy was the last great light of Greek scientific culture, the synthesis of a thousand years of wisdom accumulated by the peoples who lived between the Indus and the Mediterranean. Rome, preoccupied with quite different matters and more interested in the government of men than in the domination of nature, added little to this heritage and took much away from that drive of Odysseus[54] thanks to which, in a handful of centuries, primitive creatures had become men and women capable of converting Aristotle's "astonishment" into knowledge.[55]

The barbarians — blond and braided Germans driven by the Huns — were pressing on the borders of the empire. It was the dawn of a strange period in history, a fertile ground for racial crossbreeding, poor in science but very rich in spirituality. From that period, almost a thousand years later, a new world would be born, built on the recovered heritage of the ancients and on the fruits of its slow maturation in the Middle Ages. A new consciousness, of which the thought of Isidore, a very learned bishop of Seville who lived at the turn of the sixth and seventh centuries, is an example.

---

[53] R.R. Newton, *Astronomy, Astrology, Ptolemy, and us*, in "Johns Hopkins APL Technical Digest", 1, 1, 1982, pp. 77–80.

[54] The expression "drive of Odysseus" remains for the thirst for knowledge. It comes from the verses of Canto XXVI of Dante's *Inferno*, where the Greek hero urges his companions to cross the Pillars of Hercules and enter the Great Sea just to know what is there: « *Call to mind from whence ye sprang: / Ye were not form'd to live the life of brutes, / But virtue to pursue and knowledge high*» (vv. 118–120).

[55] Cf. L. Russo, *The Forgotten Revolution. How Science Was Born in 300 B.C. and Why it Had to Be Reborn*, Springer, Berlin 2004.

Astronomy is the study of the stars, which deals with the reason
for the course, and the behavior of stars with respect to themselves and to the Earth. In fact, there is a difference between
astronomy and astrology. Astronomy deals with celestial revolutions, the rising and setting of the stars, or the reasons why
such motions occur the way they do. Astrology, on the other
hand, is loosely based on natural observation but with superstitious conclusions. It is a science of nature when it investigates
the course of the Sun and the Moon or the stops, at regular
intervals, of the stars. It is different from the superstitious line
of thought followed by astrologers, who draw omens from the
stars and establish correspondences between the twelve signs of
the zodiac and individual parts of the body or soul. It is from
the course of the stars that they try to prophetically derive a
person's nativity and character.[56]

There is little to say about the progress of astronomy in the many centuries that elapsed between the time of Ptolemy, when the Antonines, the
emperor-philosophers, ruled in Rome, and the fall of Constantinople at the
hands of the Turks in 1453, and beyond. We shall confine ourselves to mentioning the role of Arab civilization in rescuing the knowledge accumulated
by the Greeks, and some truly original contributions. For example, the
introduction of the decimal and the zero in Europe, which led to an arithmetic system within reach of all and to an algebra that finally triumphed
over geometry. And again the studies on refraction; the improvement of the
astrolabe, a true portable planetarium of incomparable fascination; the long
series of accurate astronomical observations and the further complications
made to Ptolemy's astronomical model, which improved its performance;
and the return to the Aristotelian purity of uniform motions, abandoning
the idea of the equant. From Iraq to Morocco, after the fall of the Roman
eagles, the descendants of the Prophet's warriors tried to unite the world
under a different banner, a new faith, and a single language. In the long
run, they too failed, but their greatness served to carry the culture and science of the ancient peoples of the Mediterranean into the modern world.[57]

---

[56]Isidore of Seville, *Etymologiarum sive Originum libri XX*, III, 24.
[57]G. Saliba, *Greek Astronomy and the Medieval Arabic Tradition: Medieval Islamic
astronomers were not merely translators. They May Also Have Played a Key Role in
the Copernican Revolution*, in "American Scientist", 90, 4, 2002, pp. 360–367.

As the Roman Empire waned, another actor appeared on the astronomical scene, destined to play a major role in a momentous clash between science and faith. The new religion, born from the preaching of Jesus of Nazareth, sought a definitive order after the liberalization of the cults brought about by Constantine the Great's Edict of Milan in 318. To ensure its survival and affirmation, it was necessary to establish order among the various souls of a now universal faith, which nevertheless suffered from partisan jealousies and profound cultural differences among its followers. In the year 325, the emperor took matters into his own hands and convened an ecumenical council in Nicaea, Anatolia. The great theological conundrum to be resolved concerned the nature of Christ in relation to the Father, which had been challenged by the Egyptian presbyter and theologian Arius and his followers. This was only the tip of the iceberg of a clash between different powers and civilizations.

The Council also established a rule for calculating the date of Easter, the most important of the Christian holidays, which celebrates the resurrection of Christ and, in continuity with the pagans, the end of winter and the rebirth of nature. In order to distinguish it from Passover, the similar holiday by which the "deicid Jews" commemorated the liberation of the people of Israel from slavery in Egypt, the Christian Easter was established on the first Sunday following the first full moon after the vernal equinox. This choice meant that the date of Easter[58] would always fall between March 22 and April 25. This harmless rule involved three key elements of Christian symbology, but it turned a religious matter into a complicated astronomical problem: finding the date of a movable feast each year and communicating it sufficiently in advance to the faithful scattered throughout the vast territory of the empire. The Archbishop of Alexandria, who was in a strong position having won the theological dispute, took on this task. He could still count on the scholars of the famous Library, which had been founded after the death of Alexander the Great by the ancestor of the dynasty of the Greek pharaohs, Ptolemy I the Savior. Called into question, the heirs of Claudius Ptolemy began with a sensational proposal. They suggested moving the date of the vernal equinox to March 21, four days earlier than

---

[58]The rule established by the Nicaea Council was not immediately clearly understood by those who had been celebrating the resurrection on Passover. It was also difficult to apply uniformly everywhere, since Rome and Alexandria had different methods for determining the vernal equinox. Eventually, however, it was accepted by all of the Western churches.

the convention adopted in the time of Julius Caesar. It was the fledgling Church's first attempt to bring the calendar back in line with the seasons.

The facts are as follows.[59] The seasonal cycle lasts a full tropical year, which is the time it takes the Sun to pass through the vernal equinox twice in a row. If we measure this time interval using the mean solar day as the unit, which is the other fundamental cycle for life on Earth, we get a bad number, of which 365.2422 is only an approximation. The Babylonians rounded it to 360, so that one day in the celestial protractor was equal to one degree. The Egyptians, who had a natural and rather precise metronome in the flooding of the Nile due to the simultaneous appearance of the star Sirius just before dawn, realized that the solar calendar in vogue in the Fertile Crescent advanced the seasons by more than 5 days a year. So they increased the year to 365 days. The addition was considered by the priest-astronomers as a non-time, a suspension of ordinary life to pray and, why not, to bring offerings to the temple. However, with this substantial correction, the Egyptians continued to lose almost 1/4 of a day per year, equal to three months in 360 years; enough to experience a real reversal of the seasons with respect to the tropical points, with the solstices in spring and fall, and with the equinoxes when the temperature reaches extremes of hot and cold. Far too much! So, from time to time, the pharaohs intervened with impromptu corrections to bring the calendar back in line with the cycles of nature on Earth. This was, of course, an arbitrary management of time, suitable to meet the needs of an informed oligarchy.[60]

Finally, in 46 B.C., on the recommendation of the Egyptian astronomer Sosigenes, a protégé of Cleopatra, Julius Caesar modified the civil calendar of Rome and its provinces by adding a fourth year, one day longer, to each cycle of three years of 365 days. This leap year (*annus intercalaris*) was also called *annus bissextus* in Latin, since the extension to 366 days was achieved by doubling the day of February 24, *sexto die ante Kalendas Martias* (sixth day before the first of March). Caesar, now at the height of his military glory and power, rested in Alexandria in the welcoming arms of Cleopatra, but he had not forgotten his former role as *pontifex maximus*, that is, as we would say today, the holder of the office of the calendar. It could not have

---

[59] E.G. Richards, *Mapping Time: The Calendar and its History*, Oxford University Press, Oxford 1998.

[60] To this day, the international organization responsible for ensuring the synchronism between the time of atomic clocks and that marked by the movements of the Earth, imperceptibly variable in an unpredictable way, "tweaks" the time when necessary (about once every 18 months) by adding a leap second.

been better at that time. However, the problem that the Julian reform had sought to solve was still alive, albeit in a much less conspicuous way. Among other things, the cycle of the seasons now lasted less than the year. In fact, the Julian reform had introduced an average calendar year of 365.25 days, longer than the solar year, but only by 0.0078 days, or 11 minutes and 14 seconds. It looks like nothing but ... "*gutta cavat lapidem*" (dripping water hollows out stone), as Ovid wrote in a letter from his exile in Tomis on the Black Sea. After almost 4 centuries, at the time of the Council of Nicaea, the advance of the calendar in relation to the annual cycle of the seasons had already exceeded 3 days. Therefore, the wise heirs of Ptolemy moved the vernal equinox to March 21, curing the symptom but not the cause. Even today, after 17 centuries, the equinox falls on this date, but this is only possible as a result of another reform, known by the name of the pope who promoted it, Gregory XIII, born Ugo Boncompagni from Bologna.

The need had been in the air for some time;[61] so much so that already in 1514, on the occasion of the V Lateran Council convened by Pope Julius II della Rovere, his successor Leo X, a cultured and refined scion of the Florentine Medici family, had begun an investigation of the problem involving experts from all over Europe. It is likely that Copernicus himself was contacted to give his opinion, although there are no documents to prove it. It was a flash in a pan. The pope, preoccupied with far more concrete matters of political balance and the beginning of Martin Luther's rebellion against the sale of indulgences, lost quickly interest in the operation. Perhaps he was also discouraged by the uncertain knowledge of the length of the solar year, *condicio sine qua non* (a condition without which it could not be) for a calendar to work. The question was revisited in 1582. The Calabrian Aloysius Lilius and an influential Jesuit of the Roman College, the German Christopher Clavius, demonstrated to Gregory XIII that, in order to recover the time stolen by the Julian year and to obtain an almost perfect agreement of the calendar with the seasonal cycle,[62] 3 days should be removed every 4 centuries. But how? The two mathematicians proposed a strategy as elegant as it was simple, aimed at evenly distributing the correction. It was enough that the secular years were no longer promoted to leap years unless they were multiples of 400. This provided excellent synchronicity

---

[61]C.H. Genung, *The Reform of the Calendar*, in "The North American Review", 179, 575, 1904, pp. 569–583.
[62]The Gregorian rule satisfies the (approximate) decomposition of the number 365.2422 into simple fractions: $365.2422 \simeq 365 + 1/4 - 3/400 = 365.2425$.

with the seasonal cycle, and Easter was saved! But in the meantime the damage had been done. In order to return the hands to the Nicaea dial and recover the time that had "slipped away", Gregory XIII ordered that Thursday, October 4, 1582, the day of St. Francis — but also of Bishop St. Petronius, protector of Bologna, his hometown — be followed by Friday, October 15, the day of St. Teresa of Ávila, with all due respect to the saints and blessed who would not see their feasts celebrated that year.

In addition to getting its hands on the calendar, as every new power on Earth has always done, the Church of Christ intervened almost immediately, and heavily, in science and astronomy in particular, causing a rapid decline there where its armed wing could strike.[63]

During the long Middle Ages, the Western world, preoccupied with other matters, had lost, partly deliberately, the memory of its past, including the scientific achievements of the Greeks. But at the beginning of the second millennium, much of what then seemed lost forever came back into circulation in Latin and in the languages spoken in Europe, thanks to translations from Arabic and Greek; and while the thought of Aristotle — «master of those who know», says Dante Alighieri in Canto IV of the

---

[63] In 1846 the German mathematician Carl Jacobi made this brief and personal reconstruction of the transition between the Middle Ages and the Renaissance while speaking *About Descartes' Life and Method of Reason* (quoted by T. Dantzig, *Number: The Language of Science*, Macmillan, New York (NY) 1930): «History knew a midnight, which we may estimate at about the year 1000 A.D., when the human race lost the arts and sciences even to the memory. The last twilight of paganism was gone, and the new day had not yet begun. Whatever was left of culture in the world was found only in the Saracens, and a Pope eager to learn studied in disguise in their universities [Gerbert of Aurillac (c. 946 – 1003) who took the name of Sylvester II], and so became the wonder of the West. At last Christendom, tired of praying to the dead bones of the martyrs, flocked to the tomb of the Saviour Himself, only to find for a second time that the grave was empty and that Christ was risen from the dead. Then mankind too rose from the dead. It returned to the activities and the business of life; there was a feverish revival in the arts and in the crafts. The cities flourished, a new citizenry was founded. Cimabue rediscovered the extinct art of painting; Dante, that of poetry. Then it was, also, that great courageous spirits like Abelard and Saint Thomas Aquinas dared to introduce into Catholicism the concepts of Aristotelian logic, and thus founded scholastic philosophy. But when the Church took the sciences under her wing, she demanded that the forms in which they moved be subjected to the same unconditioned faith in authority as were her own laws. And so it happened that scholasticism, far from freeing the human spirit, enchained it for many centuries to come, until the very possibility of free scientific research came to be doubted. At last, however, here too daylight broke, and mankind, reassured, determined to take advantage of its gifts and to create a knowledge of nature based on independent thought. The dawn of the day in history is know as the Renaissance or the Revival of Learning».

Inferno (v. 131) — once revived by Muslim scholars, was attacked by hosts of warlike religious fundamentalists,[64] it found a new champion in Thomas Aquinas.

With great dialectical skill, *Doctor Angelicus* managed to combine Christian theology with the thought of the Greek philosopher: a synthesis that Dante would immortalize in poetry in the *Divine Comedy*.[65] It may have been the beginning of a virtuous marriage[66] and was instead the harbinger of a series of misfortunes for science, which lost its freedom and momentum, and for the Church, which, in order to defend its own entrenchment against the attacks of more or less interested reformers, had to abandon its message of love. While in Europe, literature and art had again flourished in the shadow of the excessive powers of the theological institutions, the philosophy of nature remained at stake, frozen by a password: *ipse dixit* (he said it himself).[67]

It would be astronomy, once again more advanced than other sciences, that would pay the highest price in blood and suffering and then lead the revolution that would open the door to the modern age. The fact is that it

---

[64] A strong blow to Aristotelianism (and Averroism) was given in 1277 by Étienne Tempier, Bishop of Paris, with the famous condemnation of 219 propositions of theological interest but with cosmological implications. An event which, according to the epistemologist Pierre Duhem at the end of the 19th century, was the authentic beginning of the scientific revolution. Cf. T. Siegfried, *The Number of the Heavens: A History of the Multiverse and the Quest to Understand the Cosmos*, Harvard University Press, Cambridge (MA) 2019.

[65] Cf. S. di Serego Alighieri, M. Capaccioli, *The Sun and the Other Stars of Dante Alighieri. A Cosmological Journey through the Divina Commedia*, World Scientific Press, Singapore 2022.

[66] Note, however, the modernity of St. Thomas in distinguishing between principle and model: the first is intangible, while the second is falsifiable: «You can bring an argument for two purposes. First, to rigorously prove a given principle: as in the natural sciences rigorous arguments are brought up to demonstrate that the motion of the heavens always has a uniform velocity. Second, you can bring an argument not to scientifically demonstrate a given principle, but only to show how the resulting effects are intimately linked to the principle, posited [as an axiom]: as in astronomy, eccentrics and epicycles are admitted because once this hypothesis is accepted, we can give an account of the irregularities that appear to the senses in the motion of celestial bodies; however this argument is not binding, since perhaps [such irregularities] could also be explained by admitting another hypothesis.» (Thomas Aquinas, *Summa Theologiae*, I q32a2, p. 390.)

[67] Assertion without evidence, or dogmatic expression of an opinion. Marcus Tullius Cicero used the Latin translation of this Greek locution in his *De Natura Deorum* (On the Nature of the Gods) to describe the attitude of blind trust of Pythagoras' followers toward their master. It was later reused for Aristotle by the Arab philosopher Averroes, and from there re-imported to Europe.

was not important for the believer to examine the Lord's work in order to understand it — it was argued — but rather to know and follow the will of the Creator. «It is clear from an ecclesiastic who has been elevated to a very eminent position [Cardinal Cesare Baronio?]», Galilei would write to Christine of Lorraine, Grand Duchess of Tuscany, in 1615, «that the Holy Spirit's intention is to teach us how to go to heaven, and not how heaven goes.»[68] Even a figure of the stature and genius of Augustine of Hippo had warned in his *Confessions* not to neglect man for the sake of Ulysses' drive to know the world: a call of profound wisdom that would not have been difficult to mystify.

The plaster cast given to astronomy by Rome and also by the Reformed Churches should not lead us to believe that people no longer looked at the sky or measured it. The positions of the stars, fixed and wandering, were indispensable to travelers and especially to astrologers, who were respected even in ecclesiastical circles. Astronomy was one of the disciplines of the *quadrivium*,[69] which was part of the standard curriculum for the education of a cleric about to enter the university. It was also taught to aspiring physicians in the belief that the stars played an important role in determining the functioning, good and bad, of the human body. A practice that was already in vogue at the beginning of the 17th century, so much so that the main pool of Galilei's students in Padua were among the apprentice surgeons who had to learn the rudiments of astrology.

The book of the heavens that astrologers had to read was written in the tables that gave the position — ephemeris (from the Greek ἐφημερις, daily) — of the Sun, planets, and stars, and the dates of eclipses. Famous, after those of the *Almagest*, were the *Tabulae alphonsinae*, commissioned in the mid-thirteenth century by Alfonso X the Wise, king of Castile and Leon, who for this work used the skills of about fifty astronomers of the Arabic school of thought. Thinking back, the quotation from Horace (*Epistle*, II, 1, 156) comes to mind: «*Graecia capta ferum victorem cepit et artes intulit*

---

[68] G. Galilei, *Lettera a Cristina di Lorena*, in Id., *Scienza e religione. Scritti copernicani*, M. Bucciantini, M. Camerota (eds.), Donzelli, Roma 2009, p. 47 (in Italian).

[69] In medieval time, *quadrivium* ("four ways" in Latin: arithmetic, geometry, music, and astronomy), along with *trivium* ("three ways": grammar, logic, and rhetoric), formed the liberal arts training program that prepared students for the study of theology and philosophy. Both terms were coined in the VI century by the Roman senator and scholar Severinus Boethius, a precursor of the Scholastic movement.

agresti Latio».[70] It was the time when, to bring the arts to the Europeans were those same Muslims whose descendants we are now rejecting as they seek asylum.

The Alfonsine tables improved on the now obsolete tables produced of Ptolemy, although they were based on the same world model for calculations. In turn, they were taken up and perfected again and again, until in the 15th century, with the German Regiomontanus, alias Johannes Müller of Königsberg,[71] the practice of using the Copernican heliocentric model prevailed (see the next chapter). The astrologers took advantage of this, as did the navigators who would soon begin the epic of geographical discovery. A scientifically and sociologically instructive anecdote concerns the most famous of these explorers. During his fourth transoceanic voyage, Christopher Columbus had to dock on the coast of Jamaica to repair his ships, which had been damaged by a storm, and to replenish his supplies of fresh food and water. Faced with open hostility from the natives, the admiral resorted to a ruse. The ephemeris tables of Regiomontanus he had brought with him predicted that the Moon would go into eclipse on the evening of February 29, 1504. He then threatened the natives to make the star disappear, and in the midst of the eclipse, he made a bargain with them for its reappearance and so he obtained the satisfaction of all his demands.[72]

Another crisis of Aristotelian scientific thought, following the more exquisitely metaphysical ones of the Middle Ages and early humanism, began as early as the 14th century.[73] The first impulse came from the Frenchmen Jean Buridan and Nicole Oresme, who took a stand against the immobility of the Earth. It was a time when the philosophy of nature was still in search of a working method, and subtle reasoning was worth more than sensible experience. The players of this risky game, closely watched by the Church, which had specialized the Holy Office in this task, often wore partisan shirts, the colors of which were superstition, fear, narrow-mindedness, and ignorance. One light among many was that of Nicholas Krebs of Kues, a German cardinal of encyclopedic culture known by his Latin name, Nicholas of Cusa. Around the middle of the 15th century, he

---

[70] Greece, once conquered, in turn conquered its uncivilized conqueror and brought the arts in the rustic Latium.

[71] Latinized term for Königsberg, Müller's hometown.

[72] W.F. Rigge, *The Columbus eclipse*, in "Popular Astronomy", Vol. 31, 1923, pp. 506-509.

[73] P. Rossi, *The Birth of Modern Science in Europe*, Blackwell Publishers, Hoboken 2000.

promoted an innovative vision of the world that disregarded the classical confinement to the sphere of the fixed stars. In this infinite universe, the Earth was a star *mobilis*, albeit *nobilis*. These were still mere creaks of the ancient foundations of the Ptolemaic heavens. But the times were rapidly ripening for a true revolution. To ignite it, fate chose a quiet Polish canon, Nicolaus Copernicus.

Chapter 3

An introvert, a reactionary, and a visionary: Copernicus, Kepler, and Brahe

> Therefore, mathematical Truth
> prefers simple words,
> since the language of Truth
> is itself simple.
>
> Tycho Brahe,
> *Epistolarum astronomicarum liber primus*

> Citizens, do you want a revolution
> without a revolution?
>
> Maximilien de Robespierre,
> *Address to the National Assembly,*
> *5 November 1792*

Whoever thinks of the revolutionary Copernicus[1] as a daring guerrilla of knowledge — a kind of Che Guevara *ante litteram* (ahead of his time) ready to fight to the point of the utmost sacrifice for the noble cause of scientific truth — could not be further from reality. Nicolaus was a reserved person, attached to the relatively quiet life he had built to cultivate his talents and especially his great passion for astronomy. He was arguably the most conservative of all the innovators.

The youngest of four brothers, he was born in 1473 in Toruń, an important trading center on the Vistula River, now the capital of the Kuyavian-Pomeranian Voivodeship, one of Poland's first territorial subdivisions. The region had come under the Polish crown only seven years earlier, at the end

---

[1]D. Sobel, *A More Perfect Heaven: How Copernicus Revolutionized the Cosmos*, Walker & Co., New York 2011; O. Gingerich, *Copernicus: A Very Short Introduction*, Oxford University Press, Oxford 2016. For an overview of the Copernican revolution, see A. Koestler, *The Sleepwalkers: A History of Man's Changing Vision of the Universe*, Hutchinson, London 1959; A. Bettini, *Da Talete a Newton*, Bollati Boringhieri, Turin 2019 (in Italian).

of the victorious uprising of the indigenous Prussian population against the Teutonic Knights who had ruled these lands for two centuries. The Hanseatic cities, tired of a harsh yoke that limited personal, entrepreneurial, and commercial freedom, had appealed to the King of Poland, Casimir IV, for protection. It was more a provocation than anything else. However, it had triggered a war that lasted thirteen years and ended with the capitulation of the monk-soldiers and the loss of their eastern territories. Thus, the great astronomer was born within the geographical borders of Poland and under a Polish ruler. But the family background and the cultural context in which he was formed in his youth, although opposed to the Teutonic Order, had German roots, so much so that Nicholas' mother tongue was German. It is reasonable to assume that he also spoke Polish, although there are no documents to prove it. He certainly knew Greek and was fluent in Latin, a must for any educated man of his time. He also learned Italian during his long period of study in the *Bel Paese*.[2]

The Copernicus family was a wealthy and prominent one. The father, also Nicolaus, traded in copper, a metal called *Kupfer* in German, from which the surname Kopernik may have been derived. The mother, Barbara, belonged to the influential Watzenrode family, originally from Lower Silesia. The war against the Teutonic Knights had further enriched the Koperniks, partly because of the position Father Nicolaus had gained among the insurgents in Toruñ. Then their business began to deteriorate, and with it their standard of living. To make matters worse, both Barbara and Nicolaus died in 1483, leaving their four children, two boys and two girls, alone and in poverty. It could have been the beginning of a nightmare for the future architect of an epochal scientific revolution[3] had it not been for the providential intervention of her maternal uncle. A cultured and wealthy man, Lucas Watzenrode took the four orphans into his home and provided for their education.[4]

The course of history thus resumed in the right direction. In fact, only six years after these events, Uncle Lucas was unexpectedly elected Bishop Prince of Warmia. Secular power in this small ecclesiastical state

---

[2]"Beautiful country"; expression coined by Dante to refer to Italy: «*del bel paese là dove 'l sì suona*» (of the fair land where the 'si' [yes] doth sounds; *Inferno* XXXIII, v. 80).

[3]T. Kuhn, *The Copernican Revolution: Planetary Astronomy in the Development of Western Thought*, Harvard University Press, Harvard 1992.

[4]W.M. Stachiewicz, *Copernicus and His Work: A Biographical Sketch*, in "The Polish Review", 17, 4, 1972, pp. 62–81.

in northern Poland was exercised by the bishop of the cathedral at Frombork, a town overlooking the Vistula Bay on the Baltic Sea. The investiture of Watzenrode as prince of Warmia,[5] which was strongly wanted by Pope Innocent VIII Cybo, put an end to a long struggle between the assembly of priests of Warmia and King Casimir IV, which had exploded after the transit of the region under Polish sovereignty. For a time, it also opened up unexpected prospects of prosperity for the Copernicus brothers. It was really a good chance for those who did not despise an easy life.

Nicolaus studied both astronomy and astrology while attending the Jagiellonian University in Cracow, supported by the money of his uncle, who wished to make him a lay clerk in the service of the Chapter of Frombork Cathedral.[6] He did not take a degree there, but during four years of relentless reading he accumulated an authentic heritage of scientific and philosophical knowledge and some familiarity with mathematics. Meanwhile, Watzenrode prepared the ground for his nephew's position in the diocese. But his brazen nepotism met with the opposition of the chapter. The winds of revolt against abuses were already blowing in its pews. As everywhere in Europe, they found their ultimate *raison d'être* (reason for being) in the shameful behavior of the new Roman pontiff, Alexander VI Borgia, elected in 1492: a fateful year, because in the short span of a single revolution of the Earth, besides the ascension of Don Rodrigo de Borja to the throne of Peter, Christopher Columbus discovered America,

---

[5] In the Middle Ages, an investiture was an act by which a lord granted a possession or right (the *beneficeum*) to another person (the *vassus*). In the 11th century, lay rulers considered it their prerogative to appoint bishops and abbots of their choice, both as an affirmation of their superiority over the Church and because religious subjects, being in principle childless, were less likely to transform the *beneficeum* into a permanent possession of their family. This custom led to a conflict between the empire and the papacy, which reached extreme peaks, such as the deposition of Pope Gregory VII, born Hildebrand of Sovana, by Henry IV and the subsequent excommunication of the emperor, who was forced to come to terms with the pontiff. In 1122, Pope Callixtus II and Emperor Henry V reached an agreement. The Concordat of Worms provided that the election of bishops would be a matter for the Church, and that bishops would swear an oath of allegiance to the secular monarch. The emperor retained the right to preside over the election of all high ecclesiastical offices and to settle disputes, but gave up the right to elect the pope. From these premises arose the appointment of Watzenrode as Prince of Warmia.

[6] Organized on the model of the Chapter of Meissen, the Guild of Frombork consisted of 16 canons, only five of whom were to be prelates. Cf. E. Rosen, *Copernicus was not a priest*, in "Proceedings of the American Philosophical Society", 104, 6, 1960, pp. 635–661.

Granada, the last Arab stronghold in Western Europe, surrendered to the very Catholic Kings of Castile and Aragon, and Lorenzo the Magnificent died in Florence. Three events that would mark the history of the Western world, added to the fall of Constantinople at the hands of Sultan Mehmed II in 1453 and the completion of the *42-line Bible* by Johannes Gutenberg in 1455, the first movable type printing book.

Hearing the creaking of his comfortable chair, the prince-bishop decided to postpone his plans until a better time and to send his nephews, Andrew and Nicholas, on his own educational path: a *cursus studiorum* that passed through the renowned northern Italian universities. Thus, beginning in 1494, the two young men were sent to the *Bel Paese*[7] where the vigorous renaissance of the arts and sciences was accompanied by a widespread paralysis of consciousness and a pervasive sense of individualism. It was a kind of long Erasmus+ program[8] that Nicolaus divided into four periods, with as many trips home, first at the expense of his uncle and then of the Chapter of Warmia. He was in Bologna, later in Rome, Padua, and Ferrara, where he studied various disciplines, including medicine. Finally, he graduated in canon law in Ferrara. It was the academic title he needed to legitimately take on the job his powerful uncle had reserved for him. During the eight years he stayed in Italy, he also found a way to meet and work with astronomers. In particular, with Domenico Maria Novara da Ferrara, who was responsible for the annual astrological predictions for the city run by the House of Este, mainly concerning the fate of Italian princes and their enemies.

In 1503, he returned to his homeland permanently and began his service in Frombork as a canon of the cathedral. This job was the appropriate sinecure for him and for his life project. In the following years, he served his uncle Lucas as a doctor and secretary and performed various other tasks and duties, but he concentrated mainly on the project of a radical revision of the Ptolemaic system of the world. He did not even allow himself to take a wife and, as far as is known, his love life was limited to a late and much resented affair with a younger governess, Anna Schilling, of whom Copernicus was a great uncle.[9] Nicolaus had an obsession: he was convinced that the Sun

---

[7] Z. Wardeska, *Nicholas Copernicus' Italian Studies: Their Chronology and Scope*, in "The Polish Review", 31, 1, 1986, pp. 5–11.

[8] Erasmus+ is the EU's programme to support education, training, youth, and sport in Europe; cf. https://erasmus-plus.ec.europa.eu/it.

[9] http://copernicus.torun.pl/en/biography/1503-1543/6/.

was stationary and that the Earth revolved around it. He was not the first to think this, but he proved to be the most convincing.

By the time of Eudoxus, another disciple of Plato, Heraclides Ponticus, and a century before him, the Magno-Greek Pythagorean Philolaus, had already speculated that the daily motion of the heavens was the consequence of a rotation of the Earth, which was therefore not completely immobile. Perhaps Eudoxus himself, or some other philosopher of the Pythagorean school, argued that Mercury and Venus were primarily orbiting the Sun. This moderate geocentrism became total heliocentrism with Aristarchus.[10] We know relatively little about his life. He probably studied in Alexandria, alongside people of the caliber of Archimedes, on the campus of the Library created by the Ptolemies in the city on the Nile delta, and died in Samos, his native island off the coast of Anatolia, where Pythagoras himself had been born more than two and a half centuries earlier. Only one work remains of him; it reports in a few pages the ingenious methods for measuring the distances and dimensions of the Moon and the Sun (we will speck of them in chapter 12).

Aristarchus is also celebrated for having created a theory that, leaving the Sun and the stars motionless, explained the phenomena of the heavens by a diurnal rotation and a heliocentric revolution of the Earth. Not only that, but in order to account for the seasonal cycle, he had correctly speculated that the axis of rotation of our planet was significantly tilted on the orbital plane. This was an insight that went beyond pure geometric models and implied bold physical assumptions. We read about it in *The Sand Reckoner*, a kind of letter written by Archimedes to his king Gelon II in the 3rd century B.C. The passage concerning Aristarchus served the Syracusan mathematician as a kind of bibliographical reference to the (maximum) volume of the cosmos with which he could estimate, for the benefit of the tyrant, the number of grains of sand needed to fill it. It was an incredible calculation «for most of those unfamiliar with mathematics»,[11] which forced this absolute genius to invent new names for numbers:

> You [King Gelon] are aware the universe is the name given by most astronomers to the sphere the centre of which is the centre of the earth, while its radius is equal to the straight line

---

[10]W.H. Stahl, *The Greek Heliocentric Theory and Its Abandonment*, in "Transactions and Proceedings of the American Philological Association", 76, 1945, pp. 321–332.

[11]Archimedes, *Arenarius*, in *Id.*, *Opere*, A. Frajese (ed.), UTET, Turin 1974, p. 448.

between the centre of the sun and the centre of the earth. This is the common account as you have heard from astronomers. But Aristarchus has brought out a book consisting of certain hypotheses, wherein it appears, as a consequence of the assumptions made, that the universe is many times greater than the universe just mentioned. His hypotheses are that the fixed stars and the sun remain unmoved, that the earth revolves about the sun on the circumference of a circle, the sun lying in the middle of the orbit, and that the sphere of fixed stars, situated about the same centre as the sun, is so great that the circle in which he supposes the earth to revolve bears such a proportion to the distance of the fixed stars as the centre of the sphere bears to its surface.[12]

Aristarchus' cosmology met with opposition from his contemporaries, for both ideological and rational reasons. The motion of the Earth — it was correctly argued — should be reflected in a periodic oscillation of the stars due to the progressive drift of the observer's point of view: an annual variation of the position that had never been observed. Aristarchus defended himself by arguing that the stars were too far away to produce significant parallax effects. His vision of the universe implied much larger spaces than the narrow sky of the geocentric model, for which it was enough that the sphere of fixed stars could contain the orbit of Saturn. In *The Sand Counter* Archimedes had estimated, without taking the position of any of his competitors, that the heliocentric model could accommodate up to a trillion universes forged in the manner of Eudoxus.

In addition to the purely astronomical arguments, there were also physical and aesthetic reasons for rejecting heliocentrism, related to Aristotle's theory of natural places[13] and the belief that the Earth couldn't move because of its enormous mass. Reasonable mistakes that, together with religious prejudices, caused Aristarchus to be disbelieved. In fact, according to the account given by the Greek historian Plutarch in his *On the Face of the Moon*, the philosopher was accused of impiety and perhaps persecuted for this crime of opinion. He was certainly not the first — if the legend has any basis in truth — and he would not be the last.

---

[12]Ibid, p. 447. For the English translation, see: https://mathshistory.st-andrews.ac.uk/Biographies/Aristarchus/.

[13]See, for example, B. Russell, *A History of Western Philosophy*, Simon & Schuster, New York 1986.

In any case, after him there was almost complete silence on the question of heliocentrism, with the one notable but useless exception of the Babylonian astronomer Seleucus of Seleucia, who in the second century B.C. also speculated about infinite horizons. Not even a genius at the level of Hipparchus was tempted by the idea of setting in motion the "hearth of the world". The authority of Ptolemy, the long sleep of science in the Middle Ages, and the subsequent marriage of Greek-Alexandrian cosmology with Christian theology, guaranteed by the doctrine of the doctors of the Church and protected by the stakes of the Inquisition, would have been enough to dampen any ambition. After all, the geocentric solution was endorsed by the Bible. «Sun, stand still over Gibeon, and you Moon, over the Valley of Aijalon», was Joshua's plea to his God, not an impartial arbiter, to give the leader of the Israelite tribes more time to exterminate the Canaanites. «So the Sun stood still, and the Moon stopped» — continues the *Book of Joshua* (Jos. 10: 12–14) — «till the nation avenged itself on its enemies, as it is written in the *Book of Jashar*. The Sun stopped in the middle of the sky and delayed going down about a full day. There has never been a day like it before or since, a day when the Lord listened to a man. Surely the lord was fighting for Israel!». Clearer than that! Woe to the unwary destroyers of a theory validated by the Bible! And woe they would have been, as we shall see.

Copernicus[14] had begun to have doubts about the centrality of the Earth when, as a student in Italy, he had noticed some flaws in the Ptolemaic model of the Moon's orbit. Back in Poland, he started to work on a heliocentric hypothesis.[15] His equipment for observing the night sky was no better than that of the Alexandrian astronomers, nor were the astronomical data and theoretical tools used to construct his celestial clock. To make the Earth and the other planets revolve around the Sun, Frombork's canon, faithful to the Aristotelian dogma of uniform circular motions, recycled the Ptolemaic gears of epicycles, deferents, and eccentrics, adapting them arbitrarily to better satisfy appearances. He only abandoned the equant, which nobody really liked. His new paradigm was based on three hypotheses:

---

[14] E. Rosen, *Aristarchus of Samos and Copernicus*, in "The Bulletin of the American Society of Papyrologists", 15, 1–2, 1978, pp. 85–93.
[15] T.W. Africa, *Copernicus' Relation to Aristarchus and Pythagoras*, in "Isis", 52, 3, 1961, pp. 403–409.

1. there is no single center for the celestial spheres (therefore the Copernican model is not a perfect heliocentrism);[16] 2. the Sun occupies the center
of the universe; 3. the Earth is central only to gravity and to the sphere of
the Moon, and its motions about the Sun affect the apparent positions of
the planets, but not of the stars, because they are too far away. A clear
cut, then, to the claim that man occupies a privileged position. This is
why cosmologists today call it the "Copernican Principle"[17] the postulate
of a very democratic universe, free of privileged observers. This is actually
an extrapolation to a situation completely unknown to Copernicus, whose
merit, while enormous, is limited to unhinging a preconceived promotion of
mankind's role.

By the beginning of 1514, when Niccolò Machiavelli published his
earthly recommendations in *The Prince*, the foundations of the new cosmos
had already been laid. But Copernicus wanted to avoid any clash with the
Church, from which he indirectly received his bread and butter. So he had
no intention of communicating his work except to a very few friends and,
above all, of making it public. Thirty years earlier, to justify his silence
about potentially dangerous inventions, Leonardo da Vinci had written to
Ludovico il Moro, Lord of Milan: «this I do not publish or disclose because
of the bad nature of men».[18] He was thinking of others. Copernicus was
thinking of himself. These were hard times, and a simple lapse in communication could bring dramatic costs. Thus, prey to his chronic scientific
autism, he reluctantly drafted a *Commentariolus* of his speculations, but
ruled out the possibility of having it printed.[19] The plan was to circulate
the manuscript only among a few trusted readers. But in its slow wanderings through Europe, the dangerous text even touched the court of Pope

---

[16] In the *Astronomia Nova*, Kepler identified two different positions of Copernicus regarding the center of the world, one ideological and one practical, when the model had to be
confronted with appearances: «Now, as regards the body that is at the center [of the
world]; if there is none — as Copernicus wishes when calculating, as does Tycho, within
certain limits; or whether it must be the Earth — as Ptolemy and Tycho too want,
within certain limits; or if it must be the Sun, as I myself believe, and also Copernicus,
that is, when he theorizes. I begin to discuss this with physical arguments»; cf. R.S.
Westman, *The Melanchthon Circle, Rheticus, and the Wittenberg Interpretation of the
Copernican Theory*, in "Isis", 66, 2, 1975, pp. 164–193.
[17] A term coined by the English cosmologist Hermann Bondi in 1960. Cf. E. Harrison,
Cosmology: The Science of the Universe, CUP, Cambridge 2000, p. 140.
[18] Leonardo da Vinci, *Codex Leicester*, f. 15A–22v, ca. 1506.
[19] E. Rosen, *The Commentariolus of Copernicus*, in "Osiris", 3, 1937, pp. 123–141.

Clement VII de' Medici, though without causing any consequence. It would not be printed until after the author's death.

The amazement that these forty pages caused in various contexts convinced the reluctant canon to insist on his research. The cosmic clockwork he constructed was as complicated and difficult as those of his Greek and Arab predecessors, but it allowed the planets to be placed on a scale of distances from the Sun that was directly constrained by observations. It also eliminated the bizarre architectures invented to account for retrograde motion. Nicolaus then went back to work and within thirty years produced what would become the watershed between ancient and modern science, *De revolutionibus orbium coelestium* (On the revolutions of the celestial spheres): a monumental work that might never have seen the light of day without the decisive intervention of Georg Joachim Rheticus, the humanist name of Georg Joachim von Lauchen.[20]

The story of this young mathematician, the only true pupil of Copernicus, is emblematic of those times. He was born in 1514, the year of the *Commentariolus*, in a village at the crossroads of Austria, Switzerland, and Liechtenstein, in the bosom of a wealthy family that came from northern Italy on his mother's side. But when he was only fourteen years old, he had lost his father, a medical doctor sentenced to death for sorcery. For this dramatic story, he even had to change his father's family name to his mother's Germanized surname (von Lauchen) and finally, as was the humanist fashion of the Renaissance, adopt the Latin name of his native region, Raetia.[21] Then, not at all discouraged by his poor debut in the world, he rolled up his sleeves with the intention of making his way through life. Enthusiasm was in the air. At the University of Wittenberg, the center of the Protestant revolt, he was noticed by Martin Luther's intellectual arm, the humanist and theologian Philipp Melanchthon. A humanist and theologian, Melanchthon was committed to the project of school reform in Germany and, like the brothers of Ignatius of Loyola on the Catholic front, was convinced of the importance of education in the transmission and preservation of the faith. He hired Rheticus as a professor of mathematics, but allowed him to postpone taking up the position until after

---

[20]D. Danielson, *The First Copernican: Georg Joachim Rheticus and the Rise of the Copernican Revolution*, Walker & Co, New York 2006.

[21]Former province of the Roman Empire, comprising the territories occupied in modern times by eastern and central Switzerland, southern Germany (Bavaria and most of Baden-Württemberg), Vorarlberg, Tyrol in Austria, and part of northern Lombardy in Italy.

a long *peregrinatio academica* (academic mobility) to visit the greatest astronomers: a kind of multi-level "Erasmus+ program" to complete his education.

Copernicus' fame had now spread beyond the borders of Poland. Rheticus had heard of him in Nuremberg, perhaps from the publisher Johannes Petreius, who would later print the first edition of *De revolutionibus orbium coelestium*; and perhaps he could also consult a copy of the *Commentariolus* in his print shop. So he decided to go and meet the famous astronomer. After a three-week journey, mostly on foot and over a terrain made difficult by the raids of the Teutonic Knights, the young Lutheran reached Frombork and introduced himself to the Catholic Copernicus. George Joachim was literally bewitched by him and became his only disciple. He stayed with his idol for two years, helped him to complete the monumental treatise on the motions of the heavens by working until exhaustion, persuaded him to give it to the press — the elderly scientist feared the reaction of the Roman Church — and in 1540 he himself wrote a preview of the work in a *Narratio prima* (First account), which was also a paean to the Maestro.

Then, unable to prolong his stay, he returned to Wittenberg to resume his work at the university. Unfortunately, he failed in his plan to supervise the printing of Copernicus' manuscript in Nuremberg, the publishing capital of northern Europe. Forced to move to Leipzig at Melanchthon's behest, he transferred the curatorship to the theologian Andreas Osiander, a cultured Bavarian Lutheran, who created "the case". On his own initiative — as Johannes Kepler would later reveal — the new editor added an anonymous preface to the text in which he argued that the heliocentric model presented in the book was merely a mathematical exercise useful for calculating the planetary ephemeris, without any claim to truth.

> For it is the duty of an astronomer to compose the history of the celestial motions through careful and expert study. Then he must conceive and devise the causes of these motions or hypotheses about them. Since he cannot in any way attain to the true causes, he will adopt whatever suppositions enable the motions to be computed correctly from the principles of geometry for the future as well as for the past.[22]

---

[22]See N. Copernicus, *On the revolutions*, trans. by E. Rosen, The John Hopkins University Press, Baltimore, https://math.dartmouth.edu/~matc/Readers/renaissance.astro/1.1.Revol.html.

This was indeed a sudden slowdown compared to the opening words of the book:

> Diligent reader, in this work, which has just been created and published, you have the motions of the fixed stars and planets, as these motions have been reconstituted on the basis of ancient as well as recent observations, and have moreover been embellished by new and marvelous hypotheses.

Perhaps Copernicus never knew it. Legend has it that a freshly printed copy of his book arrived in Frombork in May 1543, when the astronomer was in a coma on his deathbed. Better that way! If he had seen this presentation, the gentle Pole would probably have been very upset, if only in his own mind, about this downgrading of his work. There are also historians who argue that Copernicus had a chance to read Osiander's text instead, and that out of jealousy he even deleted the reference to Aristarchus by his own hand. Others even doubt that the Canon held the heliocentric hypothesis to be true, considering it merely a useful mathematical tool.[23] In any case, Osiander's preface served to protect the book for a while from the wrath of the fundamentalists and the red-hot pincers of the Inquisition; at least until May 1616, the year of Galilei's first trial, when the work of the very talented «astrologer» Copernicus was suspended *donec corrigatur*, that is, until it had been adequately purged. In the meantime, however, a few thousand copies, two bestsellers for the time, had been able to circulate freely throughout Europe, infecting minds and consciences.

The philosophical and religious consequences of Copernican heliocentrism delivered the final blow of a new world against the walls of the old one, guarded by the Church and the secular powers. Over the previous ninety years, with the fall of Constantinople on one side and the last Arab stronghold of Granada on the other, the political geography of Europe had changed. The Mediterranean-centered horizon had widened with the discovery of the Americas and the opening of new routes and old markets to the East. The power of the Roman popes had been thrown into crisis by the reformers, and Gutenberg's new technology of printing books had opened up the democracy of knowledge. Then an innocent

---

[23] Cf. T. Siegfried, *The Number of the Heavens: A History of the Multiverse and the Quest to Understand the Cosmos*, Harvard University Press, Cambridge (MA) 2019.

scientific theory created the conditions for the restoration of man's freedom to think.[24]

Initially, the Roman Curia, educated and still fairly confident of its own strength, did not oppose Copernicus, so much so that forty years after the publication of the book, the tables of *De revolutionibus orbium coelestium* were consulted by the Jesuits who were preparing the revision of the calendar on behalf of Gregory XIII (see p. 131). The Protestants, on the other hand, were totally negative from the start, with Luther in the lead.[25] The former Augustinian friar called the Polish canon mad for daring to contradict Scripture. He even dragged into anathema his own disciple, Osiander, who had joined in the alleged heresy.

> There is talk of a new astrologer [sic!] who wants to prove that the Earth moves and not the Sky, the Sun and the Moon. It would be as if someone were traveling in a wagon or a ship and imagined to be still while the earth and the trees are moving. [Luther observed] So it goes now. Whoever wants to be intelligent must not agree with anything that others value as true. He must invent something of his own. That fool wants to overturn the whole art of astronomy. However, as the Sacred Scripture tells us, Joshua commanded the Sun and not the Earth to stand still.[26]

As for Calvin, there is no trace in his writings of the admonition often attributed to him: «Who will venture to place Copernicus' authority above that of the Holy Spirit?».[27] In any case the French theologian was bitterly hostile to the theories of the Polish astronomer. Proof of this is the invective he used in one of his sermons against the advocates of the thesis that «it is the Earth and not the Sun that moves», accusing them of being «delusional, insane», and «possessed by the devil».[28] Even part of the secular world

---

[24] P. Frank, *The Philosophical Meaning of the Copernican Revolution*, in "Proceedings of the American Philosophical Society", 87, 5, *Papers on Archeology, Ecology, Ethnology, History, Paleontology, Physics, and Physiology*, 1944, pp. 381–386.

[25] J.R. Christianson, *Copernicus and the Lutherans*, in "The Sixteenth Century Journal", Vol. 4, No. 2, 1973, pp. 1–10.

[26] K.E. Föstermann, H. Bindseil (eds.), *D. Martin Luthers Tischreden, oder Colloquia*, Vol. 4, Gebauer'sche buchhandlung, Leipzig 1848, p. 575.

[27] P.Ch. Marcel, *Calvin and Copernicus*, in "Philosophia Reformata", 46, 1, 1981, pp. 14–36.

[28] Cf. E. Rosen, *Calvin's Attitude towards Copernicus*, in "Journal of the History of Ideas", 21, 3, 1960, pp. 431–441.

disputed Copernicus. The brilliant French mathematician François Viète, for example, insulted him for being «more a master of the dice than of the [mathematical] profession».[29] Perhaps he was not entirely wrong, at least in emphasizing the Pole's limited ability to handle geometry.[30]

But why, after so many centuries of consolidated geocentrism, did the heliocentric theory look so intriguing? It was certainly not because of the predictive power of the Copernican model, which seemed neither better nor simpler than its Ptolemaic antagonist. In fact, it required almost twice the number of circles used by Ptolemy.[31] The model of the Polish astronomer, however, possessed a harmony, in the Pythagorean sense, which the other theories did not. It naturally explained, without resorting to *ad hoc* interventions, phenomena such as the constant proximity of Mercury and Venus to the Sun and the fact that the reversals of path of Mars, Jupiter, and Saturn (retrograde motions) occurred only in coincidence with oppositions to the Sun.[32] For the purists, it had the merit of preserving the delicious uniform circular motions, while avoiding the geometrical tricks like the Ptolemaic equant, which seemed so little like the divine perfection. It's a safe bet that William of Ockham, the 14th century opponent of too many assumptions, would have appreciated it.

But, at the end of the 16th century, the Danish nobleman Tycho Brahe, an inflexible heir to the legendary Vikings, kept repeating to himself that the Earth was «a very heavy and dense and opaque body» that could hardly be moved.[33] And Psalm 104 — he added — is very clear where it tells us that «He [God] set the Earth on its foundations; it can never be moved». Bound by these and other prejudices, he spent his existence, his fortune along with a flood of resources from his king, and an utterly unique skill as an observer to formulate a new cosmology that could bring astronomy back in line with faith and common sense after the shake-up given by Copernicus. If the Pole was a mild reformer, Brahe was the stalwart champion of astronomical counter-reformation, this time grounded in accurate measurements and a more direct empirical derivation of planetary laws.

---

[29] P. Gassendi, O. Thill, *The Life of Copernicus (1473–1543)*, Xulon, Fairfax (VA) 2002, p. 282.
[30] Cf. R.S. Westman (ed.), *The Copernican Achievement*, University of California Press, Berkeley-Los Angeles (CA)-London 1975.
[31] O. Neugebauer, *The Exact Sciences in Antiquity*, Dover, New York 1969, p. 204.
[32] G. McColley, *The Universe of De Revolutionibus*, in "Isis", 30, 3, 1939, pp. 452–472.
[33] A. Blair, *Tycho Brahe's Critique of Copernicus and the Copernican System*, in "Journal of the History of Ideas", 51, 3, 1990, pp. 355–377.

In fact, Brahe admired Copernicus, whose portrait he even kept on a wall of his workplace.[34] He wrongly regarded his Polish predecessor as a great mathematician and rightly as a great thinker. His attitude was one of rational respect, not unlike that of Copernicus toward Ptolemy or of Aristotle toward his master: «*amicus Plato, sed magis amica veritas*» (Plato is a friend, but truth is a better friend), and by the latter for Homer: «a man should not be honored above the truth», and again from the 16th century anatomist Andreas Vesalius to the great physician Galen of Pergamum. Brahe especially appreciated the withdrawal of the Ptolemaic trick of the equant, which had the flavor of a mere artifice invented only to safeguard an ideology. He also recognized the elegance of the heliocentric model, which allowed the reproduction of the retrograde motion of the planets in a natural way, without the invention of *ad hoc* tools. But he categorically rejected the idea of a moving Earth, in part because of the lack of evidence such as the annual parallax. So he reversed the perspective and developed a cosmology in the manner of Copernicus, in which all planetary bodies except the Earth orbited primarily around the Sun, but the Sun and the Moon circled the stationary Earth. From a purely kinematic point of view, the two models were equivalent, except for the different centers assigned to the spheres of the fixed stars. The Tychonic solution seemed to satisfy all the requirements of common sense and Scripture. Brahe used all his ingenuity as an "experimental physicist" to try to validate it, to improve his observational instruments and techniques, and to estimate the reliability of his measurements. He was the first true astronomer in the modern sense, a pioneer who anticipated the methodology that Galileo would theorize half a century later in the *Dialogo sopra i due massimi sistemi del mondo* (Dialogue on the two great world systems): «what experience and sense show us, must be put before every discourse, even if it appears very well founded».[35]

Tycho was born in 1546, three years after the death of Copernicus, in the ancestral castle in a southern province of Sweden, then part of the Kingdom of Denmark.[36] The story of his early life has the flavor of a Nordic saga. He was brought up not by his parents, both heirs to rich and influential clans

---

[34] Ibid.

[35] G. Galilei, *Dialogo sopra i due massimi sistemi del mondo*, Edizioni Studio Tesi, Pordenone 1988, Giornata I, p. 70 (in Italian). Also: *Dialogue Concerning the Two Chief World Systems*, Modern Library, New York 2001.

[36] V. Thoren, *The Lord of Uraniborg: A Biography of Tycho Brahe*, CUP, Cambridge-New York 1990; J.L.E. Dreyer, *Tycho Brahe: A Picture of Scientific Life and Work in the Sixteenth Century*, Cambridge University Press, Cambridge 2014.

of the Danish nobility, but by a paternal uncle, Jorgen Brahe, Vice-Admiral of the Royal Fleet, notorious for his brutality with the crews. According to the memory of the same astronomer,[37] the rude sailor literally kidnapped him, who then was only 2 years old, invoking a promise from the parents that gave him the right to raise the child as his son (since he and his wife had none). Jorgen's program was to have the nephew study law and then direct him to become a court official. With this goal in mind, he sent Tycho to school in Copenhagen. It was there that the 14-year-old boy was introduced to astronomy by a solar eclipse. He was amazed that such an extraordinary natural phenomenon could be predicted mathematically and decided to learn more about it. Thus he began a love affair with Urania that would last a lifetime.

At first, Tycho read everything he could get his hands on, benefiting from a princely budget. In those days, despite the technological leaps made a century earlier by Johann Gutenberg with the invention of movable type printing, each book cost a fortune, so much so that mere mortals, who could not count on a religious institution to support them, had to join academies in order to have access to libraries suitable for humanistic and scientific education and research. To give a striking example, one of the 180 copies of the *42-line Bible* printed by Gutenberg could cost a clerk 3 years' salary.

Since the noble scion's educational program included a long training tour of Europe, in 1562 Tycho was sent to the University of Rostock in northern Germany with plenty of money in his pocket and a bodyguard to protect him. It was a precaution taken in vain, for four years later the impetuous young man lost his nose to the sword of a fellow student, his countryman Manderup Parsbjerg, in a duel in the dark of night among the tombstones of a cemetery next to the Church of St. Mary.[38] It seems that the reason for the quarrel between the two noble gentlemen was neither the sweet eyes of a girl, nor the kind of snub that might leave a commoner indifferent but makes blue blood boil. It was the violent conclusion of a discussion they had begun a few weeks earlier to determine which of them was the best mathematician. This accident earned Tycho the nickname "the man with the golden nose". In fact, he remedied the mutilation with prostheses that he made himself

---

[37] «[. . .] without the knowledge of my parents [my uncle] took me away with him while I was in my earliest youth». K. Ferguson, *The nobleman and his housedog*, Review, London 2002.

[38] Cf. D. Teresi, L.M. Lederman, *The God Particle: If the Universe Is the Answer, What Is the Question?*, Mariner Book, Boston 2006, p. 79.

from leather or precious metals and attached to his face with an adhesive of his own invention. It has been speculated that these mercury-laden artificial noses slowly poisoned him and led to his early death, but a recent analysis of the astronomer's remains ruled out this suspicion.

While Tycho was still in Germany, and all in one piece, another decisive episode took place. He had decided to observe a conjunction of the two highest planets, Jupiter and Saturn, attracted by the deep astrological significance of the event.[39] When he consulted the astronomical tables in circulation, he noticed that they gave different predictions. In particular, those compiled by Copernicus were off by several days. An obvious sign, he thought, of an inadequate model of the world, miscalibrated by inaccurate observations. Thunderstruck by this idea, he decided to devote himself to measuring the heavens with the greatest accuracy and completeness: a mission that required freedom of action and great technological and economic resources.

Then fate came to him resoundingly. In 1565, Uncle Jorgen died of pneumonia contracted while rescuing his king, Frederick II of Denmark, from the icy waters of a Copenhagen canal, where the sovereign had accidentally fallen after heavy drinking; or perhaps the sea dog died of a fever contracted while serving the crown. In any case, the royal gratitude turned this unfortunate event into a bonus that Tycho decided to spend in due course on the construction of a grand observatory. He pondered the project for several years while he traveled across Europe buying astronomical instruments, including a large mural quadrant: a kind of oversized and quartered protractor fixed vertically at the meridian plane and used to determine the height of stars at the moment of culmination with the aid of a mobile viewfinder. In 1572 he was ready to begin. His father's death had left him the sole heir to the family fortune. He retired to his native castle and set up his first observatory. There he immediately made the coup that in a few weeks turned him from an unknown amateur to a world-renowned astronomer.

On the night of November 11, he noticed that in the constellation named after Cassiopeia, the vain wife of the Ethiopian King Cepheus and the ill-fated mother of the beautiful Andromeda, a new star had appeared, even more brilliant than Venus. Tycho was neither the only nor the first

---

[39] K.A. Woody, *Dante and the Doctrine of the Great Conjunctions*, in "Dante Studies, with the Annual Report of the Dante Society", No. 95, 1977, pp. 119–134.

to observe the "nova"[40] (we now know that it was actually a supernova, the intense and prolonged flash of light that follows the catastrophic collapse of a high-mass star). Chained to the postulate of the immobility and perfection of the heavens, the Aristotelians dismissed the event as an exhalation of the atmosphere, such as was supposed to give rise to comets. They too were transient phenomena, but their existence was perfectly legitimate, since they were officially assigned to the corruptible sublunary world: irrelevant for astronomy, therefore, even if they might intrigue for their possible astrological significance.

Tycho was not enchanted by dogmas. With a dense series of measurements, he showed that the new firefly was situated farther away than the Moon and therefore belonged to the heavenly world. In fact, he proved that, unlike the Moon and the other planets, the star had no parallax, just like all the other fixed stars in the firmament. It was a stunning discovery, the kind that today would cause one to rush his results to a prestigious journal such as *Nature*, and which might be worthy of a nomination for the Nobel Prize in Physics, perhaps awarded after a few years of decanting. We can also imagine the resounding title of the discovery paper: *The Aristotelian Paradigm Falsified by a Danish Amateur Astronomer*, and the subtitle: *The Finding Calls for a New Cosmology. No comment from the Vatican spokesman.* More prosaically, Tycho titled his anti-Aristotelian pamphlet *De nova et nullius aevi memoria prius visa stella* (On the new and never before seen star in anyone's life or memory).[41] In fact, it sold out. As time passed, the dazzling intruder that had come out of nowhere lost its brightness and disappeared from sight in March 1574. All the while, Tycho was busy collecting data to tell posterity about the ordeal of the mysterious object that openly violated the postulate of the immutability of the heavens. On this occasion, he also made an important methodological innovation regarding photometric estimates. First, when the source was very bright, he compared its luminosity and color with those of Venus and Jupiter. Then he turned to nearby, fainter stars. This technique of comparing with stellar standards to estimate magnitudes by eye survived until the mid-20th century.

Thanks to the fame acquired with the study of the nova in Cassiopeia, Brahe was invited by the University of Copenhagen to teach a short

---

[40]P. Morrison, *Dissonance in the Heavens*, in "Bulletin of the American Academy of Arts and Sciences", 30, 2, 1976, pp. 26–36.
[41]Brahe was the first to use the term "*nova*", which means new in Latin, to refer to the phenomenon of the sudden appearance and subsequent slow disappearance of a star.

course in mathematics. On that occasion, after praising the «incomparable» Copernicus, calling him «the second Ptolemy» and the one who «speculated the most accurate way than anyone before him», he attacked the
heliocentric system, guilty of being «contrary to any physical principle».[42]
The time had come to get serious and, playing the card of sovereign gratitude, to prepare to re-establish cosmology on a solid observational basis.
He then accepted Frederick II's invitation to build an astronomical facility
on a strip of land jutting out in the narrow channel separating Denmark
from Sweden: an island about 15 kilometers from the castle of Elsinore,
which would shortly thereafter be chosen by William Shakespeare to set
the tragic story of Hamlet, the sad prince. Thus, in August 1576, the last
great observatory of the pre-telescope era was erected on the island of Ven,
immediately named Uraniborg in honor of the muse of astronomy.[43]

Tycho went big, spending a fortune on turreted buildings, guesthouses,
offices for astronomers, majestic instruments for observation, workshops for
craftsmen to make his little kingdom self-sufficient; but also on salons to
welcome the visiting Danish nobility and the many foreign scholars passing
through, to whom he administered science, good beer, and the thrill of the
prophecies of a psychic dwarf, Jep the Jester. Tycho was cultured, curious,
as refined as a descendant of the Vikings could be, and he was definitely
in love with Urania. An energetic and seductive manager — his motto was
"*not haberi sed esse*" (do not appear, but be) — he became a harsh master,
almost ruthless with his subordinates and the inhabitants of Ven, treating
them like serfs and punishing them for a trifle with imprisonment in the
dungeons of his castle-observatory. Apparently, the teachings of the rude
sailor who raised him had not fallen on deaf ears.

Shortly after completing the elegant construction of Uraniborg, Tycho
realized that the fierce north winds were causing intolerable vibrations in
the instruments hosted by the observatory towers, but he did not lose heart.
He immediately built a new complex at ground level, sheltered from the
wind and much more Spartan than the first, which he named Uraniborg,
City of Stars.[44] He could afford all this thanks to the family fortune and
the king who protected him with huge sums of money. It is estimated that,

---

[42] Blair, *Tycho Brahe's Critique of Copernicus and the Copernican System*, cit, p. 356.
[43] The first astronomical establishment in Europe had been founded a century earlier, in
1471, in Nuremberg, Franconia, by Johannes Müller (Regiomontanus) and the merchant
and humanist Bernhard Walther.
[44] J. Shackelford, *Brahe, Laboratory Design, and the Aim of Science: Reading Plans in
Context*, in "Isis", 84, 2, 1993, pp. 211–230.

during his stay in Ven, Tycho received the equivalent of 4 billion euros, enough to build another LHC at CERN (from the French *Conseil Européen pour la Recherche Nucléaire*).

The man with the golden nose stayed in Ven for 21 years, mistreating his peasants and fathering eight children with the beautiful Barbara Kirsten, whom he took with him despite her humble origins, without marrying her and without recognizing her offspring. Tycho was definitely a stubborn macho. For example, while he had a good relationship with his sister Sophia, whose intellectual growth he encouraged, he considered her an inferior creature: «I strongly advise you to stay away from astrological speculations, because I am convinced that you should not go into too abstract a subject for a female mind».[45] Sophia, however, did not listen to him and became a famous astrologer and a recognized expert in the natural sciences.

This was not the only flaw of the great astronomer. He drank a lot and made his animals drink too. The French abbot Pierre Gassendi, who was his first biographer, tells of a moose Tycho kept as a pet that died falling down the stairs because it was drunk on beer. He also cultivated astrology, like everyone else in those days. Man of his time, more than anything else he racked up one astronomical success after another. For example, by cleverly measuring the diurnal parallax, he proved that comets move outside the sphere of the Moon. It was a second slap in the face to Aristotle and his followers, who saw their eternal and unchanging heavens not only abused by transient stars, the novae, but even invaded by capricious stars with tails: changeable-looking travelers, wandering dangerously between crystal spheres like mad elephants. A little further south, in Saxony, in the same years, Michael Maestlin, who would later become Kepler's astronomical mentor, achieved the same result.

Brahe's astronomical observations often reached the limits of the human eye, and sometimes, unfortunately, went beyond them. It happened, for example, when the Dane thought he had measured the angular diameters of the stars.[46] But he was wrong: stars remain point-like sources even for the gigantic pupils of modern telescopes, unless appropriate and sophisticated

---

[45] G. Bernardi, *The Unforgotten Sisters: Female Astronomers and Scientists before Caroline Herschel*, Springer, Cham 2016.

[46] C.M. Graney, *Stars as the Armies of God: Lansbergen's Incorporation of Tycho Brahe's Star-Size Argument into the Copernican Theory*, in "Journal of History of Astronomy", 44, 2013, pp. 165–172.

stratagems are used. Why speak of it then? Because this error would have serious consequences, as proven by Galileo shortly afterwards. Using these meaningless data, Tycho arrived at a false conclusion which, by giving substance to his prejudices, ended up conditioning his entire cosmology. It must be said that if the measurements were wrong, the way in which the Dane used them was impeccable. He began by assuming that the stars were so many replicas of the Sun. This working hypothesis allowed him to derive their distances by comparing their apparent dimensions, which he thought he knew, with the true diameter of the Sun, which he really knew, albeit approximately. He found exactly what he hoped: the stars were so close that their apparent positions should be sensationally affected by one annual revolution of the Earth, if at all. In short, the phenomenon of parallax should have manifested itself so clearly that it could not escape observation, despite a limited ability to measure small angles.

For Tycho, the stellar distance test was the crystal clear evidence he needed to show that Copernicus' heliocentric circus was not working. It was therefore necessary to take a step back and return to a cosmological paradigm based on an immobile Earth. But the solution could not be the classical Ptolemaic model, which was too often defeated when confronted with its accurate observations. Brahe was then forced to soften his rigid geocentrism and admit that, with the exception of the Moon and the fixed stars, all the other wandering bodies revolved primarily around the Sun. This was a reworking, not all that new, of the models of Heraclides Ponticus and the late antique Latin writer Martianus Capella (4th–5th century), who, however, had placed only Mercury and Venus in circumsolar orbit. A proposal similar to Tycho's was made, among others, by his contemporary Nicolaus Reimers Baer, known by the Latin name Reimarus Ursus, a mathematician at the court of Rudolf II of Habsburg. He famously quarreled with the Dane over who had come up with the idea first. Kepler himself was caught between them in the unpleasant aftermath, albeit fleetingly.[47]

The Tychonic model[48] had elements that satisfied all parties at the time (except Aristotle). It improved the theoretical facsimile's adherence to the phenomena observed in the original, without weakening the primacy of an

---

[47]E. Rosen, *Kepler's Defense of Tycho against Ursus*, "Popular Astronomy", Vol. 54, 1946, pp. 405–412; N. Jardine, A.-P. Segonds, *La guerre des astronomes. La querelle au sujet de l'origine du système géo-héliocentrique à la fin du XVIe siècle*, Vol. 1, Les Belles Lettres, Paris 2008 (in French).

[48]Blair, *Tycho Brahe's Critique of Copernicus and the Copernican System*, cit.

Earth to which the Dane did not even grant the benefit of a diurnal rotation (a property rescued in Ursus's model, where, among other things, obvious flaws such as the embarrassing intersection of the orbits of Mars and the Sun occurring in Tycho's scheme could be resolved). The motivation for this total paralysis was twofold: to adhere to the letter of Scripture and to resolve what seemed to him to be a major inconsistency. How is it — he wondered — that the mane of a galloping horse is not touched by the much faster movement of the Earth? Brahe could not grasp what would become clear to Newton about a hundred years later: that our planet rotates on its own axis, carrying the air and the horse with it.

Once the Earth had been immobilized, it remained to be understood how the stars could revolve around it, covering immeasurable distances every day. Tycho asked this question without venturing an answer. He could have invoked divine power as the reason, but instead he was cautious. He was, after all, a true physicist. Instead, the Dane focused on measuring the apparent orbit of Mars, because it was the one that posed the most problems in the epicycles he had inherited from Copernicus.

Science flourished in Uraniborg, also thanks to the favor of Frederick II. But when the king's life was cut short by alcohol, fortune turned against Tycho. The new king, Christian IV, who was eleven years old when he came to the throne, had a personal grudge against the astronomer from Ven. There were rumors that Tycho was having an affair with Christian's mother, Queen Sophia. Someone has even speculated that the young king was Tycho's biological son and that he was the instigator of the astronomer's alleged murder in Prague, where, according to the story, Brahe was supposedly poisoned with mercury by a cousin. Perhaps this legend, true or false, inspired William Shakespeare's tragedy *Hamlet*, set in the castle of Elsinore, a few miles from Ven.

With his vengeful nature, Christian tightened the purse strings more and more as he grew elder, if only to support his obsessive and ruinous penchant for arms: a very expensive hobby even for a monarch. It would have led to his disastrous involvement in the Thirty Years' War,[49] which ended in a severe defeat at the hands of Albrecht von Wallenstein, the supreme commander of the Catholic army. For the sake of truth, however, it must be said that Uraniborg cost the king a real fortune: a sum estimated at

---

[49]P.D. Lockhart, *Denmark, 1513–1660: The Rise and Decline of a Renaissance Monarchy*, Oxford University Press, Oxford 2007.

about 1–3% of the nation's gross product,[50] which is more than an average European nation spends on armaments (in peacetime)[51] today. Brahe lived like a Renaissance prince, sparing no expense. When the measure was filled and the disagreements became more heated, perhaps fearing for his own life, in 1597 Brahe packed his bags and moved to Germany to look around and, as we would say today, put himself on the market.

He did not remain unemployed for long. From his golden lair in Prague Castle, the Holy Roman Emperor seized the opportunity to hire a great astronomer who could provide him with high-quality horoscopes. Rudolf II of Habsburg was a complicated man, a patron-collector with a personality bordering on madness, obsessed with art and the occult sciences. He made an offer that the Dane could not refuse because it was so generous: the title of Imperial Court Astronomer and sufficient funds to build a new observatory, which Tycho promptly set in a castle in central Bohemia, about 50 km from Prague. Once he had resumed his work, the Dane needed an assistant to help him analyze the measurements from Mars. He found this in a German teacher from the Protestant school in Graz, Johannes Kepler. The young man had made a name for himself by publishing a somewhat bizarre book in which he suggested that the Great Architect, in determining the different proportions of the planetary spheres, had resorted to Plato's five regular solids, using them as separators between one sphere and another. "It doesn't matter", Tycho must have thought, weighing the fact that the German was a convinced Copernican. What did matter was that Kepler mastered the mathematics, and he was indeed the best. Keeping the young man in line would have been no problem for an aristocrat accustomed to command.

But Tycho did not complete the project of validating his cosmological model. He died in 1601 after a sudden and brief illness — a mysterious end about which historians have speculated much, going so far as to fantasize that it was, as we have said, a premeditated murder. It seems that the unyielding Dane, in a fever, wondered if his life had not been lived in vain: «*Ne frusta vixisse videar*».[52] His main goals were two. The first had been achieved, providing science with a *corpus* of astronomical measurements of

---

[50] Cf. Thoren, *The Lord of Uraniborg*, cit.; cf. also https://thonyc.wordpress.com/2014/10/30/financing-tychos-little-piece-of-heaven/.

[51] This sentence was written before the war between Russians and Ukrainians triggered a frenzied arms race.

[52] Dreyer, *Tycho Brahe*, cit., p. 309.

extreme accuracy and a method for obtaining and evaluating their accuracy. No one could have done better with the naked eye. As for the second goal, the re-founding of a rigorously geocentric cosmology on the ruins of the Aristotelian world he himself had demolished, there was still one piece missing, the analysis of the precious database of Martian positions. On his deathbed, the man with the golden nose passed the baton to his young German assistant. A gesture that led to a catastrophe for his cosmology, which would eventually be falsified, but turned out to be a blessing for mankind, as we will now see.

Kepler[53] came from a town in Baden-Württemberg, in southwestern Germany, on the northern edge of the Black Forest. He was born in 1571, three months after the Christian triumph over the Turkish fleet at Lepanto, into a modest and messy family of Lutheran faith. The child was actually baptized Catholic, but was later educated according to the canons of the Reformation: a setting to which Kepler remained more or less faithful for the rest of his life. His father was a rough and violent man who earned his living as a mercenary. He disappeared in the war in the pay of the Duke of Alba, the "butcher of Flanders", when Johannes was still a child. The mother, ignorant, petulant, and quarrelsome, according to her son's own words, ran an inn and practiced alternative medicine. It was a dangerous activity at the time of the witch hunts, especially for those who, like her, had a bad family history: the aunt who raised her had died at the stake.

Johannes grew up frail, sickly, very nearsighted, tormented by scabies, fundamentally insecure, in constant need of reassurance, and even emotionally unstable. One of the many to whom nature did not seem to have given much — and even less fate, which had made him born poor and derelict. The only way to escape his fate and the clutches of a degraded cultural and social environment was through a career in the church (or the army, but it required good health). Fortunately, the new course of the Lutheran reform gave the boy the opportunity to study in the seminary. The school was especially hard. The body was humiliated in order to open the mind to religion. It was there that Kepler developed his deep faith in a God who bestowed perfection; and it was there that he planned a future as a shepherd of souls. Fortunately for all of us, fate thought otherwise.

After receiving his baccalaureate, he was allowed to enroll at the University of Tübingen, which was not far from home, in the heart of the land

---

[53]M. Caspar, *Kepler*, Dover Books, New York 2003.

that had raised the Hohenzollerns. The university, which by now had a more than secular history, had the reputation of being an authentic bulwark of Lutheran orthodoxy. As elsewhere, the basic training for theological studies included classical languages and Hebrew, as well as the disciplines of the *trivium* and *quadrivium*, especially mathematics and astronomy. One of Kepler's professors was Deacon Michael Maestlin, whose duties included teaching Ptolemaic cosmology. In private, however, he openly professed to be a follower of Copernicus, despite the anathema to the Polish canon issued by Luther himself and by Melanchthon.[54] Kepler, who still intended to be a preacher, inherited a passion for astronomy and an absolute faith in the model of the world centered on the Sun, the source of life and all motion.

Fate, which, in making him meet Maestlin,[55] had played his first ace, spent the second creating the opportunity to divert on sciences the genius of the young man who chiefly wanted to devote himself to the care of his faith. In the newly opened Lutheran college in Graz, the mathematics teacher had died. When the regents of the University of Tübingen were asked to name a replacement, they chose Johannes Kepler, a theology student, although he had not yet graduated. Perhaps the choice was due to the fame he had earned among students and professors for his ability to make horoscopes;[56] or possibly it hid the intention to remove from the University a subject with the stench of Copernican heresy, who was also suspected of Calvinist sympathies. Thus our hero was sent to southeastern Austria, a journey made mostly on foot to a predominantly Catholic community, with a salary barely enough to live on.

In Graz, in addition to teaching, which did not take up much of his time anyway, Kepler also carried out other tasks, including making some fortune-telling horoscopes. Most importantly, he had time and opportunity to reflect and develop his search for God in things, and to conceive his first major cosmological work, the *Mysterium cosmographicum* (Cosmographic Mystery). What obsessed him most was the need to understand what mathematical rules the Supreme Architect had used to define the

---

[54] R.S. Westman, *The Melanchthon Circle, Rheticus, and the Wittenberg Interpretation of the Copernican Theory*, in "Isis", 66, 1975, pp. 165–193.

[55] C. Methuen, *Maestlin's Teaching of Copernicus: The Evidence of His University Textbook and Disputations*, in "Isis", 87, 2, 1996, pp. 230–247.

[56] S.J. Rabin, *Anti-Astrology Polemic*, in "Renaissance Quarterly", 50, 3, 1997, pp. 750–770.

number of the wandering stars, their orbital sizes, and their motions.[57] This was an entirely new approach to science, more akin to the Renaissance's focus on creation than to humanism's attitude toward admiration and imitation. Anticipating the Newtonian revolution, the young German sought to identify causes rather than note their effects, in the context of a prejudicial leaning toward the heliocentric approach.

He had a sudden intuition while explaining the periodic conjunctions of Jupiter and Saturn to his students. Diving into the calculations and tweaking things a bit to get them back — a shortcut that did not much upset his neo-Platonic spirit — he experienced an extraordinary coincidence. The planetary distances in the Copernican model, which he believed in for physical if not metaphysical reasons, were proportional to the radii of a particular sequence of spheres tangent internally and externally to the Platonic solids: octahedron, icosahedron, dodecahedron, tetrahedron, and cube. These are convex polyhedra that meet a number of requirements: each has all its faces composed of the same regular polygon, all its edges are the same, and so are the angles between two faces. At the end of his treatise on geometry, Euclid proved that only five of them can exist. They are also called "Platonic solids" because Socrates' disciple had studied these polygons from a topological point of view and used them in the *Timaeus* in connection with the four primordial elements (earth, air, water, and fire) and with the shape of the universe in the *Phaedo*. Perfect figures, therefore, in the sense of harmony and simplicity, which apparently served, all and alone, as connecting elements between successive pairs of planetary spheres. This fantastic geometric construction gives us a glimpse of a mystical soul intent on attributing to creation the perfection of the Creator: a Pythagorean belief in number and proportion as divine instruments. In this vein, and in the context of extravagant exercises on the affinities between music, astronomy, and nature, Kepler would later make his greatest discoveries, the eponymous three laws of planetary motion.

The *Mysterium* attracted the attention of the Imperial Astronomer, who invited the author to Prague for an interview. Recently married, the young scientist was not inclined to move. But Brahe had managed to intrigue him by writing about Mars. So Kepler took advantage of a carriage ride

---

[57] P. Barker, B.R. Goldstein, *Theological Foundations of Kepler's Astronomy*, in "Science in Theistic Contexts: Cognitive Dimensions", special issue of "Osiris", 16, 2001, pp. 88–113.

to reach the Dane at his castle observatory in Bohemia. He knew that his host was a great observer, perhaps the greatest of all time, but he ignored how authoritarian and grumpy he was. It did not take Johannes long to realize that he was incompatible with the aristocratic astronomer, both in character and in scientific belief. Too bad he had to leave behind that wonderful legacy of Martian measurements, a disappointed Kepler must have thought on the difficult journey back to Graz. But enough is enough!

A year passed, and the young German was forced to reconsider his decision. In Austria, the most ardent Counter-Reformers had begun to meticulously apply the formula *cuius regio, eius religio* (whose kingdom, their religion), which, loosely translated, means that subjects must adopt the religion of the one who rules the place. A Solomonic commandment, intended fifty years earlier by Emperor Charles V to calm tempers in the tangled conflict between religion and power, and which would shortly thereafter serve as the trigger for the deadly Thirty Years' War. In order to remain in Catholic Graz, Kepler risked having to conform to the state confession, an unacceptable and hateful conversion for a staunch Lutheran. So he decided to accept the position of assistant on Brahe's salary and move with his family to Bohemia, where Catholic orthodoxy was less virulent. He had to work under a master on Martian data and also on a project of a stellar catalog and new planetary ephemeris that Rudolf II had urged. The emperor demanded from his astronomers a state-of-the-art astrological instrument, an updated and corrected re-edition of the *Prutenic Tables* (where Prutenic means Prussian) compiled half a century earlier by the astronomer Erasmus Reinhold on behalf of Albert I, Duke of Prussia, and based on the Copernican model. In turn, they had improved and superseded the *Alfonsine Tables* sponsored by Alfonso X of Castile and published in Toledo in 1483. The new ephemerides, the last based on astral positions obtained with the naked eye, would later be called the *Rudolphine Tables* in honor of the sovereign who had sponsored them.

Kepler arrived in Prague with all his belongings and took up residence in the New Town in Brahe's house; after a year spent in the castle outside Prague, Tycho had been asked by Rudolf II to move to the imperial capital, as he wanted his astrologer next to him. There the young German assistant was guarded at sight. Under these conditions, the hidden purpose of using Mars observations to refine the *Mysterium* model was not feasible. As a staunch geocentric, the inflexible Dane would never allow it. But fate had a third ace in store, which it played with surprise. As we have said, in October 1601 Tycho fell unexpectedly ill and died within a few days; he was only 55 years old. *Mors tua vita mea* (Your death is my life). Kepler, who had already made a name for himself at court, took over the position and prerogatives of his late boss. He was appointed imperial astronomer, advisor to the superstitious Rudolf, and his personal astrologer, and above all he inherited *de facto* all the material concerning the astronomical measurements of the Dane. «Tycho has the best observations in the world», he wrote to his old master Maestlin, «he only needs an architect to make a building».[58] He would have no qualms about using this legacy to his advantage.[59] Perhaps he too consoled himself by thinking, like Machiavelli, that "the end justifies the means".

Over the next decade, the German experienced a period of intense scientific activity, but also one marked by grief and constant illness. Prague, the capital of the Holy Roman Empire, was a melting pot of culture, fueled by the mad genius of Rudolf II and a magnificent imperial court, where astrologers, magicians, idlers, and artists of the caliber of the Milanese Mannerist painter Giovanni Arcimboldo roamed. Lutheran and Roman Catholic communities still coexisted in relative peace in the gritty alleys of the Old Town and the beautiful streets of the New Town across the Vltava River, and the Jews flourished after centuries of persecution. Kepler worked on the astrometry of Mars without losing sight of the sky and its signs, which guaranteed him the sympathy of the sovereign and the amazement of the people. Thus, while he was dealing with a great triple conjunction of Jupiter and Saturn, one of those that occur in close groups of two or three every 800 years, he observed the appearance of a new star *in pede Serpentarii*, that is, in the constellation of "the one who carries the snake" (*Ophiuchus* or *Serpentarius* in Latin). It was the year 1604. Friar Ilario Altobelli was the first to notice the phenomenon from Verona and also told Galileo about it. A series of coincidences, combined with a boundless astrological creativity, convinced the German scientist that the star of the Magi mentioned in the Gospel of Matthew[60] was indeed a nova star of the same nature

---

[58] O. Gingerich, J.R. Voelkel, *Tycho and Kepler: Solid Myth versus Subtle Truth*, in "Errors: Consequences of Big Mistakes in the Natural and Social Sciences", special issue of "Social Research", 72, 1, 2005, pp. 77–106.

[59] E. McMullin, *Kepler: Moving the Earth*, in "HOPOS: The Journal of the International Society for the History of Philosophy of Science", 1, 1, 2011, pp. 3–22.

[60] «After they had heard the king [Herod the Great], they went on their way, and the star they had seen when it rose went ahead of them until it stopped over the place where the child was» (Matthew 2:9).

as the one he had discovered in the constellation Ophiuchus.[61] «If this is not true, it is well founded», Giordano Bruno would probably have commented, had he lived, taking a sentence from his *De gli eroici furori* (The heroic enthusiasts).

There was no limit to Kepler's eclecticism and his desire to interpret every celestial accident as the result of a numerical coincidence. But how could a man so imbued with mathematical harmony give up the beauty of uniform circular motions in which equal angles are democratically covered in equal times? Since these motions did not work in practice because they were inconsistent with observations, he searched Brahe's data for another physical quantity that would remain linear with time, and finally found one. It was the remarkable ellipticity of the Martian orbit that favored the discovery. In 1609, the year of Galilei's happy encounter with the telescope, Kepler published two of his famous three laws. The first states that the planets describe elliptical orbits in which the Sun occupies one of the foci, and the second that the line connecting the planet to the Sun sweeps equal areas in equal times.[62] The Aristotelian paradigm was now shattered.

Happy days! But nothing lasts forever. When his patron Rudolf II was forced to pass the crown of Bohemia to his brother Matthias, a Catholic fundamentalist, Kepler began to think that the time had come for him to leave Prague. He was now a famous scientist and could choose prestigious places to go, including Padua, where the university chair of astronomy was vacant after Galileo moved to Florence. But he still wanted to live in a German-speaking country. Thus, after the death of Rudolf II, he moved to Linz with the title of "Territorial Mathematician", granted to him by the

---

[61] Comparing Brahe's 1572 nova with that of 1604, Kepler expressed with remarkable lyricism his conviction that the second apparition was not accidental but rather the consequence of God's will: «*Itaque prior illa Mundo non praemonito supervenit, et velut improvisus hostis, occupatis urbis moeniis, prius in foro comparuit, quam cives expeditionem eius fama percepissent: nostra vero, vulgo expectata a longo tempore, cum multa solennitate et triumphali pompa, ad diem constitutum est ingressa*» (And so the former came into the world without warning, like an unexpected enemy who, after occupying the city walls, appears for the first time in the marketplace without the citizens having any knowledge of his campaign: our [star], on the other hand, generally expected for a long time, appeared on the scene on the appointed day with much solemnity and triumphal pomp), J. Kepler, *Gesammelte Werke*, M. Caspar, W. Von Dyck (eds.), Munich, 1938, I, p. 272.

[62] E.J. Aiton, *How Kepler Discovered the Elliptical Orbit*, in "The Mathematical Gazette", 59, 410, 1975, pp. 250–260.

new Holy Roman Emperor on the condition that he complete the *Rudolphine Tables*: an arduous but very important undertaking. Today we are accustomed to downloading to our personal computers sophisticated programs that display the starry sky with attractive graphics at any (reasonable) time and from any observation point on Earth or position in space. In the past, however, there were considerable problems in predicting celestial events. The art of constructing ephemeris tables was reserved for the few skilled mathematicians who could tame first the Ptolemaic model and then the no less complicated Copernican model, and for the tireless observers who could measure the positions of the stars with the necessary precision using elementary instruments for calibration and verification. It was to these celestial magicians that sailors and travelers entrusted their lives and fortunes as they wandered the vast expanses of water or sand, guided only by the stars.

Completed in 1623 and published four years later, the *Rudolphine Tables* relied heavily on Brahe's observations. For this reason, the Danish astronomer's heirs, who had been absent for a good quarter of a century, now claimed a share in the profits of the work, demanding a kind of inheritance of their relative's scientific legacy. For once, Kepler showed the same determination in practical life that had made him so good in science. He stood his ground and won the case, managing to keep the proceeds of his work to restore his own finances, which had been ruined by the expenses of helping his mother. She had been tried for witchcraft ten years earlier and fortunately acquitted after a long, painful, and expensive trial.[63] Johannes himself was not sure if the old woman was really above suspicion, but he had gone to great lengths to save her from being burned at the stake, again demonstrating unparalleled skill. It was said that no lawyer could have argued the defense better than he did. At the same time, the mother was said to have been saved not by the logic of his arguments, but rather by his desire not to strike an icon of science. More or less 10 years later, Galileo would receive a similar treatment, as we will see in a moment.

Meanwhile, planetary studies had also come a long way. In 1615, a Kepler reassured by a new marriage to a young woman whom he liked much more than his late first wife, published the first volume of *Epitome astronomiae Copernicanae* (On Copernican astronomy), his most popular work. It was essentially an astronomical manual based on the heliocentric

---

[63] U. Rublack, *The Astronomer and the Witch: Johannes Kepler's Fight for His Mother*, Oxford University Press, Oxford 2015.

theory of the Polish canon, domesticated by his own discoveries, including, in the last volume, the famous three laws. We have already mentioned the first two. The third and last one, the most difficult to discover and in some ways the most important, was first announced in *Harmonices mundi* (The Harmony of the World), in the context of a research of concordances transversal to music,[64] geometry and physics. He established the direct proportionality between the cube of the semi-axis of the planetary orbit and the square of the time needed to complete a full orbit, that is, the year of that particular planet. The masterpiece was complete. The mathematical prowess and boundless imagination of a die-hard dreamer met the extremely precise data collected by Brahe on Mars, a body whose elongated orbit was particularly conducive to falsifying the classical dictum about the circularity of motion. The three empirical laws would soon become the litmus test for validating Newton's theory of gravitation, for the first time based on causes rather than merely artfully constructed to reproduce effects.

Having completed its design, fate allowed twilight to fall upon this story. Even Linz had become inhospitable to the Lutherans. So Kepler set out on the last and most painful pilgrimage of his life. Aging and increasingly tired, he began to wander the war-torn Palatinate, dividing his time between study, work, and caring for his family. He also agreed to serve as astrologer to a superstitious warrior, Albrecht von Wallenstein, a *homo novus* (self-made man) who became very powerful fighting the Lutherans in the Thirty Years' War. Though famous and much sought after, Johannes was unable to find peace because peace had left his Germany, which was being brutalized by armies from all over Europe, leading to massacres and plunder to the usual blasphemous cry: "God is with us". He died while traveling in Regensburg in 1630 at the age of 59. That same evening, the Moon eclipsed for a while, as if to accompany this man who had measured the heavens to the underworld. Dictating his epitaph: *Mensus eram coelos, nunc terrae metior umbras / mens coelestis erat, corporis umbra iacet* (I measured the heavens, now I measure the shadows of the earth / the spirit was heavenly, here lies the shadow of the body), Kepler had somehow foreseen it. It was one of the many incredible coincidences that accompanied the life and scientific thought of this absolute genius.

In the space of a century, a quiet man, a refined observer, and a magnificent visionary had contributed together, though for different reasons and

---

[64]D.P. Walker, *Kepler's Celestial Music*, in "Journal of the Warburg and Courtauld Institutes", Vol. 30, 1967, pp. 228–250.

in different ways, to the demolition of the old cosmology and the founding of a new one. Nobody could do more with the naked eye. It was at this point in our history that Galileo's glass eye providentially entered the scene. The last authoritative record of the cycle that was about to close was Johann Bayer's *Uranometria*, published in August 1603. The monumental star atlas contained the visual cartography of the entire sky in 48 plates, as many as the constellations of Claudius Ptolemy, plus a table for the deep southern sky, almost unknown to the author of the *Almagest*. The title was allusive: *Measure of Urania*, which means Measure of the Sky. Bayer, a German lawyer with a passion for astronomy, had supplemented Brahe's formidable catalog with data from other sources. Then, constellation by constellation, he measured the brightness of the stars, introducing a nomenclature he had borrowed from the mid-sixteenth-century Sienese scholar Alessandro Piccolomini. Bayer named each star with a letter of the Greek alphabet followed by the genitive form of the constellation's Latin name. For example, Sirius ("the shining"), the brightest star of Canis Majoris, became α Canis Majoris, and Mirzam ("the herald" announcing Sirius), the second brightest, became β Canis Majoris, and so on and so forth; even for those who did not have a Greek-Roman or Arabic proper name. *Uranometria* also listed four nebular-looking sources, two of which were in fact such. All these were signs of a growing interest in the objects of the stellar sphere beyond purely practical purposes. What was still missing was the tool to study them better.

Chapter 4

The glass eye:
Galilei's telescope and its discoveries

So [easy to understand] are all truths,
once they are discovered;
the point is in being able to discover them.

Galileo Galilei,
*Dialogue Concerning
the Two Chief World Systems*

Equipped with his five senses,
man explores the universe around him
and calls the adventure Science.

Edwin P. Hubble,
The Exploration of Space,
in "Harper's Magazine", 1929

Some may wonder why the *perspicillum*, which is a rather simple instrument, was not invented earlier. The question is far from idle, since some 4,000 years ago the inhabitants of the Mediterranean coasts were already familiar with the optical elements that make up the telescope.[1] This fact is evidenced by the discovery of some lenses in Crete and the Near East, although they appear too crude to be of any practical use. They were made of quartz, a natural crystal, and of glass, a material that, according to Pliny the Elder, the encyclopedic writer who perished in the eruption of Vesuvius in 79 A.D., was accidentally discovered by the Phoenicians more than a thousand years before the founding of Rome.[2] Technically, glass is an amorphous solid obtained by rapidly cooling (quenching) a liquid to stop

---

[1] A.L. Oppenheim, *Towards a History of Glass in the Ancient near East*, in "Journal of the American Oriental Society", 93, 3, 1973, pp. 259–266.

[2] Cf. I.C. Freestone, *Pliny on Roman Glassmaking*, in "Archaeology, History and Science: Integrating Approaches to Ancient Materials", M. Martinon-Torres, T. Rehren (eds.), Left Coast Press, Walnut Creek (CA) 2008, pp. 77–100.

crystallization and give the product a disordered microscopic structure. But vitrification is not a true change of state.[3] Rather, it is the creation of a state of high viscosity in the fluid so that the material retains its shape, just as a solid would. In short, the glass from which you are sipping your favorite beverage as you read these pages is itself a liquid, even though you would have to wait an extremely long time to see it drain away.

The basic ingredient of ordinary glass is silica, a sand made from sedimentary rock that must first be melted and then, as mentioned, rapidly cooled. The biggest obstacle is the high temperature required for melting, about $1,800°C$, which ordinary furnaces cannot reach. But don't panic: to lower it, just add some soda and lime to the silica. The first artificial glasses were probably made in this way, from the accidental presence of these ingredients in bonfires that sailors or camel drivers lit on the sand and then quickly extinguished with a few buckets of water. But in addition to man-made glasses, there are also natural glasses, the most exotic of which were formed billions of years ago in some corner of the cosmos before being brought to Earth by meteorites. The indigenous ones, instead, originate from the discharge of lightning on sand or, more commonly, from the rapid cooling of lava. Everyone knows obsidian, the shiny black volcanic glass used since the Neolithic to make weapons, cutting tools, mirrors, and even those figurines of Quetzalcoatl, the feathered serpent, sold in Mexico City's street markets.

Natural glasses and crystals are generally opaque and highly pigmented materials, so they are not very useful for optical purposes. There are, however, some notable exceptions. The inexhaustible Pliny tells us that the Emperor Nero observed gladiatorial fights through an emerald,[4] but does not clarify whether he did so out of habit or necessity because he was shortsighted. In his *Naturalis historia*, the admiral-scientist also speaks of transparent spheres that were used to light fires with the Sun's rays. The practice had been known for a long time. Five centuries earlier, an old Athenian — the character named Strepsiades in *The Clouds*, the sparkling comedy in which Aristophanes pokes fun at the Sophists — suggested to Socrates a bizarre stratagem to thwart the results of a trial: melt the wax from the

---

[3] Historically, there are three states of matter: solid, liquid, and aeriform. In the first, a body retains both shape and volume; in the second, only volume; and in the last, neither. Other states are defined in modern physics, including plasma and exotic condensed states.

[4] Pliny, *Naturalis historia*, XXXVII, 64.

tablet on which the court clerk was taking notes. How? By concentrating the sunlight through «that beautiful stone you can see right through, the one they use to start a fire». Another cultured Roman, Lucius Annaeus Seneca the Younger, tutor of the Emperor Nero, said he had learned that a glass ball filled with water magnified things, and confessed that he himself had used this device to read.[5] The remarkable fact of this story is that the device described by Seneca represents a real technological leap compared to the full glass sphere mentioned by Pliny. Water is much clearer than the glass pastes of the time and provides better magnification. All of which the Roman philosopher experienced and appreciated, but did not understand. Refractive optics — the science and technology based on transparent materials capable of deflecting the path of light rays — was in its infancy.

In the early Christian era, the Alexandrian philosophers were quite familiar with the mathematical model of refraction, a phenomenon that makes a rudder submerged in water look broken and governs the functioning of lenses and prisms. Ptolemy also tried his hand at the subject, but in vain, as did later Arabs in the madrasas of Morocco and Iraq. It was not until the early 17th century that the empirical formula known as the law of refraction, which relates the angle of an incident ray of light to that of an emergent ray in the transit from one transparent medium to another of different density, was correctly formulated in Europe. The discoverer was a Dutchman, Willebrord Snell,[6] who is also remembered for building a huge mural quadrant which was then installed in the Leiden University Observatory. Snell was an unusual fellow: talented, versatile, but rather eccentric. He had worked for a few months in Prague with Tycho Brahe, knew Kepler, and dabbled in difficult geometrical and geodesic problems. Because he did not bother to publish his refraction formula, it was rediscovered some 15 years later by the French philosopher, mathematician, and scientist René Descartes, who made it his own and because of that was later accused of plagiarism. The heavy insinuation rested on the suspicion, supported by the recollection of another mathematician and astronomer, the Dutchman Christiaan Huygens, that Descartes had met Snell during a visit to Leiden and even procured a copy of his notes. Moreover, the Frenchman's

---

[5] Lucius Anneus Seneca, *Questiones Naturales*, Book I, c. 5, 6.
[6] J.A. Vollgraff, *Snellius' Notes on the Reflection and Refraction of Rays*, in "Osiris", 1, 1936, pp. 718–725.

proof was based only on theoretical arguments, without any direct measurement,[7] and contained two errors that fortunately compensated for each other. Kepler himself, stimulated by some unexpected phenomena during lunar and solar eclipses, had guessed the law on his own, thus proving that great scientific discoveries need genius, but blossom only when their time is ripe. *Scientia non facit saltus* (science does not make leaps), one might say, reworking a famous phrase of the naturalist Carl von Linné and (only apparently) in spite of the American philosopher of science Thomas Kuhn, who was an advocate of the striking "paradigm shifts" caused by revolutions in thought.[8] It is not surprising, then, that some discoveries have more than one father, although the not always Solomonic simplifications of historiography tend to give all the credit to one person only, at least in the context of a homogeneous cultural and political area. People like heroes! But Bertold Brecht's Galileo argues: «Unhappy is the land that needs a hero».[9]

The law of refraction was the result of experience alone and the ability to transform experimental data into simple empirical formulas. Its interpretation in terms of physical properties would come much later, thanks to the theory describing the classical electromagnetic interaction. It was formulated in 1865 by the Scotsman James Clerk Maxwell (p. 481), another one of those superior minds obsessed with finding the Creator in creation. When light travels in a material medium, the phase velocity decreases with respect to the vacuum, and thus the wavelength increases while the frequency does not change. This fact is due to a resistance that is characteristic of the medium itself and is technically known as "refractive index". Therefore, when light passes from one material to another with a different refractive index, it is forced to change pitch, which is the cause of an abrupt change in the direction of the light rays. All refractive optical systems, starting with lenses, are based on this phenomenon.[10]

---

[7] A. Mark Smith, *Descartes's Theory of Light and Refraction: A Discourse on Method*, in "Transactions of the American Philosophical Society", Vol. 77, No. 3, 1987, pp. 1–92.

[8] T.S. Kuhn, *The Structure of Scientific Revolutions*, The University of Chicago Press, Chicago 1962.

[9] B. Brecht, *The life of Galileo*, LA Theatre Works, Los Angeles (CA) 2008.

[10] The law of refraction is the natural consequence of the principle formulated by the French mathematician Pierre de Fermat in 1666, according to which the distance travelled by light from point A to point B is that covered in the shortest time. This, in turn, derives from the principle postulated by Christiaan Huygens in 1678 that every point of a light wave is a secondary source of a spherical wave.

The properties of lenses are said to have been discovered around the turn of the year 1000 by an Iraqi with an impossible name and many talents. In the West he was called Alhazen.[11] His curiosity about the functioning of the human eye, which is a refractive system, led him in the right direction, but his brilliant speculations did not yield any practical results. More than two centuries passed, and interest in the optical instrument resurfaced far from the Fertile Crescent and the Nile Delta, where Alhazen had spent most of his life. It was an Englishman, Roger Bacon, a thirteenth-century Franciscan with an extraordinary vision of the future, who decided to take up the problem again. Although the theory was still dormant, the time was ripe to transfer the accumulated knowledge to technological applications.

The leap in perspective took place where geography, nature, and millennia of history had prepared the cradle of a new era: the Italian peninsula. There, at the end of the 13th century, the first spectacles appeared and began to spread among the educated bourgeoisie and the clergy, complete with frames that held the lenses in front of the eyes rather than on the object.[12] Initially, they were only able to correct presbyopia, a pathology caused by the loss of elasticity of the crystalline lens, and not the defects due to the refraction of ocular fluids. These primitive devices, used mainly for reading, performed rather poorly because of the rough and opaque lenses used to make them, and they cost a lot of money. In short, they were seen primarily as a badge of the wearer's intellectual role and rank; almost like a luxury item. But as time went on, eyeglasses improved and gained a larger and larger market, and, as is usually the case, the growing demand favored the transformation of a simple craft into a coordinated production effort and commercial enterprise.

The first association of eyeglass manufacturers was founded in Venice and adopted very strict rules. The members, whose foundries were

---

[11] D.C. Lindberg, *Alhazen's Theory of Vision and Its Reception in the West*, in "Isis", 58, 3, 1967, pp. 321–341.

[12] The first suggestive iconographic evidence of the spread of spectacles in the Western world dates from the mid-14th century. The famous portrait of Cardinal Hugues de Saint-Cher is part of a large fresco in the chapter house of the Dominican convent of San Nicolò in Treviso, Italy (now the Episcopal Seminary). The High Priest, who is depicted in curial garb at his desk in the act of writing, wears heavy glasses with a bridge that allows him to hold them on his nose. Tommaso Barisini of Modena completed the painting in 1352, when Hugues had already been dead for a century. Therefore, we do not know whether the glasses he wore are a license from the painter rather than an accurate representation of the technology already available in the 13th century.

concentrated mainly in the ancient settlement of Rialto and in the *sestriere* called Dorsoduro (because the land there had "hard ridge, i.e. it was more stable), were held to secrecy about the techniques, and violations were punished by death. The lenses were made of rock crystal or beryl (a mineral that, in the presence of external colorants, can become an aquamarine or a very precious emerald), because glass was considered a mediocre material. This is always the case: novelties struggle to establish themselves because of the known devil, which, according to the well-wishers, is better than the unknown. But things soon changed as a result of actions driven more by interest than by foresight.

In 1295 the government of the Serenissima ordered the transfer of all glassmaking activities to the island of Murano. The pretext was to remove all the furnaces from the city, which were often the cause of fires. In reality, the measure concealed the intention of arming, in an enclave surrounded by the sea and easy to guard, the precious secrets of an art for which Venice held a monopoly. In fact, no one else in Europe produced glass of the same quality and artifacts such as vases, cups, and candelabras as elegant and colorful as in the lagoon workshops, where the ability to create amalgams was combined with the ancient mastery of the glassblowers. Despite the careful supervision of the Council of Ten, a sort of Venetian Gestapo, the concentration of all glassmaking activities in Murano did not prevent industrial espionage — proof that the strategy of the Chinese Wall is not a win-win solution for anyone — but it certainly favored a more rapid technological development. The result was a new production line designed to provide lens grinders with homogeneous and transparent glass, free of bubbles and impurities, and relatively cheaper than natural crystals. Even the glass used by Galileo for the optics of his first telescopes would come from Murano.

In the 16th century, with the leakage of technical information and the desertion of some of its masters, Murano lost its leadership in glassmaking to the Netherlands, where the flourishing commercial activities of a people in symbiosis with the sea had initiated a period of extraordinary prosperity. The Dutch bourgeoisie, though Calvinistic in dress and public morals, was in search of lightness and beauty. Artists, architects, teachers, and even craftsmen were needed to produce the goods the wealthy merchants demanded, including eyeglasses. Thus, a Flemish school of skilled lens grinders quickly developed. A profession that the philosopher Baruch Spinoza, a Sephardic Jew, would turn to in the mid-17th century to make ends meet after his religious community excommunicated and banished him.[13] It is this shift in the center of gravity of the glass world that perhaps explains why the birth of the telescope occurred in Flanders rather than Venice.[14]

But how were the lenses produced? They were made by grinding the glass with abrasive powders of increasing fineness, typically metal oxides, until the desired shape was achieved. The result was verified by simple optical tests as the process progressed. A rotating lathe provided axial symmetry. The final step was polishing, which removed surface roughness to limit the light scattered by the remaining imperfections (see also p. 115).

The growing demand for ocular prostheses created a real boom throughout Europe, despite the protectionist ambitions of the Venetians and Florentines. Nevertheless, the glass lens was slow to become a tool for astronomy. By itself, it was not enough to explore the heavens, or even to approach distant objects on Earth. One had to combine at least two of them, by accident if not by design, as probably happened in a Dutch laboratory. Or, in order to bring the stars closer, concave mirrors should have been considered.[15] In fact, the Greek and Alexandrian philosophers — beyond refraction, which they mostly played with without any technological result — had fully mastered the operating principles of curved mirrors and had learned to apply them to instruments of various kinds. Observing the behavior of sunlight, they had introduced the concept of a ray of light, which they modeled as a straight line emanating from the eye; a proposition attributed to Empedocles and widely accepted, with the notable exception of Lucretius.[16] They

---

[13]Here is a dramatic passage from Baruch de Spinoza's excommunication, proclaimed by the the united congregation of Portuguese Jews in Amsterdam: «Cursed be he by day and cursed be he by night; cursed be he when he lies down and cursed be he when he rises up. Cursed be he when he goes out and cursed be he when he comes in. The Lord will not spare him, but then the anger of the Lord and his jealousy shall smoke against that man, and all the curses that are written in this book shall lie upon him, and the Lord shall blot out his name from under heaven».

[14]D. Whitehouse, *Glass: A Short History*, Smithsonian Books, Washington (DC) 2012.

[15]The term "mirror" is derived from the Latin verb *miro*, meaning "I look at", and denotes any shiny surface capable of reflecting light in an orderly fashion. While this may seem obvious, it is useful to point out an important difference with lenses. They transmit light; mirrors do not. Therefore, lenses must be made of some transparent material, typically glass, while mirrors require some kind of substrate that can be polished and made capable of reflecting light back; so not necessarily glass. In fact, early mirrors were made of white polished metal such as silver, copper, or tin.

[16]T. Lucretius Caro, *De Rerum Natura*, IV, vv-8183–8190:

persaepe levis res atque minutis
corporibus factas celeris licet esse videre.

knew the law of plane reflection — the incident and reflected rays have the same inclination and lie on the same plane as the perpendicular to the reflecting plane[17] — and had generalized it to curved surfaces by considering the corresponding tangent plane at each point. Then they thought, correctly, that the rays of light coming from a distant source could be treated as parallel lines. In short, they had reduced the optics of reflecting surfaces to a purely geometrical problem, bounded only by obeying Euclid's five postulates.[18] They discovered and exploited the magical properties of so-called "conic surfaces of rotation", obtained by rotating an ellipse, a parabola, or a hyperbole about the symmetry axis.

The figure we are most interested in is the paraboloid.[19] As first demonstrated by the Greek mathematician Diocles, who lived between the 3rd and 2nd centuries B.C., a mirror with a concave surface reflects bundles of

---

in quo iam genere est solis lux et vapor eius,
propterea quia sunt e primis facta minutis,
quae quasi cuduntur perque aëris intervallum
non dubitant transire sequenti concita plaga;
suppeditatur enim confestim lumine lumen
et quasi protelo stimulatur fulgere fulgur.

These Latin verses read: "first of all, it is very often seen that light things and things made of minute bodies are fast. Of this kind are, of course, the light of the sun and its heat, because they are made of minute elements which are almost beaten and do not hesitate to cross the intervening air, urged on by the next blow. Immediately, in fact, light follows light, and, as in an unbroken series, splendor is stimulated by splendor".

[17]Dante Alighieri in *Purgatory* XV, vv. 16–23, gives a magnificent description of reflection:

Come quando da l'acqua o da lo specchio
salta lo raggio a l'opposita parte,
salendo su per lo modo parecchio

a quel che scende, e tanto si diparte
dal cader de la pietra in igual tratta,
sì come mostra esperienza e arte;

così mi parve da luce rifratta
quivi dinanzi a me esser percosso.

It reads: "As when the ray of light is reflected by water or by a mirror, so that the ascending ray makes the same angle as the descending one with respect to the perpendicular to the plane, according to what experience and science teach us; so I seemed to be struck at that point by a reflected light, which made me quickly look away".

[18]L. Mlodinow, *Euclid's Window. The Story of Geometry from Parallel Lines to Hyperspace*, The Free Press, New York 2001.

[19]The parabola is the set of points on a plane that are equidistant from a line (the directrix) and a point (the focus). The paraboloid is the surface obtained by a complete rotation of a parabola about its axis of symmetry.

straight lines parallel to the axis of symmetry in such a way that they converge at a single point. Not coincidentally, Kepler would later call it "focus" (Latin for fire), because there the sunlight collected by a lens is capable of igniting a fire. Now, replace the bundle of straight lines with the geometric model of light rays, and the game is over. We have a converging mirror, a sort of funnel for light rays coming from distant sources. Of course, the reverse is also true. The rays coming from a light source placed at the focus of a paraboloid are reflected in a parallel beam. This is how car headlights work: their directionality allows us to see into the darkness of the night.

In addition to discovering the tricks of the paraboloid, the ancient Greeks had also learned to make practical use of this geometric figure. It is said that in 202 B.C. the Syracusan Archimedes set fire to Roman ships besieging his city from the sea by concentrating the sunlight on them through burning mirrors. For five years, Rome had fought Carthage to the death. While Hannibal roamed Italy at will, the Senate had sent the consul Marcus Claudius Marcellus to Sicily with three legions and fifty quinqueremes to punish the Syracusans for turning to the enemy. Plutarch reports that when «the Romans attacked them by sea and land, the Syracusans were dumbstruck with terror: they thought that nothing could withstand such a furious assault by such forces.». So the siege lasted 18 months, thanks to the wonders of the 70-year-old philosopher: machines made with those levers he mastered, with catapults and more. But the matter of the burning mirrors is probably just a legend, perhaps a reworking, either artistic or by inaccurate translation, of the story of throwing incendiary bullets. There is no trace of it in the classical sources. We also tend to believe that Archimedes had no reflecting parabolas, or even spherical bowls of sufficient size. If he really tried to set fire to Marcellus' battleships, he must have done so with a battery of flat mirrors, unwittingly anticipating the solution of the tessellated mirrors and the mosaics that we will present in the last chapter. Again, it is highly doubtful that his actions could have caused much more than a few burns to the enemy's skin, no worse than a simple sunburn.

In Alexandria, however, a concave mirror actually projected a powerful beam of light at night from a tower on the island of Pharos. This first lighthouse in history, commissioned in the third century B.C. by the founder of the Macedonian dynasty of the Ptolemies, was designed to guide sailors to the harbor in the Nile Delta, avoiding sandbars and reefs. It is believed that the reflector element was made by assembling bronze segments and that the light source placed on the focus was oil-fired. However, there is no accurate record of this "wonder of the world", which would have been second only to the Great Pyramid of Giza in durability. The Greeks, while generous with information about their ideas, were stingy with data about their technological achievements, probably because they were concerned about protecting their strategic interests.

The invention of all these optical machines is the fascinating result of the theoretical knowledge and first hypotheses that the ancients had about the behavior of light. However, with so much knowledge, they did not develop anything useful to look at the sky. Some believe that this lack of advancement was due to the Roman conquerors' disinterest in the sciences.[20] Incidentally, the descendants of the Latin shepherds who became masters of the world did not need the firmament to read the future. All they required were the entrails of animals, the flight of birds, and a handful of augurs astute enough to predict what the people or the powerful wanted to hear. These *feri victores* (the savage victors of whom Horace speaks; see p. 46) would thus be those who, more by ignorance than by choice, blocked the maturation of Greek science. Or perhaps there was not enough time for the seed to bear fruit. For while Rome was dying and laughing — *"[Roma] moritur et ridet"* were the scathing words of the 5th century Christian writer Salvianus to describe the moral and military disintegration of the Empire as seen from his observatory in Roman Gaul — the Middle Ages were knocking powerfully at the door of history. From the steppes of Central Asia and from the cold North, a river of humanity was pouring in torrents beyond the now fragile borders of a decaying civilization, helping to destroy the old order and fueling that extraordinary crucible from which a new world was to emerge, the so-called Western civilization. Despite the commendable efforts of the Islamic scholars, who salvaged what could be saved and added some original contributions, it would be necessary to wait until the 17th century to see optics become a complete science, with a solid connection to the emerging modern physics. The first task was to resume and refine the hypotheses on the properties of light radiation sketched by the Greeks.

Among the post-Renaissance scientists, Kepler was the first to take a vigorous initiative.[21] In *Ad Vitellionem paralipomena, quibus astronomy*

---

[20]Cf. Russo, *The Forgotten Revolution*, cit.
[21]S. Dupré, *Kepler's Optics without Hypotheses*, in "Synthese", 185, 3, Special Section: "Seeing the Causes: Optics and Epistemology in the Scientific Revolution", 2012, pp. 501–525.

and pars optica traditur,[22] published in 1604, an ideal continuation of the work of the Silesian monk Erazmus Ciolek Witelo[23] more than three centuries earlier, he claimed that light propagates without limit, with an intensity decreasing inversely with the square of the distance (law of conservation of flux), and at an infinite speed. The latter is a false conjecture that Galilei would not have accepted without proof (see p. 138). Kepler also considered visions and recovered some theses dear to the Arabs. The Platonic school, and later Euclid, imagined mysterious rays emerging from the eye to hit the target. The German reversed perspective, arguing that it was light that reached the eye and formed the image on the retina through the crystalline lens; an intuition already present in Islamic culture and used by Filippo Brunelleschi to lay the scientific foundations of perspective.[24] He then correctly described myopia and presbyopia, and finally, in the *Dioptrice*, illustrated the working principles of converging and diverging lenses and proposed a new optical scheme for the telescope, without putting it into practice. This was in 1611.

In Padua, "trying and trying again", Galileo Galilei, out of sheer instinct of genius, had already experimented with an astronomical use of the telescope, proving its power with facts and giving prompt communication to the world in the *Sidereus Nuncius*, which was almost a press release that one read in one go.

> Indeed, with the glass you will detect below stars of the sixth magnitude[25] such a crowd of others that escape natural sight that it is hardly believable. For you may see more than six further gradations of magnitude. The largest of these, which we

---

[22] Freely translated: Additions to Witelo's work, in which the optical part of astronomy is treated. The subtitle reads: *Potissimum de artificiosa observatione et aestematione diametrorum deliquiorumq[ue] Solis & Lunae. Cum exemplis insignium eclipsium. Habes hoc libro, lector, inter alia multa nova, tractatum luculentum de modo visionis, & humorum oculi usu, contra Opticos & Anatomicos,* that is: First of all concerning the observation and estimation of the apparent diameters of the Sun and the Moon made in a workmanlike manner. With excellent examples of eclipses. You will find in this book, oh reader, among many other new things, a clear treatise on the mode of seeing, and on the use of the humors of the eye, against opticians and anatomists.

[23] He wrote *Perspectiva*, a treatise largely based on the work of the Muslim polymath Alhazen.

[24] Leon Battista Alberti, another perspective pioneer and influential intellectual, argued instead that this science of representation was completely independent of the physics and physiology of light.

[25] Galilei refers to the scale of magnitudes of Hipparchus, which places the stars barely visible to the naked eye in the sixth magnitude.

may designate as of the seventh magnitude, or the first magnitude of the invisible ones, appear larger and brighter with the help of the glass than stars of the second magnitude seen with natural vision. But in order that you may see one or two illustrations of the almost inconceivable crowd of them, and from their example form a judgment about the rest of them, I decided to reproduce two star groups. In the first I had decided to depict the entire constellation of Orion, but overwhelmed by the enormous multitude of stars and a lack of time, I put off this assault until another occasion. or there are more than five hundred new stars around the old ones, spread over a space of 1 or 2 degrees. [...] In the second example we have depicted the six stars of the Bull called the Pleiades (I say six since the seventh almost never appears) contained within very narrow limits in the heavens.[26]

They are words of wonder with which the Pisan scientist recounts his experiences and documents them with the first deep maps of the Orion and Pleiades star clusters, sketched with the dexterity of someone who, in his youth, had frequented the workshops of Florentine artists.[27] He goes on to explain the importance of a good telescope:

Accordingly, on the seventh day of January of the present year 1610, at the first hour of the night, when I inspected the celestial constellations through a spyglass, Jupiter presented himself. And since I had prepared for myself a superlative instrument, I saw (which earlier had not happened because of the weakness of the other instruments) that three little stars were positioned near him — small but yet very bright. Although I believed them to be among the number of fixed stars, they nevertheless intrigued me because they appeared to be arranged exactly along a straight line and parallel to the ecliptic, and to be brighter than others of equal size.

---

[26]The quote, as well as the following, is taken from the digital version of the Latin translation of the *Sidereus Nuncius* in: http://people.reed.edu/~wieting/mathematics537/SideriusNuncius.pdf.

[27]When Galileo was young, he thought of becoming a painter. He studied perspective with Ostilio Ricci, the court mathematician in Florence. He was a close friend of the late mannerist and early baroque painter Lodovico Cardi, known as Cigoli, and an advisor to Agnolo di Cosmimo, known as Bronzino. Cf. H. Bredekamp, *Galilei the Artist: The Drawing, the Moon, the Sun*, Akademie-Verlag, Berlin 2007 (in German).

This is the vivid and dry account of the first encounter with four faint stars that, after a few nights of feverish observation, revealed themselves to the astonished Galileo to be the many moons of the planet Jupiter. Literally a bombshell! Indeed, although the discovery did not promote Copernican cosmology *tout court*, it falsified the preconception of an Earth chosen as the center of all celestial motions and also demolished an intriguing objection to the heliocentric model.[28] According to the Ptolemies, with the revision of the roles and the consequent downgrading of the Moon to the rank of a satellite of the Earth, the planet inhabited by humanity acquired a curious primacy. It seemed to be the only one with such a vassal. A peculiar and suspicious feature, no doubt! But the new moons of Jupiter proved that this feature was only the fruit of short-sightedness.

In short, one man's genius and courage had turned the telescope into the avenger of the *ipse dixit*, and into a picklock capable of breaking the yoke of uncritical subservience to Aristotle — a gregarious attitude that had slowed the progress of scientific knowledge even in the period of the revival of letters and the arts.[29] It was an extraordinary achievement, perhaps one of the greatest in the history of science, aided by a technological leap, the construction and skillful use of a «superlative instrument».

But how is it possible that a simple leather tube and a pair of glass lenses have changed the way we see the world? Let us help ourselves with an example. Imagine quenching your thirst by trying to fill a glass with water by exposing it to rain. If only a few drops fall, you will not be able to collect enough liquid for a good sip unless you are willing to be patient for a very long time. To avoid dying of thirst while waiting, you need a tool that will speed up the process and make it plausible; a kind of prosthesis that can be attached to the glass. For example, a very large funnel and *voilà, les jeux sont faits*. Cheers! Galilei's telescope does the same thing as the funnel. The lens collects and focuses the dim light from the stars, and a microscope projects it into the eye, which is the glass of the metaphor, filling it just enough to "water" the retina, and from there, the brain. A conceptually simple stratagem that has remained essentially unchanged from the Galilean telescope to today's terrestrial and space giants, in which the observer's eye has been replaced by much more sophisticated and powerful detectors. But the funnel analogy ends there. In fact, there are important differences

---

[28] C. Crombie, *The History of Science from Augustine to Galileo*, Dover, New York 1995.
[29] The approach to knowledge was no longer to climb on the shoulders of giants to see farther, but rather to rest uncritically on the shoulders of a single giant, Aristotle.

between an ordinary rain and a rain of light, and there are several issues that need to be addressed regarding how the radiation is intercepted and focused by the optics, and finally collected by the detector. We will see them as we go along.

Galilei knew almost nothing about all this technical devilry.

> And first I prepared a lead tube in whose ends I fitted two glasses, both plane on one side while the other side of one was spherically convex and of the other concave. Then, applying my eye to the concave glass, I saw objects satisfactorily large and close. Indeed, they appeared three times closer and nine times larger than when observed with natural vision only. Afterwards I made another more perfect one for myself that showed objects more than sixty times larger. Finally, sparing no labor or expense, I progressed so far that I constructed for myself an instrument so excellent that things seen through it appear about a thousand times larger and more than thirty times closer than when observed with the natural faculty only. It would be entirely superfluous to enumerate how many and how great the advantages of this instrument are on land and at sea. But having dismissed earthly things, I applied myself to explorations of the heavens.

So he wrote again in *Sidereus Nuncius*, emphasizing that he had started from scratch in this process, with the aim of bringing out all his ingenuity, together with the «labor and expense»: a subliminal message to recommend that his efforts be adequately rewarded in hard currency.

Aware of the extraordinary nature of his astronomical discoveries, Galileo intended to make them public in record time. He chose a patient and trustworthy Venetian publisher, Tommaso Baglioni, who was willing to accept delivery of the manuscript in installments. In the grip of anxiety, the Paduan professor had decided to begin the preparation of the printing plates while the observations were still in progress. He had two good reasons for doing so: on the one hand, he was in a desperate hurry and was afraid of being scooped if he revealed all his cards to an outsider at once. On the other hand, he also needed time to complete the work. In the absence of any precedent, he had to first see, then convince himself that he had not been deceived by a mirage, and finally construct an interpretive code for what he had discovered.

The difference between the appearance of planets and fixed stars also seems worthy of notice. For the planets present entirely smooth and exactly circular globes that appear as little moons, entirely covered with light, while the fixed stars are not seen bounded by circular outlines but rather as pulsating all around with certain bright rays.

In a matter of weeks, he destroyed classical cosmology, which for millennia had been confined to the boundaries of the Solar System, and opened to humanity a window to a boundless and unknown universe, proposing a new archetype:

What was observed by us in the third place is the nature or matter of the Milky Way itself, which, with the aid of the spyglass, may be observed so well that all the disputes that for so many generations have vexed philosophers are destroyed by visible certainty, and we are liberated from wordy arguments. For the Galaxy is nothing else than a congeries of innumerable stars distributed in clusters. To whatever region of it you direct your spyglass, an immense number of stars immediately offer themselves to view, of which very many appear rather large and very conspicuous but the multitude of small ones is truly unfathomable.

These observations ideally mark the beginning of the exploration of our star system, a titanic endeavor that will keep astronomers first, and astrophysicists later, busy for the next four centuries, and is still ongoing.

The discoveries announced by the *Sidereus Nuncius* received the unconditional appreciation of many. Prince Federico Cesi was enthusiastic and hastened to associate Galileo with the newly founded *Accademia dei Lincei*. Pope Paul V Borghese, who had also taken on the Venetians, welcomed the astronomer to Rome with all honors. Cosimo de' Medici, Lord of Florence, was very pleased with the dedication of the book and the Jovian moons, which Galilei had cleverly christened *Medicea Sidera*.[30] (the Medici stars) As early as the late spring of 1610, he called his fellow citizen, offering him

---

[30] Kepler was the first to use the term "satellites" for the bodies orbiting Jupiter, which Galilei had continued to call planets. The etymology of the word is interesting. *Satelles* is the Latin name (from an Etruscan root) for the bodyguards of the last kings (*Tarquinia gens*) of the monarchical period of Rome: a term mostly derogatory (like that of thugs), but understood by the German astronomer in its best meaning of subject or servant.

the title of «Primary Mathematician of the Studium of Pisa and Philosopher of the Most Ser.mo Grand Duke, with no obligation to lecture and to reside neither in the Studium nor in the city of Pisa, and with a salary of one thousand scudi a year, Florentine coin»:[31] about 100,000 euros, adding the weight in gold and converting to today's value. It is interesting to note that Galileo himself requested that the title of "Mathematician" be supplemented with that of "Philosopher".[32] We can explain this singular claim by reading what he writes in the *Istoria e dimostrazioni intorno alle macchie solari: Letter I*, of 1612:

> Those deferents, equants, epicycles etc., placed by pure astronomers to facilitate their calculations, but not to be considered as such by astronomers philosophers, who, in addition to the care of saving in any way the appearances, try to investigate, as a maximum and admiring problem, the true constitution of the universe, since this constitution is, and in only one way, true, real and impossible to be otherwise, and for its greatness and nobility worthy to be placed before any other knowable question for the speculative minds.

After 18 years spent in the shadow of the reassuring banner of the *La Serenissima* Republic of Venice, «the best years of my life», he later confessed,[33] the Paduan professor was able to return to his Tuscany: a stone's throw from Rome, however, and therefore more at the mercy of the Inquisition, since his friend Saverio Francesco Sagredo[34] had feared. Confident in the beginning of a new life, Galileo decided to get rid of his old one, leaving in Padua his partner, Marina Gamba, and his son Vincenzio, whom he would recognize only ten years later. He left the other two daughters with their grandmother Giulia in Florence until they were old enough to enter

[31] G. Galilei, *Opere*, *X. Corrispondenza 1574–1610*, Barbera, Florence 1968, p. 400, *Letter to Belisario Vinta*, June 6, 1610.

[32] «As for the title and pretext of my service, I would like, in addition to the title of Mathematician, that H.H. [His Highness] adds that of Philosopher, professing to have studied more years in philosophy, than months in pure mathematics». Id., *Opere*, *XVIII. Corrispondenza 1639–1642*, Barbera, Florence 1968, p. 353, *Letter to Belisario Vinta*, May 7, 1610.

[33] Id., *Opere*, *XVIII. Corrispondenza 1639–1642*, Barbera, Florence 1968, p. 209, *Letter to Fortunio Liceti*, June 23, 1640. See also A. De Angelis, *I diciotto anni migliori della mia vita*, Castelvecchi, Roma 2021 (in Italian).

[34] Giovanni Francesco Sagredo was a Venetian nobleman. Interested in physics, he was an experimenter and builder of scientific instruments.

the convent. A somewhat cynical way, we have to admit, to get himself out of trouble and to secure support in his old age. But such was life at that time. Almost immediately, however, the events took a different turn. Along with the highest honors, the first serious troubles began to rain down on Galileo.

There were those who accused him of plagiarism. It was not the first time, and he had always been able to defend himself by scratching: an attitude similar to that of the Southern general Robert Lee, who, recalling the Battle of Chancellorsville, won despite a clear numerical inferiority, is said to have boldly declared: «I was too weak to defend myself, so I attacked». Others, however, argued that the Medicean planets were nothing more than a blunder. Among the latter were honest and competent people, such as the Paduan Giovanni Antonio Magini, who twenty years earlier had bypassed him for the professorship of mathematics at Bologna, and the powerful Jesuit Christopher Clavius.[35] Both had to change their minds after what Galileo would have called a «sensible experience», that is, after observing Jupiter with a good telescope and being definitively convinced that the instrument was not producing mere illusions, perhaps manipulated by the devil.

One of the most bitter confrontations was with Christoph Scheiner, a German Jesuit, who demanded priority for the discovery of sunspots. The sanguine Pisan resolved it with a pamphlet printed by the Lincean Academy, in which he literally tore his opponent to pieces, and with good reason. Contextually, he also imprudently took the opportunity to affirm «*apertis verbis*» (openly) his belief in the physical reality of the heliocentric model. In keeping with the Aristotelian vision of the perfection of the heavens, Scheiner had elaborated an extravagant explanation of the spots with which he saved the ideal integrity of the celestial body. In fact, the Jesuit, showing uncommon intelligence and intellectual honesty, later recognized Galilei's reasoning and agreed with him that the features were pustules on the surface of the star. He then continued to observe the Sun for a long time with instruments he built himself, and obtained important results. But Galilei's worst enemies were some obtuse fundamentalists, exalted and ignorant, who appealed to the blasphemous conflict between the Copernican theory, held as true by the great scientist, and the *Sacred Scripture*.

---

[35] Cf. J.M. Lattis, *Between Copernicus and Galileo: Christoph Clavius and the Collapse of Ptolemaic Cosmology*, The University of Chicago Press, Chicago (IL) 1994.

The attack started from the Dominican monastery of San Marco[36] in Florence, from where Girolamo Savonarola had launched his tragic crusade 130 years earlier. The case escalated to the point where it reached the Vatican tribunal. The boil had not yet burst, but the infection was advancing inexorably.

Prominent among the secular voices of dissent was that of the Florentine Ludovico delle Colombe.[37] This irreducible Aristotelian had already shown himself prone to the most brazen logical acrobatics and reckless physical theories in order to protect his beloved Peripatetic philosophy. One example was when, in order to safeguard the postulate of the immutability of the heavens, he had denied that the nova of 1604 was really a star that had appeared out of nowhere. It was the publication of the *Sidereus Nuncius* that triggered his open hostility towards Galileo. Cornered by observational evidence, he too argued for the apparent contradictions with the *Scripturs*, of which he listed all the passages that, taken literally, should have demonstrated the immobility of the Earth and thus the falsity of the Copernican model. Galilei dismissed him with his usual ferocity, giving him the name of *"pippione"*, a derogatory term in the Florentine dialect meaning fool. Perhaps he could have spared his compatriot if he had taken into account the critical position delle Colombe had taken against the judicial astrologers, the horoscope makers who claimed to ascribe to the heavens, and therefore to God himself, the propensity for evil. But mercy was not Galileo's best virtue, especially when the fault involved an alleged outrage against reason.

On a continent worn down by its long history and made "old" by geographical discoveries, atavistic rivalries between nations and powerful men had taken on the guise of religious fundamentalism, setting the stage for an imminent global war in the shadow of the Cross of Christ himself. The Church of Rome, for its part, was determined to implement at all costs

---

[36]The Ferrara-born Girolamo Savonarola, a Dominican from the convent of San Marco in Florence, preached a model of popular government for the Republic that had been estab lished in Florence after the Medici were driven out, dreaming of making the city a "New Jerusalem". Excommunicated in 1497 by Pope Alexander VI, in 1498 he was hanged and burned at the stake along with two brethren as a «schismatic heretic preaching new things». His works were later included in the Index of Forbidden Books. Yet, when he had refused him the purple, Pope Borgia had called him «great servant of God» and then claimed after his death that he would gladly have included him in the list of saints. Truly a singular contradiction. Cf. C. Vasoli, *I Processi di Girolamo Savonarola*, in "Rivista di storia della Chiesa in Italia", Vol. 58, No. 2, 2004, pp. 550–563 (in Italian).

[37]M. Biagioli, *Galileo, Courtier: The Practice of Science in the Culture of Absolutism*, The University of Chicago Press, Chicago (IL) 1993, pp. 170–183.

and by all means the Tridentine strategy, which centered on unity around the pope, discipline in the hierarchy, communal life of the clergy, and the restoration of tradition. It entailed a massive action to recover values and powers, while within its own bosom the fierce dispute over theological issues between Dominicans and Jesuits continued, later leading to the fratricidal controversy over Chinese rites (the opposition to Jesuit openings to new converts on issues of the cult of the dead). Not surprisingly, in a climate of ideological entrenchment, even those who had come to the aid of the Copernicans by seeking a synthesis between science and faith had done nothing but pour gasoline on the fire. Emblematic is the case of the Calabrian Carmelite Paolo Antonio Foscarini, author of the *Letter on the Opinion of the Pythagoreans and of Copernicus* (1615), who was judged «daring» by Cardinal Robert Bellarmine and then added to the Index of Forbidden Books. Fortunately for the naive monk, he died while the trial of him and the imprudent Neapolitan publisher who had printed the book without verifying the existence of the required *imprimatur* (let it be printed) was still in progress. He was thus spared a world of trouble and suffering, for the harshness of the times had dulled minds and clouded consciences.

In such a situation, a great deal of prudence would have been required, a quality with which Galilei was poorly endowed. This is clear from the tone of a letter written by Bellarmine to the venerable Father Antonio Foscarini on April 12, 1615:

> First, I say that it seems to me that your Paternity and Mr. Galileo are proceeding prudently by confining yourselves to speaking in a supposed and not absolute way, as I have always believed Copernicus spoke. For there is no danger in saying that by assuming that the Earth moves and the Sun stands still, one saves all appearances better than by postulating eccentrics and epicycles; and this is sufficient for the mathematician. But it is another thing to want to assert that in reality the Sun is in the center of the world and only turns on itself, without moving from east to west, and that the Earth is in the third heaven and revolves around the Sun with great speed; this is a very dangerous thing, which is likely not only to irritate all scholastic philosophers and theologians, but also to harm the Holy Faith by making false Sacred Scripture [...] I say that, as you know, the Council [of Trent] forbids interpreting Scripture against the common consent of the Holy Fathers; and if your Paternity will

read not only the Holy Fathers, but also the modern commentaries on Genesis, the Psalms, Ecclesiastes, and Joshua, you will find that they all agree in the literal interpretation that the Sun is in heaven and revolves around the Earth with a great speed, and that the Earth is very far from heaven and sits motionless in the center of the world. Now consider, with your sense of prudence, whether the Church can tolerate giving Scripture a meaning contrary to the Holy Fathers and to all the Greek and Latin commentators. Nor can one answer that this is not a matter of faith, since it is not a matter of faith "as to the subject", but it is a matter of faith "as to the speaker" [...] I say that if there were a true demonstration that the Sun is in the center of the world and the Earth in the third heaven, and that the Sun does not revolve around the Earth but the Earth revolves around the Sun, then we should proceed with great caution in explaining the Scripture that seem to contradict it; we should rather say that we do not understand them than that what is demonstrated is false. But I will not believe that there is such a demonstration, until it is shown me.[38]

Nothing new. The same concepts had been expressed about the *Koran* by the Muslim philosopher Averroes in the 12th century.[39]

In a few words, the subject that Galilei had treated so impetuously went beyond the boundaries of science and became a serious political issue at a difficult time for the papacy. Paul V was forced to act for reasons of state: "Nothing personal, just business", one could almost say, lamenting this unfortunate mixing of Caesar's business and God's matters. In February 1616, the scientist, already in Rome «to respond spontaneously to some accusations, or rather slanders, that had been made against him by his emulators»,[40] was summoned to the private residence of Cardinal Bellarmino. There, in the presence of five ecclesiastics as witnesses, he was warned not to teach, defend, and treat as true the two propositions of the

---

[38] G. Galilei, *Opere, XII. Corrispondenza 1614–1619*, Barbera, Florence 1968, pp. 183–184. All quotations below are taken from *I documenti vaticani del processo di Galileo Galilei (1611–1741)*, S. Pagano (ed.), "Archivio Segreto Vaticano", Città del Vaticano 2009 (in Italian).

[39] A. Wohlman, *Al-Ghazali, Averroes and the Interpretation of the Qur'an. Common Sense and Philosophy in Islam*, Routledge, London 2013.

[40] Id., *Opere, XII. Corrispondenza 1614–1619*, Barbera, Florence 1968, p. 205.

scandal: the immobility of the Sun and the mobility of the Earth. Indeed, this was a trifle compared to what could have happened to him. The Pisan had narrowly avoided the accusation of heresy, which had been raised in the first acts of the trial and which would probably have cost him severe penalties. He had also saved his works from the guillotine of the Sacred Congregation of the Index of Forbidden Books. A wise friend of his, Piero Guicciardini, had warned him in vain that Rome was not «a country for coming to dispute over the Moon, nor for wanting, in the running century, to support or bring new doctrines there».[41] The pope's arm was again the Jesuit Robert Bellarmine, a man of learning, intelligence, and piety enough to deserve a place among the saints of the Church, but also the one who had delivered Giordano Bruno to his lay executioners sixteen years before. Copernicus' *De revolutionibus orbium coelestium* had instead ended up in the grinder of the Inquisition, and the thesis on the centrality and immobility of the Sun had been censored as «foolish and absurd in philosophy and formally heretical». Its publication was therefore to be suspended «*donec corrigatur*», until corrections were made. The writings of the contentious Pisan, instead, remained on probation.

Galileo stayed in Rome for some time. He would not give up in the stubborn hope of overturning the verdict on the Copernican theory, in which he now firmly believed. He talked too much and risked too much, more out of presumption than recklessness. His friends urged him to hurry back to Florence. «Staying out of this country would be of great benefit and value to him», said Guicciardini, a profound connoisseur of men and things. «[Galilei] is inflamed by his opinions, he has extreme passion in him, and little strength and prudence in knowing how to keep it under control, he does not notice and does not see what he should, so that, as he has done hitherto, he will remain deceived and put himself in danger»,[42] he wrote to Cosimo II in Florence; and again in a later letter to Curzio Picchena, secretary to the Grand Duke: «He is in a fixed mood to tame the friers and to fight with those with whom he can only lose». Finally Galilei surrendered. Back in Florence, he simply kept a low profile for a while, heeding Bellarmine's warning not to interfere in a matter the Church did not intend to discuss. Then, also the victim of a devious trap set by fate, he let himself go again.

---

[41] Ibid., p. 207.
[42] Ibid., p. 242.

After the death of Paul V and the brief pontificate of Gregory XV, the conclave of 1623 had placed on the papal throne a Florentine, Maffeo Barberini, scion of a rich merchant family, who took the name of Urban VIII. It was an unexpected choice that mediated between the positions of the French and the Spanish. Their crossed vetoes had narrowed the list of favorites to the advantage of an outsider. Galilei, who had known Barberini for a long time, felt that he could count on his benevolent protection. After all, they were fellow citizens, a condition that mattered then and still matters today. With this conviction, he went to Rome to present his theory of the tides to the Holy Father. We know it was wrong, but in his opinion it proved the validity of the heliocentric model. Inspired by the reaction to the movement of water in a vessel, he was convinced that tides were caused by the combined motion of the Earth's rotation and revolution. The water, he speculated, had to rise by inertia as it was left behind by the planet.[43] The mission to the Vatican palaces was a calculated gamble and perhaps a way to test the pulse of the Pope. The response was good-natured. Urban VIII disputed the thesis of the Pisan physicist, but the matter ended there, despite the blatant violation of the 1616 warning. At the papal court, the issues on the table were of far greater magnitude than a physical theory (albeit with a certain load of theological implications), and it did not seem convenient to attack a man now famous for matters of natural philosophy.

The bomb exploded when Galileo published a book expounding, in classical dialogue form, the "Two Chief World Systems", the Ptolemaic geocentric system and the Copernican heliocentric system.[44] Well aware that he was playing with fire, the Pisan had taken all the necessary steps to secure in advance the approval of the Sacred Congregation of the Index, in other words, to guarantee himself the *placet* of ecclesiastical censure. But when the Pope had the book in his hands, he believed, or perhaps he was led to believe, that his figure was overshadowed in the guise of Simplicio, Ptolemy's last peripatetic supporter of the three participants in the *Dialogue*.[45] Needless to say, he flew into a rage. Perhaps, however, without reason, since Galilei had counted on his goodwill and did not intend to

---

[43]Cf. L. Russo, *Flussi e riflussi: indagine sull'origine di una teoria scientifica*, Feltrinelli, Milan 2003 (in Italian).

[44]S. Drake, N.M. Swerdlow, T.H. Levere, *Essays on Galileo and the History and Philosophy of Science: Volume 2*, University of Toronto Press, Toronto 1999.

[45]They were the Copernican Salviati, the intelligent and neutral layman Sagredo, and Simplicio, a follower of Aristotle.

offend him. In any case, the reactions that followed were anomalous compared to the sin, if it was a sin.

The year was 1632. Europe was experiencing the third phase of the very bloody Thirty Years' War, marked by the entry into the battlefield of the Swedes led by King Gustavus II Adolphus Vasa, called the Great. Galilei, now old and shabby, was summoned to Rome and put on trial for having disregarded Bellarmine's precept. His was a political trial, conducted according to the rigid schemes of consolidated procedure, where the patina of law masked practices worthy of the harshest absolutist regimes.[46] Accused of heresy, the scientist was intimidated by all means, even the threat of torture. He was led in the presence of the executioner who, as usual, displayed his own paraphernalia: tools such as ropes for the terrible strappado — a torture to which the philosopher Tommaso Campanella had been subjected for forty hours in a Naples prison in 1601 — and red-hot pincers. A kind of illustrated warning to break the defendant's resistance and make him submit. The strategy of terror, which had not worked on Giordano Bruno, achieved the desired effect on Galileo. Frightened, tired, and disconsolate, the now aged warrior finally gave in and decided to recant. In the biography of the scientist written by Giuseppe Marco Antonio Baretti for the English public in 1757, we read that after the humiliating act of submission, he whispered the famous «And yet it spins». If so, we should feel even more compassion for such human weakness in this giant of thought. The Church had won, but in the manner of Pyrrhus.[47]

The surrender earned Galilei a relatively mild sentence, inflicted with the relief of a court called upon to judge a defendant too famous to be treated with the usual severity. The scientist was placed under house arrest, but he was guaranteed "four-star accommodations", first in Rome, then in Siena, and finally in a villa on the hill of Arcetri, south of Florence. He was also ordered to read the penitential psalms three times a week: a duty he passed on to his daughter, a nun. The suspension of harsher punishments was contingent on good behavior. The judges counted on the deterrent effect of fear, which must have been really great, because from then on Galilei acted as if he understood the enormity and gravity of his mistakes. He

---

[46] W.A. Wallace, *Galileo's Science and the Trial of 1633*, in "The Wilson Quarterly (1976)", 7, 3, 1983, pp. 154–164.

[47] «The armies [of the Romans and King Pyrrhus of Epirus] parted [after the Battle of Asculum in 280 B.C.]; and, it is said, Pyrrhus replied to one who congratulated him on his victory that another such victory would utterly destroy him». From Plutarch's *Life of Pyrrhus*.

stopped being a cosmologist, at least publicly, and devoted himself entirely to physics, which at the time did not seem to concern the heavens (the Aristotelian division of the world into two spheres still persisted, though not for long).

Thus began the twilight of a genius,[48] Sweetened by the love of his eldest daughter, Sister Maria Celeste, and of his disciples, especially Vincenzo Viviani and Evangelista Torricelli, but also by the admiration of eminent visitors such as Giusto Sustermans, who painted his famous portrait, the English philosopher Thomas Hobbes, who made his pilgrimage to Arcetri in 1638, and perhaps the poet John Milton. Nine years illuminated by a final turn, the publication of the *summa* of his scientific method in the *Two New Sciences*: another dialogical work with the same characters of the one that had led Galilei to a clash with the Pope. It is true that a leopard cannot completely change its spots! In 1638 he went totally blind. There has been much debate about the nature of this infirmity. Today it seems entirely excluded that the scientist burned his eyes while looking at the Sun. More likely it was glaucoma.

Galileo Galilei died in early January 1642, just one year before that other giant of our history, Isaac Newton, was born in a Lincolnshire hamlet. Interestingly, England's failure to adhere to the Gregorian calendar reform, due in large part to Anglican hostility to Roman Catholics, meant that the Julian date of Newton's birth fell in the last days of the same year. This false coincidence suggested to the Italian poet Giacomo Leopardi the romantic image of a "nature" that, orphaned by one genius, immediately provides another. According to Leopardi, the chain began with the death of Michelangelo, which occurred in the year of Galileo's birth. This is a very touching conjecture, but unfortunately it is also chronologically incorrect.

---

[48] A. Righini, *Gli ultimi anni di Galileo*, in *Il colle di Galileo*, Firenze University Press, Florence 2012, pp. 41–54 (in Italian).

Chapter 5

The heroic enthusiasts:
The dawn of the telescopic sky survey

I would never die for my beliefs
because I might be wrong.

Bertrand Russell,
from an interview with the New York Post.

I know that I am mortal by nature,
and ephemeral; but when I trace at my pleasure
the windings to and fro of the heavenly bodies,
I no longer touch the earth with my feet:
I stand in the presence of Zeus himself
and take my fill of ambrosia.

Claudius Ptolemy, *Almagest*

The Dutch telescope model that Galileo had perfected and would not abandon until his last years, marked by blindness, consisted of a plano-convex objective and a plano-concave eyepiece.[1] However, this combination of lenses is not the only one possible for a telescope. In fact, Kepler, stimulated by reading the *Sidereus Nuncius*, which had rekindled his old passion for optics, soon discovered another. This is the story.

In the early spring of 1610, Kepler's friend Matthäus Wacker von Wackenfels, a diplomat of many interests, had told him briefly about the exciting discoveries published by the Paduan professor. At first, Kepler feared that the Medici stars mentioned by von Wackenfels were themselves planets. That would have been a disaster for his theory of the five Platonic solids, in which there was no room for other tenants. But his fears did not last long. A few days later, the Tuscan ambassador to the Prague court presented him

---

[1] H.I. Brown, *Galileo on the Telescope and the Eye*, in "Journal of the History of Ideas", 46, 4, 1985, pp. 487–501.

with a copy of the book. Kepler literally devoured it and, on impulse, wrote a *Dissertatio cum Nuncio sidereo*[2] (Conversation with the Sidereal Messenger), a kind of letter-document full of strange geometrical proposals and religious digressions, which he sent to the author, congratulating him and declaring himself his faithful squire. A few pages would have been enough for anyone else to accomplish the task, but not for the German scientist, who loved to naively put down on paper all his thoughts, no matter how twisted, and every movement of his soul.

His firm stance in favor of Galilei immediately drew some criticism, and he perhaps began to fear that he had been too hasty in blindly endorsing the disruptive claims of a shrewd Italian who up to that point had been juggling the opposing sides of the Aristotelians and Copernicans. So, without even waiting for the reply from the Pisan genius who, as usual, did not bother to answer, he wrote to him again begging for a copy of the instrument — an instance embellished with praises that belied the need to match words with deeds. After a few months, Galilei finally showed up. The imperial astronomer's appreciation was a precious support for him now that, after his initial excitement, he found himself dealing with organic skeptics, such as the Paduan Aristotelian Cesare Cremonini, to whom «the look through those glasses [stunned] the head»,[3] and especially the influential Magini. He then replied to Kepler, thanking him for the precious praise, but being careful not to send what the German had asked him to send.[4] Probably in his heart he distrusted this crazy dreamer who sought God in numbers. But above all, he had no intention of giving up to a potential competitor one of the telescopes from the small lot he had set aside for the mighty men of earth, princes and cardinals. He was aware that he had just touched the Pandora's box of heavenly secrets and did not want to lose the advantage of his temporary pole position.

Faced with a refusal communicated to him in rather blunt terms, Kepler did not surrender.[5] As in other circumstances, he was greeted by good

---

[2] L. Simoni Varanini, *"Dissertatio cum Nuncio Sidereo" between Galileo and Bruno*, in "Bruniana & Campanelliana", 9, 1, 2003, pp. 207–215.

[3] Cf. G. Galilei, *Opere, XI. Corrispondenza 1611–1613*, Id, p. 145, *Lettera di Paolo Gualdo a Galileo Galilei, 29 luglio 1611* (in Italian)

[4] A. Postl, *Correspondence between Kepler and Galileo*, in "Vistas in Astronomy", 21, 4, 1977, pp. 325–330.

[5] Cf. O. Gingerich, *How Galileo and Kepler Countered Aristotle's Cosmological Errors*, in "*Cosmology Across Cultures*", ASP Conference Series, 409, 2009, pp. 242 ff.

fortune. At the end of the summer, the Elector[6] Ernest of Bavaria, Archbishop of Cologne, came to Prague with the task, later proved to be in vain, of settling the dispute between Rudolph II and his brother Matthias.[7] He carried with him a telescope given to him by Galileo, which he immediately put at the disposal of the imperial astronomer. At last! Kepler lost no time. Not trusting his own battered eyes, he organized a consensus experiment: he asked several people to certify the existence of Jupiter's moons. When he received confirmation, he wrote again to the chief mathematician of the Studium of Pisa, praising the virtues of the telescope but lamenting the absence of any theoretical study of the properties of this formidable new instrument. Then he thought about it and decided to take care of the problem himself.

We know how this Parsifal of science loved to approach nature and its phenomena from religious preconceptions inspired by the beauty of geometry, order, and symmetry. Once again, Kepler was true to himself. Unlike his Italian colleague whom he nevertheless admired, he attacked the problem theoretically, using his model of light and the laws of refraction. In short, he succeeded in finding a combination of lenses that was different from the Dutch telescope and yet equally capable of performing the task of bringing distant objects closer and magnifying them. He simply replaced the divergent lens of the Galilean eyepiece with a convex one, without changing the geometry of the objective, which had to remain convergent. Although aware of the superiority of his optical design, Kepler merely described it in *Dioptrice*,[8] published in 1611. He saw with his mind where others saw only pitch black, but he had little talent for practical things, partly because of severe nearsightedness and deformed hands left by chickenpox that had nearly killed him as a child. The industrious Scheiner[9]

---

[6]Member of the college of German princes responsible for electing the emperor of the Holy Roman Empire since the 13th century.

[7]V. Press, *The Habsburg Court as Center of the Imperial Government*, in "The Journal of Modern History", 58, *Supplement: Politics and Society in the Holy Roman Empire, 1500–1806*, 1986, pp. S23–S45.

[8]Full title: *Dioptrice seu demonstratio eorum quae visui et visibilibus propter conspicillanon ita pridem inventa accidunt* (On the refraction or demonstration of the things seen and visible as a result of the telescopes that have not too lately come into use). Cf. A. Malet, *Kepler's Legacy: Telescopes and Geometric Optics*, in A. Van Helden, S. Dupré, R. van Gent, H. Zuidervaart (eds.), *The Origins of the Telescope*, KNAW, Amsterdam 2010, pp. 281–300.

[9]L. Ingaliso, *Filosofia e cosmologia di Christoph Scheiner*, Rubbettino, Soveria Mannelli (CZ) 2005 (in Italian).

was the first to test this variant of the telescope, giving a detailed account in the *Rosa Ursina*,[10] printed in 1630.

The instrument proved inconvenient for terrestrial applications because it returned inverted images. However, it had a wider field of view than the Galilean telescope — a valuable bonus for finding and tracking objects on the moving sky — and a larger exit pupil, which made it easier to align the eye with the eyepiece. It also possessed a crucial added value that had eluded even Kepler's vivid imagination: the ability to focus simultaneously on the observer's retina both the target and an object placed in the common focus between the two lenses. A young Englishman was the first to realize this property, which the Galilean model has not because the common focus lies outside the interval between the concave and convex lenses. Legend has it that William Gascoigne was accidentally helped in his discovery by a spider that had decided to weave its web between the objective and the eyepiece of one of his Keplerian telescopes. This happened in the same years that the old Galilei, in the gilded cage of Arcetri, spent the rest of his life in the gray of incipient blindness, no longer able to look at the sky.

Gascoigne was born in 1612 to a family of small landed gentry in northeastern England; he would die 32 years later, killed by an arquebus bullet at the Battle of Marston Moor, outside York, while serving his king, Charles I Stuart, who was unsuccessfully trying to put down Oliver Cromwell's Parliamentarian rebellion. These were difficult times for a nation grappling with religious and political tensions that had resurfaced strongly after the epic period of Elizabethan absolutism. The throne had passed to the Stuarts, in whose veins ran the blood of the unfortunate Mary: rulers very different from the Virgin Queen who had led the country to conquer the seas. Nevertheless, the fertile climate that had produced William Shakespeare, Christopher Marlowe, and William Gilbert under Elizabeth I, and that would soon give the world Isaac Newton and Edmund Halley, remained on the island across the Channel.

To promote the scientific figure of Gascoigne,[11] along with those of his correspondents Jeremiah Horrocks and William Crabtree, was the

---

[10]The work, whose full title is *Rosa Ursina sive Sol ex admirando facularum & macularum suarum phoenomeno varius [. . .]* (Rosa Ursina, or the Sun, from the wonderful phenomenon of its faculae and spots), is dedicated to Paolo Giordano II Orsini, Duke of Bracciano, scion of one of the most influential princely families in medieval Italy and Renaissance Rome.

[11]D. Sellers, *The "Discovery" of William Gascoigne*, in Id., *In Search of William Gascoigne: Seventeenth Century Astronomer*, Springer, Berlin 2012, pp. 15–22.

Astronomer Royal John Flamsteed, founder and first director of the Royal Greenwich Observatory.[12] It happened in 1660. Charles II Stuart, the son of the beheaded Charles I who had regained the English throne after the collapse of the Commonwealth, had graciously granted the establishment of an academy of science — the Royal Society of London for the Improvement of Natural Knowledge. Chemistry, physics, and medicine excelled: disciplines that were firmly rooted in the British Isles. The astronomer Flamsteed[13] needed native school leaders to promote his science. He chose the *"Nos Keplari"*, as the three young men called themselves, fully identifying themselves with the thought and work of the great German scientist. With a clever marketing operation, Flamsteed launched Gascoigne into the Olympus of the great, arguing that the young man should even be considered the ideal successor to the work of Galilei and Kepler: ideal because a premature and bloody death had denied him the opportunity to put his discovery to good use. It is an extravagant theory, if not a true counterfactual, that survives in the Anglo-Saxon world and on the Web. In fact, the figure of Gascoigne is more like that of Lippershey. He too, like the Dutchman, introduced a new technology that others would perfect and use in many different fields of human activity.[14] Nothing more can be added, because history is not made with ifs and buts. Anyway, what is this technology all about?

---

[12] Ordered by Charles II Stuart in 1675 and built the following year on the hill in Greenwich Park overlooking a loop of the Thames, the first English public observatory was entrusted to the direction of a Astronomer Royal with the task of «correcting with the utmost care and diligence the tables of the motions of the heavens, and the positions of the fixed stars, in order to find the much desired longitude of the places to perfect the art of navigation» (S. Newcomb, *Popular Astronomy*, CUP, Cambridge 2011). Famous for having become the reference for the Earth's Prime Meridian in 1884, the observatory was moved to Herstmonceux Castle in East Sussex in 1998. A curiosity: John Flamstead suggested the date for the inauguration of the activity on the basis of a horoscope he had made himself, probably inspired by Tycho Brahe who had used the same approach with Uraniburg.

[13] For an overview of the royal British astronomers, see E. Winterburn, *The Astronomers Royal*, National Maritime Museum, London 2003. On Horrocks, the greatest of the *Nos Keplari*, see. A. Chapman, *Jeremiah Horrocks, The Transit of Venus, and the "New Astronomy" in Early Seventeenth-Century England*, in "Quarterly Journal of the Royal Astronomical Society", 31, 1990, pp. 333–357; P. Aughton, *The Transit of Venus: The Short, Brilliant Life of Jeremiah Horrocks, Father of British Astronomy*, Weidenfeld & Nicolson, London 2004.

[14] F.J. Dyson, *Two Revolutions in Astronomy*, in "Proceedings of the American Philosophical Society", 140, 1, 1996, pp. 1–9.

The instrument invented by the young Englishman, which would later evolve into the "micrometer", consists of two thin wires (originally made of spider slime) arranged crosswise in the plane perpendicular to the optical axis of the telescope, in the focal position common to the two optical elements, between the objective and the eyepiece of a Keplerian telescope. Their intersection marks a precise and repeatable position in the instrument's field of view. To improve their visibility against the black background of the sky, the wires can be illuminated by a parallel beam from a lamp placed on one side, so that the rays intercept them and are scattered towards the observer without dazzling him. An example will help us understand why such a seemingly insignificant tool represents a real technological and methodological revolution, not only in astronomy. Imagine going hunting with a rifle equipped with a telescope — a weapon like those used by camouflaged snipers in war movies. Thanks to the power of the optical scope, you can see your quarry better and farther than with the naked eye; but without a reference to tell you where to point your gun, you are doomed to fail at every shot. The viewfinder will not help you because you cannot focus it together with the target. This problem is brilliantly solved by Gascoigne's micrometer, which, thanks to its design, always keeps the viewfinder in focus.

When astronomers were able to replace the simple viewfinder of a mural quadrant with a Keplerian telescope equipped with a micrometer, they found an instrument dozens of times more sensitive and much deeper than that available to the king of naked-eye observations, Tycho Brahe. In its most advanced version, the device could be used to directly measure angles and relative distances on the celestial sphere, for the benefit of quantitative astronomical mapping. Here is how it works. A diametral wire, crossed perpendicularly by two others, one fixed and equally diametral and the other movable, is mounted by means of a rotating ring at the focal plane common to the objective and the eyepiece. By means of a worm screw, the latter wire can be displaced parallel to itself by a measurable amount. To determine the angular separation of pairs of stars close to each other on the celestial sphere, one should first align the sources to the single wire by rotating the ring. Then, after pointing the telescope so that one of the stars falls at the intersection with the fixed cross, the moving wire is shifted until it intercepts the other star. With proper calibration of the instrument, the rotation provides the angle that the line connecting the two stars makes with the direction of North, while the angular distance between them is evaluated by the translation that the moving wire undergoes. A similar strategy can be used to estimate the size of extended celestial objects, such as planets or details of the lunar surface.

These conceptually simple operations are complicated by the diurnal motion of the celestial sphere, which forces the observer to constantly track the source — a line-of-sight adjustment that becomes more frantic the farther the target is from the poles. The telescope cannot simply be held by the astronomer as a soldier holds his rifle, because the tremors of the body would render it completely useless for looking and especially for taking measurements. Anyone who has tried to look at the sky with high magnification binoculars without at least resting his elbows somewhere knows this well. A rotating mount is needed to stabilize and operate the instrument. Do not be fooled by movies in which bearded sea wolves, planted on the deck of a ship rocked by the waves, scan the horizon for enemy vessels, fearlessly wielding a long marine binocular. In this context, the need to compensate for the oscillations caused by the motion of the waves, reactivity and experience played an important role, along with a modest magnification, which would have been too small for astronomical purposes. In addition, these brave captains only had to see, not measure.

The needs of the arquebusier who wanted to hit a target were different. To aim his heavy weapon (by eye), the feathered and picturesque soldier used a fork-like rod planted on the ground that allowed him to swing the rudimentary gun left and right, that is, in azimuth, as astronomers would say, and to vary its elevation. One can do the same thing with a telescope by using a mount that enables the instrument to rotate smoothly about two axes, one of which is vertical (azimuth)[15] and the other horizontal (elevation). For example, the telescope tube can be suspended on a fork that rotates about the vertical to permit an experienced astronomer to follow the target along its diurnal motion. This crude altazimuth mount has no installation problems. A plumb line or level (e.g., a flat surface of liquid mercury) is sufficient to align the azimuth axis with the vertical of the site. But guidance is very tricky and requires a trained intelligence, human or artificial. The reason is that the direction around which the sky rotates uniformly coincides with the azimuth axis only at the poles. Elsewhere,

---

[15]"Azimuth", from an Arabic word meaning directions, is a term used in spherical reference systems. After identifying the plane passing through the observer and perpendicular to the vertical of the place ("zenith", which in the Arabic language means direction of the head), it is the angle between the North and the projection made on such a plane by any direction outgoing from the observer.

this mount implies a non-uniform rotation of each of the axes to keep the target fixed with the field of view — an operation that requires natural endowments and, above all, long experience of the operator; and it never comes perfect!

This drawback was overcome by an ingenious stratagem devised by the same Jesuit who first experimented with the Keplerian model of a biconvex telescope. To support his helioscope, an instrument with an aperture only slightly larger than the telescope used by Galileo for his first discoveries, Scheiner tilted the azimuth axis of the mount so that it coincided with the Earth's polar axis. In this way, he could follow the Sun, which was the target of his observations, while keeping almost[16] unchanged the angle of elevation of the instrument during the guidance. In short, his idea was to cancel out the motion of the celestial vault by rotating the telescope about a single direction, parallel to the terrestrial polar axis, with the same uniform angular velocity as the diurnal motion, 360° per day, but in the opposite direction to that of the stars. The Jesuit had invented the equatorial mount, which would later be well suited for the motorization of telescopes in the 19th and 20th centuries. Only today, in the age of computers and automation, has it become possible and advantageous to return to the primitive altazimuth mounts, entrusting the management of the complex accelerated movements around the two axes to specialized electronic controllers. We will see this at the end of our story, when we meet the modern giants.

The introduction of the ocular micrometer made it possible to observe a target for some time, and this greatly simplified the work of the astronomer. But seeing deeper and easier was not enough to improve the accuracy of traditional angular measurements without a parallel refinement in the mechanical characteristics of the instruments. Mounts, both altazimuthal and equatorial, were then equipped with precision goniometers on both axes: large graduated circles for reading the instrument's rotations relative to fixed references. This was no different from what had been done in Tycho Brahe's wall quadrants, which had set the standard for quality in the pre-telescopic era, but with increasing accuracy. The technological challenge was the usual one: to divide the circle into equal parts so finely that the astronomer could discern tiny angles while battling gravity-induced deformations and thermal expansions.

---

[16]To be precise, the Sun is not the best target for an equatorial mount because, in addition to the diurnal motion common to all other celestial bodies, it has a slower (but systematic and peculiar) annual motion along the ecliptic.

Neither Gascoigne nor his two friends had the time or means to draw important astronomical results from the micrometer. The fame of Jeremiah Horrocks,[17] another of the *Nos Keplari*, who also died very young of unknown causes, is rather based on an observation made with a now classic technique pioneered by Galileo himself. William had been introduced to astronomy at the age of 16 and, like many personalities in our long history, fell madly in love with it. Despite his humble farming origins, his precocious genius had gained him access to the University of Cambridge, where he supported his studies by working, as Newton would do shortly afterwards. However, he soon abandoned his studies, probably for lack of funds, without a degree but with a good knowledge base. In a short time, he had devoured and metabolized the now classic authors, especially Brahe, Galilei, and his beloved Kepler, and had begun an observational campaign that led to his most famous achievement, the observation of the transit of Venus across the sun's disk.

The event, a sort of micro eclipse, is relatively rare because it requires a near-perfect alignment of the two celestial bodies with the Earth. If the orbits of our planet and Venus were coplanar, the favorable condition would occur regularly every 584 days (synodic period). However, an inclination of only $-3°.5$ of Venus' orbit on the ecliptic is enough for the blue planet, like a capricious operetta bridesmaid, to miss the appointment most of the time. To know if and when the transit will occur, one needs an accurate model of the heliocentric motions of the two planets involved in the carousel: geometries and numbers such as distances, orbital flattenings, and inclinations that only an experienced astronomer can handle, now as then (with the difference that today we can resort to pre-packaged programs that run on a modest PC and provide all the ephemerides even to those who are completely ignorant of the subject).

With the data at his disposal, albeit not very precise, Kepler had correctly predicted a transit in 1631. However, no one had been able to confirm the phenomenon because in Europe the event had taken place after sunset, and in the Americas both the natives and the Spanish invaders had more pressing matters than observing the night sky. This sounded very unfortunate because, according to the German astronomer's calculations, it would be 120 years before the transit of Venus' shadow across the solar disk could

---

[17] A. Chapman, *Jeremiah Horrocks, William Crabtree, and the Lancashire Observations of the Transit of Venus of 1639*, in D.W. Kurtz (ed.), "*Transits of Venus: New Views of the Solar System and Galaxy*, Proceedings IAU Colloquium", 196, 2004, pp. 3–26.

be spotted again. But the 21-year-old Horrocks disagreed with his favorite astronomer. His data and calculations suggested that a new transit would occur on Sunday, November 24, 1639, just eight years later. So he prepared to observe the event. The young man was out for glory, and the occasion seemed the most propitious. He had a modest instrument in terms of aperture and magnification, but, as he later claimed, of excellent quality. Unable to look directly at the Sun through the telescope on pain of irreversible eye burns, he resorted to the stratagem devised by the industrious Scheiner to view and record the famous spots. The procedure had been successfully tested eight years earlier by the French abbot Pierre Gassendi during the transit of Mercury across the Sun. The image produced by the eyepiece was projected onto a white surface perpendicular to the optical axis and held tightly to the instrument; basically a sheet of graduated paper on which, among other things, it was possible to draw what was seen and make fairly accurate heliographic measurements.

When the moment of truth arrived, the sky was cloudy; a rather common occurrence in foggy Albion. To make matters worse, there was also an obligation to sanctify Sunday with customary religious practices, which took up some of Jeremiah's time. But the stubborn young man did not give up. Like a good Englishman, he was determined to play his game until the referee blew the final whistle. So he stuck to his instrument and was rewarded. In the afternoon, for just over half an hour and until the Sun set, he was able to observe the black shadow of the planet between the clouds as it passed over the reddish disk of the Sun. It was a masterstroke and a proof of how the telescope had enabled any person of good will to perform great astronomical discoveries. Horrocks' hastily made estimates allowed him to drastically reduce the value of the diameter of Venus and to bring the radius of the Earth's orbit to 95 million kilometers, closer and closer to the modern measure of the Astronomical Unit. His correspondent Crabtree, who traded in textiles for a living, was the only other person to observe the transit, if for moments. In short, only two were the witnesses who could vouch for the reality of the phenomenon and reassure the skeptics. All the others had been persuaded by Kepler's ephemerides to desist.

We must now return to the design of Kepler's telescope. Because of its unquestionable merits, the convex model with two lenses was gradually adopted by astronomers all over Europe, and at the end of the 18th century, thanks to the addition of two other lenses capable of correcting the orientation of the image, it was also adopted for terrestrial use. Almost everywhere, artisans and small shops flourished for the production of Keplerian telescopes and microscopes, which were increasingly requested by princes and cardinals for their games and their *Wunderkammern*, and as endowments of colleges and academies.[18] Astronomy enthusiasts, often penniless, could also build both the optics and the mounts themselves. In fact, the paraphernalia of a lens grinder was simple enough to be easily replicated at home. In the more technological version of the brothers Constantijn and Christiaan Huygens,[19] active in Holland in the second half of the 17th century, the rough piece of glass was attached with mastic to the end of a rod as long as the desired radius of curvature. The other end of the rod was constrained to a fixed point. A skilled operator pressed the surface of the glass blank to be formed against a rotating iron disk. Abrasive powders of increasing fineness and plenty of water helped to give the lens the desired shape and then gloss. When both sides were finished, the operator used sunlight and a set of diaphragms to check that the different rings of the lens all gave the same focal length. If there were any defects, the blank was reassembled on the grinding machine and retouched. During the 17th century, this process enabled expert opticians to produce lenses with diameters of up to 20 cm and beyond. The main problem was not only the processing of the glass, but also the procurement of a homogeneous raw material free of bubbles and impurities.

In short, with a little money and relatively modest effort, it was possible for anyone to plunge into the darkness of the night sky in search of known objects, for the sheer pleasure of admiring the firmament; or to go in search of unknown bodies and phenomena for the sake of knowledge and, why not, in the secret hope of making a splash and gaining eternal fame. The *Sidereus Nuncius* had unleashed a veritable gold rush in the vast and unexplored Far West of the heavens, with sleepless pioneers armed with simple tools, great enthusiasm and infinite patience. All it took was a clear night, a smattering of astronomy, keen eyesight combined with a good dose of luck, and the rare ability to understand what the telescope was showing.

---

[18] A. Van Helden, *The Telescope in the Seventeenth Century*, in "Isis", 65, 1, 1974, pp. 38–58.

[19] J.G. Yoder, *Unrolling Time: Christiaan Huygens and the Mathematization of Nature*, CUP, Cambridge 2004.

To be at the forefront, however, one also needed a high-quality instrument, which was not easy to come by, and a certain amount of caution in its use.

While it is true that Galilei had charted the paths of the firmament for all to see, his condemnation, especially in Italy, had discouraged most from following his example. The great scientist was saved by his fame, which made him almost untouchable. But what would have happened if someone less known, or even Mr. Nobody, had taken his place? Also for this reason the most brilliant among his disciples[20] practiced in the most diverse fields of knowledge, but abandoned astronomy and especially cosmology. Thinking about the heavens was too risky. To examine it with the telescope was much less so.[21] The Jesuits, the Pope's soldiers, had quickly become avid observers and expert instrument makers. To be on the safe side, it was enough to refrain from risky interpretative hypotheses: an attitude that Galilei had failed to maintain, to his great detriment.

The stars of the firmament were not yet the targets of observers, too faint, motionless, and anonymous. To these dim lights, professional and amateur astronomers preferred the Moon, the planets, and the Sun — extended objects from which drawings and maps could be made, possible companions could be flushed out, and intriguing mysteries revealed or unraveled. Such as the unique appearance of Saturn. Galilei had described it as «composed of three [stars] which almost touch each other, never change or move relative to each other, and are arranged in a row along the zodiac, the middle one being three times larger than the lateral ones.»[22] It happened in June 1610. The same telescope that had brought him triumph now betrayed him by denying the possibility of disentangling the planetary disk from the belt of rings. With his extraordinary scientific insight, Galileo warned that his description of Saturn's peculiar morphology was implausible and wanted to hold out until he had a better idea. At the same time,

---

[20]The mathematician Benedetto Castelli (1578–1643), a monk from Cassino, an early pupil and a very faithful friend; Bonaventura Cavalieri (1598–1647), who pioneered infinitesimal calculus; Evangelista Torricelli of Faenza (1608–1647), a mathematician and physicist known for his studies in hydraulics, who succeeded Galilei as the Grand Duke's court mathematician; and the youngest of all, Vincenzo Viviani (1622–1703), who was also the Maestro's first biographer.

[21]O. Gingerich, *How Galileo and Kepler Countered Aristotle's Cosmological Errors*, in J.A. Rubiño-Martín *et al.* (eds.), "*Cosmology across Cultures*, Cosmological ASP Conference Series", 409, 2009, pp. 242–249.

[22]Galilei, *Opere*, XI, cit., p. 327, *Lettera di Galileo Galilei a Belisario Vinta, 30 luglio 1610* (in Italian).

he prudently decided to guarantee his authorship by communicating the discovery with the anagram[23] of a dry Latin phrase, «*altissimum planetam tergeminum observavi*», which loosely translated means: "I have observed the distant planet and found it three-bodied". It was a trick to protect the secret, like putting your will in a safe to make it public in case of need. He then sent the unintelligible string of characters to Giuliano de' Medici, a Roman Catholic prelate belonging to a cadet branch of the famous Florentine family. At the time, he was serving as the Grand Duke's ambassador to the Habsburg Imperial Court in Prague and acted as an intermediary in the correspondence with Kepler.

When the German genius held the anagram in his hands, he saw it as an open invitation. Solving puzzles was his passion. He attacked the problem with the same determination as Jean-François Champollion with the Rosetta Stone and Alan Turing with the Enigma machine, and in a short time thought he had found a plausible solution, even if he had to sacrifice one of the characters that make up the string: «*Salve umbistineum geminatum martia proles*», which means "be greeted, double knob, children of Mars". But that was not the correct answer. Galileo had not discovered a pair of Martian satellites (twins). The astonishing fact is that the red planet actually has two moons, Phobos and Deimos; names that come from the Greek words for fear and terror. These two rocky bodies have remained unknown until 1877, when the American Asaph Hall saw them through a telescope with an aperture of 66 cm. The ineffable Kepler made no observational verification of his solution of the anagram. He liked his because it restored order to the numerical sequence of the satellites: no moon for Venus, one for Earth, two for Mars, and four for Jupiter, as his god-architect had surely wanted. He did not change his mind even when Galileo, at the urging of Emperor Rudolph II, decided to withdraw his reservations. Although the authentic interpretation of the anagram removed all meaning from the story, the belief that Mars was accompanied by two satellites became so widespread that Jonathan Swift included it in his *Gulliver's Travels* more than a century later. Magic? Not really; just chance and the fact that people tend to remember what they like or are curious about and forget the rest.

The Pisan scientist had a habit of encrypting his letters. In the fall of 1610, having completed his move to Florence, he took up the telescope again

---

[23]The string of the anagram was: *smaismrmilmepoetaleumibunenugttauiras*.

and, observing Venus, realized that «*Cynthiae figuras aemulatur mater amorum*», that is, "the Mother of Love [an epithet for the planet Venus] copies the forms [phases] as the Moon", poetically renamed with the epithet of the goddess Artemis, Cynthia.[24] This was a sensational discovery and a real stone thrown at the geocentric model which, instead of a complete cycle as observed for the Moon, did not grant more than a small sickle to Venus (the mother of love), trapped between the Earth and the Sun, as she moved on her epicycle. Before revealing his discovery, Galilei had sent Giuliano de' Medici the customary anagram of a Latin phrase that Kepler had then uselessly tried to unravel. Exhausted, the German had resigned himself to writing to Galilei: «I beg you not to keep me long in the dark about its hidden meaning, and to be frank with me. Meanwhile, eager to solve the riddle, I tried various anagrams without deriving anything that makes any sense at all except this: "it is indeed proved that Jupiter, with a spot oh! red, rotates"».[25] Incredible as it may seem, while barking up the wrong tree, the imperial astronomer had once again hit the nail on the head, anticipating two real phenomena, the rotation of the planet and the existence of the Red Spot,[26] the great anticyclonic storm clearly visible on the surface of Jupiter. The fruit of the genius and the curiosities of the anagrams! As for the Jesuits, they remained unperturbed. Cynthia's whims did destroy Ptolemy's cosmology, but they also fit perfectly into Tycho's geocentric clockwork.

Only two years later, when the first poisoned arrows arrived from the Florentine monastery of San Marco (see p. 98), Galileo returned to the mysterious Saturn and saw it single; then again «*tergeminum*», tricorporeal. This apparently extravagant behavior is due, as we now know, to the considerable inclination of the plane of the rings to that of the planet's heliocentric orbit, so that as Saturn proceeds on its slow path, the perspective from which these slender structures are viewed by the terrestrial observer changes. When the Earth crosses the plane on which they lie —

---

[24]Cynthia, the Latinization of the Greek name for the Moon, was one of the epithets of Artemis. According to myth, the goddess of the hunt and her brother Apollo were born on Mount Cynthus, on the island of Delos. Their mother Leto had taken refuge there to escape the vengeance of Hera, who systematically took it out on the poor girls seduced by her lascivious husband Jupiter.

[25]Galilei, *Opere*, XI, cit., p. 11, *Lettera di Giovanni Keplero a Galileo Galilei, 9 gennaio 1611* (in Latin).

[26]But see G.D. Goehering, *Kepler's solutions to Galileo's anagrams*, in "Journal of the British Astronomical Association", 92, 1981, p. 41.

an event that occurs on average every fifteen years — the rings appear so sharp as to be invisible. Unable to figure out where the trick lay, Galileo felt an unusual sense of uncertainty after the excitement of his sensational success in Rome. In December 1612 he wrote to Marco Velseri, one of his German correspondents, rich, cultured, and very powerful, who was also a Lincean:

> Now what can we say about such a strange case? Maybe the two minor stars consumed themselves, the way sunspots do? Maybe they disappeared and suddenly ran away? Maybe Saturn devoured his children? Or was it illusion and fraud the appearance with which the lenses have for so long deceived me and many others like me who have seen them quite a few times? Perhaps time has now come to revive the hope, already close to withering, in those who, guided by deeper contemplations, have felt all new observations to be fallacious, nor they can in any manner subsist? I cannot say a decisive thing in such a strange, unexpected and new case; the shortness of time, the unprecedented fact, the weakness of intellect, and the fear of error make me very confused.[27]

The bizarre morphology of Saturn[28] was correctly interpreted mid-century by Christiaan Huygens,[29] a Dutch scion of a wealthy and cultured family. His father, a diplomat in the service of the Orange family, reconciled his hatred of the Spaniards with a passion for poetry, music, and natural philosophy. He corresponded with Galileo and Marin Mersenne, a famous French theologian and mathematician, and was a personal friend of Descartes, whom he occasionally entertained at his home. An excellent educational context[30] for his five children in a strict and meritocratic society undergoing rapid economic, political, and cultural expansion. The United Provinces — seven counties, duchies, and lordships that had rebelled in 1568 against

---

[27]Id., *Lettere*, F. Flora (ed.), Ricciardi, Milan 1953, *Lettera di Galileo Galilei a Marco Valseri, 1 dicembre 1612*. See also E.A. Partridge, H.C. Whitaker, *Galileo's Work on Saturn's Rings*, in "Popular Astronomy", Vol. 3, 1896, pp. 408–414; A. Van Helden, *Saturn and his anses*, in "Journal for the History of Astronomy", 5, 1974, pp. 105–121.

[28]N. Howard, *Rings and Anagrams: Huygens's System of Saturn*, in "The Papers of the Bibliographical Society of America", 98, 4, 2004, pp. 477–510.

[29]H.J.M. Bos *et al.* (eds.), *Studies on Christiaan Huygens: Invited Papers from the Symposium on the Life and Work of Christiaan Huygens*, Swts & Zeitlinger, Lisse 1989.

[30]B. Stoffele, *Christiaan Huygens: A Family Affair*, Master's thesis, Utrecht University, 2006.

Philip II of Spain and his cruel emissary, the Duke of Alba — lived their golden age in the shadow of Calvin and the Orange family, uncrowned rulers of a republic ruled by an oligarchy of merchants. Amsterdam was reminiscent of Venice for its canals, but also of Florence for its painters, writers, philosophers, and scientists.

It was in this Para-Renaissance atmosphere that Christiaan grew up, a boy in poor health and prone to depressions, but of extraordinary ingenuity and insatiable curiosity. He received a thorough education, as befitted a nobleman: from physical activities (dancing, riding, and fencing) to those of the intellect (law, philosophy, mathematics), of course without neglecting the *bon ton*, first at home until the age of 16, then touring the best schools in Holland. The father, with the usual parental tendency to reproduce clones, wanted to make him a diplomat, but willingly gave up his purpose in favor of his second son's inclination for scientific studies, supported by Mersenne's judgment of him: «a new Archimedes».[31] Christiaan was then 26 years old and had a good reputation as a mathematician when, together with his brother Constatijn, he discovered a passion for grinding glass. This was not just an idle hobby to pass the damp Dutch winters or to put into practice the studies of optics they had begun a few years earlier. The two young men wanted a perfect lens to scan the planets, hoping to emulate Galileo's achievements. They wanted glory and soon became true experts in the design and manufacture of optical instruments. Such self-motivation deserved a reward, and so it was.

In March 1655, Christiaan had a telescope with an aperture of 42 mm, a focal length of 380 cm, and an eyepiece capable of 50 magnifications. The instrument did not match the qualities of those made a few years later in Rome by the master Giuseppe Campani, but its performance was good and it had a large field of view. While pointing it at Saturn, the young Dutchman saw a star next to the giant planet that could have been a satellite. He needed time to be sure that there was a physical connection between the two bodies, for the distant planet moves slowly among the stars of the firmament. But he could not delay to communicate the possible discovery, for fear of being overtaken by someone alerted by a leak or aided by chance. So, following the lesson of Galileo, he announced his finding to a small group of friends by means of a Latin anagram. This wise precaution, however, caused him some headaches. An English correspondent with a

---

[31] Ibid, p. 87.

penchant for jokes, after revealing the meaning of the coded message, made him believe that the fact was already known across the Channel. Apparently, Galilei had been better at inventing his anagrams. In any case, after a few months of uninterrupted research, Huygens was sufficiently certain of his discovery to make an ecumenical announcement. He had spotted the first *Luna Saturni*, a new tenant of the Solar System that almost two hundred years later John Herschel would name Titan. This had not happened for almost half a century, since the time of the Medicean planets.

Since appetite grows with food, Christiaan tried to double his catch by pointing his telescope at Venus and Mars, but found nothing. Neither planet showed the presence of any satellites. Puzzled by the result, the Dutchman worked out a reasoning that led him to an astonishing conclusion: there were six known planets, including the Earth, read as one planet in the Copernican system, and six satellites. Perfect parity! Therefore, he concluded, there was no other object to discover. Pythagoras and Kepler would have enthusiastically agreed (the latter even though he saw his solution to Galilei's anagram contradicted). Disappointed by this thought, Huygens resumed his study of Saturn. He had built a more powerful telescope and was able to confirm an earlier intuition. The two "moon crescents" seen by Galileo were nothing more than the projection of an immense disk around the planet. He immediately implemented the usual communication strategy: first an anagram and, after certainty, a printed report entitled *Systema Saturnium* (1659), which he dedicated to Prince Leopoldo de' Medici. No doubt, the Pisan's method had spread.[32]

Huygens' discoveries irritated Roman astronomical circles, which for some years had been considered the first in Europe in the production of optics. In particular, Eustachio Divini[33] did not take well to the news that an outsider had built a telescope good enough to discover the moon of Saturn that he had searched in vain for with his prestigious instruments. Divini was indeed famous for the quality of his clocks and the lenses for telescopes and microscopes that he began to produce in Rome after he had laid down his sword and met scientists of the caliber of Castelli and Torricelli. He was

---

[32]Cf. G. Anzini, *La forma del pianeta Saturno in un'esperienza dell'Accademia del Cimento*, in "Bollettino della Società di Studi fiorentini. Rivista di Studi storici", 2, 1998, pp. 1–12 (in Italian).

[33]The biographical information is taken from M. Muccillo, *Divini, Eustachio*, in *Dizionario biografico degli italiani*, Vol. LX, Institute of the Italian Encyclopedia, Rome 1990 (in Italian).

talented and innovative. He may even have been the first to make practical use of Gascoigne's micrometer. His mood turned to real anger when he read the *Systema Saturnium*. A decade earlier, the Italian artisan, who also liked to explore the heavens, had published a drawing of the planet that the Dutchman now dared to criticize. The diatribe took on harsh tones. Divini was supported by a French Jesuit who blew on the fire, hoping to demolish the heliocentric approach used by Huygens to explain the phenomenon of the disappearance of the rings. After an exchange of vitriolic pamphlets in which, besides scientific arguments, accusations and epithets such as «*vilem vitrarum artificem*» (dirty glass grinder) were applied to poor Eustachio, the dispute reached the Accademia del Cimento, the Florentine institution founded by the disciples of Galileo to promote the experimental method. It was not an undue interference, but the consequence of an initiative of Huygens, who had challenged Leopold of Tuscany.

The academicians decided to test the hypothesis of the rings with a simulation and assigned the task to the Neapolitan astronomer and physiologist Giovanni Alfonso Borelli. In 1660 he procured a scale model of the *Systema Saturnium* made of wood and plaster. After placing the simulacrum at the end of a long, dark tunnel, Borelli asked a group of «idiot»[34] witnesses, unaware of the true shape of the model, to observe it in the dim light of a few torches and to report their impressions. The test was repeated several times, gradually changing the perspective of the rings. The conclusion was that the Dutch aristocrat was right and the Italian craftsman was wrong. Divini resigned and returned to his technical work. In the course of this dispute, the Neapolitan Francesco Fontana was brought into play. He was also a maker of Keplerian telescopes, which, «although reversed, allow one to see the Moon and the stars.».[35] Divini had presented him as the first and true discoverer of not one but a handful of Saturn's satellites. This nonsense, invented to win the battle with Huygens, was one of the shadows that obscured the profile of the great Neapolitan optician.

---

[34]This is not an insult, but rather the use of the adjective in its pure Latin meaning of "ignorant", i.e. a person who does not know.

[35]*Il carteggio linceo della vecchia Accademia di Federico Cesi (1603–1630)*, G. Gabrieli (ed.), "Atti della Reale Accademia nazionale dei Lincei. Memorie della Classe di Scienze morali, storiche e filologiche", s. 6, VII, fasc. 4, 1942. On Fontana, cf. P. Molaro, *The Neapolitan Francesco Fontana and the Birth of the Astronomical Telescope*, in "Journal of Astronomical History and Heritage", 20, 2, 2017, pp. 271–88; M. Capaccioli, G. Longo, E. Olostro Cirella, *L'astronomia a Napoli dal Settecento ai giorni nostri. Storia di un'altra occasione perduta*, Guida, Naples 2009 (in Italian).

Fontana,[36] who was only a young man when Galileo inaugurated the telescope era (he was born around 1585), spent his life making instruments for himself and for noble clients, earning great praise from men of the caliber of the German encyclopedic Jesuit Athanasius Kircher. Torricelli said of one of his lenses that it was «the best [...] made in a thousand glasses over a period of 30 years».[37] An excellent craftsman, Fontana also knew how to use his telescopes effectively. He made the first drawing of Mars and captured the planet's daily rotation — a discovery fully supporting the Copernican model. He drew a valuable map of the Moon, the most beautiful of all those produced to date, and discovered the characteristic bands on Jupiter's disk, a sign of the violent turbulence that characterizes the planet's atmosphere. What a prize! But he himself, perhaps misled by false reflections in his instrument, announced some other celestial phenomena that did not actually exist: "witty inventions", in the words of the Venetian playwright Carlo Goldoni, which provoked the harsh criticism of an influential Jesuit astronomer, Giovanni Battista Riccioli.

About fifteen years younger than Fontana, Riccioli was a fierce conservative and a staunch geocentric. What was most important to him was to keep the Earth at the center of the world. Since the Ptolemaic model had difficulty explaining appearances, he adopted Tycho's mixed model. But he had to adjust it by positioning Mercury, Venus, and Mars as satellites of the Sun, while keeping the orbits of Jupiter and Saturn centered on the Earth, obviously along with the sphere of fixed stars. He even arrogated to himself the right to baptize lunar craters, shamelessly favoring the exponents of geocentrism over Aristarchus and Copernicus and, most singularly, even over his more illustrious brothers.

Fontana's bad press was certainly due to his scientific bluffs and gaffes, but also to the vehement reaction of Galileo's fans. Believing that he was Della Porta's heir, the Neapolitan presented himself as the authentic inventor of the telescope. This claim must have sounded blasphemous to men of culture who had promoted the genius of Pisa throughout Europe as a symbol of the struggle for scientific freedom. It must have seemed even more odious to them that a mythomaniac optician was waging war against

---

[36] G. Arrighi, *Gli "occhiali" di Francesco Fontana in un carteggio inedito di A. Santini*, in "Physis", 6, 1964, pp. 432–448 (in Italian).
[37] *Le opere dei discepoli di Galileo Galilei. Carteggio 1642–1648*, Vol. 1, P. Galluzzi, M. Torrini (eds.), Giunti-Barbera, Florence 1975 (in Italian).

the Galilean attempt to separate science from faith. It was probably this conservative and pro-Jesuit position that put Fontana out of the game of history. His compatriots should have thought about reevaluating his work, but in the "*Bel Paese*" the losers and the minor figures have always been condemned to oblivion, to the great waste of the maieutic value of memory. A striking example of this self-destructive attitude concerns a semi-unknown Sicilian priest, Giovanni Battista Odierna (Hodierna, in the spelling of the time). His story is also emblematic of the sudden and widespread diffusion of the telescope in every corner of Europe and among the most diverse social classes.

Born at the turn of the 16th century in the southern Sicilian town of Ragusa, a land suffocated by the heat and the Spaniards, Giovanni had taken to self-study to cultivate the sciences. His humble origins gave him little chance of intellectual advancement outside of a career in the Church: a rule more or less true everywhere in Italy among the poor. Thus, at the age of 25, he became a priest. Dierna — that was his name, to which he had added a "ho" to underline his enthusiasm for modernity[38] — alternated religious duties with a passion for science, particularly astronomy. A staunch admirer of Galileo, he was particularly attracted by the Pisan's forays into the vast universe of stars, still unknown and yet to be explored: a Far West *ante litteram* for brave young minds. The opportunity to change his life came in 1637, when the brothers Carlo and Giulio Tomasi,[39] the scions of an ambitious Sicilian family, pervaded by an obsessive religious fervor, received from Philip IV, King of Spain, the authorization to build a new city in their estates, in the barony of Montechiaro, not far from Agrigento, the ancient Akragas. Odierna was called to be in charge of the town planning and the health of the souls, with a small salary, service quarters, and some time to be spent for his studies. He later became the city's first archpriest. But Palma was truly a lost oasis in the desert for a man burning with curiosity to explore the infinite nature of things. He had few means of communication with the outside world, and the inefficiency of those he had made his isolation all the more apparent. «I don't have a partner or a friend or someone close to me who can support me

---

[38]"*Odierno*" in Italian means "of the present".

[39]The last descendant of this noble family is Giuseppe Tomasi di Lampedusa, who became famous in the 1960s for his novel *Il Gattopardo* (The Leopard), in which he told the story of his ancestor, Prince of Salina and amateur astronomer, at the time of the unification of Italy.

a little. My teacher is my mind, and I have no one with whom I can share my problems»,[40] Odierna complained to a co-religionist, the Cistercian monk Domenico Plato, professor of philosophy at the Monferrato monastery, in regards to the eclipse of January 1656. His intellectual loneliness was alleviated by a few trips to Palermo, Naples and Rome, where he met Fontana and Kircher, and by a few letters exchanged with great personalities such as Huygens, Riccioli, and the Polish astronomer Johannes Hevelius: mail that, if it ever arrived, could take several weeks to reach the addressee.

Although he was far from the beating heart of Europe, Odierna had one advantage: he could count on a magnificent southern sky, which he used to make some important observations. He collected data on the Medicean planets and came very close to discovering the true nature of Saturn's unique appearance. But above all, he was the first man in the world — as far as we know — to systematically aim his modest instrument at diffuse celestial objects and to catalog them.[41] Like almost everyone else, after Galileo's observations of the Milky Way, he believed that nebulae were poorly resolved or totally confused clusters of stars. He even tried to give them a morphological classification. In addition to star clusters and gas nebulae, Odierna's compilation included at least one galaxy, the great spiral galaxy in the constellation Andromeda, which astronomers refer to as M31 (see chapter 9). Obviously, the Sicilian priest had no idea what that flickering ball of white light might represent. He certainly did not imagine that it was the signal of a distant association of hundreds of billions of stars accompanying the Milky Way in the Local Group. His scientific merit lies in having identified a class of celestial objects, albeit an extremely heterogeneous one. Nor is it diminished by the fact that the nebula had already been seen with the naked eye by the Arabs before the year 1000, and with the telescope in 1612 by Simon Marius (Latinized name of Simon Mayr). This German astronomer is one of the many who claimed to have observed the Medicean planets before Galileo and who gave them the names of the lovers of the Lord of Olympus: Io, Europa, Ganymede, and Callisto. Three women and one boy! In any case, Odierna's discovery would have

---

[40] G.B. Hodierna, *De admirandis phasibus in Sole et Luna visis*, Nicolai Bua, Panormi 1656, p. 38. Cf. also G. Piazzi, *Sulle vicende dell'astronomia in Sicilia*, G. Foderà Serio (ed.), Sellerio, Palermo 1990, p. 64 (in Italian).

[41] G. Fodera-Serio, L. Indorato, P. Nastasi, *G.B. Hodierna's Observations of Nebulae and his Cosmology*, in "Journal for the History of Astronomy", 16, 1, 1985, pp. 1–36.

been enough to make him a small national hero. Instead, even the little fame he had gained in life quickly vanished.

The story of Odierna is not unique. In the second half of the 17th century, the creative impulse in "*Bel Paese*" was fading as a result of the disintegration of the mosaic of powers and competencies built up by the Renaissance. Architecture and the visual arts still held up well compared to the more advanced countries of Europe, driven by a glorious tradition and the Counter-Reformation need to demonstrate the power of the Church. But the sciences languished, stifled by ecclesiastical censorship and worn down by a general political, economic, and social decline that followed the shifting center of gravity of history. What would have happened if space and power had been given instead to the school that had sprung up around Galileo? We will never know. Certainly, the best Italian astronomer of the 17th century, Giovanni Domenico Cassini, found it convenient to emigrate to the France of the Sun King to express the full potential of a brilliant mind and the inspiration of a great scientific manager.

Chapter 6

The heavenly laboratory: Cassini, Huygens, Bradley, and the measures of time and light

> If I had a time machine
> I'd visit Marilyn Monroe in her prime
> or drop in on Galileo
> as he turned his telescope to the heavens.
>
> Stephen Hawking, *MailOnline*

> But in science the credit goes
> to the man who convinces the world,
> not to the man to whom the idea first occurs.
> Not the man who finds a grain of new
> and precious quality but to him who sows it,
> reaps it, grinds it and feeds the world on it.
>
> Sir Francis Darwin, *First Galton Lecture*

Giovanni (Giovan) Domenico Cassini was born in 1625 to a wealthy bourgeois family in Perinaldo, a village in the hills of the Ligurian hinterland, a fief of the Marquises Doria.[1] His long life spanned the entire 17th century, a period of contrasts between reality and thought, nature and sentiment, ethics and customs, and a fertile ground for the seeds of a new world and the progressive disintegration of the *ancien régime*. As was usual in those days, his education began at home and continued in Genoa. There, within the austere walls of the College of the *Societas Iesu* (Society of Jesus), Giovan Domenico came into contact with the world of science that the Jesuits knew and practiced with mastery. His approach to astronomy was further encouraged by his early interest in astrology, which he later disavowed. The horoscope that made him famous as an interpreter of heavenly wills predicted the military victory of Pope Innocent X Pamphili over Duke Ranuccio II

---

[1]G. Bernardi, *Giovanni Domenico Cassini: A Modern Astronomer in the 17th Century*, Springer, New York 2017.

Farnese in the conflict sparked by the murder of the Bishop of Castro,[2] of which Ranuccio was accused of being the chief instigator.[3]

In 1648, at the age of 23, Cassini was lured within the borders of the Papal States by Cornelio Malvasia, Marquis of Bismantova, who was looking for an astronomer to take charge of his private observatory built in the main tower of the family castle at Panzano, a village near Bologna. The young man proved himself so excellent that the following year, thanks to the good offices of his patron, he was appointed to the chair of astronomy at the Studium of Bologna, one of the most prestigious universities in Europe. He took the place of a great scholar, Bonaventura Cavalieri, the Jesuate[4] who had laid the foundations of calculus.

In fact, Cassini's promotion was not Malvasia's only success as a talent scout. When not waging war on behalf of some third party, the nobleman dabbled in the humanities, science, and astrology. Among his protégés was Geminiano Montanari,[5] a brilliant, depressed, and quarrelsome fellow from Modena, with whom Malvasia produced the *Ephemerides novissimae motuum cœlestium* (New ephemerides of the motions of the heavens) and a meticulous map of the Moon, drawn by equipping the telescope with a grid of shiny silver threads. For his part, Montanari is rightly remembered for having reported in 1670 the peculiarities of the second brightest star in the constellation of Perseus, β Persei, which the Arabs had named Algol, the "head of the Medusa". The star periodically changed its brightness with a regular cycle of about three days, testifying to «an instability of the firmament»[6] that Montanari wanted to validate by collecting other suspicious cases: known phenomenologies — of which it was risky to speak — such as transient new stars (*novae* in Latin) and Mira Ceti, the "wonderful star" that appeared and disappeared within a span of about a year, discovered in 1609 by the Frisian theologian David Fabricius in the constellation of the Whale.

---

[2] A small feud in central Italy, elevated as a vassal of the papal state but actually independent, ruled by the Farnese family.

[3] H. Gamrath, *Farnese: Pomp, Power and Politics in Renaissance Italy*, L'Erma di Bretschneider, Rome 2007.

[4] A religious order founded in 1360 by Giovanni Colombini, a wealthy banker from Siena, as a fraternity of laymen inspired by the spirituality of St. Jerome, it had this name because all sermons of Jesuates started and ended with the work "Jesus". Reformed two times, the mendicant order was suppressed in 1668.

[5] M. Cavazza, *Bologna and the Royal Society in the Seventeenth Century*, in "Notes and Records of the Royal Society of London", 35, 2, 1980, pp. 105–123.

[6] Title of a work by Montanari that remained unfinished due to the author's death.

A hundred years later, the English amateur astronomer, John Goodricke,[7] a deaf-mute boy who died at the age of 21 from pneumonia contracted while observing the sky in the cold of night, would have correctly modeled the mysterious behavior of β Persei pointed out by Montanari. Goodricke imagined Algol as a system of two stars orbiting each other in a plane containing the observer in such a way as to produce reciprocal periodic eclipses. This was an extraordinary insight for its implications, as it extended the question of the centrality of motion to astronomical environments other than the Solar System alone, causing this fundamental pillar of Aristotelian cosmology to lose its anthropological and psychological force. Later, eclipsing binaries would become a laboratory for measuring the masses and radii of stars and a clue to a technique for searching for extrasolar planets.

Let us return to Cassini. During his stay in Bologna, Giovan Domenico achieved some brilliant results that made him famous throughout Europe. He skillfully managed his moderate preference for Copernicus without arousing the resentment of the Pope, who was in a sense also his employer. As early as 1652, for example, he had shown, on the basis of Brahe's earlier work, that comets move across the sphere of the Moon; but he did not insist on the consequences of a result contrary to the unchanging and perfect heavens. At the telescope, he was a master with an innate ability for observation that allowed him to highlight some important characteristics of the giant planets. He discovered the Red Spot on the surface of Jupiter — the same one predicted by chance by Kepler when he was trying to solve one of Galileo's anagrams (see p. 118) — so large that it contained two or three Earths. His priority, however, was challenged by Robert Hooke. The English scientist claimed to have discovered the "eye of Jupiter" a year before the Italian astronomer. It is not entirely clear which of the two men holds the palm of the discovery. The matter is as uncertain as the biography of the giant, long-lived anticyclonic disturbance that has plagued the Jovian atmosphere for centuries:[8] another of the many mysteries with which the history of astronomy is littered. After the first sighting, the Red Spot was closely followed by astronomers for about fifty years, but during the next 120 years the feature disappeared from the chronicles. Whether

---

[7]M. French, *John Goodricke, Edward Pigott, and Their Study of Variable Stars*, in "Journal of the American Association of Variable Star Observers", 40, 2012, pp. 1–13; M. Hoskin (ed.), *The Cambridge Concise History of Astronomy*, CUP, Cambridge 1999, p. 172.

[8]J.H. Rogers, *The Giant Planet Jupiter*, CUP, Cambridge, 1952, pp. 5–9.

this was due to a momentary weakening of the storm or to a lack of good observations is unknown. It was sighted again in 1830 and has since become the logo of the Jupiter, as rings are for Saturn.

Just by observing these rings, Cassini identified the division, which now bears his name, as a deep groove that bisects the annular structure around the planet. We now know that this trench was carved out and is kept clean by the gravitational action of the satellite named Mimas.[9] With true scientific creativity, Giovan Domenico also proposed that the thin annular structure was made up of small fragments. The hypothesis would be validated in the mid-19th century by the 25-year-old James Clerk Maxwell. Using Newton's mechanics, the Scottish genius proved that if the rings were rigid, they would not be able to withstand the tidal forces[10] of the gigantic primary. The result also provided valid support for the hypothesis that the Solar System was formed by the collapse of a gaseous cloud into a rotating disk (the so-called "nebular hypothesis"; see p. 249). For this work Maxwell received the Adams Prize, one of the most prestigious awards conferred by the University of Cambridge. There he candidly stated that he knew of «no practical use for Saturn's rings», but that when these structures were examined from a purely scientific point of view, «they became the most remarkable bodies in the heavens, except, perhaps, those even less useful bodies — the spiral nebulae».[11]

Cassini showed the same skill in the management of a great project that directly concerned the Church: the reconstruction of the meridian line of the Basilica of San Petronio in Piazza Maggiore (Main Square) in Bologna, which had been destroyed during the renovation of the building. This story deserves a digression because it leads us along the path of two great scientific feats: the measurement of the speed of light and the discovery of aberration, both made possible by the clever use of the telescope. The idea of using a church to trace the annual ascending/descending motion of the Sun had come eighty years earlier to Ignazio Danti,[12] an Umbrian Dominican and

---

[9] A satellite of Saturn, Mimas was discovered by William Herschel in the late 18th century. Its mythological name refers to one of the giants, the offspring of Gaia (Earth), who, according to the Greek Hesiod, was born from the drops of blood of Uranus (Heaven) when he was mutilated by his son, the Titan Cronus (the Time).

[10] They stretch a finite body when the force field has a nonzero gradient (i.e., is not uniform).

[11] Cf. J.C. Maxwell, *On the Stability of the Motion of Saturn's Rings*, 1859; http:// mathshistory.st-andrews.ac.uk/Extras/Maxwell_Saturn.html.

[12] F.P. Fiore, *Danti, Ignazio*, in "Dizionario Biografico degli Italiani", XXXII, 1986; http://www.treccani.it/enciclopedia/egnazio-danti_(Dizionario-Biografico)/.

cosmographer at the court of Cosimo I de' Medici, whom Pope Gregory XIII Boncompagni had involved in the preparation of the calendar reform.[13] To suppress the drift of the seasons with respect to the calendar, the length of the tropical (or solar) year should be known to an accuracy of at least one part in 100 million. Simply put, the tropical year is twice the time interval between two successive solstices, that is, between the two epochs in which the Sun meets the "tropics" (Cancer and Capricorn), stops and reverses its motion (along the meridian at noon). In fact, the adjective "tropical" comes from a Greek word meaning "turn". The numerous perturbations superimposed on the pure motions of the Earth's rotation and revolution — most notably the precession of the equinoxes — make the concept of a year susceptible to different definitions corresponding to as many different durations. For example, the sidereal year (about 365.2564 mean solar days) identifies the time interval in which the Sun makes a complete revolution relative to the fixed stars; the anomalistic year (365.2596 mean solar days) instead uses the moment of minimum or maximum distance between the Earth and the Sun as its endpoint. The two definitions would coincide if the apse line of the Earth's orbit, i.e., the major axis of the ellipse that best represents our planet's orbit around the Sun, maintained an invariable direction and did not rotate due to planetary perturbations in an attempt to coincide with the equinoctial line.

But the year that has the greatest impact on human life is the tropical year, associated with the seasonal cycle. To determine its length, astronomers since ancient times have used the shadow of a pole planted vertically in the ground, which has the pompous name of gnomon, from the Greek word meaning "indicator". It is a very simple and rudimentary instrument to use. Not surprisingly, the measurements it provides are not very accurate. Danti realized that in order to improve the results, the difficult-to-read shadow of a pole stuck on the ground *en plain air* would have to be replaced by a sharp beam of light traveling within a tall, stable, dimly lit environment, much like what could be found in large cathedrals. It would be enough, he thought, to drill a small hole as high as possible in the south wall of a church and follow the projection of the thin beam of solar light onto the floor, taking advantage of the semi-darkness of the building and its great solidity. Note that until the 16th century, the floor plan of churches

---

[13]C.H. Genung, *The Reform of the Calendar*, in "The North American Review", 179, 575, 1904, pp. 569–583.

was usually aligned with the cardinal points so that the apse wall was "oriented", that is, it appeared to the faithful in the direction in which the Sun rises (in Latin *"oriri"* (to rise), from which orienting acquires the meaning of directing). Therefore, looking away *versus Solem orientem* (in the direction of the rising Sun) has a clear symbolic meaning, for it is from the East that salvation (that is, Christ, represented by the Sun) comes. This practice is much older than Christianity. For example, the facades of Greek and Roman temples were oriented in this way.

Danti carried out his first experiments in Florence, in the Church of Santa Maria Novella, between 1572 and 1575. He took up an idea of the astronomer and cartographer Paolo Dal Pozzo Toscanelli,[14] who a hundred years earlier had made in the dome of Santa Maria del Fiore, also in Florence, a gnomonic hole at the record height of 86 m. Then, having moved to Bologna to teach mathematics at the university, he applied the idea to San Petronio with good results, although the peculiar orientation of the basilica, imposed by the constraints of the front square, prevented him from accessing the meridian line. Hampered by the rows of columns that ran at an angle to the usual arrangement of churches, he was forced to record the times of the solstices along a direction more than half an hour away from that of noon, a complication with consequences for the accuracy of the measurements.

Danti's monumental sundial, the pride of Bologna, survived for almost a hundred years, but had to be demolished after another renovation of the cathedral. Thus, in 1655, the *Fabbriceria di San Petronio*, a committee of four citizens elected by the city government to oversee the construction and repair of the Basilica, commissioned Cassini to build a new one. The astronomer accepted the challenge and in a short time created a masterpiece of scientific engineering.[15] Taking advantage of a new wing of the temple, he conceived a meridian line 67 m long, fed by an "eye" in the vault at a height of 27 m. The solar beam penetrated the forest of columns and imprinted itself on the ground in the exact direction of noon, allowing the Sun's height to be measured throughout the year: a true jewel! When it was inaugurated with great pomp, in the presence of the professors of the

---

[14] E. Garin, *Ritratto di Paolo Dal Pozzo Toscanelli*, in "Belfagor", 12, 3, 1957, pp. 241–257 (in Italian).

[15] Gio. Domenico Cassini, *La meridiana del tempio di S. Petronio, Bologna*, 1655 (in Italian); see https://www.liberliber.it/mediateca/libri/c/cassini/la_meridiana_del_tempio/pdf/cassini_la_meridiana_del_tempio.pdf.

Bolognese Studium, the San Petronio heliometer was the largest sundial in the world. It was characterized by an unprecedented precision in the measurement of the tropical year, the obliquity of the ecliptic, the refraction, but also the apparent motion of the Sun, whose variations during the year were a matter of dispute between the Ptolemies and the Copernicans.

The former, advocating uniform circular motion, attributed any change in the Sun's apparent speed to a variation in distance from the Earth. The others, based on Kepler's third law, argued instead that the motion of the star on the celestial sphere depended on the square of the distance. The stage was set for a decisive experiment. In fact, the helioscope simultaneously provided both the variation of the distance, traced by changes in the diameter of the solar disk projected by the instrument, and that of the Sun's speed along the meridian. Giovan Domenico confirmed that Kepler was right, but he was careful not to make a case. It was too dangerous for a subject of the Papal States and very distasteful to the palate of a true conservative.

Cassini's fame as an astronomer and manager of major scientific projects attracted the attention of Jean-Baptiste Colbert. A pragmatic disciple of Cardinal Mazarin, he dreamed of modernizing France by promoting its industrial and colonial development and enriching the country despite the enormous waste, in war as in peace, of his sovereign, Louis XIV. Colbert's interest in the sciences was purely instrumental. He had encouraged the construction of a state astronomical observatory in Paris[16] to support the growth of maritime traffic, in open competition with England, Holland, and Spain. At that time, he was looking for a director capable of dealing with the various problems of orientation at sea, and in particular that of measuring geographical longitude. The rampant professor from Bologna seemed to be the right person. Invited to Versailles for an interview, the too spontaneous and direct Cassini made a bad impression on the Sun King, who certainly could not tolerate the contradictory. But the astute Colbert, who knew his man well, maneuvered to persuade the sovereign to choose the impertinent Italian, and he never regretted it. Giovan Domenico had the qualities of a leader and brought with him his great experience in measuring the eclipses of the Galilean satellites, which seemed to be the key to opening the longitude chest. What followed is an emblematic example of the use of the sky as a physical laboratory.

---

[16]For more information, see R. Barthalot, *The story of the Paris Observatory*, https:// sites.google.com/site/histoireobsparis/a-short-history-of-paris-observatory.

To understand the problem posed by longitude,[17] imagine for a moment a sailor crossing the ocean and stumbling upon an uncharted island. To find it again in the future, he must measure the geographical coordinates of the place.[18] The first coordinate is the longitude, which is an angle associated with the semimeridian where the island is located. The second is also an angle associated with the parallel that contains the destination. The intersection of the two circles marks the location of the place. Our sea wolf must be able to measure these angles if he wants to find his paradise island again after sailing away from it. But how?

Latitude is roughly given by the height of the North Star above the horizon. All our navigator has to do is find this angle with a sextant on a clear night. Measuring longitude — the angle between the local meridian and a reference meridian, such as that of Greenwich, counted in the direction of the Sun's motion — is far more difficult. But the Earth's diurnal rotation can help us. In fact, longitude is also the difference in time between the moment of astronomical noon at the prime meridian and the same event at the place of interest. In short, if you want to know your longitude relative to a friend — after he has moved to Greenwich to make your life easier — call him and ask what his solar time is (the time to or from his astronomical noon). Then take the difference between that time and your local time at the exact moment of the call, paying attention to the signs. The result is your longitude, expressed in hours (if you want it in degrees, just multiply it by 15). This operation is conceptually simple, though not very accurate in practice; but what if there is no real-time communication available? The trick is to have a clock that is either accessible to all observers wherever they are, or transportable from one place to another without losing its rhythm, so that everyone can deal with a kind of absolute time scale.

In the 16th century, with the proliferation of ocean crossings and geographic explorations, the need for accurate longitude was complicated by the inability of pendulum clocks to withstand the stress of travel. In 1567, the very powerful eldest son of Charles V of Habsburg, Philip II of Spain, whose galleons sailed almost blindly across the Atlantic at great risk to men and goods, vainly offered a substantial reward to anyone who could present him with a satisfactory solution to the problem to of taking stock at sea.

---

[17]D. Sobel, *Longitude: The True Story of a Lone Genius Who Solved the Greatest Scientific Problem of His Time*, Walker & Company, New York 1995.
[18]For sake of simplicity, here we will not make any subtle distinction between terrestrial and celestial coordinates.

The solutions devised by early explorers, such as Amerigo Vespucci, were mostly based on lunar occultations of planets or bright stars: events that occurred more or less simultaneously in all parts of the world, and therefore lent themselves well to the synchronization of clocks, provided that the corresponding ephemeris was sufficiently precise. The Spanish monarch's initiative was taken up thirty years later by his son Philip III, who added an attractive income to the generous prize. Galilei tried to secure the rich premium by offering the four newly discovered satellites of Jupiter as a universal clock.[19] Indeed, with their very uniform motions around the primary, the Medicean Moons behave like the hands of a clock, quite precise and even universal, but certainly impractical. It was Galilei himself who admitted that a beat of the observer's heart would be enough to take Jupiter out of the telescope's field of view, even if the feet were firmly planted on the ground. Imagine being on the deck of a ship, subject to the action of the waves. With this argument, the Spaniards rejected the proposal and left the Pisan scientist high and dry. More than two decades later, when the Dutch decided to dust off the idea and seek a meeting with the now elderly scientist, Galileo was already a recluse at his villa in Arcetri, virtually inaccessible,[20] and there the matter ended. But as telescopes progressed, the Jovian moon clock, though impracticable at sea, gained increasing credit for longitude measurements on land. For example, the topographic map of France was redrawn using this tool. The result surprised Louis XIV because his reign turned out to be not as extended as he had thought. His half-serious comment was that he had to concede to the astronomers a part of his homeland of which not an inch had ever been given to his enemies.

In order to fine-tune the celestial clock, the astronomers began to measure with the greatest precision, for each of the four Medicean moons, the epoch of immersion in the shadow of the primary and that of the subsequent emersion; especially for Io, which took the least time to complete an orbit, being the closest to Jupiter. Our Cassini was among the most assiduous and accurate observers, as he had begun to accumulate data since his stay

---

[19] R. de Grijs, *European Longitude Prizes. I. Longitude Determination in the Spanish Empire*, https://arxiv.org/ftp/arxiv/papers/2009/2009.12778.pdf.

[20] In 1631, Galileo had rented the villa "*Il gioiello*" (The jewel) to be closer to his daughters, who were nuns at the nearby convent of San Matteo in Arcetri. After his trial and conviction in 1633, it was decided that the villa would become his prison. A special permit, issued by the ecclesiastical authority, had to be granted both to people who wanted to go to the villa to meet the old scientist, and to Galilei himself when he had to leave his house and go to the city, mainly for health reasons.

in Bologna. He was looking for one thing and found another (not uncommon in astronomy!). He expected Io's eclipses to follow each other with a uniform timing, like the ticking of a metronome. Instead, he found that they advanced as Jupiter approached the Earth to reach opposition to the Sun, and delayed as the distance increased to reach conjunction. Cassini argued that this peculiar "inequality" could be explained by assuming that light has a finite velocity.[21] He also attempted to estimate its value, but in time his strongly conservative spirit gained the upper hand. Then a distinguished director of the *Observatoire*, he gave up, perhaps because he did not want to be forced by facts to believe in what he had rejected *a priori*. He entrusted a young associate, the Dane Ole Rømer, with the routine task of measuring the timing of the eclipses of the Medicean Moons. It was a wise choice. Rømer was familiar with this type of work. He had done it in his native country when he assisted Abbot Jean Picard, a French astronomer whom Cassini had sent to Denmark to determine the difference in longitude between Paris and Copenhagen. The Jovian satellites had posed the challenge of a more accurate cartography of Europe, and the Paris Observatory had taken up the challenge.

In 1675, combining his data with Cassini's, Rømer became convinced that the series of Io eclipses was modulated by advances and delays in the light signal. Traveling at a finite speed, light took less time to reach the observer as Jupiter gradually approached, and took longer in the opposite case.[22] The Danish astronomer merely noted that the signal took about 22 minutes to travel the diameter of the Earth's orbit: a time interval that was overestimated by 30%. Had he done the full exercise with the numbers at his disposal, he would have found a velocity of 135,000 kilometers per second: an astonishing result, though still less than half the modern value, in part because of a seriously flawed Astronomical Unit (see also p. 306). But he would not, or perhaps could not, openly draw the consequences that his measurements might entail. His boss would not have forgiven him. So it was Huygens who did it. Although he, too, was a regular at the Paris Observatory, his shoulders were sufficiently covered to protect his considerable independence of judgment. He reported the value of the speed of light in a letter and, being the gentleman he was, he did not fail to emphasize the role and merits of his colleague Rømer (a practice not so common in

---

[21]C.B. Boyer, *Early Estimates of the Velocity of Light*, in "Isis", 33, 1, 1941, pp. 24–40.
[22]M. Romer, I.B. Cohen, *Roemer and the First Determination of the Velocity of Light (1676)*, in "Isis", 31, 2, 1940, pp. 327–379.

science, yesterday as today). What an extraordinary achievement! With an intelligent use of a Galilean moon, the new astronomy had made it possible to assign a value to a speed enormously greater than any movement perceivable by man. It solved once and for all an ancient dilemma and showed how the heavens could serve as a powerful laboratory of physics (the dichotomous universe of Aristotle had meanwhile been reunited by Newton). The matter is of such importance that it requires another, albeit brief, digression.

Although our senses and daily experience suggest that light travels instantaneously, reason, even in its most elementary forms, does not. In fact, as early as four and a half centuries before Christ, Empedocles of Akragas, a man of many minds and curious habits, speculated that light traveled at a finite speed. Aristotle disagreed, despite admiring the Sicilian philosopher so much that he borrowed his entire table of elements to explain the chemistry of the sublunar world (see p. 34). Who was right? Claudius Ptolemy accepted the Aristotelian thesis. He endowed it with an emissionist model of vision, already proposed by Pythagoras and Euclid, in which light is emitted from the eye and immediately reaches the object being viewed. In this framework the justification for an infinite speed was supplied by Heron of Alexandria, a brilliant mathematician who lived in Roman Egypt in the first century A.D., working as a teacher and researcher in the Library. When a man opens his eyes on a starry night — he argued — he can see both nearby objects and very distant sources at the same time.

At first, the Arabs passively accepted Ptolemy's "tactile" thesis; then they changed their minds and proposed the "intramissionist" theory, according to which light travels from the source to the eye. However, they continued to believe that the journey took place in an instant; a view shared by Kepler and Descartes, among others, and which the French philosopher also tried to prove. His argument was that even if light were fast enough to travel the distance between the Moon and the Earth in an hour, this incredible feat would not be enough to synchronize the perception of a lunar eclipse with the actual occurrence of the phenomenon. In principle, he was right, but only because he assumed a speed of light nearly three thousand times less than its real value. Without any solid evidence, it seemed utterly foolish to believe that something could travel as incredibly fast as light, covering a distance of 300,000 kilometers in a tiny second of time.

Like St. Thomas the Apostle, who wanted to see for himself before believing something, Galilei instead doubted and wanted to subject the matter to an experimental test. In *Discorsi e dimostrazioni matematiche intorno a due nuove scienze* (Discourses and mathematical demonstrations concerning two new sciences), his last work and masterpiece, published in 1638, the Pisan scientist had his friend Sagredo say: «But what and how much shall we estimate that this speed of light to be? Perhaps instantaneous, momentary, or, like the other movements, temporary? Will we be able to determine, by experience, what it is?». Simplicio, a prudent conservative, had a ready answer: «Everyday experience shows that the expansion of light is instantaneous; whereas, seeing an artillery fire in the distance, the brightness of the flame reaches our eyes without interposition of time, but not the sound our ears, except after a considerable interval of time». At which Sagredo, *homo novus* of science, unleashed the victorious weapon sharpened by Galilei, the recourse to the arbitration of nature:

> Ah, Mr. Simplicio, we infer nothing more from this well-known experience than the fact that sound is brought to our ear in no less time than that of light; but it does not assure me that the coming of light is therefore instantaneous, rather than in a finite but very rapid time. Nor is this observation any more conclusive than the other of those who say: "As soon as the Sun reaches the horizon, its splendor reaches our eyes"; but who assures me that its rays have not reached the said limit earlier than our sight?[23]

To untie this knot, Galileo conceived an experiment that remained on paper until 1676, when the idea was implemented by the *Accademia del Cimento* (Academy of Experiments). Two observers, A and B, each armed with a shielded lantern, were to be placed one mile apart on two hills so that they could see each other. The first of them, say A, was to uncover his lamp, and the second, B, upon seeing the light, was to respond immediately by sending a signal with his own lamp. The plan was to estimate the total time of flight of the light signal from A to B and then back to A and to relate it to the distance $\overline{AB}$ between the two operators. But with which clocks? Assuming, and not conceding, that a fairly accurate one existed at the time, the measurement was still compromised by the prevalence of human reactions on the total time of the operation. The only conclusion reached was that the speed of light, if not infinite, should be fairly large. A substantial failure that in 1657 pushed Giovanni Alfonso Borelli, a talented

---

[23] G. Galilei, *Discorsi e dimostrazioni matematiche intorno a due nuove scienze attenenti alla mecanica e i movimenti locali*, in Id., *Opere*, F. Brunetti (ed.), UTET, Turin, 1980, Vol. 2 (day I), pp. 636–637.

Neapolitan, to propose the use of a system of reflecting mirrors to extend the base to 35 km,[24] which is equal to the «distance between Florence and the city of Pistoia». In this way he hoped to dilute the effect of the errors introduced by the human factor. The expedient was still totally inadequate, but the scheme of the academician of *Cimento* would serve as a track for the French physicist Hippolyte Fizeau[25] for his famous and definitive experiment of 1849. It may be interesting to note that the method proposed by Galilei is echoed in the modern technique of measuring planetary distances by means of radar pulses fired at and reflected by a celestial body. In this case, the speed of light is known and the distance is deduced from the time of flight of the signal.

This digression allows us to understand that Rømer's result came in the context of differing opinions about the speed of light, where most people continued to believe that it was infinite. To clarify and confirm the Danish astronomer's result, another independent measurement was needed. Once again, and quite unexpectedly, the sky and the telescope came to the aid of physicists. This fact does not surprise us at all today, but it was not at all obvious at a time when the Aristotelian dualism between the terrestrial and celestial worlds was still very much alive. Here is what happened.

The precision achieved by observational instruments in the late 17th century had revived the question of parallax. Giovan Domenico Cassini, while triangulating Mars from Paris and Cayenne on the occasion of the Great Opposition of 1672 (p. 357), succeeded in measuring the distance of the Red Planet from the Earth with remarkable accuracy, despite the difficulties posed by atmospheric refraction, which bends the path of light rays. To determine the length of the base, that is, the difference in longitude between the observatories in France and Devil's Island in French Guiana, he had used the Medicean Moons. The result was a number large enough to produce, when entered into Kepler's third law,[26] a value of the Astronomical

---

[24]R. Foschi, M. Leone, *Galileo, measurement of the speed of light, and the reaction times*, in "Perception", 38, 2009, pp. 1251–1259.

[25]J.H. Rush, *The Speed of Light*, in "Scientific American", 193, 2, 1955, pp. 62–67. Cf. also L. Patton, *Reconsidering Experiment*, in "The Journal of the International Society for the History of Philosophy of Science", 1, 2, 2011, pp. 209–226.

[26]The proportionality constant between the cube of the radius of the planetary orbit and the square of the revolution time (period) becomes unity when distances are expressed in Astronomical Units (AU) and time in Earth sidereal years. Thus, from Kepler's third law, by measuring the sidereal periods of the planets — which is relatively easy — it is possible to construct the map of the Solar System in AU. Consequently, it is sufficient to know any relative distance between two planets in meters (or any other linear unit) to derive the value of the Astronomical Unit (in the same units).

Unit of 140 million km, still slightly underestimated, but 50 times larger than Copernicus', 17 times larger than Brahe's, and 10 times larger than Kepler's.

Thanks to the telescope, the diurnal parallax had begun to put the Solar System on the right scale and, more importantly, had calibrated the basis of the annual parallax, that is, the radius of the Earth's orbit. However, no one had yet been able to prove the existence of the cyclic oscillation of the nearest stars predicted by the annual orbit of the observer around the Sun, much to the delight of the stubborn geocentrists led by the Jesuits. Confident that sidereal distances were significantly shorter than they actually were, the conservatives argued that if the phenomenon was real, it should be of such magnitude that it could not escape telescopic observation. It seemed that the moment of truth had arrived. To validate the new cosmology, the burden of proof fell on the Copernicans. They had to demonstrate the existence of the annual parallax. Among the first to undertake this genuine and desperate crusade was Robert Hooke, a prolific English scientist who lived in the second half of the 17th century; we will meet him again in the next chapter. Being in London, he chose to observe γ Draconis, a bright star that culminated at the zenith of the city.[27] In this way he could report the position of the source directly to the vertical defined by the plumb line, without even having to correct for the insidious effects of atmospheric refraction. He knew what to expect from parallax. In its annual cyclic motion around its "true" position, the star should have shown its maximum excursion to the south in December and to the north six months later. The amplitude of the oscillation was unknown. The bet was that it would be measurable.

Hooke felt like a titan, determined to remove the Earth from the Aristotelian center of the world. He needed the famous lever. So in 1669, fleshing out an idea of Galileo's, he built his "Archimedes Machine", a telescope

---

[27] γ Draconis is actually the brightest star in the constellation of Draco. In 1603 Bayer, author of the *Uranometria* (see p. 79), mistakenly placed it in third place. Instead, he promoted the first Thuban (Basilisk in Arabic), which, due to the precession of the equinoxes, had served as the "pole star" for 2,150 years beginning four millennia before Christ. γ Draconis, which the Arabs called Eltanin ("Great Serpent"), is almost 4 times brighter than Thuban. It is located on the head of the Dragon, of which it is one of the eyes (the other is Rastaban). It is an orange giant with almost twice the mass of the Sun and 100,000 times the volume. It is moving straight towards us and will reach its minimum distance in 1.5 million years to become the brightest star in the night sky for a while. Its other name is Zenith Star, because it passes near the vertical at the latitude of the Royal Greenwich Observatory.

pointed steadily at the zenith through a hole drilled in the roof of his house. However, bad weather and increasing health problems severely limited his observations, which eventually proved inconclusive. So Hooke left. The founder of the *Académie des Sciences*, Jean Picard, along with Giovan Domenico Cassini, also attempted the exercize, but were equally unsuccessful due to the incompleteness and inaccuracy of the data collected. The matter was embarrassing to the heliocentrists, to say the least. If the Earth really revolved around the Sun, why could they not see its motion reflected in the stars? A quarter of a century later, John Flamsteed stepped in. The director of the Royal Greenwich Observatory, he decided to focus on the Polar Star, whose position, due to the effect of parallax, should describe a small circumference around the north celestial pole each year. A skilled and tireless observer, the Englishman was on the hunt for the scientific achievement that would elevate him in the eyes of his king and put him on a par with the other scientists of the Royal Society, with undoubted moral and material benefits. He threw himself into the enterprise and, when he was sure of the reliability of his measurements, told the world that he had measured the first sidereal parallax in history. But he was wrong, as Jacques Cassini was quick to point out with obvious glee, given the rivalry between the French and the English in every field, including science. The oscillation of the Polaris, while real, wrote the son of Giovan Domenico and his successor at the helm of the Paris Observatory, was not in the right phase with the parallax. The stellar "tremors" reported by the English astronomer had to be something else. Although reluctantly, Flamsteed finally agreed.

Scans of the position of Draco's third brightest star were made in 1725 by a young Oxford professor and future Astronomer Royal, the Reverend James Bradley.[28] Following in Hooke's footsteps, the thirty-two-year-old scientist decided to build a dedicated instrument, as we would call it today: a telescope wedged vertically into a fireplace in the home of a wealthy friend and astronomy enthusiast. The telescope was well made and had only one degree of freedom, a small north-south excursion that could be controlled and measured by a micrometer. The system guaranteed an accuracy of one arc second in relative position, which was 60 times better than Tycho's dials. For two years, whenever the sky permitted, Bradley measured the displacement of γ Draconis with respect to the zenith. Finally, to his surprise, he

---

[28] A. Chapman, *Pure Research and Practical Teaching: The Astronomical Career of James Bradley, 1693–1762*, in "Notes and Records of the Royal Society of London", 47, 2, 1993, pp. 205–212.

found again the amplitude measured by Flamsteed for Polaris, but also the same phase shift of a full quarter circle from the predicted parallactic oscillation. The star's maximum southward drift occurred in March instead of December, and its northward drift in September instead of June. Confused, he repeated the measurements with a better instrument and more stars. Nothing changed. How could this phenomenon be explained? It almost seemed like an elaborate joke of fate.

At first, Bradley speculated about possible oscillations of the Earth's rotational axis, which he called "nutations" from the Latin *annuere*, which means to nod: an idea that stuck in his head anyway and would eventually lead him to another important discovery. Then he had an intuition worthy of a Nobel Prize. He noticed that the oscillation he had measured had an annual cycle, but its amplitude was independent of the distance of the source. The cause, therefore, should not be sought in the variation of the observer's position, but in the change of direction of the velocity vector of the Earth during its revolution around the Sun. In this way, he found a natural explanation for the intriguing quarter-turn phase difference in the annual cycle. In fact, the velocity is perpendicular to the vector radius and thus anticipates the position by 90°. Legend has it that he got the idea during a boat trip on the Thames, when the shaft of the mainmast flag responded to a turn of the rudder without a change in wind direction. Bradley had killed three birds with one stone: he had discovered aberration[29] as an apparent deviation of the position of a celestial source due to the composition of the velocity of the emitted radiation with that of the annual motion of the earthly observer; he had falsified geocentric cosmology once and for all; and he had also found an independent and accurate way to measure the speed of light.

A clarification is in order. Everyone knows that, when walking fast under a deluge of water, the umbrella must be held at an angle forward, because the drops are not falling vertically, but at an angle. The reason for this rain aberration lies in the composition of the two velocities involved. The same thing happens with light when the observer moves in a direction different from the line that connects him to the source. This statement seems to violate one of the cornerstones of special relativity, the invariance of the modulus of velocity of light, but it does not. In Einstein's theory, the

---

[29] A.B. Stewart, *The Discovery of Stellar Aberration*, in "Scientific American", 210, 3, 1964, pp. 100–109.

phenomenon retains the same geometric properties of the classical formulation, while the speed of light does not change its value (which is always the same in a vacuum). Of course, Bradley could not worry about all this.

The discovery of light aberration was announced at the Royal Society in 1729. In little more than a century, the telescope had won countless victories while remaining quite primitive. The greatest obstacle to the progress of this fundamental astronomical machine was another defect of refractive optics, chromatic aberration. In order to understand what it is, we must start with some considerations on the nature of light. To begin with, we consider a point-like celestial source, such as a star. Its light reaches the observer as a beam of parallel rays, which can be described according to a wave or corpuscular model, depending on our convenience. In the first case, we will use the concepts of wavelength and frequency, representing each monochromatic component, perceived as a single color, like a wave (flat, given the postulated large distance) advancing at the speed of light and pulsing at a very specific rhythm. We can imagine this wave as an oscillation propagating on the surface of a pond hit by a stone. Its frequency is given by the number of wave peaks that pass through a given and fixed position in a unit time interval. The wavelength, on the other hand, is the distance between two successive peaks at a given time. It is not difficult to understand that the product of wavelength and frequency equals the speed at which the disturbance advances. Think of a train whizzing by. If the wavelength is the length of each car, and the frequency is the number of cars, all identical, that pass you in a second, you see that the distance traveled by the train in a unit of time, which is the wanted speed, is given by the number of cars that have passed (frequency) multiplied by the length of each car (wavelength). In the particle model, we use instead the photons introduced by Albert Einstein: grains of energy that also move at the speed of light and are characterized by a frequency proportional to the energy that each of them carries. Quantum mechanics, a new science born at the beginning of the 20th century to interpret the strangeness of the infinitely small, perfectly explains the wave-particle dualism which, as far as we are concerned, is only operational. In the following, we will casually refer to one or the other model as needed.

The plane wave scheme works for a single star. But in an astronomical image there are usually multiple sources, both point-like and extended. What do we do in this case? We settle for a simplified scheme and try to represent the part of the sky where the sources are projected as a mosaic of small tiles, the famous picture elements (pixels). Each of these pixels combines and represents the flux contribution of the sources (or fractions of sources) that it ideally covers and, as far as we are concerned, becomes a point source itself. In summary, the image that reaches us can be thought of as a bundle of plane waves, as many as there are pixels, each perpendicular to the direction in which the fraction of the source that generated it is located in the sky. Astronomers call the focal plane the place where the optical device forms the image of these pixels. Unfortunately, the flat surface to which the name alludes does not exist, except as a compromise between conflicting needs, as we shall see.

Focusing a single plane wave, i.e., a single point-like source, is not a big problem. A concave paraboloid mirror, as described by Diocles, ensures perfect convergence of an axial ray to the focus. Off-axis, however, a star will look like a comet, with its coma toward the center of the field and its tail out in the radial direction. This is why this aberration is called "coma". In essence, it limits the useful range of a parabolic mirror to a small cone around the axis, with a clear waste of the field of view. Coma is just one of the possible distortions that lenses and mirrors can impose on images. There are aberrations that occur only when the light is a cocktail of many colors, and others that persist even in monochromatic light; aberrations that occur on the axis of symmetry of the optical system, and others that occur only off-axis.

Mirrors are free of chromatic aberration because reflection is independent of the color of the radiation. In refractive optics, focusing is instead achieved by the property of light rays to change direction as they pass through media of different densities. The amount of deviation depends on the refractive index of the lens material, which varies significantly with color. As a rule, blue light converges faster than red light. Other things being equal, a lens will therefore have different focal planes for different light frequencies. By favoring one color, the other colors project blurred images onto it. The effect is similar to that of a defective trichrome, in which the blue, yellow, and red arrays are not well aligned. The technological trick to overcome this serious drawback was not found until the 18th century in England, as we will read in the next chapter. Thus, in the second half of the 17th century, in order to increase the resolution while keeping the iridescence of the images under control, astronomers were induced to build instruments with ever longer focal lengths. They led to mechanical complications (increasing magnification by shortening the focal length of the eyepieces was impractical for technical reasons). As a rule, the focal length quadrupled when the objective diameter doubled. In short, the need to gather more light in order to see deeper and in more detail had forced the use of improbable mounts: tubes up to 50 meters long that bent miserably under their own weight, causing the optics to lose alignment. Maneuvering them required systems of ropes and pulleys that would have frightened a longtime sailor.

Robert Hooke tried to shorten these monstrous frames by bouncing the light back and forth through flat mirrors, but his ingenious zigzag never caught on. Christiaan Huygens and his brother Constantijn then came up with the idea of mechanically separating the lens from the eyepiece and maneuvering the two elements independently to orient them according to the astronomer's needs while maintaining their alignment.[30] The advantage of this solution is obvious. A telescope without a tube is by definition a bend-free instrument. The disadvantages lay in the complexity and extreme difficulty of operation. The objective, hinged on a high support, was moved by a system of levers and rods, while the free and independent eyepiece had to be kept on axis by hand, the observer being assisted by a string that connected it to the objective and acted as a viewfinder. Trying is believing, as the saying goes. Although the device worked well, it was rather inefficient and a real ordeal for the observer. But great science is also about sacrifice, and in some cases, heroism. That is why the crazy air telescopes found more than one lover and produced some important discoveries. Huygens, for example, boasted that he and his brother worked on lenses with focal lengths up to 64 m, also using the technology developed by Baruch Spinoza. Giovan Domenico Cassini, with a telescope of 11 m, discovered Rhea, the fifth satellite of Saturn, and then two other moons with 30 and 44 m telescopes equipped with optics worked by Campani. The lens mount reused, after being taken to the gardens of the Paris Observatory, one of the wooden towers of the incredible machine created to supply water to the fountains of the Château de Marly, the retreat of the Sun King. And Bradley measured the diameter of Venus with a 65-foot refractor in 1722. But the madness record belongs to Adrien Auzout, a Frenchman who, at the end of the 17th century, built aerial refractors with focal lengths of 90 and even 180 m, and dreamed of constructing one of 300 m to try to spot animals on the Moon! Science was already living a revival of the ancient pluralist problem of Epicurean memory: the question of the many worlds

---

[30] J.G. Yoder, *Unrolling Time: Christiaan Huygens and the Mathematization of Nature*, CUP, Cambridge 2004; S.A. Bedini, *The Aerial Telescope*, in "Technology and Culture", 8, 3, 1967, pp. 395–401.

reformulated in terms of the habitability of the Moon and the planets of the Solar System.[31] It was necessary to build telescopes powerful enough to allow astronomers to see what life forms were colonizing our satellite.

The 17th century, which had opened with the *Sidereus Nuncius*, closed with its burden of contradictions, torn between the excitement of the merchant class and the persistent dismay of a devastating global war that lasted 30 years and of repeated epidemics, between the impetus of the conquests of new lands and the horrors of the extermination of the indigenous peoples and the annihilation of their cultures. A century that marked the end of feudalism and the beginning of capitalism in the great nations of the Old Continent, and that changed the map of power. The monarchical absolutism of Louis XIV had given Europe to France at the expense of the Habsburgs of Spain and Austria. Germany had been reduced to an archipelago of small states, while Holland and England, where the parliamentary monarchy was born, remained sheltered to compete for the routes and new shores of the vast sea. Art, driven by the need to restore the image of the Counter-Reformation Church and the economic well-being of the bourgeoisie, had experienced a new flowering.

With the abandonment of magic and alchemy and the adoption of the Galilean method, science had developed considerably, both on the side of natural philosophy, which became increasingly mathematical, and on the side of medicine. And it became more and more clear to the powerful that pure knowledge was not only an end in itself, isolated in its "*turris eburnea*" (ivory tower, from the biblical *Song of Songs*), but the key to technological advances useful for everyday life and, above all, for the domination of men. In this sense, the first public astronomical observatories of the telescope age were established in the 17th century, not so much to interrogate the cosmos for astronomical or astrological purposes, but to use the sky for practical purposes. The first such facility was built in Copenhagen on the roof of a large round tower, 35 m high, accessed by a spiral staircase. It was part of the Royal Trinity complex, a concentration of academic resources, including a library and a large church, and had been ordered by King Christian IV, the same one who had forced Tycho into exile. It is evident that the seed of astronomy planted by Tycho in Denmark was paying off. In fact, it was one of his students and followers, Christen Sørensen Longomontanus, who

---

[31] M.J. Crowe, *The Extraterrestrial Life Debate, 1750–1900: The Idea of a Plurality of Worlds from Kant to Lowell*, CUP, Cambridge 1986.

proposed the great institution to the sovereign, which was inaugurated in 1642.

This was followed in 1667 by the Paris Observatory, commissioned by Colbert, and eight years later by the Greenwich Observatory, built at the behest of Charles II. Institutions financed by the state and dedicated to specific tasks, flanked by numerous other astronomical observatories that flourished in every corner of Europe by universities, princes, religious institutions, or simple private individuals. The emblem of this cradle of activity is the astronomical observatory built by Heweliusz, a wealthy beer merchant, in the city of Gdansk, Poland. We know him better by his Latinized name, Johannes Hevelius. In 1641, when he was barely thirty years old, he built an observatory with his own funds on the top floor of a group of houses. He called it Sternburg, "City of Stars", and among other magnificent instruments he installed a 45 m tubeless telescope. Johannes had very sharp eyesight, so that he could see stars of the seventh magnitude, two and a half times fainter than those accessible to ordinary people with good vision. He measured the correct rotation of the Sun using as a reference the spots discovered by Galileo. In his *Selenographia* of 1647, he reproduced accurate maps of the Moon based on ten years of observations, and estimated the height of mountain ranges. He also observed some comets and the variable Mira Ceti[32] which he himself named "Astonishing Star in the Whale"; discovered the nova star appearing in Cygnus in 1670, and made a catalog of over 1500 stars with positions so accurate that they rivaled the British astronomical community, specialized in precision astronomy. The volume in which they were presented, the *Firmamentum* (1690), is a formidable celestial atlas, the best after Bayer's *Uranometry* published nearly ninety years earlier, and also a fantastic work of graphic art, much sought after by collectors.

But in the meantime the reflecting telescope, which was to become the ultimate astronomical machine, had made its timid appearance. To tell the story, we must cross the Channel and enter foggy Albion to meet Isaac Newton at Cambridge.

---

[32] Mira Ceti is a symbiotic system of two gravitationally bound stars, a red giant (Mira A) undergoing mass loss and a high-temperature white dwarf companion (Mira B) accreting mass from the primary. Its variability may have been noticed by ancient Chinese, Babylonian, and Greek astronomers. Cf. D. Hoffleit, *History of Mira's Discovery*, https://web.archive.org/web/20070405082807/http://www.aavso.org/vstar/vsots/mirahistory.shtml.

Chapter 7

God said, Let Newton be!
The reflecting telescope and
light and gravity

> The growth of new systems out of old ones,
> without the meditation of a divine Power,
> seems to me apparently absurd.
>
> Isaac Newton,
> *IV letter to the Reverend Bentley*

> If Sir Isaac Newton had not been distinguished
> as a mathematician and a natural philosopher,
> he would have enjoyed a high reputation
> as a theologian.
>
> Sir David Brewster,
> *Memoirs of the Life, Writings, and*
> *Discoveries of Sir Isaac Newton*

In 1670, Isaac Newton, newly appointed professor at Trinity College, Cambridge, taught a course in optics, which he repeated the following year. It was an excellent opportunity to dust off an earlier interest in light and colors[1] and bring ideas together, according to the rule that in order to understand something, one must first have taught it.[2] The result was a work of synthesis that Newton, certainly not lacking in self-respect, judged to be «the oddest if not the most considerable detection wch hath hitherto beene made in the operations of Nature».[3] He exaggerated, but not entirely, as we will see soon.

---

[1] P. Fara, *Newton Shows the Light: A Commentary on Newton (1672)* "A Letter ... Containing His New Theory about Light and Colors ...", in "Philosophical Transactions of the Royal Society. A. Mathematical, Physical and Engineering Sciences", 373, 2015, p. 2039.

[2] *Homines dum docent, discunt* (Men, while teaching, learn); from L.A. Anneus Seneca, *Epistulae ad Lucilium*, 7, 8.

[3] Letter by Isaac Newton to Henry Oldenburg, secretary of the Royal Society, January 18, 1672: http://www.newtonproject.ox.ac.uk/view/texts/normalized/NATP00236.

Isaac was not yet thirty years old. He was born on January 4, 1643,[4] in Woolsthorpe, a hamlet in the northeast of England, in a petty-bourgeois agricultural environment, relatively far from the epicenter of the English Civil War,[5] but apparently under an unlucky star.[6] Before he was born he had lost his father, a relatively wealthy but illiterate yeoman. And he was just 3 years old when his mother, Hannah Ayscough, left the farm to follow her new husband, leaving her son in the care of his grandmother. An abandonment that certainly did not help to improve the closed character of a boy bordering on autism and misogyny. In 1653, his hated stepfather died, leaving a fair inheritance, and Hannah, widowed for the second time, returned to Woolsthorpe with her three second-born children. We must believe that Isaac was uncomfortable with this cohabitation. In any case, to attend the grammar school in Grantham, a small market town not far from home, he was given room and board by an apothecary. This latter had a daughter with whom Newton seems to have established an emotional bond, the only one in his life. The girl, however, later married another man.

His studies were interrupted when Hannah demanded that, as the eldest son, he cultivate the family's small estate. But Isaac neither knew nor wanted to be a farmer, so his mother agreed to send him back to Grantham, to the King's School, which he attended until 1661. To satisfy his ambitions, or perhaps in response to the bullying of his classmates, the boy stopped being a distracted and indolent student to the point that his behavior convinced the headmaster to prepare him for university. Thus, at the age of 18, Newton entered Trinity College, Cambridge, a prestigious school for the education of the scions of the nobility and the upper middle class. Although he was not entirely without financial means, because of his low social background he was enrolled in the category of students called sizars, who had access to subsidized tuition in exchange for menial jobs. Inside and outside the strict classrooms, there was an air of rigid conservatism. Aristotle's philosophy dominated, tempered by an English-only pragmatism: from the third year on, students could open their minds to other authors. Newton's first project was an anonymous law degree. As soon as he could, however,

---

[4]This date refers to the Gregorian calendar. According to the Julian calendar, which was in use in England until 1752, Newton was born on Christmas Day, 1642.

[5]The term English Civil War refers to a series of conflicts between royalists and parliamentarians that bloodied England and Wales from 1642 to 1652 and saw the death on the gallows of King Charles I and the beginning of the dictatorship of Oliver Cromwell.

[6]M. Keynes, *The Personality of Isaac Newton: Notes and Records of the Royal Society of London*, 49, 1, 1995, pp. 1–56.

he devoted himself enthusiastically to the books of Descartes, Gassendi, Hobbes, and Boyle, the philosophers whose innovative ideas dominated scientific debate, and then to Galilei and Kepler. These readings, along with the positive influence of good teachers, led him to change course. There is no clear evidence that the first signs of his genius appeared during these years, but a kind of miracle occurred soon after graduation.

It was 1666, the year of the "number of the beast" that John's *Revelation* said would rise from the sea to ravage the earth. The plague, which had been endemic in Europe for three centuries, had regained its strength and was sowing black death throughout England. To prevent the spread of the infection, Cambridge University was closed and students were sent home. Back in his native village, instead of taking up the spade for which he seemed destined, Newton chose to give free rein to his need to know.[7] Following the pioneering work of his teacher Isaac Barrow,[8] he developed an innovative calculus tool that he called "fluxions".[9] He was also interested in mechanics and the mysterious force of gravity that causes apples to fall from trees, and he started optical experiments with glass prisms that he had purchased at a village fair where these curious objects were sold for amusement. He began almost as a game, «try[ing] therewith the celebrated phaenomena of color» with a very spartan instrumental apparatus. «[H]aving darkened [his] chamber, and made a small hole in [his] window-shuts»[10] — young Isaac wrote —, he extracted a beam of sunlight intercepted by the prism, taking care to project the outgoing beam onto a white wall a few feet away. What he saw was the opening of the range of colors that he later called "spectrum". He counted seven colors, as many as the planets, the days of the week, the wonders of the world, or the branches of the menorah, in deference to an emerging interest in esotericism: throughout his life he would devote himself to alchemy and biblical studies.

As he recounted in the introduction to his first scientific contribution, entitled *A New Theory about Light and Colors* (1672), he paused for a

---

[7]R.S. Westfall, *Newton's Marvelous Years of Discovery and Their Aftermath: Myth versus Manuscript*, in "Isis", 71, 1, 1980, pp. 109–121.

[8]A. Malet, *Isaac Barrow on the Mathematization of Nature: Theological Voluntarism and the Rise of Geometrical Optics*, in "Journal of the History of Ideas", 58, 2, 1997, pp. 265–287.

[9]Instantaneous rate of change of a time-varying quantity (fluent, from the Latin *fluere*, to flow) at a given point.

[10]I. Newton, *A Letter of Mr. Isaac Newton ... containing his New Theory about Light and Colours ...*, in "Philosophical Transactions", No. 80, 1671/72.

while to admire the scene: «It was at first a very pleasing divertissement, to view the vivid and intense colors produced thereby».[11] Then scientific curiosity took over from sheer amazement. Where did these magnificent pigments come from that the prism had dispersed? No one knew. Some thinkers argued that they were a kind of modulation by the shadow on the otherwise white sunlight: a side effect caused by bodies. In fact, the shadow was considered a substance rather than a reduction or absence of the light signal.[12] Newton was not convinced and began to think about how to address the problem. He was already clear about what the approach to knowledge should be: questioning nature with «sensible experiences» in order to induce general laws from some facts. This was a *modus operandi* in line with Galilei's and contrary to that of Aristotle and his school, which relied on deducing effects from *a priori* hypotheses and aimed at the quality of phenomena rather than measurable quantities. Therefore, in order to determine whether colors were intrinsic to solar radiation or due to an amorphous component interacting with matter, the young Isaac conceived and performed a crucial experiment.[13] He used a lens to intercept the light scattered by the prism. Surprisingly, the chromatic fan closed and returned a white light. The result settled the question once and for all: no one could argue that there were colors in things.

This was the *destruens* side of Newton's foray into optics. The *construens*[14] would lead him to elaborate a model of light consisting of tiny particles, differentiated by color, which matter could absorb, reflect, or diffract;[15] not waves, as Descartes, Huygens, and Hooke had claimed. The choice of camp to side with came from the observation that light rays travel in straight lines, like material bodies. Waves, instead, propagate as spherical surfaces, the two-dimensional analog of which are the circles created

---

[11] *Ibidem.*

[12] Remember the *Genesis* (1,3) where God is said to have separated light from darkness.

[13] According to a concept explained in Francis Bacon's *Novum Organum*, Newton used the expression *experimentum crucis* (crucial experiment), coined by Robert Hooke, because he believed that his test was capable of distinguishing between two contrasting scenarios.

[14] Again, concepts from Francis Bacon's *Novum Organum*. In the process elaborated by the English philosopher to obtain certain knowledge of a phenomenon, two basic steps are distinguished: the *pars destruens* (destructive part) and the *pars construens* (constructive part). The former consists in the elimination of the prejudices of the mind (the «idols»); the latter in the application of an inductive method, improved in comparison to Aristotle, as it contemplates verification experiments.

[15] I. Sabra, *Theories of Light: From Descartes to Newton*, CUP, Cambridge 1981.

by the impact of a stone on the surface of a pond. Newton was completely right and completely wrong at the same time, as we will see later.

In 1667, at the end of a true *annus mirabilis* (wonderful year), Isaac returned to Cambridge on a small scholarship. The plague had fallen asleep and life could resume its normal course. He brought with him an extraordinary wealth of knowledge and accomplishments, yet he received his Master's degree without any praise. His supervisor, Isaac Barrow, who had sensed the qualities of his introverted assistant, endeavored to promote him as a promising mathematician in London's scientific circles. Two years later, having decided to devote himself entirely to the church, Barrow recommended Newton as his successor to the newly established chair of mathematics at Trinity College on a bequest from the Reverend Henry Lucas (unwittingly consigning the name of this minor clergyman and politician to history). It was in this capacity as a young and ambitious Lucasian professor that we meet Isaac at the beginning of the chapter: a highly respected teacher who was also allowed to escape the academic requirement of taking vows.

The nature of colors, while correct in itself, had led Newton to believe, erroneously, that there was no way to control the chromatic aberration of lenses, since, as he had correctly shown, color is an intrinsic property of light. This belief dealt a blow to the popularity of refractive optics, whose aberrations had previously been attributed mainly to the spherical shape of surfaces; or rather, it acted as a punch below the belt, coming as it did just when it seemed that a way had been found to control at least some of the spherical aberration. Descartes had done it, and then, a few years later, the Scotsman James Gregory,[16] a shy genius with little aptitude for doing (his studies on infinitesimal calculus, though disputed by Huygens, would serve Newton as a starting point for systematizing the method of fluxions). Shortly after the mid-seventeenth century, the two scientists independently demonstrated mathematically that the surfaces forged in the form of rotational conics returned images along the axis of symmetry that were completely free of spherical aberration. In short, the theory suggested that with an elliptical or even hyperbolic lens, it would be possible to focus into a single point all the light from a point source placed at infinity along the axis of symmetry. However, this beautiful result did not take into account

---

[16] E.R. Sleight, *Development of Mathematics in Scotland*, 1669–1746, in "National Mathematics Magazine", 19, 4, 1945, pp. 173–185.

the fact that the refractive properties of lenses change with the color of the signal, including focal distance.

This chromatic intolerance is the reason why, despite the partial defeat of spherical aberration, refractive optics continued to produce unsatisfactory results. Gregory was well aware of the problem. In his *Optica Promota* (The Progress of Optics), published in 1663 when he was only 25 years old, he proposed replacing objective lenses with concave mirrors to exploit the achromaticity of reflection. But by his own admission, due to his lack of manual skills, he never put into practice his idea, which, by the way, was not even entirely original. In fact, the Scottish astronomer had been inspired by a treatise on optics written by the Italian Niccolò Zucchi,[17] in which the Parma Jesuit reported that he had built a reflecting telescope as early as 1616 and used it to observe the sky from the roofs of the Roman College, the Jesuit educational institute in Rome. It all had started with the chance discovery of a bronze spherical mirror in a cabinet of scientific curiosities at the College. Zucchi, who was familiar with the geometric properties of the sphere, thought of using it as the objective of a telescope. An application that four years later Galileo too would discuss with his friend Sagredo; but neither of them did anything about it.

At that time, the Jesuit teacher had not yet developed his deep interest in astronomy that, seven years later, would be awakened in him by his meeting with the Lutheran Kepler. He, however, pointed the mirror at Mars and Jupiter and observed their images through a concave (Galilean) eyepiece that he simply held in his hand. William Herschel would follow the same procedure two hundred years later, with very different results (see p. 211). What Zucchi saw was not interesting enough to convince him of the soundness of his intuition, and so he let it go. The idea was just brilliant, but the implementation was ineffective. Beyond the quality of an optical element — about which the only information we have is a naive statement by Zucchi boasting that the mirror was the work of an experienced and very careful craftsman — the makeshift astronomer had to face a practical problem: how to access the focus without interfering with the path of the incident ray? So as not to obstruct the modest aperture of the instrument with his head, Zucchi resorted to looking at the sky obliquely with respect to the geometric axis of his mirror in order to move the focused image out of the path of the incoming ray. It was a clever solution, but it introduced

---

[17]D. Bartoli, *Della vita del P. Nicolo Zucchi della compagnia di Giesu [sic]. Libri due,* il Varese, Rome 1682 (in Italian).

an intolerable aberration typical of spherical surfaces. In the end, the cure was worse than the disease!

The fact is this one. A paraboloid is a very demanding objective. It returns (almost) point-like images only for stellar sources very close to the axis of symmetry. Those further away appear gradually more and more elongated and blurred (comatic), with severe loss of resolution. The sphere, on the contrary, is strictly democratic. Thanks to its infinite symmetry axes, it is not affected by coma aberration and maintains the same image quality everywhere in the field of view, however large this may be. This property is retained even when the sphere is sawn and reduced to a spherical cap (hereafter mirror). The new symmetry introduced by the cut creates aperture effects, but does not alter the properties related to reflection. Unfortunately, the optician's blanket is always short: if you pull at one end, you strip at the other. Spherical mirrors do not cause coma, but they cannot focus a parallel beam to a single point. If you look at a point source, you will see it diffuse due to spherical aberration. There is no other way than to accept a compromise and resort to geometric shapes that are somewhere between paraboloids and spheres. The analogous case of spherical lenses will help us understand why this strategy pays off.

The spherical aberration present in lenses is due to the fact that rays closer to the center are deflected less than those at the edges. It is necessary to induce them to bend just enough to catch the angle of the others, and this is achieved by increasing the curvature in the central part of the refracting element. The result is lens profiles that, to exaggerate, look more like lemons than lentils. *Mutatis mutandis*, a similar strategy is applied to mirrors; or, in the logic that united we win, spherical mirrors are coupled with refractive elements (objective lens) that behave opposite to the sphere, so as to produce a mutual cancellation of spherical aberration. Other combinations of mirrors and lenses will be considered later (chap. 15).

In Zucchi's off-axis observing strategy, the reflecting telescope, although achromatic, returned elongated and confused images. The problem was how to access the on-axis focus without obstructing the incoming beam too much. In addition to that, there was the challenge of obtaining perfectly polished surfaces by skillfully machining a material that could reflect light; and this step did not seem to be an easy one at all. The material in question is named "speculum", known since ancient times and used to make drip mirrors for women's toilets. It is a bronze alloy enriched with tin and bleached by adding arsenic or silver; it is difficult to grind and rather inefficient, reflecting no more than 65% of incident light even under the most favorable conditions. Worse, it oxidizes quickly when exposed to air, quickly rendering polished surfaces useless for optical purposes. But beggars can't be choosers! It would be nearly two centuries before the technique of depositing a thin layer of silver on a polished glass plate was discovered,[18] later replaced by the most modern aluminizing. Operations that can be repeated several times in the life of an astronomical mirror, without having to rework it each time, as in the case of the speculum, and with a minimum waste of time and money (see p. 328).

After the early failures of Zucchi, and then of Cavalieri, Mersenne, and some others,[19] the idea of an instrument with a reflecting objective was taken up by Newton downstream of his experiments on light and the reading the work of the Italian Jesuit. The Cambridge professor's aim was to remove the obstacle of chromatic aberration, which, as we saw, he believed to be insurmountable in refractors.

Isaac was a skilled craftsman. He secured a speculum disk, worked it into a polished spherical surface two inches in diameter, and made a short, easy-to-handle telescope. The beam reflected by the primary optics was broken by a flat mirror with an elliptical contour positioned centrally at an inclination of 45° from the axis, so that its shadow on the light collector was a circle. This design kept the instrument less than 20 cm long. The secondary mirror was suspended above the first by four tie-rods hooked to the telescope tube in the manner of a large spider. It sent the beam sideways so that the latter could be viewed with great ease by an externally placed eyepiece. In short, for the price of a small obstacle, Newton had succeeded in using the parabolic mirror along its axis, thus eliminating some of the problems encountered by Zucchi. An invention that, if not worthy of the Nobel Prize, would today warrant a patent of the type that significantly changes income (although easy to copy and therefore difficult to protect).

---

[18] In the mid-17th century, James Gregory had actually imagined using glass mirrors, but he had given up. The technique used since the Middle Ages of depositing an amalgam of tin and arsenic on the back of a transparent plate to obtain a mirror capable of reflecting light, which had made the Murano mirrors famous, was not suitable for astronomical use because the light had to pass through the glass layer twice in order to be reflected, with obvious complications.

[19] On July 7, 1626, Cesare Marsigli, a cultured gentleman from Bologna, wrote to Galilei about two fellow citizens, heirs of a certain Cesare Caravaggi, who claimed to have discovered new burning phenomena by refractive methods and to have made a *perspicillum* with mirrors (probably in the manner of Zucchi); G. Galilei, *Opere*, XIII, *Corrispondenza 1620–1628*, Barbera, Florence 1966, p. 266.

The small reflector worked so well that the Cambridge professor was able to observe the satellites of Jupiter and the phases of Venus.

> With the Telescope which I made I have sometimes seen remote objects & particularly the Moon very distinct in those parts of it which were neare the sides of the visible angle. [...] One of the Fellows of our College is making such another Telescope with which last night I looked on Jupiter & he seemed as distinct & sharply defined as I have seen him in other Telescopes. When he hath finished it I will examin more strictly & send you an account of its performances, For it seemes to be somthing better then that which I made.[20]

The news spread, and in 1671 the Royal Society asked for a copy of the instrument to compare with a refractor of equivalent aperture. The consequences of this story could have dramatic implications for the progress of science due to a quarrel between the paranoid Cambridge genius and another peculiar character, the physicist Robert Hooke,[21] staff member of the Royal Society.

Hooke was seven years older than Newton. He, too, was born to a modest family far from the nation's beating heart, on Wright Island in the English Channel — the place of the rock music festival and the famous Dik Dik song. Despite these unfavorable circumstances, compounded by poor health, Robert, who combined boundless ambition with exceptional skills in painting and mechanics, never gave up. When his father died, he used a modest inheritance to emigrate to London, where he took a job in a painter's studio. The city was a human anthill, smelly and unhealthy for those without adequate means of livelihood. Hooke was determined to break out of the ghetto of pariahs that surrounded the palaces of the aristocracy and wealthy merchants, plagued by hunger, fleas, and Cromwell's police. He managed to gain admission to the prestigious school associated with Westminster Abbey, where he was noted for his unusual wit. He then took a circuitous route to Christ Church College, Oxford. It was at this university, as famous and exclusive as Trinity, Cambridge, where Newton

---

[20]Letter to Henry Oldenburg, 16 March 1671; http://www.newtonproject.ox.ac.uk/view/texts/normalized/NATP00307.

[21]P.E.B. Jourdain, *Robert Hooke as a precursor of Newton source*, in "The Monist", 23, 3, pp. 353–384; S. Inwood, *The Forgotten Genius: The Biography of Robert Hooke 1635–1703*, MacAdam-Cage, San Francisco (CA) 2003; L. Jardine, *The Curious Life of Robert Hooke: The Man Who Measured London*, HarperCollins, New York 2004.

had studied, that his boat passed: the one that crosses a man's life at most once and is not to be missed. He met Robert Boyle, an Irish nobleman who was a reference point for natural philosophy in England, and became his assistant, starting a precious training with a powerful and brilliant scientist, rightly considered one of the fathers of modern chemistry.

In the early 1680s, Hooke returned to London as a permanent member of the newly formed Royal Society, of which he would later become secretary. He was charged with organizing and managing the experiments that were presented to the members on a weekly basis: practical demonstrations of a wide variety of branches of science and technology. Hooke carried out his work diligently, but he did not stop there. Above all, he was a tireless, imaginative, sharp, and versatile researcher. His skill in drawing proved invaluable in the drafting of his plans and representations of his discoveries (a talent that Leonardo da Vinci despised to the utmost). So he had great gifts, but also a moody and shadowy temperament, aggravated by poor health. Jealous of his work to the point of encrypting his notes (Leonardo also did this), Hooke spent much of his life fighting over the priority of discoveries and inventions. He clashed with everyone, especially Newton. The Lucasian professor, stimulated by the Royal Society's interest in his prototype reflecting telescope, had printed his notes *Of Colors* in 1671, then expanded them into the *Opticks*, a work published only after Hooke's death — later we'll see why. Still chasing fame, Newton counted on triumphant success. Instead, his study was met with lukewarmness and skepticism. In particular, the corpuscular hypothesis was sharply criticized by Hooke, who supported the wave model promoted by Huygens. Newton did not take this well. Sensitive and vindictive, he decided to withdraw from public debate altogether; and might have died without communicating to the world the extraordinary results of his studies of mechanics and gravitation, had it not been for another of those all-round scientific figures with whom 17th-century England was richly endowed.

Edmund Halley[22] had argued extensively with Robert Hooke, the architect, mathematician, and astronomer Christopher Wren, and other learned men in his circle on the controversial question of the model of the world, without getting much out of it. For this reason, although he was not yet thirty years old, he had dared to approach the gruff Lucasian professor holed up at Trinity College. He had heard of his genius and hoped to gain

---

[22]M.V. Fox, *Scheduling the Heavens: The Story of Edmond Halley*, Morgan Reynolds, Greensboro (NC) 2006.

some enlightenment from him. Newton appreciated the exuberant young man enough to tell him about an application of his mechanical theory to a special law of force by which he believed that bodies attracted each other, on Earth as in the heavens. The idea that this force depended on the inverse of the square of the distance was not new: the last to mention it had been Hooke himself.[23] But Newton had done much more with it. In full compliance with the Galilean dictate, he had found a way to put this insight into a physical model constructed in the language of mathematics and resistant to experimental proof. The problem was that this singular character, an explosive cocktail of genius and recklessness, had lost his notes! Halley understood the momentous scope of a scenario that identified in a characteristic property of every physical body, the mass, the reason why it is up to the heavy Sun to remain (almost) immovable in the center of the world and not to the much lighter Earth. But that was not all. In contrast to Aristotle and his cosmic dualism, Newton's theory of gravity unified the heavens and the Earth for the first time after many centuries, identifying in a single universal force the cause of both the weight in the sublunar world and the motions of the stars in the heavens. Since then, the history of physics has been a never-ending quest for the unification of all the laws of nature: a scientific monotheism justified by the brilliant results obtained and which, as we shall see, has also infected cosmology.

Angry at the whole world for the insult he felt he had received, Newton was reluctant to publish his jewels. He hated some of his colleagues at home and abroad with all his heart, and he was also quite concerned about the law of action at a distance, which he himself had introduced but whose essence he did not understand. «You sometimes speak of gravity as essential & inherent to matter», he would have written in 1693 to the theologian Richard Bentley who had stimulated him on this point. «Pray do not ascribe that notion to me, for the cause of gravity is what I do not pretend to know & therefore would take more time to consider of it».[24] How is it possible, he pondered, for two distant bodies to exchange a force without the action of an intermediary? Nevertheless, Halley encouraged him in his doubts — Newton would not endure the humiliation of being discovered and ridiculed in a grave error —, urged him to reconstruct his

---

[23] A. Cook, *Edmond Halley and Newton's "Principia"*, in "Notes and Records of the Royal Society of London", 45, 2, 1991, pp. 129–138.

[24] Newton's II letter to Bentley, dated January 17, 1693; in http://www.newtonproject. ox.ac.uk/view/texts/normalized/THEM00255.

lost notes, offered to help him finish the calculations and set up the experimental tests. He also guaranteed him, thanks to the money inherited from his father, a wealthy London soap importer,[25] economic support for the printing of the work, which the Royal Society was unable to finance. The entire budget of 1686 had been invested in the publication of John Ray and Francis Willughby's lavishly illustrated book *De Historia Piscium* (Latin for On the History of Fish). This was probably the greatest achievement of the man who discovered the cyclic nature of comets, beginning with the one that bears his name, and who, in the early 18th century, was able to persuade his colleagues on the thorny issue of a more accurate measurement of the Astronomical Unit (p. 306).

Thus, in 1687 the *Philosophiae Naturalis Principia Mathematica* (Mathematical Principles of Natural Philosophy) finally appeared, printed with the permission of the Royal Society. The first two books presented the ideas, rules, and tools of a mathematically based mechanics capable of representing the kinematic phenomena observed on Earth. The Newtonian formulation of the "new science" was to some extent a critical synthesis of current ideas in the vast sea of many different opinions about the causes of motion. It worked spectacularly well and was enthusiastically received by colleagues in the Royal Society and by natural philosophers throughout Europe, despite the nationalism that intoxicated the academies then and now. In the third and final book, entitled *De mundi systemate* (On the System of the World), Newton turned his interest to the heavens, applying his mechanical instruments to a single law of force to describe, as we have said, both the fall of bodies on the Earth, including apples, and the motions of wandering stars. The Lucasian professor's gravitational theory had the ability to represent and predict multiple celestial appearances of the same body, just as Halley wanted. It found and justified Kepler's laws, even correcting an inaccuracy in the third law,[26] could account for the anomalies in the motions of the Moon, and modeled the orbits of the Medicean Moons

---

[25] From the minutes of the Council of the Royal Society, June 2, 1686: «It was ordered that Mr. Newton's book should be printed, and that Mr. Halley should take charge of it, and print it at his own expense, which he undertook to do.». Cf. A.N.L. Munby, *The Distribution of the First Edition of Newton's "Principia"*, in "Notes and Records of the Royal Society of London", Vol. 10, No. 1, 1952, pp. 28–39.

[26] The ratio of the square of the period of revolution to the cube of the major axis is proportional to the sum of the masses involved. It appears to be (almost) constant within the Solar System only because the mass of the Sun dwarfs that of any planet. The giant Jupiter, for example, is only ∼0.1% of the Sun.

and comets. For the latter, Newton used data provided by Flamsteed. Subsequently, the good relationship of the Greenwich director with Newton and Halley in particular[27] would be marred by accusations that the two used some data without citing the source.

When Hooke read the *Principia*, he became furious and basically accused the author of plagiarism, if not for the mathematical treatment of the problem, at least for the idea concerning gravitation. As we have already said, a few years earlier he had hypothesized a universal gravitational force dependent on the inverse of the square of the distance, and had pointed to it as the cause of the curved orbits of the planets around the Sun: circles or ellipses instead of the rectilinear motion which, inspired by Galileo, he believed to be the natural motion even in the sky in the absence of forces. Hooke's explicit aggression hurt Newton, who reacted to him with unusual hostility, also because of the modest warmth with which the third volume of the *Principia* had been received. He discredited his opponent in every way, abusing his position and influence, even going so far as to have the canvas representing his rival in a hall of the Royal Society destroyed.

Petty, but also pavid, he did not dare to publish his treatise on *Opticks* until 1704, a year after Hooke's death, crushed by illness. Perhaps the twisted genius of Trinity College feared the outbreak of other accusations of plagiarism, such as that levelled at him by the German polymath Gottfried Wilhelm Leibniz concerning the theory of fluxions: an international dispute in which Isaac once again proved his lack of scruples. In fact, the question of scientific priority was brought before a committee of the Royal Society

---

[27]Still an undergraduate, Halley had made contact with Flamsteed and, hoping to be taken under the wing of the Astronomer Royal, sent him one of his works: «these Sr. as a specimen of my Astronomical endeavours I send you, being ambitious of the honour of being known to you, of which if you deem me worthy I shall account myself exceedingly happy in the enjoyment of the acquaintance of so illustrious and deserving a person as your self. I am Sr. Yours and Urania's most humble Servant tho' unknown. Edm. Halley» (10 March 1674/5). The Astronomer Royal had been seduced by this «ingenuous youth, versed in calculations and almost all parts of mathematics», and had made him his assistant «de facto» (in practice). As time passed, the collaboration became less and less close, until it broke down completely. In a manuscript preserved at Greenwich, a handwritten note by Flamsteed reads: «About E.H. that I am very much mistaken in him: that I never found anything so considerable in him as his craft and forehead, his art of filching from other people, and making their work his own; as I could give instances, but that I am resolved to have nothing to do with him, for peace sake». From S. Chapman, *Edmond Halley, F.R.S. 1656-1742*, in "Notes and Records of the Royal Society of London", Vol. 12, No. 2, 1957, pp. 168–174.

during the years Newton was president.[28] The arbitrators ruled in favor of the British scientist with a report written in Newton's own hand. In any case, the British power and the fame of the Cambridge professor left no chance to anyone, neither in England nor abroad. This can be understood from a passage in a letter written by Leibniz in December 1715 to a learned nobleman from Padua, Antonio Schinella Conti, who had been chosen as the arbiter of the dispute.

> Here, sir, is the letter, of which you may make whatever use you wish. I come at once to the matter that concerns us. I am very glad to hear that you are in England, where you will profit greatly by your sojourn. It must be admitted that in this country you will find very accomplished people; but they will want to attribute all, or nearly all, of the discoveries to themselves, but they will probably fail to do so.[29]

Leibnitz believed he had an impartial friend in the Italian abbot, but apparently he was mistaken. After the German polymath's sudden death in 1716, Conti abruptly reversed his hitherto very balanced position and fully embraced the English thesis.[30] "Paris is well worth a mass", one would be inclined to repeat with Henry IV of Bourbon.

In conclusion, although the genesis of the *Principia* has relatively little to do with the invention and use of the telescope, the theory of universal gravitation has become one of the main driving forces in the development of the observation of the cosmos. But Newton's name is also directly linked to the technological development of this instrument. As Galileo had done with the refracting telescope, the Englishman was the first to build a prototype reflecting telescope capable of observing the sky. Unlike the Tuscan, however, he made little astronomical use of it, limiting himself to observing the four moons of Jupiter and the crescent phase of the planet Venus, without even attempting to make any new discoveries. The brightness of the instrument, with a pupil of only 6 cm, was still too modest to compete with the large objectives then in circulation. Moreover, the optical configuration

---

[28] D.V. Schrader, *The Newton-Leibniz controversy concerning the discovery of the calculus*, in "The Mathematics Teacher, Vol. 55, No. 5, 1962, pp. 385–396.

[29] Cited by G. Cantelli, *La disputa Leibniz-Newton sull'analisi. Scelta da documenti degli anni 1672–1716*, Boringhieri, Turin 1960, p. 200 (in Italian).

[30] Cf. M.R. Antognazza, *Leibniz. An Intellectual Biography*, CUP, Cambridge 2009.

adopted by Newton allowed a maximum magnification of 40, too little for planetary detail hunters.

Soon, however, it was found that other optical configurations existed besides the one devised by Newton, which were equally capable of making light reflected by a mirror accessible for observation. Following in Gregory's footsteps, in 1673 Hooke made a telescope with a rear sight. The reflected beam from the parabolic primary was caught and returned by a second, smaller concave element with an elliptical cross section. The light then passed through a hole in the center of the primary mirror to be collected by the eyepiece. Primary and secondary were kept coaxial and tuned by empirically matching the foci of their conical surfaces. The combination caused an elongation of the resulting focus, to the benefit of magnification,[31] but required equally long tubes, which were bulky and subject to unwanted bending. This was a major problem because the performance of an instrument deteriorates when the support structure flexes under its own weight, causing the optics to lose alignment.

The obstacle had already been overcome a year earlier by Laurent Cassegrain,[32] a French priest of whom not much is known,[33] with the introduction of a convex secondary with hyperbolic cross-section. As in the Gregorian project, the combined focus was located behind the holed primary. The effect was an instrument with a good equivalent focal length, and therefore a considerable scale, but also quite compact and, at least along the axis, free of spherical aberration. In short, a handy and optically (almost) perfect telescope, albeit with a very narrow field of view. As was his wont, Newton ridiculed the gimmick, perhaps fearing that the paternity of the reflector invention was in jeopardy. Such was his power that the poor Laurent did not even try to retaliate. But history would prove him right. Today, large optical telescopes have a Cassegrain-type configuration, modified with a slight deviation of the primary from the perfectly spherical shape (aspherization) to reduce astigmatism and coma and increase the field of view.

---

[31] Even today, the Gregorian mount is used to control the heating of the secondary mirror in solar telescopes, invested by a torrent of energy.

[32] A. Baranne, F. Launay, *Cassegrain: un inconnu célèbre de l'astronomie instrumentale. Cassegrain: A famous unknown of instrumental astronomy*, in "Journal of Optics", 28, 4, 1997, pp. 158–172.

[33] Laurent Cassegrain was born in the region of Chartres around 1629 and died in Chaudon (Eure-et-Loir) in 1693. In his last years, he earned his living by teaching science at the Collège de Chartres, a French high school.

The main instrument of modern and contemporary astronomy, the reflector did not win out immediately. Refractors had a larger field of view and a better market. Indeed, many craftsmen were able to develop excellent lenses — which also had many applications in both peace and war — but few or none could produce mirrors of optical quality. Lenses, however, were left with the original sin of chromatic aberration, which, as we have seen, was kept in check by incredibly long focal lengths in the so-called "aerial telescopes". But even the epic of tube-less instruments, like the era of the mammoth and dangerous airships in the early days of aviation, could not last long. In both cases, technology had exceeded reasonable limits rather than seeking alternative solutions to problems. The time was ripe for a breakthrough, and the first step was taken by Chester Moor Hall.[34] In 1730, this British lawyer with a hobby for research had an idea to control the effect of the colors of light and thus reduce the length of telescopes without sacrificing magnification. He had thought about it for a long time and was convinced that if it worked for the human eye, where a single biconvex lens of protein substances, the crystalline, is free from chromatism, it should also work for glass lenses. What could not be achieved with a single lens, he concluded, could perhaps be obtained with two, selected and forged so that the sum of their chromatic properties would produce less total aberration. In short, it was necessary to somehow establish a virtuous cooperation between two equally undesirable but opposite behaviors. Of course, it is not always true that two wrongs make a right, but in this case it did!

To understand Hall's gimmick, recall that a biconvex (positive) lens focuses the different colors of a parallel beam of white light at different points on the axis, with blue converging faster than red. A biconcave lens does the same thing, but because it is divergent (negative), it diverts blue more than red. By combining them — the brilliant magistrate reasoned — and by carefully choosing their refractive indices, it would be possible to eliminate chromatic aberration for two colors and reduce it for the others. It seemed a plausible idea that needed to be tested with a prototype. For the biconcave lens, Hall chose the common glass used for windows and eyeglasses: a material with a low refractive index called "crown" because of

---

[34]R. Willach, *New Light on the Invention of the Achromatic Telescope Objective*, in "Notes and Records of the Royal Society of London", 50, 2, 1996, pp. 195–210.

the technique used to make it.[35] For the biconcave lens, he instead used "flint", a type of lead glass with high refractive index and great brightness, used in furniture and chandeliers, but difficult to produce in the quality required for optics. Unable to make the lenses himself, he turned to two different artisans, assigning one to work on the biconvex lens and the other on the biconcave lens. In this way, he hoped to keep his project a secret. But, incredible as it may seem, the two turned to the same third craftsman, George Bass, and subcontracted the work to him. Sensing something was up, Bass began experimenting with the two lenses he had been making. By combining them, he obtained an achromatic doublet that was essentially free of the annoying phenomenon that turned white light sources into iridescent spots. Eureka!

The secret was no longer secret, but Hall did not care. As a prominent gentleman with a good personal fortune, he did not think it appropriate for his rank to patent the invention. With typical British detachment, he simply used the achromatic doublet to make various telescopes for personal use. In 1758 a London telescope maker, John Dollond, who had been tipped off by Bass, took up the idea and made a fortune for himself and his family; and we probably would never have heard of Hall had the owners of Dollond & Son, a leading telescope manufacturing firm, not been involved in a plagiarism lawsuit (see below).

John Dollond[36] is another of the many in our history for whom fate had reserved a difficult start. He was born in London in 1706, the son of a Huguenot refugee who had fled persecution by French Catholics and earned his living weaving silk. While helping his father with his work, John found a way to get an education. He was 46 when he decided to join his son, Peter, who had opened an optical instruments shop in 1752. The Dollonds thought big. They understood that to fulfill their ambitions for growth, they had to keep abreast of developments in the debate over the nature and properties of light and color that was fueling correspondence between scientists in different parts of Europe and animating the meetings of scientific societies. This could lead to innovative technological solutions, the control of which would make a difference to their business.

---

[35] The hot glass mass was blown into a hollow sphere (the "crown") and then, after being reheated, stretched by applying a strong rotation.

[36] R. Sorrenson, *Dollond & Son's Pursuit of Achromaticity*, 1758–1789, in "History of Science", 39, 2001, pp. 31–55.

Thus, when the great Swiss mathematician Leonhard Euler proposed an achromatic solution for converging lenses based on water placed between glass menisci, John Dollond wrote an essay to challenge him, taking on the role of the untouchable Newton, who at the time was the new paradigm for the classic *ipse dixit*. In response, Euler deepened his study and produced a new model in which light was represented by a vibration excited by luminous bodies in an elastic medium and colors corresponded to different frequencies of oscillation. Perhaps even this extraordinary insight, which anticipated the studies of James Clerk Maxwell by more than a century, would not have been enough to convince the stubborn and patriotic craftsman. Instead, a work critical of Newton succeeded. It had been written by the Swede Samuel Klingenstierna, formerly a lawyer and later a brilliant natural philosopher, whose greatest limitation was that he published his studies in his native language, unknown to most. As a practical man, Dollond set out to clarify the problem empirically and threw himself headlong into experimenting with new solutions and new materials, inspired by the ideas he gathered while listening. This led him to patent the achromatic doublet, which made his company the first in Europe to produce telescopes for civil, military, and astronomical use. A millionaire's business! In 1766, the invention was allegedly plagiarized by a competitor, James Champneys, who was in fact just one of a long list. Taken to court by John Dollond Jr, he defended himself by arguing that the patent had no reason to exist because the achromatic doublet had been invented nearly forty years earlier by Hall. This was true, but the judge agreed with Dollond and upheld the argument that «it was not the person who locked his invention in his scritoire that ought to profit by a patent for such invention, but he who brought it forth for the benefit of the public».[37] You can learn something from this!

With the Dollonds' technology, the exploration of the heavens took on a new pace. In 1763, John's son, Peter, invented an apochromatic triplet, which, as the etymology of the adjective suggests, further improved lens performance by eliminating most aberrations. The British company's strength was in producing flint glass of a much higher quality than its competitors. Techniques for working with materials and production tools were jealously guarded and passed on to workers with great care to prevent the leakage

---

[37]H.C. King, *The History of the Telescope*, Dover, New York 2003, p. 154. See also J.E. Greivenkamp, D.L. Steed, *The history of telescopes and binoculars: an engineering perspective*, in *Novel Optical Systems Design and Optimization*, XIV, R.J. Koshel, G.G. Gregory (eds.), Proceedings of SPIE, 8129, 81290S, 2011.

of valuable information (as the Venetians had tried unsuccessfully to do a few centuries earlier with glass). Indeed, regulations were inadequate to protect ideas from plagiarists and thieves, and sometimes it was considered wiser to keep an invention hidden than to patent it and make it public. The Dollonds became suppliers to the royal household and the world's leading exporters of telescopes. Among the few competitors was a Jesse Ramsden, of whom we will hear more (p. 189); he was somehow part of the family, having married a sister of Peter Dollond. Thanks to these craftsmen, refractive optics, which seemed to be on its way out, made a comeback.

London was a genuine Eldorado for telescope makers because of the rich demand for refractors coming mainly from military and commercial activities at home and abroad. There were also those who, like Scottish craftsman James Short, had made money by producing reflectors with Newtonian and Gregorian mounts. As is the case today with some famous clothing brands that simply put their logos on garments produced in the Third World at bargain prices, the big British manufacturers mostly assembled components made on their behalf by small artisans. This ploy reduced costs and increased production. Short, for example, bought complete instruments from smaller companies, replaced the optics with his own mirrors and lenses, which were definitely of higher quality, and resold the products at much higher prices, passing them off as his own. His company's brochure offered minimal telescopes, with apertures of a few centimeters, a few dozen magnifications, and affordable prices, along with 60 cm giants, with a focal length of more than 20 m, capable of hundreds of magnifications but very expensive: nearly 1,000 guineas, equivalent to 150,000 euros today. Both reflectors and refractors existed, but only a few astronomers preferred mirrors to lenses. The exceptions included a Frenchman, Charles Messier[38] and a German naturalized English named William Herschel, perhaps the greatest observer of all time.

Like many others, Messier came to astronomy by accident. As a 20-year-old, he had moved to Paris in 1750 from his native Lorraine in northeastern France to earn a living. His older brother, who had raised and educated him after their father's untimely death, had given him a small selection of positions to choose from. Simple jobs, considering Charles' only skills were beautiful calligraphy, some ability in drawing, and an attitude toward order and precision instilled in him by his brother, who was an intendant.

---

[38]See chapter 1 of S.J. O'Meara, *The Messier Objects*, Cup, Cambridge, 2014.

Remembering the excitement that a comet and a partial solar eclipse had aroused in him as a child, Charles chose to work for the astronomer Joseph-Nicolas Delisle.[39] The latter, a protégé of the Cassinis, had enriched himself at the tsar's court in St. Petersburg and now ran his own observatory at the Hôtel de Cluny, a former college of the Cluniac order, in a splendid location in the Latin Quarter, and served as astronomer-geographer for the French Royal Navy. At first, Messier was assigned to copy some drawings. Then Delisle introduced him to celestial observations and gradually taught him the tricks of the trade.

At the time, one of Charles' contemporaries, Joseph-Jérôme Le Français de Lalande, was attending the Hôtel de Cluny. He was an ambitious and eclectic man who would later become director of the Paris Observatory and an opinion maker during a difficult and glorious period for his country and the world at large. The two were brought together by Delisle in the project to find the comet that Halley had predicted would reappear around 1758, after an absence of three quarters of a century. Indeed, the great English astronomer was convinced that even the feathered stars participated in the cosmic merry-go-round of planetary motion along very long and yet periodic orbits. In 1705, in a note entitled *Synopsis astronomy cometicae* (A brief summary of cometary astronomy), he had speculated that the comet appearing in 1682 might correspond to the transit of the same object observed by the Saxon Petrus Apianus (Latinization of Peter Bienewitz) in 1531 and by Kepler in 1607. Hence his prediction of the star's reappearance around 1758.[40] Halley was right, so much so that the comet now proudly bears his name. This was chosen by the French astronomer Nicolas-Louis de Lacaille, in an act both just and generous in view of the rivalry between the two nations. The wandering star reminds us of this every 75 years or so when it completes its more or less spectacular approach to the Sun in its eccentric orbit around the star, before plunging back into the dark depths of the outer Solar System, beyond the orbit of Uranus. But Halley did not live enough to verify his conjecture. He died in 1742 at the age of 85; a long life was not enough for him to determine whether his star would indeed return in time.

---

[39] N.I. Nevskaja, *Joseph-Nicolas Delisle (1688–1768)*, in "Revue d'histoire des sciences", 26, 4, 1973, pp. 289–313.
[40] R.E. Wilson, *The Story of Halley and His Comet*, in "Popular Astronomy", Vol. 8, 1910, pp. 357–366.

As the date of the object's supposed new perihelion passage approached, the anticipation in the astronomical community became feverish. Everyone wanted to be the first to see the wisp of light emerge from the darkness after so much silence, still without the magnificent plumage that comets don near the Sun. But where to look? This feat, which in other times would have been almost impossible, like looking for a needle in a haystack, then had a new ally in Newton's theory of gravitation; the same one that had allowed Halley to confirm his hypothesis and to estimate the date of the return of the small body, taking into account the disturbances caused by Jupiter and Saturn. No one had thought of this prophecy for decades. But at the time of the comet's supposed re-appearance, Lalande performed the mathematical exercise again and discovered that the object was traveling several months behind Halley's estimated time: a consequence of a powerful conjunction of the two Solar System giants that had pushed the tiny traveler into a new orbit. Other astronomers in France and England came to the same conclusion. Delisle then worked hard to calculate where the object would appear a few months before perihelion, when a good telescope should have allowed astronomers to spot it. But he made a mistake in his calculations and gave his collaborators the wrong indication of where to observe.

Messier, who by now had become a very experienced observer, diligently searched the area of the sky assigned to him by his director and apparently did not find the comet he was looking for. Instead, he discovered another, apparently much less intriguing one. Following it on its path between the Bull's Horns, he noticed a second streak of light not moving among the stars and carefully noted its position. He had accidentally found, without understanding its significance, the nebula that came to be called the Crab: the spectacular remnant of the supernova that had appeared in 1054 and been recorded by Chinese astronomers.[41] It was certainly not the first nebular object found after the advent of the telescope. The Great Nebula in Orion had been seen as early as 1610 by a wealthy Frenchman who had, among his many interests, that of looking at the sky. It was then rediscovered by a Swiss Jesuit in 1618 and a third time by Huygens. During the 18th century, Edmund Halley, Jean-Philippe Loys de Chéseaux (the true

---

[41] G. W. Collins II, W.P. Claspy, J.C. Martin, *A Reinterpretation of Historical References to the Supernova of A. 1054*, in "Publications of the Astronomical Society of the Pacific", 111, 761, pp. 871–880.

author of the Olbers Paradox),[42] and Nicolas-Louis de Lacaille (pioneer of the exploration of the southern sky), hunted down and cataloged over 40 diffuse-looking objects.

But let us return to Halley's Comet.[43] While the world searched in vain, the faint source was serendipitously discovered by a self-taught German amateur exploring the sky with the naked eye.[44] It was the Christmas night of 1758. The news was slow to reach France, and so Messier, who had adjusted the target after correcting the error in Delisle's calculations, had enough time to independently discover the star. It is easy to imagine his disappointment when he had to surrender the coveted record to an unknown Saxon peasant. From then on, he became the most avid and productive comet hunter. There was bad blood between him and Delisle. Fortunately, the master, who was already showing clear signs of jealousy toward his assistant, decided to leave the observatory and retire to a monastery: a voluntary abandonment steeped in devotion. Charles took his place at the Hôtel de Cluny, but it was not until 1771 that he also received the financially advantageous position of naval astronomer. Not bad, if it is true that scientists are paid little for work they would do for free!

Soon after, a young man from Picardy, a region in the north of France, joined the observatory. Like Messier, Pierre Méchain had essentially no university education, but he possessed the qualities that make an astronomer a great observer. Over the course of 40 years, the two discovered 27 comets, more than half the total number recorded in the same period. After sunset and before dawn, Charles systematically scanned the sky with his favorite telescope, a 19 cm aperture Gregorian reflector with 104 magnifications, later replaced by less bright but higher magnification refractors equipped with Dollond's apochromatic lenses. More than a hunt, his had become a true crusade, or rather a restless obsession. It is said that when his wife died, he misunderstood the condolences of his friends,

---

[42]The night sky is dark. If the universe were infinite, eternal, static, homogeneous and transparent, it would shine like the disk of the Sun. This paradox, attributed to the German astronomer Heinrich Wilhelm Olbers (1758–1840), tells us that at least one of the above properties of the cosmos is false. In fact, in the 20th century, it was realized that more or less all of them are false.

[43]R. Lyttleton, *The Comets and their Origin*, CUP, Cambridge 1953.

[44]Halley's Comet is believed to have made about 200,000 orbits around the Sun. Among its most famous perihelic passages is the one reported in 240 B.C. by Chinese astronomers of the Han Dynasty. They described the appearance of a "broom star", a star with a tail, an auspicious sign because the object had the ability to sweep away bad things. Cf. P. Maffei, *Beyond the Moon*, MIT Press, Cambridge (MA) 1978.

interpreting them as gestures of sympathy for having to give up the discovery of what would have been his 30th comet for another because of his domestic hard time. His tireless and fruitful observations eventually brought him great fame and many honors, including admission to the Royal Academy of Sciences in Paris: a seat he won with difficulty, overcoming the opposition of the academy's bigwigs. Perhaps out of jealousy, they challenged him for being at best an explorer and certainly not a man of science. The same sovereign, Louis XV, spoke of him as the «ferret of comets».[45] All this glory backfired on him when the Jacobins, after taking control of France, decided to erase all traces of the past. A kind of ethnic cleansing also involved the so-called "modern charlatans", as a furious Jean-Paul Marat called the scientists of the Academy, angry at not being admitted as a member. The heads of nobles, sans-culottes, and intellectuals fell without mercy or discrimination. «The Republic has no need of scholars or chemists — Pierre-André Coffinhal-Dubail, president of the Revolutionary Tribunal, is said to have replied to the appeal of Antoine Lavoisier's wife for the release of the great scientist so that he could continue his research; — the course of justice cannot be delayed».[46] «It took the mob only a moment to remove his head; a century will not suffice to reproduce it»,[47] the mathematician Joseph-Louis Lagrange later commented on this misdeed to his friend Jean-Baptiste Delambre. Fundamentalists, idiots, and tyrants have never shied away from beheading their academies.

These were hard years for Messier, but also for his friend Méchain, who, along with Lalande and Delambre, had embarked on the most extraordinary of scientific endeavors, the measurement of the meridian arc,[48] to give mankind an immutable unit of length, while France was bloodied by the Terror. Everyone was saved, and with the rise of Napoleon, things got better. At the end of a long life, Messier was able to reap the rewards of his discoveries. By then, his fortunes no longer depended on comets, but on

---

[45] J.-P. Philbert, *Charles Messier. Le Furet des comètes*, Pierron, Paris 2000.

[46] A. Demazière, *Encyclopédie des mots historiques vrais et faux*, Famot, Genève, 1980, p. 285 (in French): see also http://www.antimythes.fr/individus/guillaume_james/ gj_etudes_revolutionnaires_9.pdf.

[47] J.-B. Delambre, *Eloge del Lagrange*, in "Memoires de l'Institut", 1812, p. XIV, quoted in B. Belhoste, *L'École normale de l'an III, I. Leçons de mathématiques: Laplace–Lagrange–Monge*, Dunod, Paris, 1992, p. 202 (in French).

[48] K. Alder, *The Measure of All Things: The Seven-Year Odyssey and Hidden Error That Transformed the World*, Simon and Schuster, New York 2002.

those objects, then called nebulae, that had entered his life while he was searching for Halley's star.

With the discovery of the Crab Nebula in 1758, Messier had started a black book of diffuse celestial objects to prevent the feathered star hunters from being deceived by false sirens. Gradually, the *Catalog des nébuleuses et des amas d'étoiles* (Catalog of nebulae and stellar masses) took shape, enriched with the already known comet-like objects whose real existence Messier had already established. The first edition of this personal black-list,[49] published in 1771, contained 45 nebulae, which became 103 in the last edition, compiled with the help of the usual Méchain, the only colleague of whom Charles was not jealous. Even today, astronomers like to refer to the objects on this list.[50] They are identified by the letter "M" (for Messier), followed by the catalog entry number. For example, the Crab Nebula is M1; M8 is the beautiful Lagoon Nebula in Sagittarius; M13 is a rich cluster of old stars in Hercules; and M31 is the large spiral galaxy in Andromeda, a companion of the Milky Way in the Local Group. These are imposing and picturesque clouds of gas, clusters of stars of all ages, and galaxies with dimensions and distances that make them visible as tangled wisps of light.

Unknowingly, Messier had opened the door to the study of the deep sky, both galactic and extragalactic, which until then had essentially been a passive frame around the Solar System. The first to grasp the message of Messier's *catalog* was William Herschel, an amateur astronomer who had come to prominence in 1781 when he used a homemade reflector to discover another wandering star, the planet later named Uranus. Science is indebted to this German-born musician for two of the most exciting and fruitful campaigns to explore the universe. Two adventures that still attract the best young people for a journey beyond the Pillars of Hercules of our ignorance, out there in the deep sky, in search of origins.

---

[49] K.G. Jones, *Messier's Nebulae and Star Clusters*, CUP, Cambridge 1991.

[50] After revisions of Messier and Méchain's papers to correct an error and incorporate the two astronomers' latest discoveries, the total number of objects in Messier's *catalog* has reached 110. The last entry, M110, added in 1967, is an elliptical dwarf galaxy, companion of the spiral galaxy M31 in Andromeda.

Chapter 8

Beyond the Pillars of Hercules: Herschel and the new planets

> I would rather be a superb meteor,
> every atom of me in magnificent glow,
> than a sleepy and permanent planet.
>
> Jack London, *The Bulletin,*
> *San Francisco (CA)*

> A man of genius makes no mistakes.
> His errors are volitional and
> are the portals of discovery.
>
> James Joyce, *Ulysses*

On the night of March 13, 1781, as was his custom when the weather permitted, Frederick William Herschel[1] stood in the open air on the street in front of his house. Indifferent to the cold, he scanned the sky with a telescope he had built with his own hands: a magnificent Newtonian reflector with an objective made of speculum. German by birth, education, and mentality, William was 42 years old, a relatively advanced age for those times, and for the past fifteen years he had lived in Bath, a pleasant resort town in southwest England, famous for its natural thermal springs. Bath, as its name suggests, was a health resort, a meeting place for high society and the newly rich, born of the bloodthirsty post-Elizabethan commercial imperialism and a ruthless industrial revolution. A *unicum* in a land without volcanic manifestations, known and appreciated since the time of the Roman conquest of Brittany. Those who could afford it went there to heal the body and the spirit, to weave and strengthen relationships, to know and be known, to gossip while sipping Indian tea with a view lost on the soft green hills, perpetually damp, to let time pass quietly amidst parties,

---

[1] M. Hoskin, *Discoverers of the Universe: William and Caroline Herschel*, Princeton University Press, Princeton 2011; biography by the leading scholar of William Herschel.

ablutions, walks, *bon ton*, loves, betrayals, readings, hypocrisies, and even good music offered as a pastime to vacationers. It is with the music[2] that William made a living in this little paradise for a few. He performed as an organist at the Octagon Chapel, a fashionable private church, and as a conductor at the Assembly Rooms, a space devoted to listening and dancing. Over the years he had carved out a role and a reputation that could be compared, with a hint of exaggeration, to that of Herbert von Karajan in Salzburg.

A tireless worker, Herschel supplemented his income by composing music[3] and teaching it to hordes of wealthy, petulant, and desperately untalented students. A minor ordeal for such an artist, it nevertheless secured him bourgeois status and decent financial resources to help his family. In return, he allowed himself to devote every spare moment and remaining energy to cultivating his latest passion, astronomy. On the second floor of his house, he manufactured metal mirrors and glass eyepiece lenses with which he assembled telescopes of excellent optical quality. As we know, hand grinding and polishing requires the operator to perform harmonious and regular movements; and who better to fulfill this role than a musician? Some instruments were sold to cover expenses, but William kept the best ones for himself to observe the heavens. That night he used his favorite telescope, a Newtonian reflector with a spherical mirror of 6 inches and 1/2 and a focal length of 7 feet, mounted on a mahogany stand and moved in elevation by a system of cranks, pulleys, and tie rods, much like the sail of a racing boat.[4] The azimuth movements, obtained by driving a worm screw perpendicularly to the tube of the instrument, provided a modest angular excursion, sufficient for the astronomer to have a look and perhaps make some measurements, before the target, dragged by the rotation of the celestial sphere, relentlessly disappeared from the field of view. An obstacle that diminished with the polar distance of the target, as is soon realized by noticing that at the equator a celestial object describes a maximum circle in one day, while at the pole it remains essentially stationary.

---

[2]V. Duckles, *Sir William Herschel as Composer*, in "Publications of the Astronomical Society of the Pacific", 74, 436, 1962, pp. 55–59.

[3]In all, 18 symphonies for small and 6 for large orchestra, 12 concertos for oboe, violin, and viola, 2 concertos for organ, 6 sonatas for violin, cello, and harpsichord; recordings of some compositions can be found on the Web.

[4]H.C. King, *The History of the Telescope*, Dover, New York 2012, p. 135.

While surveying a region between Taurus and Gemini, where his astronomical maps showed no significant objects, Herschel's attention was suddenly drawn to a bright source surrounded by a faint haze. Now a consummate observer, he sensed that this was not an ordinary star. But how could he be sure? From consolidated experience he knew that the sizes of stellar images are not very sensitive to the magnification used to observe them: a point-like source remains so no matter how one looks at it.[5] The astronomer-musician had excellent eyepieces, so he went from the usual 230 magnifications to over 900. He hoped in his heart that the mysterious object would give a different response than the countless stars that dot the vault of the sky; and so it did. The image gradually widened as the magnification increased. "*Wunderbar!*" (Wanderful) he must have exclaimed, if it is true that in moments of maximum excitement a person thinks and speaks in his native tongue. His sister Carolina later reported Wilhelm's comment in German: "*Hier ist wahrhafting ein Loch in Himmel*" (There really is a hole in the sky), uttered a moment before he noticed the presence of Uranus.[6]

William, christened Friedrich Wilhelm, was born in 1738 in Hanover, the capital of the electorate of the same name in northern Germany. His father, Isaak, a talented musician, supported the family by playing the oboe in the royal band. Unlike his wife, Anna Ilse, who could barely read or write, he possessed ingenuity and curiosity far beyond the social role that fate had assigned him. It was he who educated his sons[7] by getting them passionate about music, as was customary in German families, all the more so if they had some Jewish blood in their veins (for example, Max Planck was a good pianist and Albert Einstein a talented violinist). In the barracks where they lived, the Herschels would play together late at night or talk about Newton and Leibniz. As soon as 14-year-old Wilhelm finished school, he was drafted into the Hanoverian Guard as an oboist: a humble and poorly paid family occupation (his older brother Jakob had

---

[5]Large distances relegate stars below the resolving power of the instruments used to view them. However, sidereal images do not appear point-like as they should, mainly due to atmospheric turbulence (seeing). Therefore, the use of more powerful eyepieces has no real effect, except to widen the seeing disk, i.e. the image on which the star's light is redistributed.

[6]J.R. Hind, *Sir William Herschel*, in "Nature", 23, 1881, pp. 429–431.

[7]Jacob, Wilhelm, Alexander, and Dietrich. Wilhelm was taught to play the violin as soon as he could hold a small one specially made for him. Taken from E. Winterburn, *Philomaths, Herschel, and the myth of the self-taught man*, in "The Royal Society Journal of the History of Science", Vol. 68, 2014, pp. 207–225.

been drafted before him) that subjected musicians to harsh martial discipline and exposed them to the same hardships as soldiers, though with less risk — or so the young man believed, until he was called to take part in military operations in defense of his homeland. This happened in 1756. France's King Louis XV and his allies had broken the delay by promoting a war against the Anglo-Prussian coalition that would last seven years.[8] The stakes were high: hegemony in Europe and the New World. The *Kurfürstentum Hannover* (Electorate of Hannover) stood between the two contenders in this global conflict — the real "First World War" according to Winston Churchill[9] — and immediately became the theatre of the French's first successful military exploits.

For a time, Wilhelm honored his warrior name ("Helm of Will" in German) by enduring the dangers and hardships of a soldier's life. He participated without injury in the disastrous Battle of Hastenbeck in 1757, where the French of Louis *le Bien-Aimé* (the Beloved) triumphed under the illusion that they had the final victory in their grasp. But after a week spent in a cold and damp trench, he unilaterally decided to put an end to the experience. With his father's blessing, he and his brother Jacob fled to Hamburg, and from there they set sail for England with the idea of making a living through music. Was it an act of desertion? Perhaps,[10] because at that moment the nineteen-year-old musician wore the uniform, albeit without a combat role, and his homeland was at war and, worse, in great distress. It does not matter that the Duchy of Hanover was then part of the Kingdom of England, due to the dual sovereignty of George III, and that Wilhelm was thus hiding within the extended borders of the nation. Anglo-Saxon historians, however, have often sought with enviable pragmatism to gloss over this youthful sin of the scientist who was to become the British icon of celestial exploration, to be pitted against the glories of the French Enlightenment in the centuries-long tug-of-war between the two great hegemonic nations of Europe.[11]

The fleeing German soldier was already familiar with England. Two years earlier, along with his father and his brother Jakob, he had followed his unit when it was sent across the Channel to Brithish Island to reinforce

---

[8]D.A. Baugh, *The Global Seven Years War, 1754–1763: Britain and France in a Great Power Contest*, Routledge, London 2011.

[9]H.V. Bowen, *War and British Society 1688–1815*, CUP, Cambridge 1998, p. 7.

[10]M. Hoskin, *Was William Herschel a deserter?*, in "Journal for the History of Astronomy", 35, 3, 2004, pp. 356–358.

[11]Cf. G. Hunt, P. Moore, *Atlas of Uranus*, CUP, Cambridge 1989.

defenses against a possible French invasion. The stay had lasted several months. Wilhelm took the opportunity to perfect his knowledge of the language and to familiarize himself with a less harsh and polarized society than the one that the looming shadow of Frederick II was building in Germany. The warnings of the Seven Years' War then brought the Herschels back home. Before leaving England, the young musician carefully tucked a copy of John Locke's *An Essay Concerning Human Understanding* into his backpack as evidence of his precocious intellectual curiosity. Perhaps, looking at the trail left on the water by the boat that brought him back to the continent, he was already planning to return soon to see the white cliffs of Dover again, and this time to stay. But he could not have imagined that in a few years he would become the most famous English scientist of his time.

For nine years after he had come back to England, Wilhelm, who now called himself William, wandered the island, struggling to make ends meet as an occasional conductor, copyist, composer, and music teacher: first in London with Jacob, then on his own in the north of the country, where competition from other musicians was less fierce. After many wanderings, he finally found a stable and relatively comfortable arrangement in Bath in 1766. He was later joined by his younger brother Alexander, also a musician but with a marked talent for mechanics, and in 1772 by his sister Caroline Lucretia,[12] 12 years his junior, to whom William was deeply attached. Short in stature due to the effects of typhoid fever contracted at the age of 10, the girl had received a mediocre education. Old Isaak had not had time to train her as he had his other children. When he died, he left his daughter in dire straits, with virtually no material resources and in the hands of an ignorant and chauvinistic mother. William's remittances had come in from time to time to ease the existence of this kind of Cinderella. Fortunately, she had a fair voice. So on arrival in Bath, she was given a crash course in singing and involved in the family's music business. Determined and indefatigable like her beloved brother, she took care of the household and performed as a lyric soprano; and above all, she fully shared William's new totalizing passion for astronomy, which had blossomed in the context of a thousand intellectual itches. The Herschel brothers had read James Ferguson's best-selling *Astronomy explained upon Sir Isaac Newton's*

---

[12] Cf. M.B. Ogilvie, *Searching the Stars: The Story of Caroline Herschel*, The History Press, Cheltenham 2011.

Principles, a book "galeotto"[13] that had awakened in them the desire to discover firsthand, without any other means than a telescope, the celestial wonders illustrated by the Scottish astronomer.

William and Carolina first experimented with some off-the-shelf instruments purchased in London. They also tried the route of assembling optical components that were commercially available. The girl, who had more time than her very busy brother, had taken on the task of making the cardboard tubes that would hold the lenses for telescopes with focal lengths up to 6 m. They also experimented with a Gregorian mounted reflector with a focal length of 60 cm (2 feet), borrowed directly from Bath. The results were disappointing. A qualitative leap was needed, but the economic resources that the Herschel brothers could devote to this new hobby were not comparable to their ambitions. It was, after all, a matter of satisfying a whim, albeit one nobly motivated by a formidable curiosity. Thus, the two convinced themselves that the only way to obtain a sufficiently powerful telescope was to do everything necessary at home. Completely unfamiliar with the subject, they followed the instructions given in *A Compleat System of Optick*, a two-volume work considered the bible of astronomical instrument makers, written by Robert Smith, a prolific professor of physics at Cambridge. William had come across this book almost by accident. He had read a treatise on music theory by Smith for professional reasons and liked it so much that he decided to buy the optic manual as well.

Now the Herschels had to acquire the equipment to carry out their precision bricolage.[14] They could have specialized in grinding objective lenses, as the Huygens brothers had done. But fate had another option in store, for, as Cicero argued, man proceeds with virtue as his guide and fortune as his partner.[15] A craftsman living in Bath, who years before had worked in the field of optical mirrors, had decided to give away his equipment:

---

[13]Expression taken from the episode of Paolo Malatesta and Francesca da Rimini in Dante's *Inferno* (VI, vv. 127–138): «One day, / For our delight, we read of Lancelot, / How him love thrall'd. Alone we were, and no / Suspicion near us. Oft-times by that reading / Our eyes were drawn together, and the hue / Fled from our alter'd cheek. But at one point / Alone we fell. When of that smile we read, / The wished smile so raptorously kiss'd / By one so deep in love, then he, who ne'er / From me shall separate, at once my lips / All trembling kiss'd. The book and writer both / Were love's purveyors. In its leaves that day / We read no more.» (*The Harvard Classics: The Divine Comedy of Dante Alighieri*, translated by H.F. Cary, 1909–14).

[14]E. Winterburn, *Philomaths, Herschel, and the Myth of the Self-Taught Man*, in "Notes and Records", 68, 3, 2014, pp. 207–225.

[15]«*Virtute duce, comite fortuna*»; M.T. Cicero, *Epistulae ad familiares*, X, 3.

molds, tools, blanks, and whatnot, for a piece of bread. William didn't let that happen again. He bought everything, unwittingly committing himself and his family to Newton's sponsored solution of reflecting mirrors. It was the high road to realizing his latest dream: to look deeper into the sky than any other man with a keen and powerful eye. So in the late summer of 1773, as the swarm of holidaymakers left Bath, the Herschel house was transformed into a workshop. Everything from turning and woodworking to grinding glass for eyepieces was made there, even in the room that William had previously used to receive his aristocratic and wealthy students for music lessons. Everything was noise, dust, fumes, and a coming and going of people so different from the London gentlemen. Probably the haughty shopkeepers of Bath, with British reserve, wondered what the three Germans were doing; and perhaps some one, more aware of the reason for all the fuss, passing by the house, must have shaken his head and regretted that the three industrious foreigners were wasting their strength and wealth on a crazy infatuation. "God save us from the artists!". But the Herschels were too caught up in their fever to care.

It took months of hard work. By 1774, William was able to put his eye into the eyepiece of a 168 cm Gregorian reflector and marvel at the observation of the Great Orion Nebula with an instrument made entirely at home, largely by the hands of the Herschels themselves. It was an achievement that opened up intriguing prospects on the commercial front as well. What had begun as an intellectual pastime could become a profitable industrial activity. No sooner said than done. During the day, whenever his musical engagements allowed, William transformed himself, somewhat as we read in Machiavelli's 1513 account to his friend Francesco Vettori of his days of exile on a mountain farm: «When evening comes, I return to my house and enter my study; and at the door I take off the clothes of the day, covered with mud and dust, and put on royal and courtly garments; and dressed accordingly, I enter the old courts of old men, where, received by them with affection, I feed on that food which is mine alone, and for which I was born».[16] Unlike the great Florentine, the German, once at home, shed the robes of the elegant conductor to wear those of the craftsman. Then, in a laboratory on the ground floor with his brother Alexander, he put his hand to speculum castings, ground and polished the optics, and assembled them into the frames designed for his own use and to meet the demands

---

[16]N. Machiavelli, *Lettere a Francesco Vettori e Francesco Guicciardini*, G. Inglese (ed.), Rizzoli, Milan 1996, pp. 194–195 (in Italian).

of a growing clientele. In addition to new projects, he was constantly busy regenerating the surfaces of the mirrors, which periodically deteriorated due to oxidation. It was a hectic and tireless job that did not even allow him to eat, so much so that Carolina sometimes had to feed him while he continued his work. To get an idea of this activity, over a period of twenty years William and Alexander would produce no less than 200 mirrors of 210 cm focal length, 150 of 3 m, and about 80 of 6 m: numbers that still evoke admiring awe today.[17]

On clear nights he would scan the skies. In Bath, his reputation as an astronomer almost surpassed his fame as a musician. During breaks, his students began to question him about the sky, even asking to look at the Moon and stars through a telescope. Astronomical practice had become an obsession. When a concert forced him to go to the theater on a clear evening, William would wait for the interval to rush home and take a look at the firmament, not wanting to miss any opportunity. Like a good German, meticulous, orderly, and efficient, he had begun a systematic and profound search for pairs of stars so close together that, in the absence of a powerful telescope, they appeared to be a single source. Today we call them binary systems,[18] using the terminology coined by Herschel himself. He was intent on this task, his eye glued to the eyepiece of the new instrument he had built for himself, the 7-foot Newtonian reflector, when he came upon the strange star. After dismissing the possibility that it was a point source, he thought he had discovered a nebulous object or a comet.[19] In the second case, it would have to move between the fixed stars. After 40 days of continuous observation, William became convinced that the object was indeed moving, albeit at a slow pace. This was clear evidence that it belonged to the Solar System. A comet,[20] then! What else? The musician-astronomer felt ready

---

[17]A. Maurer, E.G. Forbes, *William Herschel's Astronomical Telescopes*, in "Journal of the British Astronomical Association", Vol. 81, p. 284–291.

[18]In 1802 Herschel clarified the meaning of the terms: «If, on the contrary, two stars should really be situated very near each other, and at the same time so far insulated as not to be materially affected by the attractions of neighboring stars, they will then compose a separate system, and remain united by the bond of their own mutual gravitation towards each other. This should be called a real double star; and any two stars that are thus mutually connected, form the binary sidereal system which we are now to consider» (*The Scientific Papers of Sir William Herschel*, J.L.E. Dreyer (ed.), CUP, Cambridge 2013, Vol. 2, p. 201).

[19]R. Holmes, *The Age of Wonder*, Vintage Books, New York 2008, pp. 96–98.

[20]S. Schaffer, *Uranus and the Establishment of Herschel's Astronomy*, in "Journal for the History of Astronomy", 12, 1981, pp. 11–26.

to give news to the scientific world and especially to the Royal Society. The discovery of a feathered star, even if it was still featherless, was quite a coup for an amateur, but nothing really sensational in itself, especially after the bountiful harvest of Messier and Méchain in Paris.

The first person to suspect that the source found by the Anglo-German musician could be something far more intriguing than a comet was the director of the Royal Greenwich Observatory, Nevil Maskelyne, with whom Herschel had long been on friendly terms. The two met at meetings of the newly formed Philosophical Society in Bath, modeled on the Royal Society. The distinguished scientist attended when he was on vacation. The amateur astronomer, instead, was a regular attendee since the influential Sir William Watson, physician and naturalist, had sponsored his co-option. What had made Maskelyne suspicious about Herschel's star was that, despite its remarkable brightness, it lacked all the typical characteristics of a comet, such as a coma and a tail. Messier, alerted by a letter from Herschel, expressed the same opinion.[21] What if, instead of a feathery star — the Astronomer Royal ventured to speculate — it was a planet in a circular orbit around the Sun? This bold conjecture was soon confirmed by the Finnish Anders Johan Lexell, a mathematician working at the court of Tsar Alexander I in St. Petersburg, who was the first to calculate the orbit of the new object; and then by Johann Elert Bode, another acrobat of orbital calculations, at the time a professor at the Berlin Academy of Sciences. Finally, suspicion became certainty: the star was indeed another planet. It was Herschel himself, then only "Mr. William Herschel, of Bath", who officially informed the Royal Society in November 1782. Fortunately, he was a showman, used to publicity, and so he was not overwhelmed by the excitement of standing before the best minds in Britain, duly dazzled. It was also an opportunity to explain that, in his opinion, the discovery was not the result of chance, but a necessary consequence of his systematic

---

[21] «*Je suis toujours étonné de cette Comète qui ne porte avec elle aucun Caractère distintif des Comètes, et qu'elle ne ressemble à aucune de celles que j'ai observées, qui sont au nombre de dix-huit*» (I am continually astonished by this Comet, which has no characteristic Feature of Comets, and that it does not resemble any of those I have observed, which have the number eighteen). Quoted by Schaffer, *Uranus and the Establishment of Herschel's Astronomy*, cit. It is interesting to note Messier's comment about the number of comets he discovered. Perhaps he wanted to demonstrate his expertise in the field, but also to point out that if Herschel's object had been a comet, it would have been only one of many. Another small example of the constant competition between the French and the English in every field. It should be added, however, that Messier later congratulated Herschel on the discovery of the object when its nature was still unclear.

exploration of the heavens: «I had gradually perused the great volume of the Author of Nature and was now come to the page which contained a seventh Planet».[22] In short, "I crossed the Pillars of Hercules thanks to my perseverance and the greatness of my scientific project". Not bad for an outsider.

It is not difficult to imagine the excitement that this astonishing news caused in every corner of the world. Humanity, which had always counted five wandering stars in the sky, in addition to the Sun and the Moon, had to take note of the existence of another, more distant than all the others: a sixth planet, rising from the darkness of the night, which now claimed citizenship in mythology, astrology, and science. The newspapers, which for less than a century had kept the bourgeoisie, born of the Industrial Revolution, up to date, reported the news quickly and widely. The announcement did not, however, catch astronomers completely off guard. For several centuries, the possibility of unknown tenants roaming the Solar System had been the subject of extravagant thought and speculation. The dances had been opened by the usual Kepler back in the days when he pondered the cosmos in Upper Austria, where Melanchthon had sent him to "educate the new believers". The imaginative German scientist had noticed that the sequence of ratios between the heliocentric distances of successive planets showed a discontinuity between Mars and Jupiter (using modern values of planetary distances: $\Delta_i/\Delta_{i+1} = 0.57, 0.70, 0.67, 0.30, 0.56$). A leap that he considered disproportionate and incompatible with the mathematical harmony of divine creation (which would later become musical harmony in *Harmonices Mundi*). One can almost hear him obsessively repeating, as before him Aristotle and after him Leibnitz and Linnaeus: *Natura non facit saltus* (Nature does not make leaps).

Kepler didn't take long to untie the knot. It was enough, he argued, to fill the mysterious gap between Mars and Jupiter with a wandering star that had not yet been observed because of its small size and scarce brightness. But after only a year he abandoned this hypothesis when he became convinced that the distances between successive planets were the consequence of the divine architect's choice to use Platonic solids as dividers (see p. 73). Good thing men of genius do not suffer from changing their minds! Newton also thought that the gap deserved a justification,

---

[22]S. Schaffer, *Uranus and the Establishment of Herschel's Astronomy*, cit.

which he interpreted as the divine will to keep smaller bodies out of the clutches of massive Jupiter.[23] The Newtonian argument was then taken up again by Immanuel Kant in 1755, this time without an overemphasis on God's protective role; but the eplanation was not convincing enough to dampen the speculations about missing bodies almost everywhere in the Solar System. There were those who merely alluded to the question by using the expression "known planets" to refer to historical ones; those who emphasized the excessive jump between Saturn and the apogee of comets; and those who explicitly speculated about an elusive planetary object between Mars and Jupiter, perhaps no longer existing because it had been destroyed by an encounter with a comet, as we will say in a moment.

Thanks to his discovery, Herschel suddenly became a famous man,[24] and England was ready to adopt and protect him to prevent his primacy from being shared with others. In fact, it could be argued that the Anglo-German astronomer had indeed seen and reported the mysterious object, but like others before him, starting with Flamsteed, had not realized that it was a planet. More Christopher Columbus than Amerigo Vespucci! Fortunately for him, however, the decision to make him a hero was functional to the host country's image; and he certainly deserved it.

In the space of twenty years, thanks to his extraordinary ingenuity and iron will, he had risen from the barracks of the Hanoverian garrison to become a guest of honor in the scientific salons of the most advanced and powerful nation in the world. In addition to the prestigious and coveted membership in the Royal Society, medals and awards poured in, mainly due to the interest of his friend Watson. William and then his son John, the first and fourth presidents of the Royal Astronomical Society, were both awarded the Royal Guelph Order of Hanover,[25] with the title of Sir, which had no value in Britain, being associated with a foreign honor. George III, the English monarch, whose family, like that of the Herschels, came from

---

[23] M. Hoskin, *Bode's Law and the Discovery of Ceres*, in *"Physics of Solar and Stellar Coronae, G.S. Vaiana Memorial Symposium"*, Kluwer Academic Publishers, Dordrecht 1992, pp. 35–46.

[24] Cf. J.R. Hind, *Sir William Herschel*, in "Nature", XXIII, 593, 1881, pp. 429–431, for an amusing digression on the distortion of Herschel's name in the various printed sources that reported the discovery of Uranus in Europe.

[25] An order of chivalry established in 1815, a few weeks before the Battle of Waterloo, by the Prince Regent, later King George IV. It is named after the House of Guelph, of which the Hanoverians were a branch.

Hanover,[26] forgave the desertion of his overseas subject and called him to the palace to show his family the sky. Satisfied and well advised, he created for William the new position of Court Astronomer (not to be confused with the title of Astronomer Royal, which rightfully belonged to the Director of the Greenwich Observatory), with a modest pension of 200 pounds a year, comparable to four year's wages of a maid, but with no obligation except to remain at the disposal of the Crown. Not yet sufficiently experienced in the arts of the court, William hoped that the king would reward him for his discovery *motu proprio*, that is, on his own initiative. But George III could not risk that his gracious move in favor of the astronomer would be rejected by the recipient. The stalemate was broken when Herschel, duly guided by clever friends, addressed a petition to the king. Caroline was also granted a small allowance to assist her brother: an extraordinary achievement for a woman in an age of strong gender discrimination. «Never bought Monarch honor so cheap!», Watson would laconically comment.[27]

Herschel was also encouraged by Maskelyne to name his planet. He chose *Georgium Sidus* (star of George), borrowing it from the *Iulium Sidus* mentioned by the Latin poet Propertius and attributed by Horace (Ode XII, v. 47) to the blood-red comet that appeared in the summer of 44 B.C., after the assassination of Julius Caesar. William also found the argument to embellish this courtly gesture of his with an extravagant *excusatio non petita*. He explained that the name of the new celestial body should evoke the memory of the time when the discovery of the new planet had been made. By naming it after King George, everyone would know that the planet had been found during his reign. He was confident that he would fare better than Galilei with the Medicean Moons, but he was sorely mistaken. The French, seething with envy and particularly ill-disposed toward a sovereign who had just defeated them in the Seven Years' War, suggested through Lalande that the star be named after Herschel himself. Indeed, this was the practice for comets, which were christened after their discoverers. Better a German immigrant, they thought in Paris, than the hated and loathsome King of England. There were those who, to make lemonade

---

[26] George Louis of Hanover, founder of the royal house that continues today under the name changed in 1917 to Windsor, ascended the throne at the age of 54 as George I of Great Britain. Elector of Hanover in the Holy Roman Empire from 1698 and King of Great Britain and Ireland from 1714 until his death, he succeeded Queen Anne. He was born in Germany, loved to speak German, and frequently visited his beloved German lands.

[27] M.A. Hoskin, *The Herschel of Hannover*, CUP, Cambridge 2007, p. 53.

out of lemons, ventured the name of Neptune, preserving the tradition of classical mythology and at the same time alluding to Albion's power over the seas.

In the end, it won the Solomonic proposal of the German Bode[28] to resort to the divinity of the sky, Uranus, in order to give continuity to the divine genealogy which saw in Jupiter the grandson, in Saturn (Cronus) the father, and in Uranus the grandfather and progenitor. Gradually, this erudite opinion became so widespread that when it came time to name a new chemical element found in a mineral called pitchblende, the discoverer, who was a colleague of Bode's at the Royal Academy in Berlin, chose uranium. In fact, the German professor had carved out a leading role in the history of the new planet. After the discovery of Uranus, he had devoted himself passionately to study it. Among other things, he had found that the object, just fainter than those accessible to the naked eye, had been seen a number of times before, but had never been identified. The last person to do it was Flamsteed, who mistook the faint source for a fixed star. To tell this suggestive story, we must temporarily leave the Herschel epic, which we will resume in the next chapter.

In 1772, the 25-year-old Bode had published the updated second edition of *Anleitung zur Kentniss des Gestirnten Himmels* (Guide to the Knowledge of the Starry Sky), a popular book that would make his fortune. He was still living in his native Hamburg, a city of sailors, merchants, and brewers. A fortunate coincidence had allowed him to leave his father's business to give free rein to his desire for science. He read, wrote, and counted tirelessly, in spite of a precarious vision that depended on his left eye alone. The young man had sent a copy of his astronomical manual to Johann Heinrich Lambert, a distinguished member of the Berlin Academy and a pioneer of non-Euclidean geometry. Lambert was so impressed by this work that he offered Bode a teaching position: a fair way to recruit professors, if those who decide are honest, competent, and wise people, and if their future fame is tied to the quality of their choices. Instead, "in a state overwhelmed by corruption, laws multiply", bitterly judged Tacitus (*Annals*, III), to act as a smokescreen for embezzlers; and so no co-option, but very legal and twisted concourses like fig leaves for the slalom of cheaters!

In a footnote to his *Anleitung*, Bode had pointed out the existence of a correspondence between the distances of the planets from the Sun

---

[28] O. Gingerich, *The Naming of Uranus and Neptune*, in "Astronomical Society of the Pacific Leaflets", 8, 352, 1958, pp. 9–16.

and the terms of a geometric progression of natural numbers generated by a formula containing only an origin and a scale.[29] The intriguing coincidence had already been announced six years earlier by the Prussian Johann Daniel Titius, again without any particular emphasis, in a note added to the German translation of the *Contemplation de la Nature* by the Swiss naturalist and philosopher Charles Bonnet: the most appropriate context to host another amazing and mysterious example of mathematical harmony that would surely have sent Pythagoras and Kepler into ecstasy. Bode had heard of this numerical coincidence, but refrained from mentioning it in the first editions of his book; we do not know why. Only later he would acknowledge the role of his compatriot in formulating the law that now bears both of their names.

The simplicity and beauty of the empirical correlation between orbital radii and integers, produced by a single formula, made one suspect that rather than being a mere coincidence, it might have some physical meaning. But there was a flaw in the sequence of correspondences between integers and planets. The fifth element in the sequence, corresponding to a heliocentric distance between those of Mars and Jupiter, did not match any known celestial body. The inconsistency could be remedied with *ad hoc* hypotheses, but these ended up complicating the knot rather than untangling it. In fact, they introduced discontinuities or evolutionary phenomena, such as the existence of invisible dwarf planets or the traumatic disappearance of major bodies, that were difficult to digest in the absence of evidence. Just born, the law of Titius-Bode was already facing a dead end. It is at this point in story that the discovery of Uranus came into play, creating one of those extraordinary crossroads that fate occasionally composes for the joy and despair of mankind. As soon as astronomers were able to estimate the distance of the new planet from the Sun, they realized that it occupied exactly the position foreseen by the Titius-Bode law for a body orbiting the Sun externally to Saturn.

The coincidence was astonishing and removed from the ghetto of pure numerology the conjecture that there was a ghost planet between Mars and Jupiter. However, in order to turn a suspicion into a certainty, the phantom body had to be found. But how? The heavens seemed too big a haystack to

---

[29]The original formula is: $a_k = (0.4 + k)/10$, where a_k (in AU) is the size of the planetary orbit and k belongs to the geometric progression with the common ratio 3: $k = 0, 3, 6, 12, 24, 48, \ldots$; cf. M. Nieto, *The Titius-Bode Law of Planetary Distances: Its History and Theory*, Pergamon Press, Oxford 1972.

search blindly for the classical needle, and curiosity was too pressing to rely on chance alone. Like any great enterprise, the matter lay dormant, waiting for a leader to emerge. Shortly before the turn of the century, a new player appeared on the scene, a Hungarian baron born in Bratislava (Pressburg, then the capital of the Kingdom of Hungary) in the mid-18th century. Since 1786 he had been the director of the observatory that Ernest II, Duke of Saxe-Gotha-Altenburg, had built the previous year in the town of Gotha, Thuringia. It must be said that the Enlightenment had trained crowned heads for good, just as the exercise of power would teach the post-industrial bourgeoisie to do evil: but such is the world. After a reckless youth and training as a military engineer, Franz Xaver von Zach[30] had toured Europe as a tutor in the retinue of the Duke of Saxony, and during his travels he had been able to meet the most famous astronomers, such as Lalande, now director of the Paris Observatory, the famous mathematician Pierre-Simon de Laplace, and Herschel himself. From them, Francis Xavier had learned the craft of the scientist. He possessed great organizational skills, which proved invaluable when he decided to plan a full-scale crusade to find the missing planet between those of the "God of War" and the "King of Olympus". Von Zach was convinced that the elusive wandering star shared the ability of all known planets except Mercury to travel within the zodiacal belt. This working hypothesis greatly narrowed the hunting ground, which, however, remained a sea too vast for a single fisherman. In fact, the Baron's strategy was to explore the sky and catalog the position and brightness of all the stars present in the selected area, up to the observable limits of the instrument used. On this map of stationary targets, it would have been relatively easy to identify the moving stars, all potential candidates to play the role of the sought-after planet. The by-products would have been an impressive catalog of stars and a predictably rich haul of photometric and astrometric transients (i.e., variable sources), comets, and objects with high proper motions.

Once the rules of engagement were established, troops were needed. Indeed, since 1787 von Zach had been trying to compile a catalog of the stars of the zodiac on his own, but the poor results had convinced him of the need for a large team of collaborators. The turning point came just before the end of the Age of Enlightenment, when the Duke of Saxony sponsored a brainstorming session in Gotha on the future of astronomy. Lalande, among

---

[30]D. Vargha, M. Vargha, *Franz Xaver von Zach (1754–1832): His Life and Times*, Konkoly Observatory of the Hungarian Academy of Sciences, Budapest 2005.

others, attended the meeting, invited despite the fact that his political radicalism made him capable of far more dangerous revolutions than those concerning the firmament. The event appeared to be a good opportunity to launch the idea of a celestial crusade. At first the response to the call was lukewarm, but after a few months Baron von Zach returned to the attack on the occasion of a small meeting of German astronomers at the Lilienthal Observatory in Lower Saxony, near the city of Bremen. Among the participants were the founder of the observatory, Johann Schröter, his assistant Karl Harding, and Heinrich Olbers, a medical doctor who had become a professional astronomer. This handful of scientists formed an association, the *Vereinigte Astronomische Gesellschaft* (United Astronomical Society) or, for convenience, the Lilienthal Society, jokingly renamed *Himmelspolizei* (Sky Police). To make effective patrolling possible, the zodiacal belt was divided into 24 tiles of 15° per side, as many as the hours in a great circle, to be assigned to an equal number of astronomers with certified experience and the availability of an excellent instrument for observations.

A list of possible "celestial policemen" was drawn up. Half were Germans, including von Zach himself and, of course, Schröter, who had the most powerful telescope in the world, the *Riesenteleskop*: a reflector with an aperture of 51 cm and a focal length of 8.3 m. The others had been fished from the rest of Europe. The underlying idea was to unite in a common scientific enterprise the forces scattered across a continent torn apart by ancient and enduring rivalries and bloody wars. An Enlightenment ideal that would be interpreted schizophrenically by Germany for the next century and a half, and parochially by contemporary England. The list of "conscripts" also included the famous names of Messier, Lalande, and Herschel, and two astronomers from Italy, a country that for most was merely a geographical expression:[31] Barnaba Oriani, director of the Brera Observatory in Milan, and Giuseppe Piazzi, who was in charge of a small observatory in Palermo, which he had built and equipped with a magnificent meridian circle. The celestial policemen coveted the man, his instrument, and above all the beautiful southern sky, which was far more attractive than the foggy German sites.[32]

---

[31] In an 1847 note, the conservative Austrian statesman and diplomat Prince Klemens von Metternich wrote the famous and controversial sentence: «Italy is a geographical expression». The phrase was soon interpreted by Italian patriots in a derogatory sense in order to stir up anti-Austrian sentiments among the population.

[32] C. Cunningham, *Discovery of the First Asteroid, Ceres: Historical Studies in Asteroid Research*, Springer, Cham 2016.

Giuseppe Piazzi was no outsider. In the ten years since fate had transformed him from a professor of mathematics and theology into an astronomer, he had built an efficient observatory in Palermo and even managed to equip it with a magnificent altazimuthal circle. The instrument consisted of two graduated circles, a horizontal one with a diameter of 90 cm and a vertical one of 150 cm, connected to a telescope with an aperture of 7.5 cm. It was the masterpiece that the moody but very skilled English maker Jesse Ramsden, known for his irritating inconstancy, had completed under the daily pressure of Piazzi himself, in the two years during which the priest,[33] assuming the role of director of a still non-existent specola, stayed in London to train. Once moved to Sicily, not without complications from British customs that hindered the transfer abroad of high technology products, the instrument had been installed on the roofs of the oldest royal residence in Europe, the "Palazzo dei Normanni", seat of the Kings of Sicily during the Norman rule of the island. The indefatigable Cleric Regular of the Theatine Order had used it with a constancy bordering on self-sacrifice, producing a valuable catalog of star positions that would later earn him an award from the French Astronomical Society. Piazzi had blocked Ramsden's circle so that it could only move along the meridian, like Tycho's great wall circle. (cf. chapter 4). This was a precautionary measure to ensure that the transit epoch measurements always had the same reference (not guaranteed if the instrument was moved and then re-positioned on the meridian). With this restriction, the tracking of a star was limited to no more than two minutes around the transit position. Too small an angle to guarantee that the observer, under pressure, would not make mistakes, which for the strict Theatine would have been so many deadly sins. For this reason, Don Giuseppe forced himself to repeat the determination of the position of each star several times, during more than one night.

On January 1st, 1801, with the help of his assistant and future director Niccolò Cacciatore, he was observing again some of the fields explored the night before, when he realized that a star of magnitude 8 and the color of Jupiter gave him a different position than the one he had recorded. He immediately thought he had made a mistake and, probably as a form of self-punishment, said a *Pater, Ave, Gloria* and repeated the pointing the following night to clarify the matter. The star continued to move. After another

---

[33]Piazzi was ordained a priest in 1769, at the age of 23.

check, Piazzi gained confidence in the reality of the drift and decided to tell some selected colleagues that he had discovered a wandering star. «On the evening of [January] 3, my suspicions became certain after I was assured that it was not a fixed star. Before saying anything about it, I waited until the evening of the 4th, when I had the satisfaction of seeing that it was moving according to the same law that it had followed during the previous days».[34] He still did not know for sure what the wandering star was. So, writing to Bode, he preferred to be on the safe side and spoke of a comet.[35] But already on January 24, 1801, in a letter to his friend Oriani, he confessed the doubt that this could be «something better».[36] Had he continued his observations regularly, he would have surely been able to verify his suspicions within a few weeks and confirm the planetary nature of Ceres. But he was hampered by bad weather, which was not uncommon even in Sicilian winters, by an illness that lasted several weeks, and also by the characteristics of his magnificent instrument. The configuration that Ramsden had at the time, and which the astronomer was reluctant to change in order not to jeopardize the long program of position measurements that was underway, did not allow observations outside the meridian. To overcome this obstacle, the Theatine, with the help of the faithful Cacciatore, tried to use other telescopes of his observatory, but without success.

By the time the news of Ceres' discovery reached France and Germany, along with the few position measurements taken at the Palermo Observatory over a 41-day interval, the star had been swallowed up by the day light. It seemed lost forever. Piazzi was accused of negligence by Maleskyne for not reporting the discovery in time for others to observe the star. Perhaps the Astronomer Royal was also wringing his hands, fearing that the discovery of the phantom planet between Mars and Jupiter would tarnish the reputation of his protégé Herschel. But he was not alone in blaming Piazzi. Von Zach, who had published Piazzi's report on Ceres in his journal, the

---

[34] Id., *Risultati delle osservazioni della nuova stella, scoperta il dì 1. Gennaio all'Osservatorio Reale di Palermo*, Stamperia Reale, Palermo, 1801, p. 4 (in Italian).

[35] W.F. Bottke *et al.* (eds.), *Asteroids III*, The University of Arizona Press, Tucson (AZ) 2002, pp. 18–22.

[36] *Corrispondenza astronomica fra Giuseppe Piazzi e Barnaba Oriani pubblicata per ordine di S. E. il Ministro della Pubblica Istruzione*, Hoepli, Milano 1875, *Lettera XL di Piazzi a Oriani*, 24 gennaio 1801, pp. 48–49 (in Italian). In a verbal account given years later by Cacciatore to Captain Basil Hall, Piazzi is even said to have exclaimed to his assistant already on the third night: "Oh, oh! [...] we have found a planet when we thought we were observing a fixed star; let us look at it more carefully?" (Cunningham, *Discovery of the First Asteroid, Ceres*, cit.).

"Monatliche Correspondenz", also lamented the missed opportunity. For it seemed that the star sighted in Palermo, after its fleeting appearance, was to be regarded as lost forever in the vast sea of the heavens, like a new Flying Dutchman. Even Laplace, a guru of celestial mechanics and Napoleon's favorite scientist,[37] believed that with only Piazzi's limited set of data, it was simply impossible to predict where the object would reappear on its return in the dark night, after the months of oblivion in the blinding light of the Sun. In fact, all the methods used to calculate heliocentric orbits up to that point were based on some *a priori* hypothesis. In the case of Uranus, for example, the number of free parameters had been arbitrarily reduced on the assumption that the orbit was circular, which later turned out to be the correct choice. But in the case of Ceres, things were not so simple. Fortunately, the matter attracted the attention of a young German genius, Carl Friedrich Gauss.[38]

At the time, this "prince of mathematicians" was only 24 years old, but he was already known for some brilliant achievements, including the first proof of what is known as the "fundamental theorem of algebra".[39] He resided in Lower Saxony, in his hometown of Braunschweig, protected by the duke who had sponsored his education since childhood. When von Zach published in his journal Piazzi's astrometric measurements, which covered a tiny arc, only 9 degrees, Gauss was irresistibly tempted by the problem, which «recommended to mathematicians for its difficulty and elegance».[40] In a few weeks, he developed a revolutionary method to obtain all the characteristics of the orbit with astonishing precision and on the sole assumption that it was Keplerian, that is, an ellipse with the Sun at one of the foci. The young German now knew where in the sky to look for Ceres

---

[37] In 1799, Bonaparte, who admired Laplace as a very talented scientist, appointed him Minister of the Interior, only to regret it six weeks later because the mathematician «carried the spirit of the infinitely small into administration» (D.M. Burton, *The History of Mathematics: An Introduction*, McGraw-Hill, New York 2007, p. 479). After the sudden dismissal, perhaps motivated by the need of the first consul, a compulsive nepotist, to "accommodate" his brother Lucien, Laplace remained loyal to Napoleon until the end.

[38] E.G. Forbes, *Gauss and the Discovery of Ceres*, in "Journal of the History of Astronomy", 2, 1971, pp. 195–199; D. Teets, K. Whitehead, *The Discovery of Ceres: How Gauss Became Famous*, in "Mathematics Magazine", 72, 2, 1999, pp. 83–93.

[39] This is the theorem that determines the number of zeros in a polynomial. Despite the efforts of the greatest mathematicians, since the 17th century no one had been able to give a complete proof. During his lifetime, Gauss, always in search of perfection, produced four of them.

[40] G. Glaeser, H. Stachel, B. Odehnal, *The Universe of Conics: From the Ancient Greeks to 21st Century Developments*, Springer, Berlin, 2016, p. VI.

when it returned in the dark night. He passed the information on to von Zach, who, with the help of Olbers, fished out the object exactly one year after its discovery. Gauss immediately became famous. Everyone wanted to know what his trick was. It is said that in 1801 Laplace commented on the Ceres affair with these words: «The Duke of Brunswick has discovered in his country more than a planet: an extraterrestrial spirit in a human body».[41] True to the motto *pauca sed matura* (few but ripe), which he would have honored above all for quality, given the breadth of his scientific output, the great mathematician refused to publish the procedure because he did not think it was yet perfect.[42] The world would have to wait a few years to learn about the "method of least squares"[43] and the algorithm that would later be called the "Fast Fourier Transform".

Gauss had hit the jackpot twice. He had flexed his mathematical muscles and at the same time proved to the world that the star discovered in Palermo, with an average orbital radius of 2.77 AU and an eccentricity of 0.08, had the right numbers to be the missing planet between Mars and Jupiter. He also noted that luck, as well as Piazzi's skill and perseverance, had played a role in the planet's discovery. At the time, Ceres was near a node of reversal of its motion from retrograde to direct, and could therefore appear almost stationary to a distracted or simply unlucky observer. Bode gloated, the celestial policemen nibbled at their elbows, and Piazzi scoffed, but he was ready to protect his creature, which he misjudged to be much larger than Earth. He had baptized her *Ceres Ferdinandea*. The name recalled the maternal divinity of the earth and fertility, protector of Sicily, and the person of King Ferdinand IV of Bourbon, to whom the Theatine felt a sense of loyal gratitude («without his favor, who knows if this discovery would ever have come about»).[44] Perhaps he had met the sovereign in person when the big-nosed Bourbon,[45] during his first forced exile to Palermo at the time of the Neapolitan Republic, a little more than

---

[41]D. Teets, K. Whitehead, *The Discovery of Ceres: How Gauss Became Famous*, in "Mathematics Magazine", 72, 2, 1999, pp. 83–93.

[42]S.M. Stigler, *Gauss and the Invention of Least Squares*, in "The Annals of Statistics", 9, 3, 1981, pp. 465–474.

[43]The method provides the best fit of a parametric function (usually a polynomial) to an oversampled data set by assuming that the sum of the squares of the residuals between the function and the data is minimal.

[44]Piazzi, *Della scoperta del nuovo pianeta Cerere Ferdinandea*, cit., p. 59 (in Italian).

[45]Ferdinand of Bourbon, IV of Naples and then I of the Two Sicilies after the Restoration, owed this nickname, "*re nasone*", both to the morphology of his nose and to his frequentation of the poorest and crudest classes of the Neapolitan population.

a year earlier, stayed in the Royal Palace of the Normans, just below the Observatory. As in previous cases, the act of reverence did not last, and Piazzi's planet was soon called only Ceres.[46] In fact, it seems that the name Juno was preferred in Germany and France because of the proximity of the body to Jupiter's orbit. Napoleon himself, informed of the matter by Lalande, had expressed in favor of this solution. But this time the Italian priest prevailed. «If the Germans think they have the right to baptize the discoveries of others, let them call my new star whatever name they prefer: I will always use the name of Ceres with them».[47]

The object quickly lost its status as a planet. Observations with instruments more powerful than the Ramsden gave it a very modest diameter, three and a half times smaller than the Moon. A dwarf planet. With a hint of bitterness, Herschel, the discoverer of the giant Uranus, coined the neologism "asteroid", which Piazzi did not like at all.[48] It appeared to him an undeserved *diminutio* (curtailment) to which he hastily replied: «If Ceres must be called an asteroid, so it must be called Uranus».[49] Frustrations of a provincial. But for the woody Theatine, the setbacks were just beginning. Only three months after the discovery of Ceres, Olbers found another asteroid at about the same distance from the Sun, but on a very eccentric orbit, as Gauss, who had rushed to calculate it, immediately pointed out. Olbers suggested that the two objects could be fragments of a large planet that had once plowed the space between Mars and Jupiter and then, for unknown reasons, disintegrated. The hypothesis was strengthened in 1804 with the discovery of a third asteroid by Harding, Schröter's assistant at Lilienthal Observatory, and then a fourth in 1807, again by Olbers. Not bad, because in any case the law of Titius-Bode seemed to be safe, if together with the healthy planets there were those gone to smithereens.

---

[46] A. Manara, *Controversie e curiosità sulla nomenclatura dei piccoli pianeti*, in "Memorie della Società Astronomica Italiana", 68, 1997, p. 679 (in Italian).

[47] *Corrispondenza astronomica fra Giuseppe Piazzi e Barnaba Oriani*, cit., *Lettera XLVI di Piazzi a Oriani*, 25 agosto 1801, pp. 53–54 (in Italian).

[48] Herschel tried to make Piazzi digest the nickname by subtly explaining that it was more important to be the father of a new family of celestial bodies, even if small, than to be the winner of a race that the English astronomer, without explicitly saying so, made it seem he had won (Manara, *Controversie e curiosità sulla nomenclatura dei piccoli pianeti*, cit, p. 679).

[49] A. Bemporad, *Giuseppe Piazzi (commemorazione)*, in "Memorie della Società Astronomica Italiana", 3, 1925, p. 396 (in Italian). *Corrispondenza astronomica fra Giuseppe Piazzi e Barnaba Oriani*, cit., *Lettera LV di Piazzi a Oriani*, 2 luglio 1802 (in Italian), with the transcription of a letter sent by Herschel on May 22, 1802, pp. 61–63.

Safe, yes, but not for long. Paradoxically, it was Uranus itself that created the case.

Since its discovery, the planet has been closely followed by astronomers as it moved among the stars, with the goal of improving estimates of the orbital parameters, namely the size and flattening of the orbit, its inclination on the ecliptic, the length of the ascending node[50] and the perihelion, and the epoch required to establish the connection between space and time.[51] The measured astrometric positions served as input to a mechanical model of the Solar System, whose feedback was the confirmation of its own predictions. But the numbers did not add up! Uranus threw a tantrum, and it seemed as if it would almost undo a century of sensational achievements in celestial mechanics. As early as 1791, Delambre, Messier's young colleague at the Hôtel de Cluny, had reported that the planet was moving faster than mathematical models predicted. For a while, the discrepancy was attributed to small errors in the orbital parameters. But it did not work, and another thirty years passed during which Uranus continued to drift. Bode's bright idea of broadening the time base by looking for any observations of the planet made before the discovery, when the object was cataloged as a simple star, did not help either. He found a few, scattered over a century, but they did not change the result.

The issue had become embarrassing.[52] In reviewing and correcting Delambre's data, Alexis Bouvard, director of the Paris Observatory, confessed in 1821 that he «left to the future the task of discovering whether the difficulty of reconciling [the data] is connected with the old observations, or whether it depends on some external and unperceived cause which may have been acting upon the planet».[53] But another 22 years were not enough for Bouvard to see the end of the story; in 1843 "he stopped breathing and

---

[50]The position where the planetary orbit intersects the ecliptic as it moves north through the reference plane.

[51]Calculating orbits in a cluster of many bodies, such as the Solar System, does not allow for an exact solution. The obstacle is overcome by using successive approximation techniques. The first step consists of finding a solution to a simplified problem in which the perturbative terms are neglected; then one proceeds to successive iterations, which gradually bring the previously ignored effects back into the calculation.

[52]E.D. Miner, R.R. Wessen, *Neptune: The Planet, Rings, and Satellites*, Springer-Praxis, Berlin 2001.

[53]Cf. M. Littmann, *Planets Beyond: Discovering the Outer Solar System*, Dover, New York 2004.

calculating.[54] There were two possibilities. Either those who had previously made the measurements were wrong, or, like the trick that exists but cannot be seen, some unknown body disturbed the orbit of Uranus. Indeed, other possibilities could have been considered to resolve the tension between the observed position of the planet in the sky and the calculated position. One could have speculated, for example, that the gravitational force deviates with increasing distance from the simple expression that Newton prescribed for a massive point, but this would have been almost blasphemous[55] for the new Aristotle from across the Channel. Another way out might have been to consider the failure — always at a great distance from the Sun, under conditions of weak gravitational fields, as physicists say — of the law (of inertia) that defines the mass of bodies as the ratio of applied force to developed acceleration.[56] But until recently no one in history has dared to really think about this. We will return to this same line of reasoning later, when we talk about dark matter in the final chapter.

While Uranus was spitefully slamming on the brakes and slowing down the calculations, a French mathematician, Urbain Le Verrier, discovered a similar inequality regarding the precession of Mercury's perihelion, unwittingly laying the groundwork for the most classic of Einstein's general relativity tests[57] (see chapter 14). Another decade passed and in 1841 the question of Uranus was examined by a young Englishman, John Couch Adams, and almost simultaneously by Le Verrier himself.[58] Quite independently, the two scientists[59] asked themselves the following question:

---

[54] A. Clarke, *A Popular History of Astronomy during the Nineteenth Century*, Adam and Charles Black, London 1908, p. 79.

[55] It is worth remembering that any physical quantity that varies as the inverse of the square of the radius conserves the flux, and this is a property that is difficult to abandon. Today, however, there are those who have tried it, such as the Israeli Mordehai Milgrom, Professor of Physics at the Weizmann Institute in Rehovot, author of *Modified Newtonian Dynamics* (MOND). Cf. https://ned.ipac.caltech.edu/level5/Sept01/Milgrom2/Milgrom_contents.html.

[56] Newton's second law of dynamics states that $f = ma$, which in words means that the force f applied to a body causes an acceleration a that depends on a property of the body called (inertial) mass m.

[57] N.T. Roseveare, *Mercury's Perihelion from Le Verrier to Einstein*, Oxford University Press, Oxford 1982.

[58] J. Lequeux, *Le Verrier: Magnificent and Detestable Astronomer*, Springer, New York 2013.

[59] Bessel also had the idea of searching for the invisible disturbing body, but in 1845 he became seriously ill and had to give up. Death overtook him the following year at the age of 61.

if the planetary anomalies were the result of disturbances caused by an unknown eighth planet, where should the disturbing body be located and what mass should it have? The answer was entrusted to the solution of the "inverse problem" of finding the orbital parameters of the phantom body, with the constraint of obtaining the best adherence of the mechanical model of Uranus to the astrometric positions provided by the observations, remaining within the framework of Newtonian orthodoxy as regards gravity and inertia. Then, having calculated where the capricious Fantomas should have been, it would be possible to wait for it at the gate, armed with a telescope, to verify its real presence and figure out its nature.

Easy to say, but quite difficult to accomplish without the invaluable aid of an electronic computer, and above all, enormously tedious. The two brave astronomers, each on his own account on both sides of the Channel, embarked barehanded on exhausting and tedious calculations that required diligence and constant concentration. We are not surprised: the climb to the top, in science as in life, is usually a long journey, marked by fatigue and sweat, to be undertaken through gritted teeth, with tenacity and a great spirit of sacrifice. Seneca was right when he said that "the way from the earth to the stars is not easy" (*Hercules furens*, 437). Both of them, however, took the same shortcut, assuming that the eighth planet also obeyed the Titius-Bode law, with which they had freely determined its distance from the Sun. The hypothesis turned out to be wrong, but fortunately it had no consequences, except to discover in retrospect that the magic formula did not work. Adams probably completed the calculations before the French did. During a vacation in Cornwall, immediately after discussing his thesis at Cambridge, he had completed the first of six iterations: the solution to the inverse problem actually went through a cycle of successive approximations. Back at his university, he tried to contact the Astronomer Royal, Sir George Biddell Airy, to involve him in the practical part of the exercise: exploring the area around the position of the eighth planet indicated by his calculations, to try to bring the hidden body out of the cosmic darkness. But Airy was abroad at the time. The note Adams left him was devoid of details; the young man kept this information secret even when it was later requested, no one knows why. It seems that the Cambridge air favored the manifestations of Asperger's syndrome, a form of mild autism that probably affected Newton, as well as many other geniuses.

Meanwhile, Le Verrier had also obtained a first estimate of the position of the eighth planet, and in November 1845 he gave a report on it to the Royal Academy of Sciences in Paris. For a few years, after working as a chemist in the shadow of the great Joseph-Louis Gay-Lussac, he had devoted himself to celestial mechanics, gaining a good reputation with a work on the secular variations of the orbits of the planets. This research attracted the attention of François Arago, a fascinating figure of scientist, politician, explorer, and popularizer, who headed the *Bureau des Longitudes*. Gossips whispered that part of this interest was linked to a *affair* between François and Lucile, Le Verrier's wife. *Cherchez la femme?* Who knows?

The context of these events was only seemingly peaceful. A fire was burning beneath the ashes of the Restoration. Continental Europe was preparing for the springtime of the peoples that would, within three years, storm the strongholds of absolutism rebuilt by the Congress of Vienna. It would be the second violent outburst of a bourgeoisie empowered by the Industrial Revolution, imbued with ideals sown by the newspaper boom, strangled by a retarded and dull *Ancien régime*, and starved by famine. The flames would consume France, Bourbon Sicily, Austrian-occupied Venice and Milan, Great Poland, Hungary, and Austria itself, and drag Charles Albert of Savoy into a first, ruinous war for Italian independence. Across the Channel, England watched, controlled, and prospered by "ruling the waves" of all seas, made safe by the cynical genius of Nelson and the lucky star of the Duke of Wellington, and tending its image with maniacal devotion in every field of knowledge, including astronomy. Thus, when Airy realized that the position of the eighth planet calculated by Le Verrier was in good agreement with that of Adams, his skepticism about the young compatriot's research turned into a frantic interest. Unable to devote resources to the search for the planet, he turned to James Challis, director of the Cambridge University Observatory. He did him no great favor, as we now see.

Challis set to work, but lacking updated star charts, he missed his target, even though it happened to be within range of his sights. It was the summer of 1845. Meanwhile, Le Verrier had completed the cycle of iterations and pinpointed the location of the hypothetical planet with a sufficiently small margin of error.[60] He needed an operating arm and found it in a German, Johann Gottfried Galle. The latter had worked for ten years at the Berlin Observatory, directed by Johann Franz Encke, a great specialist in the orbits of comets. Le Verrier did not know him personally, but he had read his thesis and appreciated it. He also knew that Galle had access to an

---

[60] A. Danjon, *Le centenaire de la découverte de Neptune*, in "Ciel et Terre", 62, 1946, pp. 369–383.

extraordinary instrument, a 22.5 cm aperture refractor built by the firm of Germany's best optician, Joseph von Fraunhofer. He then wrote to the young German astronomer, asking him to take a look at the sky around the position he had estimated. Five days later, having obtained his boss's permission to use the observatory's best facilities, Galle began his search. He was assisted by a student, Heinrich Louis d'Arrest, who, among other things, had reported the existence of an updated star chart on which it would be easy to detect the existence of a traveling intruder; and so it was. Barely half an hour after observations began, the planet, a relatively bright star with a distinctive appearance, was spotted and recognized. It was the evening of September 23, 1846. «*Monsieur, la planète, dont vous avez signalé la position, réellement existe*» (Sir, the planet whose position you mentioned really exists), Galle telegraphed to Le Verrier with obvious emotion.[61]

We now call this body Neptune, but once again the question of its name caused a scuffle. The French saw Le Verrier's discovery as revenge for the checkmate that Herschel had inflicted on their scientific greatness. The British, for their part, did not give up — they never do, neither in war nor on the soccer field — and desperately tried to draw by playing the card of Adams, who instead called himself out with great dignity. To pay for the Waterloo of British astronomy were a contrite Challis and a responsive Airy, accused of incompetence and sloth[62] even by the daily press, that over the British Islands makes few discounts to power. Le Verrier instead became a hero, the man who had discovered a planet «*au bout de sa plume*»[63] (with the tip of his pen), as his friend Arago said with effective lyricism. He was awarded the *Légion d'honneur* and appointed an associate member of the *Bureau des Longitudes* and professor of astronomy and celestial mechanics at the Sorbonne. The Royal Society conferred on him the Copley Medal (which Herschel had already received for the discovery of Uranus). Louis Philippe of Bourbon-Orléans, the "*roi citoyen*" who had "reigned but not ruled" since 1830 and who would flee in 1848, chose him as astronomy tutor for his young son. Unfortunately, a hard temper and a certain obsession with perfection earned him the dislike, if not the hatred, of his colleagues.

---

[61] P. Descamps, *La découverte de Neptune: Entre triomphe et camouflet*, in "Revue d'Histoire des Sciences", 68, 2015, pp. 47–79.

[62] R.W. Smith, *The Cambridge Network in Action: The Discovery of Neptune*, in "Isis", 80, 3, 1989, pp. 395–422.

[63] F.J.D. Arago, *Œuvres de François Arago*, J.-A. Barral (ed.), Gide, Paris 1857, vol. 4, p. 515.

The discovery of Neptune had many consequences. The first and most remarkable was the profound impression it made on the minds of kings, prime ministers, generals, financiers, and swindlers. They experienced first-hand the enormous and mysterious power of mathematics and science in general, and became convinced that funding research and education in the so-called hard sciences could bring them great power and immense benefits. This was the fortune of polytechnics and universities, academies, and research laboratories. The explosion of chemistry and later of physics, especially in Germany, would do the rest.

Chapter 9

The Cyclopes:
The giant telescopes of Herschel and
the Leviathan of Lord Rosse

> No, our science is not an illusion.
> But it would be an illusion to think
> that what science cannot give us
> we can get elsewhere.
>
> Sigmund Freud,
> *The Future of an Illusion*

> It is excellent to have a giant's strength,
> but it is tyrannous to use it like a giant.
>
> William Shakespeare,
> *Measure for Measure*

We now retrace our steps a little to find the music teacher with a passion for astronomy in idyllic Bath in the late 18th century, nestled among the green hills of the Somerset countryside. The discovery of Uranus had brought him unexpected fame and immortality, but it had not quenched his thirst for the heavens. In fact, Herschel hoped to change his life by turning his hobby into a full-time occupation. But how? As we know, George III offered him the means and the opportunity. Although in a foul mood over his Redcoats' repeated defeats in North America,[1] the king had been persuaded to graciously confer upon his German subject the title of Court Astronomer, created for the occasion. It was a great honor, coveted by William's friends and admirers; but it came with a meager salary, about half of what William earned as a musician. Put yourself in his shoes. It was not an easy decision to take. When he discovered Uranus in 1781, Herschel was 42. He had spent 25 years sweating for his family's comfortable status; now, just as Urania had kissed him on the forehead, at the height of his fame, he was

---

[1] J. Black, *George III: America's Last King*, Yale University Press, New Haven (CT) 2006.

being asked to reduce his standard of living. In return, however, he would be able to stop giving private lessons and musical performances and concentrate solely on astronomical activities, instrument making, and observing the heavens.

William thought about it for a while and finally accepted. He had reasoned that the role of *sui generis* (of his own kind) courtier would not take up much of his time, although he knew that he would have to respond with due diligence to the sudden demands of the most powerful man on Earth, afflicted as he was with a growing mental imbalance that made him talkative and unpredictable. Since he could not serve the royal family by remaining in Bath, he decided to approach one of the residences most frequented by the king and his court. Eschewing London because of its smoke and light-polluted skies, he was left with Windsor Castle, the historic fortified palace of England's kings from the time of the Norman Conquest, built in a bend of the Thames west of the capital. Thus, in the summer of 1782, while in Virginia the rebels were still celebrating the final surrender of the regulars at Yorktown, and King George was forever robbed of his sleep, the Herschel clan bade farewell to the merry Bath, packed their bags, and went to Datchet,[2] a village near the royal castle. It was a sudden choice, reminiscent of the one, forerunner of other tragedies, that Galilei had made at the height of his exaltation for his discoveries, he too to put himself at the service *sine cura* of a master. One would be tempted to draw a moral from this, but two cases alone are not enough to make a rule.

Datchet was a flat area, rich in greenery and water. The foggy and unhealthy climate stifled William's longing for clear skies, and the dampness that seeped into his bones undermined his sturdy body, making him frequently ill. Caroline, for her part, disliked the atmosphere of the village, much less her new home, a crumbling ruin that had once been a hunting lodge.[3] Much more self-assured than when she first landed in England as a poor and lost immigrant, the girl complained about the roughness of the villagers, so different from the good manners of the gentlemen and ladies of her Bath circle, the dishonesty of the shopkeepers, and, as a careful housewife, the high cost of living. Despite all this, during the more than three years that William spent at Datchet, he worked at his usual frenetic pace:

---

[2] Cf. E.S. Holden, *Sir William Herschel: His Life and Works*, Charles Scribner's Son, New York 1881, chap. 3. *Life at Datchet, Clay Hall, and Slough; 1782–1822.*

[3] J. Herschel, *Memoir and Correspondence of Caroline Herschel*, John Murray, London 1879, p. 50.

exploring the sky on clear nights, entertaining the royal family from time to time with chats and observing evenings in the gardens at Windsor, but also designing and manufacturing quality astronomical instruments,[4] for personal use and for the many who wished to own a telescope made by the man who discovered Uranus. Over forty years his customers would have included wealthy bourgeois and crowned heads of Spain, Italy, and Russia, famous scientists such as Bode, Schröter, Piazzi, and John Pond, Maskelyne's successor as sixth Astronomer Royal, and even a Chinese mandarin.[5] All received a quality product, meticulously disassembled and packaged, with instructions, handwritten by Caroline, on how to reassemble the pieces once they reached their destination. The proceeds were used to finance the various technological activities that orbited around the observation equipment, to maintain a dignity that the royal purse would not allow, and to cover the expenses of the increasingly frequent trips to London to participate in the capital's cultural life. Fame had opened the doors of salons and academies to the former musician, but most people still regarded him as a lucky beginner and a newcomer to science. Indeed, the absence of a university education was still evident in his language, the lack of specific terms, and the style of his writings, and William's enthusiasm was often mistaken by the bigwigs for unrealistic naivety. He did not seem to care too much, obsessed as he was with his astronomical fever.

William was proud of his instruments: «I can now say that I absolutely have the best telescopes that were ever made»,[6] he wrote after visiting Greenwich and some university observatories in England. By now he was just signing off on the products of his brother Alexander and his staff. He had no time to waste. All his energy was focused on the gigantic project in which he and Caroline were engaged: surveying the stars in the sky to determine how many there were and how they were distributed, discovering double or multiple systems, and at the same time trying to solve the mystery of the puzzling objects cataloged by Messier, known as nebulae because of their diffuse appearance. It was a mission bordering on the impossible, requiring an iron will, great mental order, organizational skills, and an *ad hoc* telescope. Herschel loved the 6 inch aperture, 7 foot focal

---

[4] J.A. Bennett, *On the Power of Penetrating into Space: The Telescopes of William Herschel*, in "Journal for the History of Astronomy", 7, 1976, pp. 75–108.

[5] A. Maurer, E.G. Forbes, *William Herschel's Astronomical Telescopes*, in "Journal of the British Astronomical Association", 81, 1971, pp. 284–291.

[6] In M. Littmann, *Planets Beyond: Discovering the Outer Solar System*, Dover, New York 2004, p. 9.

length reflector with a mahogany tube and altazimuth stand, with which he had found Uranus. But to count the stars, a task he called "star-gage", he needed a larger collecting area.[7] Two new Newtonian reflectors were then made, the second of which with an aperture of 19 inches (almost half a meter) and a focal length of 20 feet: a giant of extraordinary quality, which made a fine display in the garden of Datchet's house, the object of admiration and astonishment of those who, visiting the Sovereigns at Windsor, ventured the little detour to see the unusual machine.

A slender wooden tower in the shape of an inverted V supported the telescope tube, at the base of which was the primary mirror. The instrument was moved in elevation by a system of ropes and pulleys. This gave the apparatus a familiar seafaring look that was certainly appreciated in England. The whole system could rotate in azimuth on a wide circular track thanks to the thrust of a winch. Both movements were manual, performed by one or more assistants, and only to point the telescope. They were far too crude to follow the smooth rotation of the celestial sphere and keep the telescope on target during observation. The astronomer stood on a balcony that could be slid along the front truss and raised to reach the eyepiece, which protruded from one side of the telescope tube near the top. The primary mirror was made of the usual alloy of copper and tin, with the addition of a pinch of arsenic to increase its reflectivity. Even so, the speculum returned little more than 50 percent of the incident light. The performance, already modest when the surface was freshly polished, declined inexorably within a few months due to weathering. Herschel required his telescopes to be in perfect condition at all times and had adopted the spare-tire strategy. Each instrument was equipped with two main mirrors, so that while one was in use, the other could be regenerated for replacement when needed.

The 20-foot telescope worked beautifully and was used consistently by the Herschel brothers. William and Caroline had organized themselves in order to optimize time and resources. Jean-Hyacinthe Magellan, an astronomer of Portuguese origin who had the opportunity to spend the Epiphany night of 1785 in Datchet, tells us it with great effectiveness:

> I stayed for the night of 6/7 January last at Mr Herschel's, near
> Windsor, in the village of Datchet, and had the lack of having
> a clear night. His large 20-ft Newtonian telescope is located in

---

[7]R.A. Proctor, *William Herschel's Star Surveys*, in "The North American Review" 140, 338, 1885, pp. 30–47.

his garden under the sky with a very simple and convenient mounting. A servant standing below turns a crank leaver alternatively forwards and backward, until a hammer strikes as soon as the telescope has been raised or lowered by the width of the field of view [...]. Next to this instrument stands a pendulum clock, adjusted to the sidereal time and showing right ascension. In this room sits the sister of Mr. Herschel and she has Flamsteed's sky maps in front of her. When he gives a signal, she notes in a log the declination and right ascension, and also writes down the other circumstances of the appearance. Mr Herschel is investigating the whole sky in this way, without so much as missing a single part of it [...] I went to bed one hour after midnight and by then he had so far discovered that night four or five new nebulae. The thermometer in the garden showed 13° Fahrenheit [−10.6° C], but unaffected by this Mr. Herschel observed during the whole night, except that he has a rest of a few minutes every three or four hours, when he strides up and down in the room I mentioned [...] For some years now Mr H[eschel] has missed no opportunity to observe the sky, if the weather allows it, and this always in the open air, for he is convinced that the telescopes work well only at a constant temperature equal to that of the air. He tries to protect himself with clothing against the rough weather conditions, is fortunately blessed with a robust constitution, and thinks in this world of nothing other than the heavenly objects [...] His sister is like him, fascinated by astronomy, and she has considerable knowledge of the calculations involved.[8]

This report is similar to that of the French geologist Barthélemy Faujas de Saint-Fond, who had visited the Herschels at Datchet a few months earlier:

I arrived at Mr Herschell's about ten o'clock. I entered, by a staircase, into a room which was decorated with maps, instruments of astronomy, and natural philosophy, spheres, celestial globes, and a large harpsichord.

Instead of the master of the house, I observed, in a window at the farther end of the room, a young lady seated at a table,

[8]M. Hoskin, *Discoverers of the Universe: William and Caroline Herschel*, Princeton University Press, Princeton (NJ) 2011.

which was surrounded with several lights; she had a large book
open before her, a pen in her hand, and directed her attention
alternately to the hands of a pendulum-clock, and the index of
another instrument placed beside her, the use of which I did
not know: she afterwards noted down her observations.

I approached softly on tiptoe, that I might not disturb a
labour, which seemed to engage all the attention of her who
was engaged in it; and, having got close behind her without
being observed, I found that the book she consulted was the
Astronomical Atlas of Flamstead, and that, after looking at
the indexes of both the instruments, she marked, upon a large
manuscript chart, points which appeared to me to indicate
stars.

This employment, the hour of the night, the youth of the
fair student, and the profound silence which prevailed, interested me greatly.[9]

What a difference with modern instruments, operated from a comfortable,
lighted, and heated control room, with a cup of good coffee in hand and
a few sandwiches at the astronomer's disposal, and a series of computers
busy pointing, guiding, monitoring, correcting, recording, and visualizing
the collected data, so that the observer can quickly adjust the course.[10]
Herschel stood in the cold and dark, climbing a ladder, at the risk of falling
and injuring himself.[11] Dangers lurked everywhere, as Caroline well knew.
One night she had been harpooned in the calf by a hook of the handling
system, risking lameness. She can still be heard whispering in the dark:
"*Lieber Bruder, sei um Gottes willen vorsichtig*" (Dear brother, for God's
sake, be careful). When they were alone, the two brothers often spoke in
German, just as the Fuhrer's scientists would do nearly two centuries later,
working in NASA's laboratories in Huntsville, Alabama, to help the United
States win the race to the Moon.[12]

---

[9] W.B. Ashworth, *Faujas-de-Saint-Fond visits the Herschels at Datchet*, in "Journal of History of Astronomy", 34, 2003, pp. 321–324.

[10] Cf. M. Capaccioli, *Un antico mestiere si rinnova*, in *Mille1Notte. Storie dell'altro mondo*, Mediterraneo, Caserta 2018, pp. 153–158 (in Italian).

[11] Littmann, *Planets Beyond*, cit., p. 7.

[12] Cf. M. Capaccioli, *Luna rossa. La conquista sovietica dello spazio*, Carocci, Roma 2019, pp. 187–214 (in Italian).

While at Datchet, William continued the search for double stars[13] that he had begun some years earlier in Bath, when he was an obscure provincial amateur astronomer. Inspired by Galileo, he believed that pairs of stars angularly close enough to be resolved only by telescope could be used to flush out phantom parallax. The strategic idea was based on two assumptions: that the duplicity was the result of random perspective alignments, and that in pairs of stars of very different magnitudes, the fainter of the two was also the more distant, so as to serve as a reference for the parallactic motion of the other, brighter and thus (presumably) closer.

So Herschel thought and hoped. But nature had a surprise in store for him. And what a surprise! The astronomer had thoroughly analyzed the question and identified the various sources of noise in the possible parallactic motion. He was not worried about the aberration, since the phenomenon identified by Bradley acts in the same way on angularly close stars, being dependent on position and not on distance. Rather, he feared the fact that stars that are thought to be fixed sometimes roam the celestial sphere, animated by their own proper motions, each on its own and apparently at random. Halley had already pointed out this feature in 1718,[14] just as the paradigm of a rigid, plastered sky had been shattered by the suspicion that the dynamical equilibrium on which the Solar System rests should apply, *mutatis mutandis*, to the entire cosmos. Newton's volcanic friend had compared the positions of Sirius, Arturus, and Aldebaran, estimated more than eighteen centuries earlier by Hipparchus, with those just obtained by Flamsteed, and found that the differences were too large to be attributed merely to measurement errors. He had only to admit that between the two epochs the three stars had drifted, each on its own, with respect to the pattern of the sky.[15] In other times, this idea would have stoked the fires

---

[13]For a brief review of double stars, cf. J.S. Tenn, *Keepers of the Double Stars*, in "Journal of Astronomical History and Heritage", 16, 1, 2013, pp. 81–93.

[14]In fact, in a summary of the observations made in 1677–78 from the island of Saint Helena in the South Atlantic Ocean, which he published in 1679 under the title *Catalogus Stellarum Australium* (Catalogue of the Southern Stars), Halley had already reported the evidence for the «mutability of the fixed stars». Cf. J.C. Brandt, *St. Helena, Edmond Halley, the discovery of stellar proper motion, and the mystery of Aldebaran*, in "Journal of Astronomical History and Heritage", 13, 2, 2010, pp. 149–158.

[15]E. Halley, *Considerations on the Change of the Latitude of some of the principal fixed Stars*, in "Royal Society Sec. Philosophical Transactions", 30, 1720, pp. 736–738. «Having of late had occasion to examine the quantity of the Precession of the Equinoctial Points, I took the pains to compare the Declinations of the fixt Stars delivered by *Ptolomy*, in the *3d* chapter of the *7th* Book of his *Almag.* as observed by *Timocharis* and *Aristyllus* near 300 Years before *Christ*, and by *Hipparchus* about 170 Years after

of indignation and the pyres of the Inquisition. Instead, it was metabolized with ease. A sign of the new times.

A rich database was needed to smooth the systematic errors caused in pure parallax by proper motions of binary stars. In fact, other things being equal, the magnitude of the unpredictable drift is greater the closer the stars are to the observer. To see this, imagine a flotilla of boats swarming across the sea: you will notice that the hulls closest to you appear larger and more mobile on average than those farther away. But nearby stars are also those where the effect of parallax is greater. To break this vicious circle, Herschel decided to observe «every star in the firmament»[16] to find the one that was right for him. The strategy was the same as that of the pearl fisherman, who collects as many shells as possible in the hope that at least one hides the desired black pearl. By systematically scanning the sky with his instruments, Herschel collected "binary systems", as he called them, and then, at different epochs, measured the angular separation and the (position) angle of the line connecting the two sources formed with the fixed direction of North.

In 1782, just as he was preparing to leave Bath and take on the role of court astronomer, he compiled the first catalog of nearly 300 double stars, 85% of which he discovered himself. Three years later he added a second list of hundreds more binaries, this time all his own work. But despite these efforts, the parallax would not appear. It seemed that the long search would end in a flop. Of course, it is not always true that everything comes to those who wait! The German did not give up, however, and was eventually generously rewarded, albeit with a different kind of money. For a quarter of a century he had followed the promising double systems of Castor (α Geminorum) and Algieba (γ Leonis), measuring the relative positions of the two stars of each pair with his precise micrometers. Finally, he had to give in to the evidence. The desired differential displacements were there, but they had neither the characteristic linearity of proper motions nor the expected periodicity of one year (as required by parallax). But even the random perspective association hypothesis no longer held water. These pairs of stars were physical systems — the astonished Herschel had to admit to himself

---

them, that is about 130 Years before *Christ*, with what we now find: and by the result of very many Calculations, I concluded that the fixt Stars in 1800 Years were advanced somewhat more than 25 degrees in Longitude, or that the Precession is somewhat more than 50″ *per ann.*».

[16]M. Hoskin, *The Construction of the Heavens: William Herschel's Cosmology*, CUP, Cambridge 2012, p. 78.

— maintained and governed by Newtonian gravity, which only then could rightfully claim universality.

In fact, the idea had been proposed some forty years earlier by the Reverend John Michell,[17] an English genius best known for a pioneering and naive formulation of the black hole concept. He had argued that stars naturally tend to coalesce due to mutual attraction. Not surprisingly, they also form physical pairs such as Mizar, the central star in the bar of Ursa Major, which Benedetto Castelli had discovered to be double in 1617. Herschel had turned this insight into an empirical truth. A quarter of a century later, the Parisian Félix Savary, professor at the *École Polytechnique*, would give it formal dignity by showing that the orbit of the visual binary ξ Ursae Majoris was consistent with the Newtonian model of two massive points subject to mutual attraction: another nail in the coffin of Aristotelian dualism, and a way to directly derive the masses of these distant worlds.

This amazing byproduct of the grueling hunt, which alone would be enough to make a scientist famous, concerns the role of the Sun in relation to the stars of the Milky Way. Copernicus had placed our star (almost) immovably at the center of the world: an official promotion based on a philosophical and religious bias. By measuring the proper motions of 13 sources, Herschel had instead realized that the Sun is a starship moving in cosmic space toward a point in the constellation of Hercules, not far from the brilliant Vega. Supreme contempt for a humanity that had boasted of a celestial *unicum* that did not actually exist.

The last catalog of double stars appeared in 1821, by which time the former musician was a scientist of years and fame. Professional astronomy had accorded him the honor of being named vice-president, and the following year president,[18] of the newly established Astronomical Society of London, renamed the Royal Astronomical Society by King William IV. An association as always strictly male; membership would not be opened to women until 1951, despite the prominent figure of Caroline, sister of the first president. Later, William's son John would publish his catalog of hundreds more binaries, bringing the total haul of double systems discovered

---

[17]R. McCormmach, *Weighing the World: The Reverend John Michell of Thornhill,* Springer, Dordrecht 2012.

[18]*De jure*, as William Herschel never chaired any session. Cf. R.G. Aitken, *History of the Royal Astronomical Society, 1820–1920 — A Review*, in "Publications of the Astronomical Society of the Pacific", Vol. 36, No. 211, 1924, pp. 131–134.

by the Herschels to well over a thousand. But the hunt for binaries did not stop: it continued with Friedrich von Struve in Dorpat (now Tartu), Estonia, with Admiral William Henry Smyth and Reverend William Rutter Dawes in England, and with Baron Ercole Dembowski in Italy. The goal was to collect information about the stars to be used to validate the first naive models of these celestial machines. More on this later; for now we must retrace our steps to tell the last of William's great exploits, the most intriguing and least successful, connected with his largest reflector, the 48 inch telescope.

In 1785, tired of the bad weather at Datchet, the Herschels moved to a farmhouse in Old Windsor, on the edge of the park, where the air was somewhat healthier. They stayed there for nearly a year. William already had in mind a new and more powerful telescope with which he hoped to lift the veil from the mysterious nebulae and determine their nature. He possessed, or at least believed he had, the technological capabilities to accomplish this. King George had encouraged him with a reward of 2,000 pounds (about 260,000 euros). A large sum, later doubled to allow the astronomer to complete the project, not without a rebuke for going over budget. When the landlady felt the taste of money, she thought of speculating on further occupation of her property and asked for a rent increase. Faced with this demand, Caroline, never one to be accommodating, flew into a rage, and the patient William set out to find a new home. He found it in Slough. The area was still the same, on the outskirts of misty Windsor, but the building was large and surrounded by a wide open space where instruments could be conveniently set up. It became the "home observatory" of an astronomical genius who, throughout his long career, never relied on a public institute or university to carry out his research.

William had probably sensed that Slough would be his last landing, and as if to cement the stability he had achieved, two years after his arrival in the village, and now in the 50s, he married a wealthy woman,[19] Mary Baldwin Pitt, 12 years his junior, who had recently lost her husband and only child. Four years later, Mary gave him an heir, John Frederick William, who would follow in his great father's footsteps. Meanwhile, the latest and largest of Herschel's reflectors had sprung up in Slough's garden to match

---

[19]M. Hoskin, *Mary Herschel's Fortune: Origins and Impact*, in "Journal for the History of Astronomy", 41, 2, 2010, pp. 213–224. The article's exergue quotes a caustic phrase from a friend of the Herschel's, writer Frances Fanny Burney: «Astronomers are as able as other men to discern that gold can glitter as well as the stars».

the glorious 20 feet: an instrument[20] with a focal length of 40 feet (12.2 m) and a mirror with a record diameter of 48 inches (120 cm). It took more than one attempt to make the giant speculum disk. The London foundry workers commissioned by the astronomer to cast it even risked their lives when the furnace exploded, hurling fragments of hot metal into the air. The feat tested the limits of current metallurgical technology, but was ultimately successful. William received an intact blank, though not of the quality he would have liked, which he ground and polished with his usual care and passion.

The architecture of the instrument was basically the same as the smaller model, the beloved 20 feet. As the structure grew, so did the number of visitors to the site. Even the King wanted to go there, accompanied by the Archbishop of Canterbury. It is said that Farmer George,[21] seeing the iron tube of the telescope lying on the lawn, urged the high prelate to go inside: «Come, my Lord Bishop, I will show you the way to heaven».[22] The great forty-foot telescope saw its first light on a clear night of February 1787, while the giant was still being completed. Hungry for the deep sky, Herschel could wait no longer. Not knowing where the focus was, he tried to find it by holding the eyepiece in his hand and pointing it directly at the main mirror with a slight offset. In this way he rediscovered how to observe directly at the prime focus, with a great saving of light. This method, already used for another reason by the Jesuit Zucchi (p. 154), eliminated a reflection and thus reduced losses, but it required great skill on the part of the astronomer.

«The object I viewed was the nebula in the belt of Orion, and I found the figure of the mirror, though far from perfect, better than I had expected. It showed four small stars in the nebula and many more. The nebula was extremely bright»,[23] he would later write, recalling that memorable night. Although very bright indeed, so bright that the astronomer had to look away quickly to avoid being blinded by Sirius' entrance into the field, the instrument proved to be of rather modest mechanical and optical quality. A huge but imperfect umbrella. Apart from the awkward handling of the huge metal tube, which was moved by the usual system of ropes and pulleys, the

---

[20] W. Herschel, *Description of a Forty-Feet Reflecting Telescope*, in "Philosophical Transactions of the Royal Society of London", 85, 1795, pp. 347–409.

[21] Nickname given to George III by satirists because of his lack of interest in politics compared to, for example, agriculture. The king, however, had also had a scientific education, and he enjoyed making a few minor astronomical observations.

[22] J. Mullaney, *The Herschel Objects and How to Observe Them*, Springer, New York 2007, p. 14.

[23] H.C. King, *The History of the Telescope*, Dover, New York 2003, p. 129.

biggest problem was the mirror, which Herschel wanted to be thin to keep the weight down to half a ton. The disk of the speculum bent with changing elevation and lost the optical figure. William then ordered a second cast, which turned out to be wrong. The blank broke while cooling. More money wasted! On the third try, he finally secured a flawless disk, uniformly 3.5 inches thick and weighing one ton. This would have been the official eye of the Cyclops, replaced only by the faulty twin during periodic maintenance of the polished surface.

The new mirror saw its first light at the end of August 1789, just as the insurrection was raging across the Channel, sparked by the storming of the Bastille by the *sans-culottes* on July 14th. What a difference between the industrious peace of Slough and the hell of the streets of Paris! Caroline would have remembered those days with these words:

> The garden and workrooms were swarming with laborers and workmen, smiths and carpenters going to and fro between the forge and the forty-foot machinery, and I ought not to forget that there is not one screw-bolt about the whole apparatus but what was fixed under the immediate eye of my brother. I have seen him lie stretched many an hour in a burning sun, across the top beam whilst the ironwork for the various motions were being fixed.[24]

Herschel's high expectations, and those of the king who had financed the project, were not disappointed. On the very night of the telescope's inauguration, thanks to its magnification of 1200×, William was able to discover a new satellite of Saturn and, within a few days, another. Many years later, at the suggestion of his son John, the two faint moons would be named for the mythological giants Enceladus and Mima, sons of the gods of heaven and earth. Yet despite the power of the instrument, for decades the largest in the world, Herschel continued to prefer and habitually use the 20-foot. He argued with disarming pragmatism that looking with a telescope larger than necessary was a waste of time that an astronomer could not afford on a clear night: «I have made it a rule never to employ a larger telescope when a smaller will answer the purpose».[25] In fact, the smaller instrument

---

[24] Ibid.

[25] Cf. *The Scientific Papers of Sir William Herschel: Including Early Papers hitherto Unpublished, Collected and Edited Under the Direction of a Joint Committee of the Royal Society and the Royal Astronomical Society*, J.L.E. Dreyer (ed.), The Royal Society, London 1912, p. 536.

cost less to maintain, was easier to prepare for observation, maneuvered better and with fewer assistants, and, most importantly, returned images of unparalleled quality that had enabled William to discover Titania and Oberon, the first two moons of his Uranus, in 1787. For 25 years, no other astronomer would have been able to see the two satellites, a testament to the German's skill as an observer and the unsurpassed quality of his telescope. It is worth noting that the names[26] these cold wingmen of the god of the starry skies were also chosen by John over half a century later, drawing for the first time on the mythology and literature of the Nordic peoples instead of the usual Greek-Roman Olympus. Still a sign of the shifting political center of gravity in Europe, but also of the unquestioned authority of the last of the Herschels, who could arrogate to himself the right to baptize the stars by even changing the custom.

The Solar System had given, and continued to give, fame, and honor to William and his family. But the real battleground for him and Caroline was the sidereal universe, a boundless and unknown land that they explored for half their lives with unparalleled method, continuity, and intelligence.[27] The cosmic epic began at Datchet in 1783, just as a peace treaty at Versailles sanctioned the birth of the nation, the United States of America, where the sprout of modern observational cosmology would flourish more than anywhere else. The stubborn German set out to solve the mystery of the Milky Way's inner structure. He understood that its appearance depended on the position of the observer within the system. But how could the observations be de-projected to reconstruct the true shape? It would have been necessary to count all the stars and assign the correct distance to each of them. Knowing that he could not cover the entire sky, Herschel identified 700 regions on a great circle perpendicular to the Milky Way. On clear nights, these selected areas were let to drift one by one in front of the telescope's aperture. In this way, William could carefully describe them to Caroline, who recorded the measurements in a logbook in the dim light of a candle. By the end of this titanic cosmic sounding, the two had managed to count over 90,000 stars. An impressive haul, which was not enough, though. To draw conclusions from the survey, distances were also needed. William usually evaluated them by the apparent magnitudes, estimated by

---

[26]Titania is the queen of the fairies and the wife of Oberon, the fairy king. They are both characters in William Shakespeare's play *A Midsummer Night's Dream*.
[27]B. Lovell, *Herschel's Work on the Structure of the Universe*, in "Notes and Records of the Royal Society of London", 33, 1, 1978, pp. 57–75.

eye, assuming that the intrinsic luminosity of the stars was constant.[28] A naive hypothesis that would later be resoundingly disproved by the vast differences in brightness and color of stars in binary systems. But with the means and knowledge at his disposal, it was the best he could do. In this particular case, however, even this shortcut was not applicable to the countless and very faint stars that the astronomer could only glimpse and count with his telescopes.

In one fell swoop, Herschel got himself off the hook by completely eliminating the need to measure distances. He assumed that the Milky Way was the image of a uniform distribution of stars in a volume of finite dimensions, and that he could see and therefore count all the stars along a given direction up to the extreme limits of the system. With these constraints, the total number N of objects within a cone of fixed aperture is proportional to the volume of the cone itself, which is truncated where the galaxy ends, at a distance from the observer proportional to the cube root of N. In short, all he had to do was rearrange the count to simulate identical cones, and he was done. William crossed his fingers, hoping that his simplifications would not completely distort the result. He was particularly concerned that he had assumed a uniform distribution of stars, which he knew was not true. But by then the die was cast. In 1785 he presented his findings in a paper with the suggestive title *On the Construction of the Heavens*, in which he argued that the Galaxy looked like a loaf 7,000 light-years wide and at most 1,300 light-years thick. A symmetrical structure with respect to an axis, in which the Sun occupied an almost central position. He also calculated that the system as a whole had 300 million stars: a spectacular number of Suns, but nearly a thousand times smaller than modern estimates. It would be a century before others could do better, taking advantage of rapid technological advances.

Herschel was the best, but not the first.[29] A similar model of the Milky Way had already been imagined by Immanuel Kant in 1755, but had not been verified quantitatively. It is uncertain whether the astronomer was aware of earlier speculation on the subject. Historians of science tend to believe that William became aware of them very late, after the game was

---

[28] Because of the conservation of luminous flux, the magnitude of a star decreases as the inverse of the square of the distance, unless the signal is attenuated by phenomena such as absorption by gas and/or dust.

[29] M. Capaccioli, *Quando l'uomo scoprì le galassie — When man discovered the galaxies*, in "Quaderni di Storia della Fisica", 23, 2020, pp. 1–41 (in Italian, with abstract in English).

over. The philosopher from Königsberg, in turn, was generally inspired by the writings of Thomas Wright, a native of Durham, a city in northeastern England. Five years earlier, this singular character, who had made a prosperous living tutoring noble maidens, had published *An Original Theory or New Hypthesis of the Universe*, a bizarre work that crossed brilliant intuitions with deviant mystical impulses. In effect, the author's goal was to fit the scientific worldview into a satisfactory religious framework, thereby correcting the apparent contradictions between the observed disorder and the postulated divine order.[30] His first concern was to defuse the reasons for that general gravitational collapse of the sidereal universe already feared by the classical scholar and theologian Richard Bentley in his correspondence with Newton: due to mutual attraction, the stars, left to themselves, should fall on each other.

Bentley, later Bishop of Worcester, had been appointed to deliver the first series of lectures or sermons (eight a year) established to fulfill a testamentary bequest by Robert Boyle, a noble and wealthy Irish natural philosopher. Deeply religious, Boyle had wished to provide the Anglican Church with an opportunity «to counter the atheistic currents of thought which which were posing a challenge to the faithful in the late seventeenth century».[31] Indeed, by involving Newton in the preparation of his lectures on *A Confutation of Atheism from the Origin and Frame of the World*, Bentley had hit the jackpot. In his first reply to the churchman, dated December 10, 1692, the Lucasian professor confessed to him: «Sir, when I wrote my treatise about our system, I had an eye upon such principles as might work with considering men for the belief of a deity, and nothing can rejoice me more than to find it useful for that purpose [...] I do not think explicable by mere natural causes but am forced to ascribe it to the counsel & contrivance of a voluntary Agent».[32] In short, to get out of his embarrassment, he brought up God. A peculiar move for a giant of physics, to say the least, or a strong act of faith? We find the answer in his final comment, in the second edition of the *Principia*, printed in 1713:

> This most beautiful system of the sun, planets, and comets could only proceed from the counsel and dominion of an

---

[30] M. Hoskin, *The Cosmology of Thomas Wright of Durham*, in "Journal for the History of Astronomy", 1, 1970, pp. 44–52.

[31] J.D. Dahm, *Science and Apologetics in the Early Boyle Lectures*, in "Church History", Vol. 39, No. 2, 1970, pp. 172–186.

[32] Newton's first letter to Bentley, in http://www.newtonproject.ox.ac.uk/view/texts/normalized/THEM00254.

intelligent and powerful Being. And if the fixed stars are the centers of other like systems, these, being formed by the like wise counsel, must be all subject to the dominion of One, especially since the light of the fixed stars is of the same nature with the light of the Sun and from every system light passes into all the other systems; and lest the systems of the fixed stars should, by their gravity, fall on each other, he has placed those systems at immense distances from one another.[33]

Sic! Having to respond to Bentley's keen remarks (contained in a letter now lost), he realized that his universe could not be finite, because it would be crushed by gravity; and if infinite, it had to be homogeneous to avoid catastrophic collapses. But a homogeneous infinite system is like a fragile house of cards, whose survival Newton entrusted directly to God by a «continual miracle».[34]

More secularly, Wright devised the remedy for the catastrophe in Halley's proper motions.[35] In his view, these stellar drifts resulted from a universal orbit of the stars around a common attractor, also understood as the pole of the cosmic moral order and the source of all the laws of nature. Pretending that everything was symmetrically distributed around this center, because what is good must be beautiful, and drawing inspiration from the rings of Saturn, he imagined the Milky Way as a «faint circle of light»,[36] a homogeneous ring of stars centered on Divine Providence. In this way he pleased everyone. It was enough to imagine a Sun far from the center to obtain an asymmetrical appearance of maximum regularity. For the earthly observer — Wright concluded — the part of the ring that surrounds him and dominates the view is perceived primarily as a diffuse streak of stars. A concept illustrated with unparalleled elegance by Su Shi, an 11th-century Chinese poet: «viewed horizontally, a range; from the side, a cliff. / It

---

[33] I. Newton, *Philosophia Naturalis Principia Mathematica*, sn, Cantabrigiae 1713, p. 388. Cf. http://www.newtonproject.ox.ac.uk/view/texts/normalized/NATP00056.

[34] P. Kerszberg, *The Cosmological Question in Newton's Science*, in "Osiris", Vol. 2, 1986, pp. 69–106.

[35] We will see on p. 351 how the problem of overcoming the tendency of gravity to general collapse was solved by Einstein in his model of a static universe; a solution which, however, left the problem of stability intact.

[36] M. Hoskin, *The Cosmology of Thomas Wright of Durham*, in "Journal of History of Astronomy", 44, 1970, pp. 44–52.

differs as we move high or low, or far or near. / We do not know the true face of Mount Lu, / Because we are all ourselves inside».[37]

Kant, who had read a summary of the Englishman's work in a magazine, described his reinterpretation of the model, purged of mystical intrusions, in the *Allgemeine Naturgeschichte und Theorie des Himmels* (General Natural History and Theory of Heaven), which appeared in 1755:

> Mr. Wright of Durham, whose treatise I have come to know from the Hamburg publication entitled the *Freie Urteile*, of 1751, first suggested ideas that led me to regard the fixed stars not as a mere swarm scattered without visible order, but as a system which has the greatest resemblance with that of the planets; so that just as the planets in their system are found very nearly in a common plane, the fixed stars are also related in their positions, as nearly as possible, to a certain plane which must be conceived as drawn through the whole heavens, and by their being very closely massed in it they present that streak of light which is called the Milky Way. I have become persuaded that because this zone, illuminated by innumerable suns, has very exactly the form of a great circle, our Sun must be situated very near this great plane.[38]

In the hands of the philosopher of Königsberg, Wright's intuition matured[39] in a physical model built according to the Cartesian method and with the laws of Newtonian mechanics, but without resorting to calculations, only to analogical parallels. Kant did not limit himself to the stationary solution, which accounted for appearances, i.e. effects, and even wanted to go into the merits of causes. Inspired by the speculations of the atomists, he imagined that the transition from chaos to a "well-ordered world" occurred through a mechanism of formation of stars and star systems from rotating clouds of particles. These incubators of worlds were able to gravitationally coalesce and yet resist the force that pushed them toward collapse. A similar hypothesis was developed independently by Lambert in 1761 and forty

---

[37]Zhang Longxi, *The true face of Mount Lu: on the significance of perspectives and paradigms*, in "History and Theory", 49, 2010, pp. 58–70.

[38]W.W. Campbell, *Historical quotations*, in "Publications of the Astronomical Society of the Pacific", 29, 2014, p. 87.

[39]F. Paneth, *Thomas Wright of Durham and Immanuel Kant*, in "The Observatory", 64, 1941, pp. 71–82.

years later by Laplace. It put an end to the long chapter of the unchanging and perfect sidereal universe and introduced into natural philosophy the concept of evolution, which would have as its first consequence the birth of thermodynamics in physics and the explosion of the Darwinian and Lamarckian revolutions in genetics.

In fact, the idea of creative evolution predates and even belongs to Galileo's thinking, as evidenced by this passage from the *Dialogue Concerning the Two Chief World Systems* (1632), in which, toward the end of the first day, Sagredo passionately pursues the praise of imperfection:

> I cannot without great astonishment — I might say without great insult to my intelligence — hear it attributed as a prime perfection and nobility of the natural and integral bodies of the universe that they are invariant, immutable, inalterable, etc., while on the other hand, it is called a great imperfection to be alterable, generable, mutable, etc. For my part, I consider the earth very noble and admirable precisely because of the diverse alterations, changes, generations, etc. that occur in it incessantly. If, not being subject to any changes, it were a vast desert of sand or a mountain of jasper, or if at the time of the flood the waters which covered it had frozen, and it had remained an enormous globe of ice where nothing was ever born or ever altered or changed, I should deem it a useless lump in the universe, devoid of activity and, in a word, superfluous and essentially nonexistent. This is exactly the difference between a living animal and a dead one; and I say the same of the Moon, of Jupiter, and of all other world globes. [...] Those who so greatly exalt incorruptibility, inalterability, etc. are reduced to talking this way, I believe, by their great desire to go on living, and by the terror they have of death. They do not reflect that if men were immortal, they themselves would never have come into the world. Such men really deserve to encounter a Medusa's head which would transmute them into statues of jasper or of diamond, and thus make them more perfect than they are.[40]

---

[40] G. Galilei, *Dialogue on the two greatest systems of the world*, Edizioni Studio Tesi, Pordenone 1988, Giornata I, pp. 64–65 (in Italian). For the English text, see: http://law2.umkc.edu/faculty/projects/ftrials/galileo/dialogue.html.

Kant was well aware of the danger to which his theory, so openly heretical, exposed him, and he took two countermeasures. He shielded himself
with the great Frederick of Prussia, to whom he dedicated the book, and
he hid behind anonymity. In any case, few readers were infected with his
revolutionary ideas about the cosmos until Herschel's star counts proved
that the Milky Way was indeed shaped like a millstone. Of course, the
astronomer's observation only confirmed the geometric model and not the
evolutionary mechanism of Kant and Laplace. Instead, it confirmed another
intuition of Wright's, supported by the philosopher of Königsberg, concerning the nature of nebulous stars. The matter is so rich in significance that
it deserves a little historical *excursus*.

Ptolemy had already listed seven diffuse-looking stars in the *Almagest*,
but it is likely that, with the exception of the cluster M44,[41] these were false
identifications produced by the observing conditions. In the Middle Ages,
however, these "nepheloid" (nebular) stars were taken seriously. The Arabs
added to the Ptolemaic list an object in the constellation of Andromeda,
which we now know corresponds to the nucleus of the galaxy M31. In the
year 964, the Persian Abd al-Rahman al-Sufi noticed the wisp of light from
this beautiful spiral that accompanies the Milky Way[42] and described it in
Book of Fixed Stars as a "small cloud". A similar expression was used six
centuries later by Antonio Pigafetta of Vicenza, Italy, in reference to the two
satellite galaxies of the Milky Way that we now call the Magellanic Clouds.
The first Westerner to observe the Large Cloud was again al-Sufi, who saw it
from northern Yemen, where it is visible. He called it "White Oax". Later,
the Florentine merchant and explorer Amerigo Vespucci may have spoken
of both clouds in a letter about his third voyage to "Cipango" (a term first
used by the Venetian traveler and writer Marco Polo in *Il Milione* to refer
to Japan): «three Canopes [sic], two bright and one obscure», i.e. the Large
and the Small Cloud together with the Sack of Coal in the Southern Cross;
and then, before Pigafetta, the Florentine Andrea Corsali, an explorer in the

---

[41] The Beehive Cluster, also known as Praesepe (Latin for "manger"): open cluster in
the constellation Cancer.

[42] I. Hafez, *Abd al-Rahman al-Sufi and his Book of the Fixed Stars: A Journey of Re-
Discovery*, PhD thesis, James Cook University, Townsville 2010. For a summary see: I.
Hafez, R. Stephenson, W. Orchiston, *Abd al-Rahman al-Sufi and his Book of the Fixed
Stars*, in W. Orchiston, T. Nakamura, R.G. Strom (eds.), *Highlighting the History of
Astronomy in the Asia-Pacific Region. Proceedings of the ICOA-6 Conference*, Springer,
Berlin 2011, pp. 121–138.

service of the Medici. Narrating the circumnavigation of the globe under the flag of Ferdinand Magellan, Pigafetta noted:

> The Antarctic Pole is not as starry as the Arctic. We see several small stars packed together, which are in the guise of two clouds slightly separated from each other [the Magellanic Clouds], and a little foggy, in the middle of which are two stars very big but not very bright, that move very little. And these two stars are the Antarctic Pole [that's why they don't move much].[43]

Galilei's first telescopic explorations seemed to dispel the idea that nebulae were unresolved stellar systems because of their great distances. Oddly enough, he did not see, or would not see, the milky glow that surrounds Orion's Belt.[44] A French amateur noticed it a few months later. In 1611 Simon Marius rediscovered the nebula in Andromeda, describing it as «a flame of a candle seen through horn [...] like a cloud consisting of three rays: whitish, irregular and faint; brighter toward the center».[45] The subject was taken up by a Swiss Jesuit, who dared to compare Orion's appearance with that of a comet, and then by Giovan Battista Odierna.[46] The Sicilian priest divided the nebulae into three classes: a) *luminous* if they contained stars visible to the naked eye, b) *nebulosae* if the stars were distinguishable only by the telescope, and c) *occultae* if not even the telescope could resolve them. M31 was one of the latter. A Frenchman found the nebula half a century after Marius. Ignoring Odierna's work, he assumed that the object had not been seen for a long time and speculated that it had faded similarly to the amazing Mira Ceti (p. 128), the variable star in the constellation of the Whale. The variation in brightness led to the exclusion of its stellar nature, since, with few exceptions, known stars seemed remarkably stable in brightness.

However, very few people were systematically interested in nebulae; and no one really knew what they were. In 1715, publishing a list of half a

---

[43] A. Pigafetta, *Relazione del primo viaggio intorno al mondo*, C. Manfroni (ed.), Istituto Editoriale Italiano, Milan 1956 (EBook; in Italian).

[44] Asterism of three bright stars evenly spaced in a straight line that marks the constellation Orion, also known as the Three Kings or the Three Sisters. According to astro-mythology, they define the belt of the giant hunter Orion.

[45] K.G. Jones, K. Glyn Jones, *Messier's Nebulae and Star Clusters*, CUP, Cambridge 1968, p. 128.

[46] Before 1654, Odierna described the Andromeda nebula in these words: «A very admirable nebula never seen to my knowledge by anyone [...] No multitude of close stars can be distinguished in it [...] It has a glow similar to a comet».

dozen of these strange sources, Halley took an opposite position to Galilei, describing them as «nothing else but the light coming from an extraordinary large space in the ether; through which a lucid medium is diffused, that shines its own proper luster».[47] The challenge then was to find a way to establish whether, in addition to the clusters of stars faded by distance, the clouds of light imagined by Newton's influential friend and collaborator actually existed. This is where Herschel came in.

It all began thanks to a striking drawing of Orion made by Huygens in 1656, much more detailed than what Odierna had been able to do with his modest means. William saw it reproduced in Ferguson's book, mentioned earlier (p. 177), and was so impressed that in 1774, at the opening of his first observing book, he pointed the telescope directly at the nebula. The object appeared to him different from what the Dutch astronomer had described. This fact intrigued him greatly and directed his interest to nebular objects. Loyal to his new homeland, he made his debut at Halley's party, but determined to question nature to confirm his thinking. He understood that in order to solve the problem, it was necessary to resort to ever more powerful telescopes: «the visibility of stars depends on the penetrating power of telescopes, which, I must repeat, falls indeed very short of showing stars that are many thousands of times farther from us than Sirius».[48] However, he feared that the technological effort alone would not be enough. The availability of ever more powerful telescopes would have led to the discovery of even more distant, and therefore unresolved, nebulae, which would have reopened the reasons for the hunt in a vicious circle. If so, how could stars be distinguished from similar but diffuse objects in the most distant sources?

Remembering Orion's lesson, he thought that the solution to the puzzle might be found by investigating the changes in shape and brightness that nebulae undergo over time. To remain unsolved, he reasoned, a star system would have to be very distant and therefore very large and resistant to short-term changes in both luminosity and appearance. But how could one be sure that any observed changes were not simply the result of astronomers' personal interpretations and different drawing skills? The

---

[47] M.J. Crowe, *Modern Theories of the Universe: From Herschel to Hubble*, Dover, New York 1994, p. 39.

[48] W. Herschel, *IV. On the power of penetrating into space by telescopes; with a comparative determination of the extent of that power in natural vision, and in telescopes of various sizes and constructions; illustrated by select observations*; published 1 January 1800; https://doi.org/10.1098/rstl.1800.0005.

terrain was rugged and littered with pitfalls, and it required another titanic effort: discovering, cataloging, and observing as many objects as possible. So the heroic pioneer embarked on another personal crusade. With the help and original contributions of Caroline,[49] within two decades, he increased the number of known nebulae from about a hundred listed by Messier to over 2,500. Some appeared spotted, others milky. He was convinced that the latter consisted of the luminous medium that Halley had mentioned. Instead, the former were systems of stars that looked fuzzy because of their large distance, but would reveal their true nature when observed with more powerful telescopes. «The stupendous sidereal system we inhabit, consisting of many millions of stars, is, in all probability, a detached nebula. Among the great number of which I have now already seen, amounting to more than 900, there are many which in all probability are equally extensive with that which we inhabit; and yet they are all separated from each other by very considerable intervals»,[50] he wrote in 1785.

William could have stopped there and won the jackpot again! Instead, he wanted to continue the chase and ended up confused, proving that sometimes the perfect is the enemy of the good.[51] This is how it happened. In 1784 Herschel came across M17 and M27, two nebulae that astronomers now call Omega and Dumbell because of their characteristic morphology, and saw that some stars were floating in the halo of light. He then thought that the residual nebulosity was the result of poor resolution and abandoned the luminous fluid model. He had a radical change of mind, as is supposed to happen to truly intelligent people. The consequences of this change were sensational for cosmology. If larger nebulae such as Orion were indeed unresolved star systems due to their distances, it implied that they were really large and capable of «surpass[ing] the Galaxy in splendor.».[52] Today we would say that they should be just as many galaxies, as suspected by Kant, to whom the effective image of the "island universe" is wrongly attributed. The metaphor, reported in the second volume of *Kosmos* eighty

---

[49] Caroline did her part in this deep-sky object hunt, not only as her brother's assistant, but also as the discoverer of prominent nebulae, including the stunning local spiral NGC 253 and NGC 205, the spheroidal galaxy sailing on the periphery of the nebula in Andromeda, which Charles Messier had observed but not included in his catalog.

[50] W. Herschel, *On the Construction of the Heavens*, in "Philosophical Transactions of the Royal Society", LXXV, 1785, p. 249.

[51] Aphorism by Voltaire, taken from an Italian proverb: *Le meglio è l'inimico del bene.*

[52] G.E. Chrisitanson, *Edwin Hubble: Mariner of the Nebulae*, Farra, Straus & Giroux, New York 1997, p. 140.

years after the publication of *General History of Nature*, was coined by the German encyclopedist Alexander von Humboldt to popularize Kantian thought. The universe became incomparably vast and would never stop growing since then. «I have looked further into space — Herschel would write in 1813 to the poet Thomas Campbell — than ever human being did before me. I have observed stars of which the light, it can be proved, must take two million years to reach the Earth».[53] Who knows what he would think today if he discovered that his two million years had become ten billion and beyond!

The story does not end here, for in 1790, while wandering with his telescope in Taurus, where his illustrious career as an astronomer had begun, Herschel came across NGC 1514, «a very striking phenomenon».[54] It was not the first time he had seen this kind of individual stars «surrounded with a faintly luminous atmosphere, of considerable extent».[55] Mindful of his Uranus, he had called them "planetary nebulae". But the central star of NGC 1514 was particularly bright, and therefore the luminescence surrounding it could not be attributed to a swarm of unresolved stars. "It's Halley's fluid", William thought, completely indifferent to the idea of having to retrace his steps once more. The result this time was the disqualification of Orion and the whole class of unresolved nebulae into small cocoon-like fluid stars, and the consequent promotion of the Milky Way to the dominant star system, «the most brilliant and beyond all comparison the most extensive sidereal system»;[56] perhaps even unique in the cosmos, considering that with the 40 foot telescope William had discovered countless other faint stars never seen before. In short, the Milky Way was getting bigger, while the nebulae were getting drastically smaller. Another turn in Herschel's cosmological thinking, the last and this time not very well done: a melancholic epilogue to an impressive chain of resounding successes.

In the midsummer of 1822 he who «*coelorum perrupit claustra*»[57] (broke the barriers of the heavens), as he wanted to have engraved on his tombstone, died at Slough in the loving arms of his wife and sister. Among his

---

[53] In W. Sheehan, C.J. Conselice, *Galactic Encounters: Our Majestic and Evolving Star-System, From the Big Bang to Time's End*, Springer, New York 2016, p. 28.

[54] W. Herschel, *On Nebulous Stars, Properly So Called*, in "Philosophical Transactions of the Royal Society", LXXXI, 1791, pp. 71–88.

[55] Ibid.

[56] Hoskin, *The Construction of the Heavens*, cit, p. 68.

[57] In H. von Humboldt, *Saggio del Cosmos*, C. Wilmant and sons, Lodi 1846, p. 20 (in Italian); Mullaney, *The Herschel Objects*, cit., p. 140.

great achievements, he could boast of having liberated the reflecting telescope from its gregarious role and destined it to become the main instrument of 19th and 20th century astronomy. A year earlier, Napoleon had breathed his last on Saint Helena, a rock lost in the South Atlantic. William Herschel had had the opportunity to meet him in 1802 when he was vacationing in Paris with his family. «The First Consul did surprise me — he would later recount — by his quickness and versatility on all subjects, but in science, he seemed to know little more than any well-educated gentleman, and of astronomy much less, for instance, than our own King. His general air was something like affecting to know more than he did know. I remarked his hypocrisy in concluding the conversation on astronomy by observing how all these glorious views gave proofs of an Almighty wisdom».[58] These words reveal the extent to which the German musician identified with the role of the loyal British subject. England had given him everything a man devoted to science could wish for. He had repaid his adopted country with his loyalty, torn the veils of heaven, and gifted the whole world with a harvest of discoveries and an example to follow. He bequeathed to humanity an immense universe, dynamic and evolving, of which he believed he had seen the limits. «The stars we see in the sky — he once said to the poet mentioned above, who interviewed him and was shocked by it — may no longer exist». In search of the causes, he entered the slippery and unexplored terrain of evolution, daring to draw an analogy between the alternation of forms and properties of celestial bodies and the seasons of plants in a garden, from germination and flowering to withering and decay. Confused elaborations of a thought that was still weak but destined to grow dramatically.

Prostrate with grief at the death of her beloved brother, Caroline returned to Hanover to live on memories. «I am grown much thinner than I was six months ago; when I look at my hands they put me so in mind of what your dear father's were, when I saw them tremble under my eyes, as we latterly played at backgammon together».[59] How much tenderness in those lines sent across the Channel to his nephew. When his father died, John was traveling the world: a restless thirty-year-old, consumed by the anxiety of knowing and, perhaps, the very human and unacknowledged desire to compete with a famous and fully accomplished father. «God knows how

---

[58] In *Life and Letters of Thomas Campbell*, W. Beattie (ed.), in "Gentleman's Magazine", ns, XXXI, 1849, p. 126.
[59] R. Holmes, *The Age of Wonder*, Vintage Books, New York 2008, p. 410.

ardently I wish I had ten lives»,[60] he had once written to Charles Bab-
bage, the college friend, then credited with the first invention of a digital
calculating machine, with whom he had shared his studies and dreams at
Cambridge.

Mourning was also a sign of the beginning of a new course. John had
done a little of everything up to that point, studying light and colors,
crystallography and chemistry, as well as the astronomy that Aunt Caroline
had taught him from his earliest childhood and that the young man had
practiced at Slough with his father's instruments on his father's subjects.
Left alone in the large house full of telescopes, he now had to confront the
memory and scientific legacy of the Herschel brothers. It was a difficult role
that John played admirably. «I think I shall be adding more to [my father's]
fame by pursuing and verifying his observations than by reprinting them»,
he wrote to Auntie Caroline; then he rolled up his sleeves and resumed
the survey of nebulae begun by father William. «I hope this season to
commence a series of observations with the twenty-foot reflector, which is
now in fine order. The forty-foot is no longer capable of being used, but I
shall suffer it to stand as a monument».[61] William had also kept it alive
mainly for reasons of prestige and to satisfy the curiosity of those who
wanted to see the largest telescope in the world.

Unlike his father, who enjoyed observing the sky immensely, John experienced astronomical practice as a necessary duty, and he admitted that
sometimes, «sick of star-gazing», he had the instinct «to break the telescope and melt the mirrors».[62] But his German blood gave him no escape:
a kind of *noblesse oblige*. Thus, having completed the exploration of the
skies accessible from Slough — an undertaking which had served him very
well to cut his teeth as a skilled observer — he decided that it was time
to turn his eyes to the southern skies and to conceive an enterprise worthy of his father. In November 1832, he set out from Portsmouth for the
Cape of Good Hope to explore the rest of the celestial vault that neither he
nor his father had been able to penetrate. He brought with him his young
wife, three children, a nurse, and 18 cases of instruments. The voyage took

---

[60]G. Buttmann, *The Shadow of the Telescope: A Biography of John Herschel*, Lutter-
worth, Guilford-London 1974, p. 14.
[61]Letter from John Herschel to Caroline Herschel dated 1 August 1823, in *Memoir
and Correspondence of Caroline Herschel. By Caroline Lucretia Herschel, 1750-1848*,
Edited by Mrs. John Herschel, John Murray, London 1876, https://digital.library.upenn.
edu/women/herschel/memoir/memoir.html.
[62]Buttmann, *The Shadow of the Telescope*, cit., p. 51.

over nine weeks on a three-masted teak sailing ship that was then used as a prison ship for convicts sent to Australia. What times! And what men! Young Herschel, who in truth had a fair amount of personal fortune, had preferred to pay for the voyage out of his own pocket, turning down both a grant from the Royal Society and the free passage the Royal Navy had offered him and his family on a military vessel. He wanted to be free to enjoy the best four years of his life without further obligations. In fact, his stay in Cape Town, in Feldhausen, a beautiful and comfortable house 50 m above sea level, lasted so long that the astronomer first rented it and then bought it. In the garden he had placed the 20 foot focal, 19 inch aperture reflector that he had brought with him from England.

The mission to "bottom of Africa" produced extraordinary results. John managed to catalog 1,700 new nebulae, 2,100 double stars, and a handful of other objects. He extended the census of Milky Way stars initiated by his father, and came to believe — but he was wrong — that the system should have a ring structure. He welcomed the new passage of Halley's Comet after the historic one of 1758, produced the first star maps of the Magellanic Clouds and the Orion Nebula, and practiced naming the moons of Saturn and Uranus that were awaiting designation: an activity in which he displayed an enviable classical education. In addition to three new children, he and his wife Margaret also produced delightful drawings documenting over one hundred species of native flora. Nothing more could be accomplished. Back home, this tireless activity earned him, among other honors, the title of first Baronet of Slough: much better than the knighthood bestowed on his father in 1816, with permission to be called "Sir".

In 1864, John succeeded in printing his *General Catalog of Nebulae*, containing some 8,000 objects, each identified by a serial number preceded by the initials GC. This monumental work remained unsurpassed for a quarter of a century until the publication of the *New General Catalog of Nebulae and Clusters of Stars*, which was in fact «the Catalog of the late Sir John F.W. Herschel, Bart., revised, corrected, and enlarged». We will say more about this basic reference later. The author, the Danish-British astronomer John Dreyer, adopted a new abbreviation, NGC, to be placed before the entry number of each object. With this convention, the great nebula in Andromeda, named M31 by Messier, became NGC 221, and NGC 598 the spiral M33 in the constellation Triangulum. For his work, Dreyer also took advantage of the legacy of Herschel's 40 inch reflector, the Leviathan of Parsonstown.

This was the largest telescope of the 19th century, a mammoth altazimuth reflector with an aperture of 180 cm. Its father was an aristocratic Irish landowner, William Parsons, third Earl of Rosse.[63] The Parsons had come from England in the late Elizabethan period to establish one of the paramilitary garrisons of the Protestant conquest in Catholic Ireland. They had then settled there, becoming increasingly wealthy, aristocratic, and nationalistic. The family's earldom was located around what is now Birr, then called Parsonstown: a rolling succession of pastures and woods in the center of the Clover Island, not far from a bend in the Shannon River. William was born at the dawn of the nineteenth century, in a painful year for the Irish nation because of the Act of Union, which sanctioned the dissolution of the Dublin Parliament and the formal transfer of the beloved homeland under the British Crown. He was educated at home, as was customary for the country gentry, in the fairy-tale setting of the ancestral palace and under the watchful eye of his father. The latter, embittered by the heavy-handedness of His Majesty's government and disgusted by the treatment of the Protestant minority toward the Catholic majority, had decided to retire from politics and devote himself to the care of his manor and lands and the education of his five children.

At the age of 19, William was enrolled at Trinity College Dublin, with a tacit exemption from the attendance and harsh discipline of a school designed to forge and amalgamate the Irish aristocracy. Then, having earned the easy degree, he was surprisingly sent to Magdalen College, Oxford, away from the comforts of Birr and into the heart of hated England. An unpatriotic gesture that can only be understood by assuming that the parents wanted to give their firstborn a unique opportunity to train as a scientist. In fact, the young man was completely indifferent to the lure of power and the dull idleness of the country gentleman, among dogs, horses, hunting, parties, submissive love affairs, and perhaps a few good books. He had material means, some ingenuity, and plenty of time to try to earn a prominent place on the stage of science; and no doubt he could think big. Returning to Ireland with a degree in mathematics, he decided to follow the example of William Herschel, who had now become an icon of science in the public imagination. It is likely that this choice was motivated by a patriotic claim on the British, but also by an ambition to answer the question left

---

[63] Earl of Rosse, *William Parsons, Third Earl of Rosse*, in "Hermathena", 107, 1968, pp. 5–13; C. Mollan (ed.), *William Parsons, 3rd Earl of Rosse: Astronomy and the Castle in Nineteenth-Century Ireland*, Manchester University Press, Manchester 2014.

unanswered by the great astronomer-musician's many slaloms: what are nebulae?

The first goal of William, who could then boast the temporary title of Lord Oxmantown, was to procure a large-aperture telescope. Unlike the French and Germans, who were trying to break the record for the largest lens, he chose the mirror solution already adopted by William Herschel. The challenge was to make it entirely at home, without the help of London's renowned opticians. But the Lord of Birr was totally inexperienced, and even less so were the craftsmen on his estate: laborers and manual workers whom he wanted to convert to the astronomical enterprise. Technical information on how to make a good speculum mirror was also scarce. Herschel, who had recently died, had not left many instructions on his recipes for preparing and working blanks. He had always regarded this information as a valuable trade secret to be jealously guarded. This being the case, the young lord had to start from scratch, and it was not easy. Among other things, his time was limited by his seat in Parliament, where he sat on the Whig benches.

Lord William then began his astronomical adventure by setting up a foundry on his property to experiment with different alloys and different cooling techniques for castings. In trial after trial, getting his hands as dirty as any other worker, he found his recipe: amalgamation of four parts copper with one part tin. He also built a steam machine for grinding speculum disks. And he finally felt ready to move on to actual production. It seemed a done deal, but instead it took him ten years to see his efforts crowned with success. The cast disks kept breaking while cooling. William persisted until he won. Accepting to abandon a monolithic primary, he succeeded in making mirrors of 15 and 24 inches in diameter by tightly assembling smaller speculum segments into one piece. Finally, in 1839, he reached the coveted size of 3 feet, not far from Herschel's record mirror.

It was just what was needed. Soon, on the lawn in front of Birr Castle — which the parents had left entirely in the hands of their son, retiring to private life, perhaps to avoid witnessing William's follies — a telescope with an aperture of nearly a meter and a design very similar to that of the Slough giant could be seen. Parsons had copied everything from Herschel's project except the idea of using direct vision, preferring a more relaxed Newtonian mount. Irish pride was safe, but astronomy did not celebrate. Beyond its grandiose appearance, the instrument was not much good: a cyclops with one eye seriously flawed, just like Slough's 48 inch. Besides the optics, the mechanics were also very unsatisfactory. Lord Rosse did not lose heart and rolled up his sleeves again. After a little over a year, he was almost ready to replace the cheap segmented mirror with a monolithic one of the same size. Just in time to welcome two talented astronomers, Sir James South and Dr. Thomas Romney Robinson. Lured by the undeserved fame of the giant telescope, the guests wanted to use it to check the resolution of the Orion Nebula into stars, and especially M1, which would soon be called the Crab Nebula at Birr. Who were these two?

English by birth, South[64] worked as an astronomer in Ireland with the Cooper family, owners of Markree Castle in the northern county of Slingo. There he had installed a refractor with a 36 cm achromatic lens made in France by the great Robert-Aglaé Cauchoix, which for a few years was the largest in the world. The mount for this record-breaking instrument had been made in Dublin by Thomas Grubb, a mechanic whose company would soon pass into the hands of the Parsons. Robinson, instead, was a pure Irishman. Like Lord Rosse, he had studied at Trinity College, Dublin, where he had been ordained an Anglican priest. In 1823 he joined the staff of the Armagh Observatory in Northern Ireland, which had been founded some 30 years earlier by an archbishop of the same name and is still well known today for the popular activities of its planetarium.

Arriving at Birr, South and Robinson found the telescope still equipped with the segmented mirror, which was known to malfunction. Nevertheless, they pointed it at a few stars, the planet Uranus, and then the Lyra Nebula, impressed by the instrument's brightness. Perhaps, they thought, once the primary's problems were fixed, this giant might really be the key to solving the mystery of the nebulae. In a fit of enthusiasm, they harassed their host until he mounted the new monolithic mirror, of which he was still not entirely satisfied. The instrument responded a little better, but the Crab Nebula remained unresolved. Convinced that discovery was only a step away, Robinson urged his noble friend to take on the pharaonic project that the Lord of Birr had been harboring for some time: a record-breaking telescope with a 1.8 m mirror aperture. To build such a monster, William could draw on the experience he had gained. He also decided to drastically change the design of the telescope mount: no longer a truss on rails like Herschel but a tube between two solid walls, 7 m apart, 12 m high, and 21.5 m deep, oriented north-south. In this way, the bending and most of the vibrations that had plagued its predecessors were absorbed. Thanks to

---

[64]M. Edmunds, *Founder of the RAS: Sir James South*, in "Astronomy & Geophysics", 60, 1, 2019, pp. 1–17.

a universal joint, the instrument would have total freedom of movement in elevation. Instead, the longitude excursion was limited to about one hour before and after the Birr meridian.

The first 1.8 m blank was successfully produced in 1842, but it cracked during machining. The casting was a titanic undertaking, as Romney Robinson, a friend of William's, effectively described it:

> The sublime beauty [of the moment of fusion] can never be forgotten by those fortunate enough to be present. Above, the sky, crowded with stars and illuminated by a most brilliant Moon, seemed to look down auspiciously upon their work. Below, the furnaces poured out huge columns of nearly monochromatic yellow flame and the ignited crucible during their passage through the air were fountains of red light, producing on the towers of the castle and the foliage of the trees, such accidents of color and shade as might almost transport fancy to the planets of a contrasted double star.[65]

Stubborn William withstood the blow. He found that the strength of the material improved dramatically after annealing and slow cooling of the speculum. The rescue program was going well and fast until the devil got his tail in it. A terrible famine struck Ireland between 1845 and 1849, sowing despair and death. The largely agricultural island was now economically dependent on the English market, which absorbed most of the country's produce. "What would you do if we stopped buying your hams and butters?", the English merchant asked mockingly in a story popular in Dublin salons. "We would eat them", was the Irish farmer's sarcastic reply. In fact, the conservative government of Her Majesty Queen Victoria held the reins of the market very short to extinguish any desire for revolt in the Catholic territories, with a harshness, cynicism, and ambiguity that, truth be told, it did not spare even at home, where workers and miners were exploited inhumanely. The aggravation of this strategy would produce, ninety years later, the crime of the Holodomor in Ukraine, the famine planned by Stalin to implement his agricultural policy.

The extremely poor condition of the Irish countryside was exacerbated in the fall of 1845 by the arrival from America of a fungus that attacked and destroyed the potato crop. As a result, 1846 became infamously known

---

[65]M.W. Parsons, *The Scientific Papers of William Parsons, Third Earl of Rosse, 1800–1867*, C. Parsons (ed.), CUP, Cambridge 2011, p. 21.

as the year of the Great Famine. For two centuries, the tuber had been the staple food of the peasantry. Its disappearance created a crisis that the British authorities did not want and could not solve. There were demonstrations, riots, and mass emigration from the island: a horde of weakened and typhus-stricken people that no one wanted and, especially in Canada, no one could contain. At this point, William was forced to turn his attention from astronomy to far more serious and saddening matters. With his father's death, he had also formally inherited the family estate, the title of Third Earl of Rosse, and the responsibility for his many settlers, as well as a leadership role in the militia.

In any case, despite an inevitable slowdown in work caused by the premises at this tragic juncture, in 1845 he finally managed to have two intact blanks to polish. In a short time, the telescope *monstre*, a real cyclops, was ready for inauguration. The ceremony was attended by the primate of Ireland, fully harnessed, who wanted to take a walk right into the metal tube of the instrument. He carried an open umbrella to emphasize its amazing proportions: 16.5 m long for a weight of 9 tons. The local farmers, half proud and half bewildered, nicknamed it "Leviathan Parsonstown" for its resemblance to the biblical sea monster created by God and endowed with legendary powers, both good and evil. In fact, Lord Rosse's telescope was not only the largest in the world. It was also profoundly innovative. The mirror, with a diameter of 1.8 meters, a thickness of 13 cm, and a weight of 3 tons, was housed in a cell with 81 triangular supports. Their function was to distribute the load and reduce the deformations caused by gravity when the inclination changed. It was replaced by a twin every six months or so to allow maintenance of the optical surface, which was subject to rapid deterioration. The mechanics exploited all the innovations of a technology that had grown vertically in Europe under the influence of the Industrial Revolution. Iron played the lion's share. No one could be surprised at the grandeur of the works created with this ancient material, after seeing bold metal bridges over rivers, improbable pylons supporting canopies and windows of greenhouses, railway stations, and entire buildings conceived as ephemeral containers for temporary exhibitions and later becoming an integral part of the urban landscape. A triumph of mechanical engineering that would culminate in 1889 with the Eiffel Tower and come to a tragic end with the First World War, also fought for and with steel and coal from the Ruhr. The ingenious gears, joints, and counterweights of Lord Rosse's new telescope had already found their way into shipbuilding and railway factories. In short, the Leviathan fully interpreted its time and at the same time proposed a new standard for observing the sky. The mount was of the Newtonian type, with a flat secondary mirror directing the beam laterally to an eyepiece located on the western side of the tube. The observer accessed it from a wooden platform placed between the two walls, which was moved according to the needs of pointing.

A few weeks after first light, South and Robinson returned to Birr to test the giant's muscles. Despite adverse weather conditions, they had enough time to verify the goodness of the instrument and to convince themselves once again that they were close to proving the truly stellar nature of nebulae, with all due respect to the great William Herschel. A belief they shared overseas with William Cranch Bond of Harward College Observatory, and in England with Sir William's son John.

> Should the powers of an instrument such as Lord Rosse's succeed in resolving [the nebulae] into stars, and, moreover, in demonstrating the starry nature of regular elliptic nebulae, which have hitherto resisted such decomposition, the idea of a *nebular matter* in the nature of a shining fluid, or condensable gas, must, of course, cease to rest on any support derived from actual observation in the sidereal heavens, whatever countenance it may still receive in the minds of the cosmogonists from the tails and atmospheres of comets, and the zodiacal light in our own system.[66]

So wrote Robinson in 1845. It seemed only a matter of time, but instead, the discovery of variable nebulae again threw the scientific community into doubt.

The controversy was reopened in 1862 by an article in the "Times" written by John Russell Hind,[67] who claimed to have observed the disappearance of a nebula in Taurus.[68] Superintendent of Her Majesty's Nautical Almanac Office, where essential data for the practice of celestial navigation were prepared, he also directed the private observatory that George Bishop, a wealthy wine merchant who had been captured by Urania, had

---

[66] M. Hoskin, *Rosse, Robinson and the resolution of the nebulae*, in "Journal for the History of Astronomy", 21, 1990, p. 340.

[67] *Obituary Notices: Hind, John Russell*, in "Monthly Notices of the Royal Astronomical Society", 56, 1896, pp. 200–205.

[68] In fact, the English astronomer had unwittingly discovered the first example of a newborn star, still struggling with the settling processes and therefore subject to brightness fluctuations.

built in London's Regent's Park. Hind's job was to hunt asteroids, which had become fashionable again with the discovery, after a 38-year hiatus, of the fifth such object, Astraea, by Karl Ludwig Hencke, one of the many secular "laymen" devoted to astronomy.[69] While exploring the ecliptic with a 7 inch refractor, this Prussian amateur astronomer spotted a faint, as yet uncatalogued star near the Hyades and a nebula beside it, later named T Tauri and NGC 1555, respectively. Nothing special. But ten years later the nebula disappeared. The fact was confirmed by d'Arrest in Leipzig and Otto Struve at Pulkovo, the observatory built on the outskirts of St. Petersburg at the behest of Tsar Nicholas I Romanov. It was only 30 years later that the star reappeared, also in a modified form. Exactly what Herschel had hopelessly tried to find!

The phenomenon was complex, but it clearly indicated the existence of celestial objects whose luminescence could not be attributed to an underlying and unresolved cluster of stars. A fatal blow to the Leviathan, rendered useless by the apparent disappearance of the problem that had motivated its construction. In short, while people were still starving in Ireland, Lord Rosse's telescope stood in Birr as a futile monument to human arrogance, an icon of the megalomania of a rich and extravagant Irishman.[70] Instead, with a sudden turnaround, the Parsonstown 72 inch had already straightened its situation, promoting itself and its creator to a leading role in the history of science. It was Lord Rosse himself who discovered, almost by accident, the spiral structure of M51, the "whirlpool galaxy" of the Canes Venantici inventoried by Messier, which we now know to be an interacting system. Not without some skill as a draughtsman, he made a stunning representation, which he presented in the summer of 1845 at the meeting of the British Association for the Advancement of Sciences, of which John Herschel was president. It was a triumph. Within 15 years, the Earl of Rosse produced or sponsored drawings of another handful of spirals, somehow —

---

[69] Not to be confused with the other German astronomer, Johann Franz Encke, who lived at exactly the same period of time.

[70] «The value of the instrument was not only seen in the enlarged power it gave to astronomers, but it opened the way to other instruments of equal power being constructed [...] The scientific fame of the late Lord Rosse will rest rather upon the mechanical than upon the observational branch of astronomy [...] Considering the immense power of the great telescope, the results that have been emanating from it, although startling in their nature, have been small in extent. Drawings of the most remarkable nebulae, a few sketches of part of the lunar surface, and lastly, a large drawing of the Orion nebula, are the chief fruits that are publicly known to have been gathered from it»; A. Webb, *A compendium of Irish Biography*, "Annual Register", London 1878.

and without fully realizing it — opening yet another window into the realm of galaxies.

In later years, the Leviathan was used mainly by Sir William's assistants. When he died in 1867, his eldest son Laurence inherited the title, the instrument, and the strategies. He hired the Dane John Dreyer to expand Herschel's catalog of nebulae to more than 9,000 objects and to edit the *Index Catalog of Nebulae in the Years 1888 to 1894, with Notes and Corrections to the New General Catalog of Nebulae and Clusters of Stars*, the coding of which is still part of the jargon of extragalactic astronomers. Dreyer later became director of the Dunsink Observatory near Dublin, headed by the Irish mathematician William Rowan Hamilton, famous for the homonymous equations of analytic mechanics, and then of the Armagh Observatory. The grand machine remained more or less functional until the end of the 19th century. From 1907, with the death of the fourth Earl of Rosse, it was left to rust on the vast green lawn surrounding Birr Castle until it was rebuilt for tourist purposes in the 1990s. Like the legendary William Frederick Cody, who, in order to make ends meet, went from hunting buffalo on the prairies of Montana to mock-hunting feathered and cackling Native Americans under the tent of Barnum's circus, the Leviathan now lends itself to being photographed for a handful of shillings by curious tourists who may not realize they are standing in front of an authentic ancestor. Meanwhile, in the simple private workshops of a few French and English pioneers, photography was born; a technique destined to revolutionize astronomy and the way it was practiced.

Chapter 10

Stop the time: Photography and the first all sky surveys

> I captured the light.
> I have arrested its flight.
> Louis Daguerre,
> *letter to Charles Chevalier*

> I don't trust words.
> I trust pictures.
> Gilles Peress

The desire to fix images over time is as old as mankind: just think of the Paleolithic rock paintings. Throughout the history of the various civilizations that have flourished on Earth, artisans and artists have gradually left visual evidence of their world, both real and ideal, mediated by inspiration, technique, and materials; and in the pursuit of fidelity, they invented optical instruments capable of reproducing realistic perspectives ready to be copied on some kind of support.[1] One such device is the *camera obscura*, a name given by Kepler to a closed room or, more modestly, to a box in which light enters only through an opening drilled in one of the walls. And there's the magic! The rays that penetrate the chamber form on the wall opposite the opening an inverted image of what would be seen if the eye were placed directly on the small hole. The ancestor of the modern camera had illustrious enthusiasts: Aristotle and Euclid played with it, Chinese philosophers and Arabs, Roger Bacon, and also Leonardo da Vinci as a painter and scientist. Even in the time of Dante Alighieri, it was common to safely observe eclipses with this device.

---

[1] Perspective representation, the ability to render the depth of three-dimensional objects in two-dimensional images, was developed in the early 15th century by Filippo Brunelleschi and popularized by the Genoese humanist and architect Leon Battista Alberti in his treatise *De pictura* (On Painting).

If the size of the entrance hole is reduced, without exaggeration, to limit the effects of diffraction,[2] the image gains sharpness. The reason is that, if we limit ourselves to purely geometrical considerations, i.e., if we neglect the effects of interference, a point aperture realizes a perfectly afocal optical system, completely free of aberrations. This is demonstrated by the improvement in our visual acuity and the increase in the depth of focus when the pupil contracts under intense illumination. That's why you can read the newspaper in full sunlight without prescription glasses. The downside is that as the aperture gets smaller, the image loses brightness, which also depends quadratically on the scale, i.e. the distance between the aperture and the wall on which the image is formed. Then, as is often the case in life and almost always in politics, there is no choice but to compromise between the two opposing requirements, unless an innovative expedient is devised, as a mathematician, physician, and astrologer from Pavia, Italy, did in the 16th century.

A kind of genius,[3] Gerolamo Cardano had the intuition to arm the chamber aperture with a converging lens.[4] It was a clever way of gaining brightness without losing sharpness, but it introduced a new complication, the loss of afocality. The object framed by the lens had to be brought into focus by a corresponding translation of the wall on which the image was formed. This fact was not only inconvenient. The presence of the optical element required a compromise between the needs dictated by the different colors of light, which, due to the vagaries of refraction, did not share the same focus.

But why bother with how a toy works? The fact is that the dark-room was gradually moving beyond mere curiosity to a growing techno-logical value. Painters, for example, had begun to use it. It was easy to copy a image (upside down) projected on a frosted glass, as in the case of Canaletto's meticulous views and some of Caravaggio's canvases. How-ever, no one knew how to capture this image without the intervention of

---

[2]Loss of directionality of light rays caused, for example, by the presence of an opaque obstacle in the path of the wave front. The portion of the wave that is free to pass does not simply propagate in the direction of the rays. In an optical system, this phenomenon causes a reduction in brightness and sharpness.

[3]G. Cardano, *Della mia vita*, A. Ingegno (ed.), Serra e Riva, Milan 1982, pp. 64–65 (in Italian).

[4]See the amusing analysis of the priority of discovery between Cardano and Della Porta in G. de' Sallustj, *History of the origin and progress of mathematics by several authors gathered in commentaries in the form of a chronicle*, Typography Gismondi, Rome 1846, vol. 3, pp. 15–17 (in Italian).

a copyist or an artist's hand. It was necessary to find a substance capable of transforming itself under the effect of incident light, in the same way that a copper plate is characterized under the sharp point of an engraver's burin. The key was found as early as 1727 — or so the story goes — when a Saxon scholar, the anatomy professor Johann Schulze, proved that silver salts darken in the light.[5] Half a century later, Carl Scheele, a Swedish pharmaceutical chemist whom history has unfairly relegated to a supporting role despite his many valuable discoveries, realized that the effect depended on the color of the source.[6] Salts reacted much more strongly to blue than to red light. Schulze's experiments were revived around 1790 by Thomas Wedgwood,[7] the sickly son of a well-known English potter.[8] He was able to capture the shadow of an object placed on a silver nitrate coated stand and exposed to strong sunlight for many hours. But the image was not stable. Wedgwood did not know how to stop the process of blackening the silver salt. This was no small inconvenience, forcing him to keep the rudimentary photograph in a dark place at the risk of complete blackening, and to view it in the dim, reddish light of a candle.

A few years passed, and in the second decade of the nineteenth century, the Frenchman Joseph Nicéphore Niépce, who is also credited with the invention of the internal combustion engine, devised a way to remove the excess of photosensitive material and thus fix the image.[9] His receptor, however, was not a silver salt but a mixture of hydrocarbons and clay called "Judean bitumen" or "Syrian asphalt", known and used since ancient times for very different purposes. Niépce had noticed that this liquid and viscous substance hardened when exposed to intense and prolonged illumination. After repeated attempts, he finally discovered how to take advantage of its photosensitivity to capture images. He first dissolved the bitumen in lavender essence, then smeared it on a polished pewter plate, which he exposed to light in a dark room in a manner qualitatively similar to modern

---

[5] S. Edwards, *Photography: A Very Short Introduction*, Oxford University Press, Oxford 2006, p. 73.

[6] *Encyclopedia of Nineteenth-Century Photography*, J. Hannavy (ed.), Routledge, New York 2013, p. 316.

[7] Ivi, p. 1483.

[8] Josiah Wedgwood, founder of the Wedgwood Company in 1759, was a pottery entrepreneur and a passionate abolitionist. He is best known for his medallion, a jasper-ware cameo with the image of a kneeling black man in chains and the words "Am I not a man and a brother?".

[9] T.O. Davenport, *The History of Photography: An Overview*, University of New Mexico Press, Albuquerque (NM) 1999, pp. 6–8.

photography. At the end of the exposure, which could last up to eight hours, Niépce treated the heliographic plate with white spirit and lavender essence to remove the material that had not been affected by the light. The image was created by the contrast between the dried bitumen and the shiny pewter exposed by the wash.

It sounds like a recipe for a potion of the Druid Panoramix, the characters of the French comic strip *Asterix*! But it is a false feeling. On closer inspection, it is in fact a technique conceptually similar to etching, used since the 16th century by artists such as the German Albrecht Dürer and the Italian Parmigianino. To make copper printing plates, the engraver would use a burin to remove the wax patina of the plate at the point where he intended the acid mordant to attack the metal. In Niépce's process, the engraver's function was replaced (in an inverted way) by light. Projected as a figure onto the bituminous plate, it marked the areas to be protected. So "written by light" (the term "photography", from the Greek words for light and writing, was coined by John Herschel in 1840), the resulting image was a unique piece. But it could also be reproduced by etching the unprotected copper parts with an acid and then cleaning the plate of the remaining bitumen to obtain a matrix ready for the printing press.

Niépce died in 1833, completely ruined by the investments made by his brother Claude to promote the commercialization of their internal combustion engine, the *pyreolophore*. His pioneering research in photography was continued by Louis-Jacques-Mandé Daguerre,[10] an enterprising Parisian stage designer who had been working with him for four years. He was another man "without letters"; but in this business, as in the Klondike gold rush, academic degrees did not matter. The challenge required ingenuity, courage, determination, and luck. Daguerre focused on using silver-plated copper plates treated with iodine vapors to create a surface film of silver iodide. These plates worked better than the bitumen-coated pewter plates used by his late partner, but exposure times were still too long for subjects that were not perfectly static such as panoramas or monuments. To be portrayed, people had to remain motionless for a very long time, a condition more reasonable in a studio and with the help of props than *en plein air*. The turning point came suddenly. Aided by chance — with all due respect to the epistemologist Karl Popper and to the relative satisfaction of Paul Feyerabend, the Austrian-born philosopher of science who advocated an

---

[10] *Encyclopedia of Nineteenth-Century Photography*, cited above, pp. 363–366.

anarchist vision of science — in 1837 Daguerre discovered by chance that a few minutes of exposure to light was enough to imprint a hidden image on the silver coating of a plate. This latent image could then be intensified (developed, we say today) by a chemical amplification technique until it became visible. The trick was to fume the plate with mercury vapors and complete the process by neutralizing and removing the remaining photosensitive material (fixing process) with a solution of water and common salt, later replaced by hyposulfite at the suggestion of John Herschel.

The invention transformed a curious phenomenon into a powerful tool with multiple applications. The imaginative Frenchman realized he had won the lottery and worked hard to make the most of it. For two years, he tried in vain to market it, telling potential buyers what it could do while managing to keep the core of the process a secret. Finally, he decided to break the seal and reveal his discovery to the public. To illustrate it, Daguerre chose an exceptional speaker, François Arago, a visionary scientist and influential member of parliament. The event took place at the French Academy of Sciences on January 6, 1839. Obviously, Arago limited himself to presenting the results without going into the details of the method. The curiosity aroused by the news was enormous. People wanted to know more, but Daguerre was not available to open up without an adequate counterpart. The situation became burdensome, as some began to suspect that the lack of details might conceal a fraud, and was finally unblocked in the summer of 1839. Under pressure from Arago himself and the chemist Joseph-Louis Gay-Lussac,[11] the French government itself acquired the rights to the daguerreotype, granting a substantial pension to the inventor, with a share also to Niépce's son.[12] Then, with a gesture of refined elegance reminiscent of the style of the Age of Enlightenment, or perhaps in the awareness that there was no way to protect the patent, the Citizen-King of France, Louis Philippe d'Orléans, made this "wonder of the world" a gift to all mankind[13] for any purpose, including, of course, the recording of the sky and its phenomena. It really is a case of saying that class cannot be bought!

---

[11]You may remember him for the homonymous law of direct proportionality between the pressure and temperature of a gas when the volume remains constant.

[12]R.D. Wood, *A State Pension for L.J.M. Daguerre for the Secret of his Daguerreotype Technique*, in "Annals of Science", 54, 5, 1997, pp. 489–506.

[13]With one exception: the hated British were excluded from the benefit. A week before the royal gift, an agent of Daguerre had patented the invention in England. Here, unlike in the rest of the world, it was therefore necessary to purchase a license to use or commercialize daguerreotypes.

The daguerreotype immediately caught the interest of ordinary people. Those who could afford it had their portraits taken, enduring the hassle of a few minutes of posing in exchange for the immortality of their image. After all, a photograph cost far less than an oil painting, was far more faithful — for some ladies, too cruelly faithful — and was ready in a moment. It was a resounding success that convinced improvised "gentlemen scientists"[14] in search of fortune or fame, to tackle the difficult problem of improving a technology whose chemical and physical principles were still very nebulous. As early as 1841, the Englishman William Fox Talbot devised a process for producing negative images on paper that had been made transparent and sensitive to light. Printed by contact, these photographs ("calotypes", from the Greek words for beautiful and figure) could be reproduced in a large number of positives. This was a decisive step forward compared to daguerreotypes, which were sharper and more detailed, but slower and, above all, unique: they could only be duplicated by photographing them again, with increased costs and loss of resolution. For their part, painters, driven out by the unfair competition of the new technique, which represented reality with disarming perfection, took refuge in playing with the human mind's ability to synthesize and express emotions: a new style anticipated by the English painter William Turner and baptized by the Frenchman Claude Monet with a view of the port of Le Havre at dawn, which the artist prophetically titled *Impression, soleil levant* (Impression, sunrise).[15]

Another ten years passed, during which the peoples of Europe again tried to raise their heads against the despotism of absolute monarchs, and the squares of the great continental cities were stained with blood. Finally, in quiet England, a sculptor, Frederick Scott Archer,[16] devised the way to further increase the stability and transparency of Talbot's calotypes by replacing the paper support with glass plates. To trap the light-sensitive salts, Archer resorted to a liquid and viscous substance, collodion or guncotton, which had been discovered a few years earlier in Germany in the course of what we would now call paramilitary research: another useful product of the many created by the Prussians' obsessive desire for supremacy. Wet

---

[14]Post-Renaissance expression used to describe scholars who conducted their research with funding and instruments independent of universities and public institutions.

[15]G.N.J. Makin, *Photography's Impact on the Painted Image after Manet: The Emergent Genre of Painting from Photographic References*, doctoral dissertation, University of Western Sydney, 2015, https://pdfs.semanticscholar.org/6f21/835b6c18892ece15ae9969794a774cd10a83.pdf.

[16]*Encyclopedia of Nineteenth-Century Photography*, cit., pp. 55–58.

collodion plates had the photographic quality of daguerreotypes at least ten times faster and produced better copies than calotypes.

But there was a price to pay. The collodion had to be kept moist to prevent the silver iodide from crystallizing and losing its light sensitivity. This meant that the plates had to be used immediately after preparation. In short, cooked and eaten. The inconvenience was not insignificant, as it forced photographers to move around carrying a field darkroom where they could store and then develop the material in real time. The game was worth the candle, however. The high sensitivity reduced exposures of well-lit subjects to a handful of seconds, allowing traditionally weak light sources to be captured (though not for more than a quarter of an hour, at the risk of collodion drying). This integration time limitation would have been the main reason for the relatively slow penetration of photography into astronomy, as we will see later.

Although the practice was still cumbersome, the last quarter of the 19th century saw a real surge in demand for images and materials for do-it-yourself. Industries for the production of photosensitive paper flourished, which, using egg white, gave rise to the induced activity of industrial chicken farming. Photographic studios proliferated. New camera lenses were studied to reduce aberrations and increase brightness. As exposure times became shorter, curtain shutters replaced the simple covers that the photographer operated in front of the lens, as we sometimes still see in film reconstructions. The ancestor of the flash was born: magnesium was used for the ignition, with some technical and economic difficulties due to the high cost of the metal, until it was produced on an industrial scale. By doping the silver halides, it was possible to obtain materials sensitive not only to blue, but also to green and yellow light. And photography became the faithful companion of journalists, starting with those sent to chronicle the bloody Crimean War,[17] reporters and paparazzi, great travelers, engineers, lovers, and artists.

The complications of wet collodion were quickly overcome, as were the health risks posed by the solvents used. It was Richard Maddox, an English physician interested in giving objectivity to his research with the microscope, who had the idea of using gelatin as a binder for photosensitive salts: a substance of animal origin used in the confectionery industry. The

---

[17]The conflict that began in 1853 as a dispute between Russia and France over control of Christian holy sites in Ottoman territory. Thanks to the first war correspondents, it was the first to be followed constantly by the people of the nations involved.

dry process made it possible to prepare the photographic material well in advance of its use and then to develop it without haste and in the best conditions. Only 32 years had passed since Daguerre's invention! In the version perfected in 1878 by two English photographers, Richard Kennet and Charles Bennet, the dry emulsion achieved extraordinary performance. It took 1/25th of a second to photograph a subject in bright light. The era of the snapshot had begun, whose main interpreter was the American entrepreneur George Eastman,[18] whose story is the archetype of the American Dream that continues today in the world of high-tech, information technology, software, and e-commerce.

The daguerreotype had been imported to the New World by Samuel Morse, inventor of the alphabet of the same name used in telegraphy, and had quickly caught on. But the real boom came with Eastman's dry emulsions. This visionary New Yorker came to photography by pure accident. The story goes that, on the advice of a friend, he bought the equipment to document a business trip to Santo Domingo with the wet collodion plates then in use: a camera with «the size of a soap box», a tripod that «could support a bungalow», a dark-room tent, a nitrate bath, boxes of glass plates, a heavy plate holder, and a water tank. So much stuff, too much! «It seemed that one ought to be able to carry less than a pack-horse load»,[19] he would later say. It was the year 1877. American society, scarred by a recent civil war, was living the contradiction between the spirit of the frontier, symbolized by the wild prairies where, only a year earlier, the Lakota, Northern Cheyenne, and Arapaho, led by Sitting Bull, had defeated General Custer, and an unprecedented drive for modernity. It could be said that every American, especially if he was white, carried in his backpack the possibility of starting a business. What mattered was not what a person was — as long as he had the right color — but what he did and how he did it.

It was in this spirit that the 23-year-old Eastman, despite his lack of specific training, embraced the idea of simplifying photographic technology and making it accessible and attractive to ordinary people. He had read a report on dry emulsions in an English newspaper and plunged headlong into the subject. He worked in a bank during the day and experimented with new solutions in his mother's kitchen at night, forgetting to sleep. In 1884,

---

[18] E. Brayer, *George Eastman: A Biography*, University of Rochester Press, Rochester (NY) 2011.
[19] For this and the preceding short quotes: ivi, p. 26.

he succeeded in patenting his recipe, along with the design of a machine for mass production of the plates. That same year, in Rochester, NY, the town south of Lake Ontario where he had grown up, he founded Eastman Dry Plate Co., later Eastman Kodak Co., a company that would become a world leader in the industry. Europe's strongest competitor, German AGFA (*Aktien-Gesellschaft für Anilin-Fabrikation*), had been founded 17 years earlier near Berlin at the behest of Paul Mendelssohn Bartholdy, son of the great composer Felix. Its excellent products would serve the cause of Nazi propaganda, in the Babelsberg studios in Potsdam and on battlefields all over Europe, proving once again that it is the application of scientific discoveries, not the science itself, that is good or bad.

The Kodak trademark[20] heralded the era of photography "for everyone": a fashionable slogan and populist fig leaf of 20th century industrial capitalism, which then found its symbols in Henry Ford and Volkswagen, the "people's car". In 1888, portable fixed-focus cameras were introduced, equipped with rolls of sensitized paper, which were replaced by celluloid the following year. They allowed up to 100 exposures and were marketed with the slogan: "You press the button, we do the rest". A message worth of Steve Jobs! But that's not all. Cinematography also took its first steps with Kodak. In 1889 Thomas Edison, inventor of the phonograph and the light bulb, patented the idea of intermittently advancing frames on flexible celluloid film to record and then reproduce moving images. A technology that would become the seventh art and the dream industry. George Eastman, a tycoon with a strong philanthropic streak and a complex and fragile personality, was aware that he had helped change the world. «My work is done. Why wait?»,[21] he would write to his close friends in 1932 before blowing his brains out with a gun, unable to bear the physical pain caused by an irreversible spinal disease.

Today, with the advent of digital detectors, the epic of analog photography opened by Daguerre has practically come to an end, but it has left an imperishable mark. Contemporary civilization owes much to this discovery, which, like the printed book, eventually succumbed to its own child, progress. From science to engineering, from art to medicine, from peace to war, from communication to education, nothing in the human world would

---

[20]The origin of the name "Kodak" is unknown. The best guess is that it means nothing. Eastman himself would have chosen this word playing on the sound and on the letter K for which he had a particular predilection.

[21]Ibidem, pp. 515–527.

be as it is if a small army of inventors had not learned to exploit the peculiar way in which silver salts react to light.[22] Neither would astronomy, to which we now return.

Daguerre himself was the first to try his hand at celestial photography, in 1839. Like Galileo, the Frenchman wanted to measure the power of his invention by making a portrait of the Moon. The image of the celestial body and a detailed daguerreotype of a poor, dead spider caught the interest of Arago, who, as we know, was instrumental in bringing the revolutionary photographic technique to the public: the future director of the Paris Observatory was likely to be excited by advances in science and technology as well as by a beautiful woman's face. Unfortunately, shortly thereafter, in March 1839, these first unique pieces were destroyed in a fire in Daguerre's workshop, along with his notes and his diorama, a precursor of film animation. Thus, the era of astrophotography was effectively opened by John William Draper.[23] At the age of 22, in 1833 this eclectic character had decided to leave his native England and emigrate to the New World, where he completed his higher education. A medical graduate, he studied chemistry and physics, as well as history and philosophy: in short, the classic do-it-yourself scholar who combined academic method with undisciplined curiosity. Perhaps he was the first to capture the image of a human face;[24] John Herschel, who was familiar with photography, was very impressed. And in 1840 he obtained a good daguerreotype of the Moon «17 days old», exposing the plate for 45 minutes to the (long) focus of a 13 cm aperture refractor positioned on the roof of New York University, where he taught. The image was immediately presented to the public and caused quite a sensation. In 1845 it was the turn of the Sun,[25] captured in bright light by the French physicists León Foucault, whom everyone remembers for the pendulum experiment at the Pantheon, and Hippolyte Fizeau, and, during

---

[22] A very important and little known example of the use of photography concerns James Clerk Maxwell's insight into the way colors are formed as combinations of certain bases. The scientist had a Scottish tartan photographed through three different filters. The images were then projected simultaneously using three different projectors, each equipped with the same filter used in the exposure. The result was a color image, the first in the history of photography. It happened in the middle of the 19th century.

[23] *Encyclopedia of Nineteenth-Century Photography*, cited above, pp. 437–438.

[24] R.D. Wood, *The Daguerreotype Portrait of Dorothy Draper*, in "The Photographic Journal (Royal Photographic Society)", 110, 1970, pp. 478–482.

[25] See the daguerrotype image of the Sun made by Louis Fizeau in 1845 at https://sunearthday.nasa.gov/2006/locations/firstphoto.php.

an eclipse, by an unknown operator on behalf of the director of the Prussian Observatory at Königsberg.

It is clear that astronomers immediately turned to photography with particular interest. Replacing direct observation with recorded images simplified both photometric and positional measurements and provided objective and reproducible evidence. However, it was necessary to wait for not only the sensitivity but also the speed of daguerreotypes to improve, neither of which was sufficient to capture the faint light of stars. Structurally better telescopes were also needed to meet the demand for long exposures. The low rigidity of the mechanical mounts did not guarantee the maintenance of optical alignment and focus for an adequate period of time, and the systems for tracking daily motion, when available, were essentially left to the skill of the observer. But the most critical node, the one that made daguerreotypes useless for celestial sources that were not light champions like the Sun and Moon, was the failure of what is known as "reciprocity". To understand what this is all about, we need to delve briefly into the nature of the photosensitization process.

Exposing photographic material to light can be compared to an aerial bombardment of a city. The example is bloody, especially in these times, but effective. When a building is hit by a bomb, it begins to be damaged, and if it is hit more than once, it eventually collapses. The same thing happens to a grain of light-sensitive material, which must be hit by several photons to release the metallic silver. Obviously, the effect of the bombardment depends on its duration and intensity, i.e. the number of bombs dropped per unit of time. If one of the two is reduced, the result remains the same only if the other is increased accordingly. In photography, this simple inverse proportionality law is called the reciprocity of illumination intensity and exposure time. The concept was introduced by the German chemist Robert Bunsen, whom we will meet again shortly, and theorized by Karl Schwarzschild, the man who wrote the mathematics of black holes. Now, in photography, reciprocity can fail. Let us see how and why.

For this purpose, we assume that the inhabitants of the city over which the air raid takes place in our example are particularly industrious and try to repair the damage to their houses after each bomb attack. In the case of heavy bombing, these efforts are in vain, as there is a high probability that a building that has already been hit will be struck again and will collapse permanently before it can be made safe again. But if the bombs fall sporadically, even for a very long time, their effects can be countered by the tireless work of reconstruction (if it takes less time than that between two successful bombs), and only a few very unfortunate buildings will collapse completely. The same happens with photons intercepting the grains of the photosensitive substance. If only a few arrive per unit of time, the free-running electrons, the more successful the higher the temperature, will have enough time to repair the effects of individual impacts before they become permanent. In short, in photography it is not true that a lower light flux can be traded for a longer integration time. This reciprocity failure affects any faint source, including stars. The situation is aggravated by the very poor efficiency of the primitive photosensitive materials, which were able to exploit only one photon per million, and only if it was blue (i.e. of the highest frequency among those of the optical window). As if, in our example, most of the bombs fall on the city without exploding, especially if they are of low power.

Astronomical photography[26] made its triumphal entry among the people in 1851, on the occasion of the First Great Exhibition in London: an event full of meanings and consequences, promoted by Prince Albert, cousin and husband of Queen Victoria, with the intention to celebrate the achievements of the industrial civilization. In the magnificent setting of the Victorian Crystal Palace, itself a monument to the new times and the new Iron Age, some daguerreotypes of the Moon were presented. The celestial object appeared as a disk of 12 cm in diameter, rich in details: a valuable document for lunar cartography. The images had been taken by John Adams Whipple with exposures of 40 seconds at the 15 inch aperture equatorial refractor of the Harvard University Observatory. They caused great excitement and were rightly awarded. Whipple was another inventive American. He was a photographer by trade. Aided by some notion of chemistry, he specialized in making his own daguerreotypes, and for these skills he was hired by the director of the Harvard Observatory, William Cranch Bond, also a self-taught astronomer.

Bond had followed in his father's footsteps, making clocks in Boston until the age of fifty.[27] In his spare time, he cultivated a passion for astronomy. In 1839, his hometown university decided to co-opt him with a general free assignment of astronomer-observer. Five years later, buoyed by the

---

[26] D. Norman, *The Development of Astronomical Photography*, in "Osiris", 5, 1938, pp. 560–594.

[27] C. Stephens, *"The Most Reliable Time": William Bond, the New England Railroads, and Time Awareness in 19th-Century America*, in "Technology and Culture", 30, 1, 1989, pp. 1–24.

excitement left by the spectacular Great Comet of 1843, which had nearly caressed the Sun, the former watchmaker, obviously capable of thinking big, decided to try his luck. Boston was a rich and snobbish city. Playing on the generosity and vanity of his fellow citizens, Bond managed to raise enough money to secure a prestigious instrument. Pragmatically, he set his sights on a replica of the large 43 cm aperture refractor that had been installed at Pulkovo, in the Russian Empire, in 1839 and was the largest in the world. Following the example of the Tsar's astronomers, he had turned to the renowned Bavarian firm of Georg Merz and Joseph Mahler, heirs to the company founded by Utzschneider and Fraunhofer (see next chapter), for the optics. The Cincinnatians had made a similar choice a few years earlier for the instrumentation of their fledgling observatory, but had been limited to a lens one-third smaller. The large refractor went into operation at Harvard in 1848 and quickly became famous for the discovery of Hyperion, Saturn's eighth satellite, plucked from the darkness of night.[28]

The success that Bond and Whipple had in London was the result of a combination of a modern instrument and an innovative technology: a winning formula that we will find again and again. At this point you may wonder why the twin telescope in operation at the Imperial Observatory of Pulkovo never appears as a protagonist in our story. The fact is that the director, a renowned astronomer and geodesist, had given his institute a more classical structure, in keeping with the conservative style of the Romanovs, which preferred systematic research to experimentation with novelties. The methodical Friedrich Georg Wilhelm von Struve,[29] educated in the Prussian style, lacked the frontier spirit that instead, by choice and necessity, permeated American society. What a fascinating contrast to his nephew, Petr Berngardovich, early Marxist, leading exponent of the

---

[28]William Lassell, an English merchant from Liverpool and amateur astronomer, had seen the satellite two days after Bond, but was the first to publish the news. He was also the one who suggested the name, following the strategy adopted by John Herschel in 1847, who had named the other 7 known satellites of Saturn by drawning on the list of the celestial god's children. Cf. E. Everett, *Report on the Discovery and Name of an Eighth Satellite of Saturn*, in "Memoirs of the American Academy of Arts and Sciences", 3, 1848, pp. 275–290.

[29]The son of the astronomer Jacob Struve, born near Hamburg in a region of Germany that was then part of the Kingdom of Denmark, was the second in a dynasty of astronomers spanning five generations. He was the father of Otto Wilhelm von Struve, the grandfather of Otto Struve (see p. 487), and the great-grandfather of Hermann Struve. During the Napoleonic occupation of Hamburg, Georg's father Jacob fled to Latvia, a governorate of the Russian Empire, to avoid a call to arms.

Russian Revolution, and author of the *Manifesto of the Russian Social-Democratic Labor Party*.

If the Moon was the Trojan Horse of photography's penetration of the astronomical world, it was the images of the Sun, both in daylight and eclipse, that fully demonstrated the investigative power of the new technique. In the middle of the 19th century, little was known about our star and about stars in general. Visual observation of the disk, very simple and without risk to the eyes thanks to Scheiner's invention (see p. 114), provided little information beyond the spots, the intensely bright regions called "facole", and some flares. Although rare, the total eclipses showed something more: a faint corona and protuberances that the brevity of the phenomenon did not allow for adequate documentation. In the absence of new suggestions, the model of the Sun still in vogue was that proposed by William Herschel at the end of the 18th century: a solid sphere surrounded by a complex two-layer warm atmosphere, which the first daguerreotypes seemed to confirm. In fact, the images obtained by Foucault and Fizeau in 1845 at the request of Arago showed a marked decrease in brightness near the edge of the disk. Exactly what one would expect from a gaseous atmosphere. The interpretive stalemate called astronomers to arms. They fumbled in the dark for the key to understanding where the Sun's light came from.

Although no one had realized it yet, this key had already been identified by a German physician who also cultivated physics among his many interests. Julius Robert von Mayer,[30] born in 1814 without the noble particle (*"von"*) attributed to him towards the end of his life, was the classic thought adventurer. Reckless and libertarian, he had soon fled the shores of his native Neckar for the Far East, following his insights. He was the first, together with James Prescott Joule,[31] an English physicist, mathematician, and brewer, to imagine the equivalence between heat and mechanical work. Their proposal anticipated the formulation of the more general principle of conservation of energy that the American physicist Richard Feynman explains as follows: «The physicist uses ordinary words in a peculiar manner. To him a conservation law means that there is a number which you can calculate at one moment, then as nature undergoes its multitude of

---

[30]K.L. Caneva, *Robert Mayer and the Conservation of Energy*, Princeton University Press, Princeton (NJ) 1993.
[31]D. Cardwell, *Science and Technology: The Work of James Prescott Joule*, in "Technology and Culture", 17, 4, 1976, pp. 674–687.

changes, if you calculate this quantity again at a later time it will be the same as it was before, the number does not change. For instance, suppose there is one bishop only, a red bishop, on the board, then since the bishop moves diagonally and therefore never changes the color of its square, if we look away for a moment while the gods play and then look back again, we can expect that there will be still a red bishop on the board, may be in a different place, but in the same color square».[32] The immediate effect of such a Solomonic law of nature is that every spender becomes inexorably impoverished unless something or someone keeps feeding him.

In fact, the problem of energy entered into the more general question of the structure of the star. In the 5th century B.C., Anaxagoras had imagined that the Sun was a fiery metal ball larger than the Peloponnesus, and the Moon a large rock.[33] A naive model that nevertheless purged physical thought of religious superstition, not without risk. Accused of impiety by the well-thinking Athenians, the philosopher was saved only by an excellent lawyer, Pericles. In order to have a model of the Sun that contemplates the gaseous state, we have to reach Kant and the developments of his nebular hypothesis: the star and its planetary system would form from the slow collapse of a primordial cloud of fluid matter. The German thinker imagined that the Sun drew energy from the materials of the protocloud because «these lighter and volatile parts — he explained in the mid-1800s — are also the most effective in maintaining fire, [and] we can see that by adding them, the body and central point of the system attain the advantage of becoming a flaming sphere, in a word, a sun. Conversely, the heavier and powerless material and the absence of these fire-feeding particles will make of the planets only cold and dead lumps, deprived of this quality».[34]

On the contrary, Herschel imagined a solid Sun surrounded by a liquid atmosphere. The first person to question an entirely gaseous sphere was the English philosopher Herbert Spencer in 1858, later a leading exponent of that Social Darwinism which attributes to nature the ability to select

---

[32] R. Feynman, *The Character of Physical Law*, The MIT Press, Cambridge (MA) 2017, p. 59.

[33] P. Curd, *Anaxagoras of Clazomenae: Fragments and Testimonies*, University of Toronto Press, Toronto 2007, pp. 110–112.

[34] I. Kant, *Universal natural history and theory of the heavens or essay on the constitution and the mechanical origin of the whole universe according to Newtonian principles*, p. 274, at http://users.clas.ufl.edu/burt/spaceshotsairheads/Kantuniversalnaturalhistory.pdf.

the best.[35] His idea of a fluid sphere agitated by radial motions was taken up by the French astronomer Harvé Faye, who in turn was stimulated by reading some articles by a Jesuit astronomer, the Italian Angelo Secchi (cf. p. 283), and quickly adopted it. «So then the exterior surface of the Sun, which from far appears so perfectly spherical, is no longer a layered surface in the mathematical sense of the word. The surfaces, rigorously made up of layers, correspond to a state of equilibrium that does not exist in the Sun, since the ascending and descending currents reign there perpetually from the interior to the superficial area; but since these currents only act in the vertical direction, the equilibrium is also not troubled in that sense, that is to say, perpendicularly to the leveled layers that would form if the currents came to cease».[36]

The models of the Sun that would be produced along this line, and which we will discuss later, would have to be tested again (*"riprovati"*, according to the Galilean methodology) by measurements of physical parameters. Let us stay for the moment with the observation and interpretation of the surface features of our star. First of all, the spots, rediscovered in Europe thanks to the use of the telescope, but already known in India in the third century. They were understood as perforations in the gas mantle, operated by unknown forces, exposing the relatively cooler and darker solid surface below. The scheme was consistent with the hypothesis of the formation of the Solar System formulated by Immanuel Kant in 1755 and taken up 40 years later by Pierre-Simon de Laplace in his *Exposition du système du monde* (Exposition of the System of the World): a more "popular" presentation of the theories then published in his *Méchanique Céleste* (Celestial Mechanics). An observation made as early as 1776 was used to support this idea. The Scottish scientist Alexander Wilson had noticed that the penumbra, the darkest part of a spot, loses symmetry as it moves from the center to the periphery due to the effect of the Sun's rotation. The same thing happens when you look vertically into a bowl with a black bottom. If you tilt it to simulate its participation in a rotational motion, the rim will appear more pronounced on one side.

Everything seemed crystal clear when a chance discovery by a Saxon pharmacist rekindled interest in these mysterious solar scars. Samuel

---

[35] J. Offer, *Herbert Spencer and Social Theory*, Palgrave Macmillan, Houndmills 2010.

[36] H. Faye, *On the Physical Constitution of the Sun* (1865), in "Progress in Physics", 3, 2011, p. 37. See also Robitaille, *On the Presence of a Distinct Solar Surface: A Reply to Hervé Faye*, cit., pp. 75–78.

Heinrich Schwabe had been forced to give up his career as a scientist due to family problems. He had one dream left. He wanted to find the ghost planet that was said to orbit around the Sun in Mercury's orbit. An ambitious and very difficult task, because the fish to be caught had to be very small. The only way to capture it, Schwabe reasoned, was to see its shadow silhouetted against the star's glittering disk. With this idea in mind, he had diligently recorded, day after day for 17 years, whatever signs appeared on the Sun's photosphere.[37] He did not find what he was looking for, but in 1843, while examining the very long series of daily observations, he noticed that the number, appearance, and even behavior of the spots he had surveyed showed a certain degree of regularity.[38] It was the anticipation of the 11-year cycle of solar activity. The physical knowledge attained at that time did not allow scientists to understand what was the cause of such a bizarre periodicity. It was only recognized that the fiery star, like the quieter Earth, was subject to some sort of seasonal behavior. John Herschel, who had a good nose, sensed that the matter deserved attention, and as early as 1858 he suggested to the numerous gentlemen scientists in his country that they engage in methodical monitoring of the Sun using collodion plates to document the location and extent of the spots. He was taken seriously, but the important discovery again came from visual observation.

In 1861 a retired Scottish engineer, James Nasmyth,[39] announced to the Literary and Philosophical Society of Manchester that under perfect atmospheric conditions the solar photosphere appeared to be covered with willow-leaf shaped filaments. He had unwittingly identified the superficial manifestation, later called granulation, of the radial motions by which the solar matter transfers outward the heat produced in the central furnace to compensate for radiation losses and carried up to more than halfway directly by photons.[40] Another of the sparse and still indecipherable pieces of a mosaic that was slowly being assembled and that would soon provide

---

[37]The photosphere is the apparent surface of a star. For radiation of a given wavelength, it is the lowest surface that it can be reach from the outside without being completely absorbed.

[38]Cf. H. von Humboldt, *Cosmos: A Sketch of a Physical Description of the Universe*, Harper & Brothers, New York 1868, Vol. 4, pp. 85–88.

[39]J. Nasmyth, S. Smiles, *James Nasmyth Engineer: An Autobiography*, John Murray, London 1883.

[40]In fact, up to 71% of the solar radius the energy transfer is radiative. It occurs without macroscopic drift of matter.

the basis for the first physical models of the Sun. The discovery was documented by impressive drawings made by Nasmyth himself. He must have been a truly exceptional observer, considering that neither daguerreotypes nor wet collodion plates obtained the following year by Warren de la Rue at Kew Observatory on the outskirts of London could confirm the phenomenon. The sensitivity of the photographic material was not yet sufficient to beat a well-trained eye. But it was only a question of time. Nasmyth had a genius for machine tools. He patented the pneumatic hammer and contributed to the development of railroads with his inventions. When he thought he was rich enough, he retired to private life, «letting younger men have their chance»[41] — he later explained — but also to devote himself to his hobbies, including Urania. Astronomers remember him mostly as the father of the optical system of the same name. The scheme of the "Nasmyth focus" is essentially that of the Cassegrain telescope, with the addition of a third flat mirror placed just before the combined focus, which allows beam extraction perpendicular to the optical axis. In Nasmyth's time, this technical trick was more of a convenience, paid for dearly by the light lost by the third reflection. Today, it is often an unavoidable necessity. Combined with an altazimuth mount, it allows the use of particularly heavy, bulky, and delicate focal plane instruments, since they can be held rigidly connected to the fork that moves the telescope in azimuth.

Nasmyth's communication on granulation provoked much discussion and some doubt. The opinion maker John Herschel immediately considered it «the most marvelous phenomen[on] that have yet turned up — I had almost said in all Nature — certainly in all Astronomy», venturing to speculate that these features were produced by «*organisms* of some peculiar and amazing type».[42] He did not go so far as to admit that the swarming on the Sun's photosphere had anything to do with authentic life forms, but he hinted that the life action was capable of producing heat, light, and electricity. Meanwhile, the director of the College Observatory in Rome, the forty-year-old Angelo Secchi, had joined the all-British game by finding a name and an interpretation for the granulations.[43] The first satisfactory

---

[41] Ivi, p. 364.

[42] C.F. Bartholomew, *The Discovery of the Solar Granulation*, in "Quarterly Journal of the Royal Astronomical Society ", 17, 1976, p. 266.

[43] P.-M. Robitaille, *On Solar Granulations, Limb Darkening, and Sunspots: Brief Insights in Remembrance of Father Angelo Secchi*, in "Progress in Physics", 3, 2011, pp. 79–88.

photograph of the structure of the solar disk would have come in the 1870s from Jules Janssen, an astronomer from the *Bureau des longitudes* whose fame is linked to the discovery of helium in the Sun.

It happened on the occasion of the total solar eclipse of 1868, a wonderful spectacle, but also a unique opportunity to study the atmosphere of the star, otherwise invisible because of its extreme weakness compared to the photospheric radiation scattered by the Earth's atmosphere. Janssen had gone as far as Guntur, on the east coast of India, to observe the Moon's passage across the solar disk, despite the disability in his legs left by a childhood accident. Nothing extraordinary, really, for a character who two years later would mock the Prussian siege of Paris[44] by flying in a hot air balloon to Oran, Algeria, to witness another eclipse. He had brought with him a prismatic spectroscope, an instrument recently revived by German physicists, as will be seen in the next chapter. At the moment of totality, he pointed his instrument at the corona (the "solar atmosphere") and saw a particularly intense yellow line that stood out among the familiar hydrogen lines. At first he thought the feature was due to sodium, but he had to change his mind. It seemed that the line had no father in any terrestrial element, since no one could reproduce it in the laboratory. A mystery! Perhaps a mistake?

The Frenchman was not alone in this story. At the same time, and independently of him, the yellow line had been discovered by an English amateur astronomer who was exploring the solar limb with a spectroscope from his home in the London suburb of Wimbledon. Joseph Norman Lockyer, an eclectic intellectual with a journalist's streak,[45] even took the trouble to name the mysterious substance. He chose ἥλιο (helium), the Greek name of the solar deity, as if to emphasize its extraterrestrial nature. The discovery, announced with incredible timing by the two authors on the same day at the same academy, was greeted with warmth and skepticism by the establishment. But thirty years later, the element was recognized in a sample of uranium ore and rightfully added to Mendeleev's periodic table. Janssen and Lockyer were duly rewarded. They had given the world knowledge of

---

[44]The siege of Paris, which lasted from September 1870 to the end of January 1871, and the subsequent surrender of the city to Prussian forces were the decisive events that led to the French defeat in the Franco-Prussian War, the founding of the German Empire, and, four years later, the birth of the Third Republic.

[45]W.W. Campbell, *Sir Joseph Norman Lockyer*, 1836–1920, in "Publications of the Astronomical Society of the Pacific", 32, 189, 1920, pp. 265–268.

the second most abundant element in the cosmos after hydrogen,[46] a property that would be discovered half a century later by an English doctoral student at Harvard, Cecilia Payne (married Gaposchkin), who, however, was prevented from publishing the revolutionary work. The French scientist would end his career by taking charge of the construction of a large observatory on the outskirts of Paris, in Meudon; his English colleague by editing the articles sent to "Nature", the scientific journal he had founded in 1869. Now content, Lockyer enjoyed the fame that his many interests brought him. These included archaeoastronomy, which he had helped to found with his studies of the megalithic site of Stonehenge, aiming to find the "shoulders" of the oldest giants in the Newtonian pyramid of scientific knowledge.

Let us return to eclipses, which in the second half of the 19th century had become the main laboratory for the study of the Sun. Unfortunately, the phase of totality is quickly consumed, putting the astronomer off until a later date and another arduous trip, usually a year later and in some remote corner of the world. Stuff for the rich! So it is easy to see the advantage that a technique capable of freezing images in time could offer in capturing such a fast phenomenon, provided that the exposures are suitably short. The first to try this idea was the Milanese physicist Gian Alessandro Majocchi, when in July 1842 the Sun played hide-and-seek with the Moon directly over his city; but his iodized plate showed not even the shadow of a signal. More fortunate was the German August Busch, director of the Königsberg Observatory, who photographed the black Sun from Scandinavia during the 1851 eclipse with a lens of only 6 cm. His daguerreotypes fully documented the reality of the coronal phenomena, which until then had been observed only for brief moments. Solid confirmation came in July 1860 from an English expedition based in Rivabellosa, in the region of Castile and León, and independently from Father Secchi, who had coordinated the observations of the event carried out with the 17 cm aperture Caochoix equatorial owned by the Roman College. The eclipse of 1868 in India and Malaysia, in which Janssen had participated, went badly for the astrophotographers, who made up for the American eclipse of 1869 and especially that of 1870, which interested the Mediterranean area.

---

[46] Cf. M. Capaccioli, *La scoperta dell'elio*, in "Il Nuovo Saggiatore", 34, 5–6, 2018, pp. 49–59 (in Italian); B.B. Nath, *The Story of Helium and the Birth of Astrophysics*, Springer, New York 2013.

The French astronomer Camille Flammarion, with a historical intelligence that even today laments the return of nationalism, wrote:

> In our enlightened Europe there still remain some vestiges of ancient fears, and we sometimes associate these phenomena, like the disagreeable facts of meteorology – such as storms, inundations, tempests – with the beliefs of antiquity in Divine anger. During one of the recent partial eclipses, which was well observed in our climate, that of March 6, 1867, the government of a girl's school in Paris took their pupils to prayers in order to advert the malediction of the Most High. I heard nothing similar said on the occasion of that of December 22, 1870; it is true that everyone had then other dominant preoccupations, and that this eclipse was itself eclipsed by that of common sense: two intelligent and reasonable nations were mutilating each other without anybody even knowing for what reason. Two hundred and fifty thousand men dead and ten thousand millions of franc were cast to the wind. Former, they would have associated with this international slaughter either that eclipse of the "terrible year" or the aurora borealis which then appeared in the sky; now everyone understands that it had no other cause but human folly.[47]

But the real breakthrough came with the American eclipse of July 29, 1878, which, because it coincided with the minimum of solar activity, made it possible to document tremendous coronal activity.[48]

Around 1880, with the advent of dry silver bromide emulsions, the photographic plate won the competition with the human eye, and astrophotography was promoted to common practice, especially by amateurs who could risk their reputation and their money more cheerfully than professional astronomers. It was no longer necessary to limit oneself to the light streams from the Sun and the Moon due to the mediocre efficiency of daguerreotypes. Nor was there any need to resort to the complicated maneuvers required by the wet collodion, which, among other things, shrank as it dried, thereby compromising astrometric fidelity. The stars, then, behaved like patient subjects, willing to remain diligently posed for hours if their

---

[47] C. Flammarion, *Pupular Astronomy*, Chatto & Windus, London 1907, p. 200.
[48] G.F. Barker, *On the Total Solar Eclipse of July 29th, 1878*, in "Proceedings of the American Philosophical Society", 18, 102, 1878, pp. 103–114.

daily movements were followed by well-controlled and well-guided telescopes. What a difference from the poor human eye, which is limited to
exposures of fractions of a second! By integrating for hours, it became possible to see and measure sources hundreds of times fainter. It really is true
that *gutta cavat lapidem* (the drop hollows out the stone)!

So when the Great Comet of 1881, sighted a month and a half earlier by
the Australian amateur John Tebbutt, stretched its enormous tail across
the skies of Europe, many tried to photograph it. First the usual Janssen,
then Andrew Ainslie Common,[49] an English gentleman scientist who kept
a 36 inch telescope in the garden of his house on the outskirts of London;
Henry Draper, son of William; David Gill, a Scottish student of Maxwell;
and William Huggins, whom we will soon find behind a spectroscope. A
curiosity: lacking any form of rapid communication, Tebbutt had not managed to get the news of the comet's discovery to England. So when the
intruder appeared to the inhabitants of the Old Continent, it was already
very conspicuous and made a great impression. The year before, Draper had
for the first time portrayed a nebula, M42 in Orion. In 1885, the Pleiades
showed the brothers Paul and Prosper Henry, two Parisian astronomers and
telescope makers, the nebula that chastely shelters the Seven Sisters[50] of
Greek mythology: Sterope, Merope, Electra, Maia, Taygeta, Celaeno, and
Alcyone.

In the same year, Robert Williams Wood began his systematic work
on the stars. To this American physicist we also owe the invention of the
first liquid mirror, obtained by rotating a tank of mercury: an idea that
had already been considered half a century earlier by Ernesto Capocci,
director of the Specola di Capodimonte in Naples.[51] Under the combined
effect of gravity and centrifugal force, the surface of the metallic liquid,
which has good reflectivity and high viscosity, takes on a parabolic shape.
Wood built an instrument with an aperture of 50 cm, but it gave distorted
images because of a technological ingenuity: the use of too thick a layer of
mercury.

---

[49] *Obituary Notices: Fellows: Common, Andrew Ainslie*, in "Monthly Notices of the
Royal Astronomical Society", 64, 1904, pp. 274–278.

[50] The Pleiades, so named probably to note that in the Mediterranean the season of
navigation began with their Heliacal rising (*plein* in Greek means to sail).

[51] P. Hickson, *Liquid-Mirror Telescopes: An Old Idea for Astronomical Imaging is
Undergoing a Technology-Driven Renaissance*, in "American Scientist", 95, 3, 2007, pp.
216–223.

But the application where photography should make a difference concerns the age-old problem of cataloging as many stars as possible, determining their positions and magnitudes accurately, and repeating measurements to track changes over time. The advantage over direct visual observation is obvious and concerns both the quality of the work and the quantity of the product, the repeatability of the measurements, and the possibility of parallelization. Star catalogs have a history going back thousands of years. In the Western world, the first known catalog was built by Hipparchus with a thousand stars. The original work has been lost, but perhaps it was with these data that the map of the heavens was carved on the marble globe held up by Atlas in the Roman copy of the Greek statue of the Titan that belonged to the Farnese family and is now in the Archaeological Museum in Naples.[52] In the year 137 of the Christian era, the astrometric measurements of Hipparchus were taken up by Ptolemy and reported in the *Almagest* with new values of position and magnitude. The Alexandrian astronomer claimed to have made the observations himself, but there are reasons to suspect that he was taking credit for something that was not his.[53] The systematic errors in his sidereal positions have led some historians to believe that Ptolemy simply corrected Hipparchus' observations by precession, and not too well. In the 15th century the Persian king Uluǧ Bek[54] compiled his own catalog of a thousand stars. He had sighted them with a giant sextant to increase positional accuracy (*ceteris paribus*, a larger circle implies more accurate angular measurements). This contribution to celestial geography complemented the revision of Ptolemy's tables sponsored by Alfonso X of Castile and Leon around 1250 and carried out in Toledo by his Muslim astronomers: valuable intellectual labor that the very Catholic kings of Castile and Aragon would later lose due to their religious integralism.

Brahe, Kepler, and Johann Bayer were the last to risk mapping the sky with the naked eye. Hevelius was the first to use a wall clock and a telescope in the mid-17th century. With the Polish astronomer, astrometric accuracy improved to half a minute of arc, an unimaginable value

---

[52]B.E. Schaefer, *The epoch of the constellations on the Farnese Atlas and their origin in Hipparchus's lost catalog*, in "Journal for the history of astronomy", 34, 2005, pp. 167–196.

[53]J. Evans, *On the Origin of the Ptolemaic Star Catalog. Part One*, in "Journal for the History of Astronomy", 18, 3, 1987, pp. 155–172.

[54]J.J. O'Connor, E.F. Robertson, *Uluǧ Bek*, in "MacTutor History of Mathematics archive", 1999, http://mathshistory.st-andrews.ac.uk/Biographies/Ulugh_Beg.html.

before Galileo's revolution. Then it was the turn of Flamsteed, the French astronomer Lacaille, and Piazzi, who achieved a precision of a few arcseconds. In 1845, the *General Catalogue*, containing over 8,000 stars, was printed for the types of the British Association for the Advancement of Science, under the supervision of Francis Baily, a former stockbroker who had made money from his studies of the scientific basis of insurance. Still a small thing compared to the monumental work of Friedrich Wilhelm August Argelander[55] and his students in the first half of the 19th century, all sick with precision and dedicated to work as a social mission. Argelander was half-Finnish — he was born in Lithuania in 1799 — but had trained with the great Bessel in Germany. There he became a personal friend of King Frederick William IV, brother of the future first emperor of modern Germany. A useful acquaintance: the ruler had indeed sponsored the creation of a small observatory associated with the University of Bonn, in the heart of the emerging kingdom of Prussia. The astronomer was thus able to realize his dream of making a census of as many stars as possible in order to detect transient celestial phenomena: an ambition and a strategy that Hipparchus also seems to have developed after observing a new star, possibly a supernova, appeared in the constellation of Scorpius in 134 BC.

Argelander's main goal was to discover new asteroids, the faint celestial wanderers that were very fashionable targets at the time. He was 60 years old when his gigantic project was launched with the publication of the first catalog. It was 1859, the year of the war between France and Austria, which would lead to the first unification of Italy, and the publication of Charles Darwin's revolutionary theory *On the Origin of Species*. The astronomer continued to work on the project until his death, using a modest 16 cm refractor locked in the meridian, with which he recorded the transits of all the stars brighter than the 10th magnitude, that is 40 times fainter than the faintest stars visible with the naked eye. A war machine that required two astronomers on every clear night, one to mark the moments of the meridian transits and a second to diligently record them in the observation log. When completed, the *Bonner Durchmunsterung* ("Bonn Systematic Survey") collected data on 324,000 objects, mostly boreal. One of Argelander's students arranged to complete the ball by cataloging the Australian stars as well. The census of the southern sky continued in Argentina with a deep

---

[55]D.M.F. Chapman, *Reflections: F.W.A. Argelander - Star Charts and Variable Stars*, in "Journal of the Royal Astronomical Society of Canada", 93, 1999, pp. 17–18.

catalog,[56] made by U.S. astronomers in the manner of the Germans, using the eye as a detector to estimate the positions and magnitudes of nearly 600,000 objects.

Finally, in 1896, the English David Gill[57] and the Dutch Jacobus Cornelius Kapteyn[58] carried out the first stellar astrophotometry campaign based on photographic images. The idea came to Gill while he was in South Africa documenting the transit of a comet. After obtaining financial support from the Royal Society, which made little distinction between professional and amateur astronomers and valued the quality of projects over academic merit, he began observations in 1885 with a 4 inch photographic camera made according to the recipe of the great Anglo-German optician John Henry Dallmeyer. The positions of the objects recorded on the photosensitive support were related to the absolute coordinates of a handful of fundamental stars. The latter had been measured with extreme precision, one by one, with traditional transit instruments, and several times in different epochs to derive their proper motions, so that the positional standards could be related to a single epoch. Specialized institutes such as the U.S. Naval Observatory in Washington, originally a U.S. Navy depot for maps and instruments, the Royal Cape of Good Hope Observatory in South Africa, one of the British Empire's many gateways to the southern skies, and the Pulkovo Observatory in Russia were responsible for this task. It was mainly on this terrain that 19th and early 20th century astronomy earned its reputation as a hyper-precise science. Needless to say, the first *Fundamental Catalogue* saw the light of day in 1879 by a German, Arthur Auwers. He is much better known for having sensed the presence of a companion in orbit around the star Sirius, later identified as the prototype of the family of degenerate stars called white dwarfs (see pp. 446–448). Another German followed, August Kopff, author of the *Dritter Fundamental Katalog* or FK3, published just before the Second World War. These tasks were carried out with *ad hoc* instruments following a scheme conceptually similar to Brahe's: waiting for the star to cross the meridian to measure the instant of transit and the distance from the pole, but with a technology and precision very different from that of the Danish pioneer.

---

[56] In astronomical jargon, the adjective "deep" is synonymous with great distance.
[57] J.C. Kapteyn, *Sir David Gill*, in "Astrophysical Journal", 40, 1914, pp. 161–172.
[58] P. van der Kruit, *Jacobus Cornelius Kapteyn, Born Investigator of the Heavens*, Springer, Berlin 2015.

Let us return to Gill's survey. When Kapteyn heard about the project, he asked the British astronomer if he could join the team. His offer was to do the hardest and most tedious part of the project, measuring the plates. The strategy of the joint venture was decidedly innovative. There was no need for Kapteyn to move to South Africa. To carry on his share of the work, he simply waited for the photographic material to arrive in Gröningen, where he had set up a special laboratory. This was the first step towards that *modus operandi* of modern astronomy called "service", which involves, out of necessity or convenience, a clear separation between the moment of observation, entrusted to specialized technicians, and that of reduction and analysis of the material by the astronomers who request it. In his letter to Gill, Kapteyn had stoically confessed that he knew the whole affair would cost him 6 years of hard work. But he was wrong by default: it took 13 years!

Why such a willingness to sacrifice? The fact is that soon after his appointment as Professor of Astronomy and Theoretical Mechanics at the University of Gröningen, Jacobus had to leave the muted microcosm of those who navigated theories and formulas and quickly find out how the world really works. Colleagues at the historic universities of Leiden and Utrecht, his *alma mater*, unwilling to share their already meager state funding with another diner, had opposed his plans to upgrade the observatory in the small town in northeastern Holland. The closure frustrated Kapteyn's ambition to continue with modern means along the path laid out by Herschel: reconstructing the distribution of matter in the Milky Way by measuring the positions of stars.

The association with Gill therefore seemed to him a blessing in disguise, an opportunity not to be missed at any cost. Certainly the drudgery and boredom of long and routine operations did not frighten a passionate researcher who had been brought up in the shadow of the rigid precepts of Calvinism: thrift, duty, humility, work, and discipline. Born in 1851, Jacobus was the ninth of 15 children. His father, a boarding school director, believed that his children should be treated no differently from other students. The future astronomer suffered so much from this egalitarianism that he spent his life striving to earn affection and respect, first from his parents and then from his colleagues. How different from the peoples of the warm lands of southern Europe, weakened and corrupted by Catholic tolerance of sin and the culture of forgiveness, but in greater harmony with the world. In short, Father Kapteyn had forged a tireless and determined robot, ruthless with himself, staff, and students, unable to lose himself in contemplation, efficient and tyrannical, though seemingly kind and caring. Thanks to him, however, Dutch astronomy, despite the country's small size, was able to assert itself on the world stage and secure a leading role that the Netherlands still holds today. After 1908, when Kapteyn came into close contact with the American George Hale at Mount Wilson, he used his prestige and valuable advice to sponsor the emigration to the United States of his young collaborators Adrian van Maanen, Pieter van Rhijn, and the Dane Ejnar Hertzsprung, his future son-in-law, thus creating a unique and lasting link between the small Holland and the North American giant. This is the channel through which Jan Oort, Bart Bok, Lodewijk Woltjer, Sideny van den Berg, and also Marteen Schmidt, the man who discovered quasars (see the last two chapters), would have passed. The visionary teacher had opened a new alternative destination to the road that led the rest of the world to Germany, which by then had become so dominant in all fields of knowledge that German had become the lingua franca of the sciences.

The construction of the *Cape Durchmusterung*[59] represented not only a scientific enterprise, but a veritable laboratory of the future, in which new astronomical technologies were conceived and tested for the first time. For example, the determination of the magnitude of stars by the size of the photographic image of each object: a decisive step towards quantitative photometry, repeatable and of relatively simple application. After astronomy, the nascent field of astrophysics also gained a reputation as a precise science. Among other things, Kapteyn invented an ingenious parallactic method for measuring the coordinates of stars. Each plate in Gill's survey was placed on a vertical stand at the focal distance of a small telescope with an altazimuth mount. By pointing at the stars one by one, a properly trained assistant could read their coordinates, relative to the center of the plate, directly on the axes of the instrument and then annotate them by hand: a meticulous and tedious task that for some time was delegated to prisoners willing to do it for a small fee.

In Italy, a similar task was performed by some nuns to earn a bonus for heaven instead of a handful of coins to stock up on tobacco. The good sisters were used to measure the positions and brightness of the stars in

---

[59] J.D. Galloway, *Kapteyn's Contributions to Our Knowledge Of the stars*, in "Publications of the Astronomical Society of the Pacific", 14, 84, 1902, pp. 97–102.

the photographic plates of the *Carte du ciel* (Map of the Sky), collected by the astrograph of the Vatican Observatory as part of the homonymous project.[60] It was the brainchild of Amédée Mouchez, director of the Paris Observatory and a former naval officer with a passion for precision cartography. In 1887, fresh from a congress of astronomers and a conversation with Gill, Mouchez became convinced that photography could revolutionize celestial topography. He then decided to launch, through the *Académie de France*, a campaign to catalog sources down to the limit of stars a hundred times fainter than those visible to the naked eye. Their precise positions would have provided a fundamental reference for a subsequent, deeper census of the immense sidereal realm. A network of stations was needed to cover the entire sky and to overcome the meteorological peculiarities of each observation site. Mouchez put together twenty of them, scattered over four continents, with the condition that each be equipped with a photographic refractor (astrograph) with a good aperture, about 30 cm, a focal length capable of giving a scale of 60 seconds per millimeter, and a field of 4 square degrees.[61] In this way, and with some trickery to eliminate distortions in the optics, a positional accuracy of half a second of arc could be achieved. For hardware, most of the participants in the venture turned to the French company of the Henry brothers and to Howard Grubb's famous firm in Dublin. In 1925, the latter was purchased by Sir Charles Parsons, the youngest son of the 3rd Earl of Rosse: a renowned engineer, inventor and businessman, who changed the brand name to Grubb-Parsons.

Italy took part in the enterprise with the astrograph of the young Catania Observatory, equipped with the optics of Carl August von Steinheil, mounted on the mechanics of Filotecnica Salmoiraghi of Milan, the company founded by the great optician Ignazio Porro. A young country, beset by the problem of homogenizing and restoring the varied mosaic of a peninsula that had remained shattered and divided for 1,500 years, could do no more. In fact, the Italian government could have even avoided the inconvenience of investing in astronomy if it had not been for the need to build

---

[60] D. Jones, *The Scientific Value Of The Carte du Ciel*, in "Astronomy & Geophysics", 41, 5, 2000, pp. 16-20.
[61] The entire celestial vault covers an area of 40,000 square degrees, equal to 10,000 times a 4 square degree field. It is obvious, however, that to map the entire sky, albeit with a single exposure per field and therefore in a single color, a much greater number of exposures is required because of the inevitable overlaps.

the unified identity and image of the nation in different spheres and contexts.[62] Among the participants in the *Carte du ciel* was the observatory rebuilt on the walls of the Vatican city by Pope Leo XIII Pecci after the loss of the historic seat of the Roman College. With the capture of Rome by the Italian army in 1870, the Pope needed more than ever to reaffirm his temporal presence. Some might ask what attraction such an expensive, laborious, and difficult undertaking, which at best would generate a catalog of precise stellar positions, could have in an age now pervaded by the seductive themes of nascent astrophysics. Hamlet's question was: to participate in the assembly line of the *Carte du ciel* or, just to give an example, to undertake the investigation of the structure of stars and discover the source of their energy? Let us not strive to find the answer. It was explicitly given by one of the fathers of the new science, Angelo Secchi, who in 1877 wrote: «The greatest progress of modern science consists in the exact determination of the locations of the stars; work that certainly lacks the bright flare of many other researches, but which is infinitely more solid and important».[63] As the great scientist that he was, the Jesuit knew that celestial cartography was a prerequisite for any investigation about stars and their system, the Milky Way, and he affirmed it with force and clarity. It was not a pleasure but a very important duty.

The collaboration initiated by Mouchez bore its first fruits before the Great War raised yet another barrier of hatred and blood between the peoples of old Europe and beyond. In the 1930s, the astrographic catalog already contained several million stars. But by then no one was interested in it: the set of measurements came back as an important memory of the sky in connection with the astrometric satellite Hipparcos, launched in 1989 by the European Space Agency. Once the reference network of bright stars was established, the construction of the real *Carte du ciel* was to be carried out, with exposures 10 times deeper, cataloguing all the stars up to the 15th magnitude, more than a million times fainter than Sirius appears to us. The result did not match the expectations and the energy used. By the time the *Carte du ciel* was completed, the great problem of the "construction of

---

[62] Following the same logic, in April 1855, the Prime Minister of the Kingdom of Sardinia, Camillo Benso, Count of Cavour, had sent an expeditionary force of 18,000 men and 3,500 horses to the Crimea to flank the British and French against the Russians, hoping to see his homeland recognized as a nation and to earn a place at the winners' table.

[63] A. Secchi, *The stars. Essay on Sidereal Astronomy*, F.lli Dumolard, Milan, 1877, p. 12 (in Italian).

the sky", as Herschel had put it, had already been solved, thanks largely to Kapteyn and his school.

The 20th century was just around the corner. It would bring extraordinary innovations, capable of challenging all acquired knowledge about celestial bodies and the very idea of the cosmos, but also of surpassing the records painstakingly achieved by groups of astronomers now scattered all over the world. Pioneers with few resources, still limited ideas, but much passion, incredible ingenuity, and a spirit of sacrifice that is a credit to the human species. Today, any graduate student holding an image of a globular cluster obtained with a wide-field telescope can inventory hundreds of thousands of stars with a few hours of computer work, a sandwich, and a cup of coffee. We will discuss these wonders of progress at the end of this book. For now, we must rewind the movie of history to tell of another formidable weapon of modern astronomy, spectroscopy.

Chapter 11

Harmony of light:
Fraunhofer, Kirchhoff, and Huggins

If I were not a physicist,
I would probably be a musician.
I often think in music.
I live my daydreams in music.
I see my life in terms of music.

Albert Einstein, 1929

What we are nowadays hearing
of the language of spectra
is a true 'music of the spheres'
in order and harmony
that becomes ever more perfect
in spite of the manifold variety.

Arnold Sommerfeld,
quoted by Wolfgang Pauli

What a great time the 19th century was! A melting pot of new ideas and contradictions, it gave rise to those scientific, technological, and social revolutions that gradually shaped, for better or worse, a new world. A century of conservatives and rebels, bigots, ruthless businessmen, and libertarians; brilliant to the point of demolishing the most sacred paradigms, such as that of eternity, replaced by the much more secular concept of evolution. A century of machines, trades, inventions, and monumental works, of ideologists and iconoclasts, of lonely men deluded by reason, tormented by ideals, forced to confront with themselves and their passions, and committed to exploring the psyche and its mists; of artists engaged by technological innovations, in search of new languages; of masses of desperate people at work in anthills, fighting for crumbs of *liberté*, *égalité*, and *fraternité*, always promised and never given. A century in which the idea that the Earth is not a privileged place with different rules from the rest of the universe was definitively affirmed: a very bitter morsel to take for the irreducible defenders of the primacy of man over nature, but also an indispensable step towards the fruitful fusion of physics with astronomy, heralding marvelous discoveries in the two apparently antithetical worlds of the infinitely large and the infinitely small.

Among the architects of the alliance from which astrophysics[1] was born in the second half of the 19th century, was another German, Joseph von Fraunhofer[2]: a self-taught man who became a pioneer of spectroscopy as a tool for probing the depths of matter. The story of the first steps in the short life of this brilliant "craftsman" would deserve the pen of Charles Dickens, or at least the more corny but equally seductive style of the Italian children's author Edmondo De Amicis.

The youngest of eleven children of a modest family of glassmakers, Joseph[3] was born in 1787 in Straubing, in fertile eastern Bavaria: an environment of wooden houses and potato-based meals, far from the marketplace of opportunity that the city offered even to the less wealthy. He was just 11 years old when he lost both his parents, one after the other. A tragedy that seemed to irrevocably undermine the future of a creature with a fragile complexion and completely uneducated, except for the little profession he had been able to learn by helping his father. What to do with him? His tutor found nothing better than to put him in the workshop of a glassmaker and decorator in Munich. The latter, a heartless man, agreed to take the boy with a pact, signed and registered with Germanic fussiness, that Joseph would work for him without pay for at least six years. The poor orphan was even denied the opportunity to attend Sunday school to learn to read and count. Under these conditions, there was little hope for the rest of his life. Paradoxically, it was another disgrace that brought this David Copperfield back into the game. In 1801, the master glassmaker's home workshop, a sort of Thénardier's tavern from Victor Hugo's *Les Misérables*, where the boy worked and lived, suddenly collapsed. Joseph was buried under the rubble, but unlike his master's wife, who died in the accident, he managed to escape his fate. Trapped under a crossbeam that had

---

[1] The term was coined in 1865 by the German Johann Carl Friedrich Zöllner; see J. Lankford (ed.), *History of Astronomy: An Encyclopedia*, Routledge, New York 2011, p. 72.

[2] M.W. Jackson, *Spectrum of Belief: Joseph von Fraunhofer and the Craft of Precision Optics*, The MIT Press, Cambridge (MA) 2000.

[3] C. Plicht, *Joseph von Fraunhofer*, in "Journal of the Antique Telescope Society", 15, 1998, pp. 20–21; I. Howard-Duff, *Joseph Fraunhofer (1787–1826)*, in "Journal of the British Astronomical Association", 97, 6, 1987, pp. 339–347.

fallen without crushing him, he had to wait several hours for rescuers to break through the debris; and here was the turning point.

The accident had attracted the attention of several onlookers, including Elector Maximilian Joseph of Wittelsbach, the future King of Bavaria, who even wanted to participate in the rescue effort by encouraging the shoveling teams. Pro-French and a staunch ally of Napoleon, the Catholic Maximilian had been educated in the values of the Enlightenment. He truly felt like the father of his people, and like a good father, he immediately had the young survivor transported to the sumptuous summer residence of Nymphenburg, the castle built by the Wittelsbachs on the outskirts of the Bavarian capital, famous for its park and the exploits of a child prodigy, Wolfgang Amadeus Mozart. A story worthy of a fairy tale by the Brothers Grimm. After he was refreshed, the prince wanted Joseph to tell him his experience and his feelings when he was revived. Who knows, perhaps Maximilian saw a light in Joseph's eyes that touched him. The fact is that when he dismissed his young guest, he gave him some money, and when the boy was able to resume his work, he sent orders to his master to give him Sundays off so that he could attend school.

For the unsuspecting young glazier, the fairy tale held an unusual prize: not the classic heart of a princess, but a front-row seat to the history of science. Naturally, he put his own spin on it and immediately set to work. With part of his royal nest egg, he bought a lens-grinding machine, which he planned to use in his spare time to make some money. But when his master discovered it, he ordered the immediate cessation of the business, which, among other things, competed with his own. He also claimed that Joseph did not waste time on books of mathematics; and to make sure that he obeyed, the rough boss assigned him a room without windows, and even forbade the young man to turn on the lamp at night. Nothing, however, could stop a true thirst for knowledge. Joseph, who could hardly read, mastered theoretical optics within a few years.

In the meantime, on the prince's recommendation, he had made contact with Joseph von Utzschneider, an enterprising state official who had founded a workshop in Munich in 1804 together with two experienced engineers, Georg Friedrich von Reichenbach[4] and Joseph Liebherr,[5] specialized

---

[4] C. Matschoss, *Great Engineers*, Ayer, New York 1970, pp. 127–139.

[5] G. D. Roth, *Die Grundlagen für eine Optische Industrie in München. Ein Werk von Joseph von Utzschneider und Carl August von Steinheil*, in "Zeitschrift für Firmengeschichte und Unternehmerbiographie", 5, 1, 1960, pp. 15–36 (in German).

in precision topographical instruments and called it by the pompous name Institute for Mathematical Mechanics. To make their way in a market dominated by English products, the three partners could count on a magnificent angle-dividing machine invented by Reichenbach, who was a son of the arts (his father built cannons), but they were completely uncovered on the optics front. The crux of the problem was flint glass disks, indispensable components for making achromatic doublets. Manufacturers on both sides of the Channel held on to the secrets of their recipes, and Her Majesty's Customs and Excise, as we have pointed out, hindered the export of high-tech products that were rightly considered to be of high strategic value (p. 189). Convinced that the only way to get their business off the ground was to start their own optical glass production in competition with the British, Utzschneider and his associates approached the renowned Swiss glassmaker Paul-Louis Guinand, who accepted the challenge. The furnaces were installed in Benediktbeuern, a deconsecrated Benedictine abbey south of Munich: a place of extraordinary charm where the manuscripts of *Carmina Burana* would soon be found,[6] but also a veritable fortress to protect from prying eyes and ears the industrial secrets that patents alone could not protect.

While all this was going on, Fraunhofer had run into trouble trying to start a new life free from the yoke of his harsh master. With the rest of the money Maximilian had donated him, he had bought his independence back. Then he had started a printing business that did not work because of the perpetual state of war to which Europe seemed condemned by Napoleon's ambitions. The people had neither the money nor the spirit to print visiting cards. Reduced to wretchedness, Joseph turned to Utzschneider, who hired him in 1806 as an optician at the Institute, with the task of grinding large astronomical lenses; little more than a specialized worker. For all his flair, Utzschneider probably had no idea what an extraordinary deal he had made when he took aboard Fraunhofer. The young man was no longer just a skilled and industrious craftsman. He now understood the hidden laws of optics and even wrote essays on the subject. After a short time, Joseph was promoted to director of the company, which had grown to 40 employees and concentrated all its activities in Benediktbeuern. Thus, within the ancient walls of the former monastery, the caterpillar became a butterfly. Freed from the need to make ends meet, Fraunhofer indulged his dream of

---

[6] Medieval poems partly set to music in the 20th century by the composer Carl Orff.

producing flawless glass and then processing and combining it in a way that minimized stray light and chromatic aberration. Although his motivations were technical and entrepreneurial — to do it well, faster, and at competitive cost — he could not neglect the scientific aspects of the problems he faced, such as quantifying the variation of the refractive index of glass as a function of the color of light. For him, a believer in Huygens' wave theory, this meant finding a way to measure wavelengths. But how?

In his thousands of after-hours experiments, the industrious Joseph had noticed that the spectrum of light from a candle doped with a few grains of table salt and passed through a prism showed an orange continuum crossed by a streak of light.[7] The remarkable fact was that this bright "line" always fell at the same position in the spectrum; today we would say at the same frequency. A kind of milestone that could be used as a reference for dispersion measurements. But it was not enough to cover all the colors of light, just as a single sign is not enough to mark a long path. More was needed. Fraunhofer thought about exploring the most powerful celestial sources, starting with the Sun. He did not know what to expect, but he did not want to leave any stone unturned. He ignored the fact that a decade earlier William Hyde Wollaston[8] had already discovered in the spectrum of our star the presence of black lines embedded in the range of iris colors. But the brilliant English physician, all focused on enriching himself with chemistry, metallurgy and technology in general, did not go

---

[7]In essence, the visible spectrum is a streak of colors like those of the rainbow. It is produced by refraction (which is color-sensitive), for example by passing a parallel beam of light through a glass prism and then focusing the output on a detector (a similar result is obtained using diffraction gratings). Since in the wave theory of light the term "color" is synonymous with wavelength (or frequency), the spectrum provides a representation of the energy content of the different components of a light source as a function of wavelength. The thermal spectrum produced by a hot body has a continuous base (the "continuous" in the jargon) on which are grafted some minima, typically deep and narrow; they are called "absorption lines" because they correspond to the selective subtraction of that part of the radiation of free atoms intercepted by the light beam on the way from the source to the observer. There are also some maxima, thin and pronounced, called "emission lines", which are the result of the equally selective thermal emission of free atoms. Clusters of very close lines produce the "bands" typical of molecules. The absorption lines require the presence of the continuum, while the emission lines do not, since they do not subtract but add energy (thus they are easier to detect). Finally, we must emphasize that the notion of spectrum can be extended to the whole range of electromagnetic waves, from γ rays to long radio waves, and, as we will see in chapter 18, it also applies to the so-called non-thermal emissions.

[8]H.G. Wayling, *A Short Biography of William Hyde Wollaston, MD, PRS (1766–1828)*, in "Science Progress in the Twentieth Century (1919–1933)", 22, 85, 1927, pp. 81–95.

beyond the baffling hypothesis that they marked the discontinuities between the different colors of the iris, of which Newton believed that a spectrum was composed. The theory fell into crisis with Fraunhofer; the lines of the solar spectrum were too many to perform the function assigned to them by Wollaston.

Joseph began assembling the necessary equipment. The factory was the right place to find the components for the first spectroscope powered by the light of a star collected by a telescope. The beam emerging from a long narrow slit[9] placed at the focus of the instrument had to be first collimated by a lens, then dispersed by a prism, and finally focused again to allow the range of colors to be studied. In this way, Fraunhofer was able to reveal the particular set of dark lines in the solar rainbow palette, including the one that appeared bright in the "salted" candle flame, and, unlike Wollaston, he wanted to understand more. The lines of the solar spectrum also seemed to remain motionless against the background of the rainbow colors, keeping their distances from each other (in wavelength units) unchanged. It was 1814, the year of Napoleon's first defeat and his exile on Elba. While Europe was preparing for the restoration of the old powers and, worse, the old customs, still unaware that the indomitable Emperor of the French had one last stroke in store, the brilliant German do-it-yourself scientist was putting the results of his experiments in writing. This would have brought him some earthly glory, such as co-optation into the Bavarian Academy, a honorary degree at the University of Erlangen, the title of Knight of the Order of Merit of the Bavarian Crown (instituted by King Maximilian Joseph I, which added the "von" to his family name), and a proscenium place in the history of science.

From "a man without letters"[10] that he was, he used brute force in his research. Observing everything he could lay his hands on, he discovered that the mysterious lines were ubiquitous in the celestial sources, but not always evenly distributed. For example, the star Sirius had a different

---

[9]Why a rectangular slit and not a simple circular aperture? The reason is that a slit aligned with the edge of a prism in the spectrograph can be thought of as a chain of circular apertures. The result is a series of parallel spectra, which improves the perception of the details of the continuous signal. It is this experimental configuration that has given rise to the name "lines" (of emission and absorption, respectively) to describe rapid and localized increases or decreases in the luminosity of the continuum. The lines are nothing more than (almost) monochromatic images of the slit.

[10]«*Omo sanza lettere*», Leonardo da Vinci's famous description of himself; obviously he did not mean to say that he was illiterate, but that he had not been educated in the scholarly language of Latin.

set of lines from the spectra of the Sun and Moon, which were obviously similar because they are related to the same source of radiation. Despite his short life, Fraunhofer was able to catalog more than 570 lines of the solar spectrum and design and build the first diffraction grating to quantify wavelengths. The phenomenon of diffraction had been known for a long time. In the 17th century, Francesco Maria Grimaldi had studied and even named it. The Bolognese Jesuit was followed in the early 19th century by the British polymath Thomas Young and the French civil engineer and physicist Augustin-Jean Fresnel, in the context of the diatribe between the advocates of the wave model of light developed by Huygens and the corpuscular model favored by Newton. But the Bavarian optician had also made it a fundamental measuring instrument.[11]

In 1826, at the age of just 39, while dreaming of building an achromatic telescope with an 18 inch aperture, Fraunhofer went to join the many glassblowers who died of tuberculosis contracted from inhaling the metal fumes. He did not know that he was very close to realizing that spectral lines are the fingerprints of the elementary constituents of matter, atoms and molecules, and the most reliable spies of the physical conditions of stars. But the revelation of these hidden secrets of nature was now in the air. A few years earlier, using the nascent laws of gases, Claude Pouillet, director of the *Conservatoire des Arts et Métiers* in Paris, the temple of engineering education created by the French Revolution, had estimated the Sun's external temperature at $1,800°C$, a high value, though more than three times lower than that accepted today. Since it was believed at the time that very hot gases did not produce light, this result was used to argue, erroneously, that the star should be a solid body. Pouillet himself, and independently John Herschel, had also measured, by means of a water calorimeter, the so-called "solar constant", that is, the amount of solar radiation passing through a perpendicular unit surface at the distance of the Earth (1 Astronomical Unit) in one second of time.

Fundamental as they were, these discoveries were also premature, for they lacked the support of a solid physical theory, tested by terrestrial experiments and applicable to celestial objects and phenomena, on the assumption that the cosmos is self-similar in every part, in defiance of Aristotle and his followers. In short, astrophysics was just around the corner, and yet it struggled to convince even the most open-minded. The best

---

[11] M.E. Rudd, *Joseph Fraunhofer's First Paper*, in "Journal of the Antique Telescope Society", 22, 2002, pp. 4–8.

known and most obvious example of this resistance is offered by the French philosopher Auguste Comte, who in 1835, in his *Course of positive philosophy*, declared:

> The stars are thus, of all natural things, those that we can know in the least varied ways. We conceive the possibility of determining their sizes, their distances and their movements whereas we would never know how to study by any means their chemical composition or their mineralogical structure, and even more so the nature of the organized beings that might live on their surface [...] I persist in the opinion that every notion of the true mean temperatures of the stars will necessarily always be concealed from us.[12]

This was a glaring error on the part of a great thinker who was so in love with the natural sciences that he placed them at the foundation of what for him was the highest: social philosophy.[13] But the pessimistic prediction did not last long. It was disproved only 15 years later by the research of two Germans, the towering Saxon Robert Wilhelm von Bunsen, a big man the size of a closet, and especially his younger and slimmer colleague, the Prussian Gustav Robert Kirchhoff.[14]

The two had met in 1850 in Breslau, in southwestern Poland, where Kirchhoff had received his first lectureship, and had quickly become friends. When Bunsen was called to the University of Heidelberg at the end of an academic peregrination through Europe, he wanted his colleague to join him. The context was stimulating: a "royal residence of the intellect" rebuilt on the ruins left by the repeated incursions of French armies in the 18th century. Although Germany lagged behind the great nation-states and was still fragmented into many small states, love of country was the reason

---

[12] A. Comte, *Cours de philosophie positive*, 19e leçon, quoted by J.-C. Pecker, *La détermination de la composition chimique des étoiles et du Soleil*, in "L'Astronomie", Vol. 72, 1958, p. 94 (in French). See also J. Hearnshaw, *Auguste Comte's blunder: an account of the first century of stellar spectroscopy and how it took one hundred years to prove that Comte was wrong!*, in "Journal of Astronomical History and Heritage", 13, 2, 2010, pp. 90–104.

[13] M. Bourdeau, M. Pickering, W. Schmaus (eds.), *Love, Order, & Progress. The Science, Philosophy, & Politics of Auguste Comte*, The University of Pittsburgh Press, Pittsburgh (PA) 2018.

[14] *Gustav Robert Kirchhoff*, https://mathshistory.st-andrews.ac.uk/Biographies/ Kirchhoff/.

for total devotion to work, order, and discipline. Service to the state was the watchword of the modern secular samurais, inspired by the model imposed on Prussia by Frederick the Great. This desperate nationalism — «Nationalism is an infantile disease. It is the measles of humanity», Einstein would later stigmatize[15] — also infected the universities, with extremely positive consequences for knowledge, but catastrophic for the fate of Europe.

Kirchhoff and Bunsen worked in different fields under the watchful eye of the great and versatile guru of German science, Hermann von Helmholtz.[16] The first was a physicist who combined strong experimental qualities with excellent mathematical training. He was born in 1824 in Königsberg, the capital of East Prussia and the birthplace of Kant.[17] There Kirchhoff had finished his studies under the tutelage of the most inspiring master of the time, the Prussian Jew Carl Jacobi. Later he devoted himself to the study of electrical phenomena, coming close to the discovery of electromagnetic waves, which was then Maxwell's masterpiece. Bunsen was a chemist instead. In Heidelberg, he worked with caustic and corrosive substances, which he wanted to classify according to the color of the light they produced when burned.

The spark ignited when Bunsen told his colleague about his project and the first encouraging results. Today we could imagine the meeting of the two friends in a noisy *Biergarten*, over two frothy mugs of *Weizenbier*. But back then, in the charming town on the Neckar River, amidst the green forests of Baden-Württemberg, the atmosphere was stern and staid. Everything took place within the gray walls of the university, but illuminated by the genius and enthusiasm of Kirchhoff. Listening to his colleague's presentation, the physicist pointed out that the work would be better done with a Fraunhofer-style spectroscope. It was a revolutionary insight, full of consequences for all mankind, which was accompanied in this *annus mirabilis* by the publication in London of a secular bible for the socialist view of «the mode of production in material life», the essay *Zur Kritik der politischen Ökonomie* (On the critique of political economy) by

---

[15] A. Einstein, *The Human Side: Glimpses from His Archives*, H. Dukas, B. Hoffman (eds.), Princeton University Press, Princeton (NJ) 1979, p. 38.

[16] Cf. D. Cahan (ed.), *Hermann von Helmholtz and the Foundations of Nineteenth-Century Science*, University of California Press, Berkeley (CA) 1993.

[17] In 1946 the city became part of the Soviet Union and is now part of the Russian Federation under the name Kaliningrad.

Karl Marx, also a Prussian Jew. Bunsen found the suggestion interesting and replied something like: "*Gut, mein Freud* (very good, my friend). Let's do it four-handed. I will put in a selection of very pure substances and my gas burner. You will do the rest". The burner was nothing more than a refinement of a prototype created by Michael Farady. Known as the "Bunsen burner", it produces a powerful but practically colorless flame with few and very thin spectral lines.

Kirchhoff accepted the challenge and went to work. The result of the joint venture, announced to the world in 1860 — while Garibaldi set sail for Sicily, Charles Darwin had just published his hypothesis on *The Origin of Species*, the Austro-Hungarians were licking the wounds of their first defeat in *Bel Paese*, and the United States were preparing for a bloody Civil War — was the sensational confirmation of Fraunhofer's intuition. The samples provided by Bunsen showed a spectrum without a continuous coloration, but with an apparatus of brilliant lines characteristic of the treated substance. The two Germans had shown that, despite Comte's certainties, it was possible to analyze the chemistry of even distant bodies as long as they were luminous, i.e. hot. It was only a first step on a long road, but the path was now set. During the preliminary tests for the experiment, another fundamental fact had come to light. Kirchhoff wanted to verify that the bright sodium line seen by Fraunhofer coincided with one of the dark lines of the Sun. To do this, he decided to superimpose the two spectra by passing the star's light through the burner flame before entering the spectroscope. By attenuating the contribution of the powerful solar rays, the two lines compensated for each other until they disappeared. All as expected. But in bright light, something happened that at first seemed absurd: the sodium line became even darker than it had been in direct sunlight. Why this violation of the additive property? After a few weeks of total concentration on the problem, the German physicist arrived at the solution to the riddle. He understood that a gas absorbs radiation passing through it only at the wavelengths at which it emits the lines of its spectrum when heated. Conversely, a hot solid produces a continuous spectrum without lines. There remained the problem of determining the ratio between the emission power and the absorption coefficient of different materials. By a *reductio ad absurdum*,[18] Kirchhoff succeeded in showing that the relationship between these two quantities depends solely on the

---

[18] Demonstration by absurdity; it consists in proving the validity of a certain proposition by showing that its denial leads to a contradiction.

color of the radiation (wavelength) and the (absolute) temperature,[19] but not on the nature of the body (that is, it was a property of matter as such!). He then devised an ideal omnivorous physical entity capable of absorbing all incident radiation, regardless of wavelength: a "blackbody", so called because it absorbs all the light that falls on it.

Without looking for it, Kirchhoff had serendipitously found the "meter" to measure the temperature at the surface of stars. Just a third of a century later, another Prussian, Wilhelm Wien[20] — using the methods of a new branch of physics called thermodynamics,[21] a science that studies heat and its relationship to the energy of bodies and the ability to produce work, which developed overwhelmingly in the time of thermal machines — would have proved that the spectral continuum of this theoretical body reaches its maximum at a wavelength inversely proportional to the temperature. Simply put, the peak of the blackbody radiation spectrum shifts progressively from red to blue as temperature increases. Therefore, a red star is cooler than a blue one, assuming, of course, that these celestial objects can be compared to blackbodies (which is almost true for normal, not-too-cold stars). At first glance, the analogy does not seem legitimate, since stars lose energy as they radiate. Instead, it is plausible because the loss is replenished from an internal source; much as parents do with the wallets of their spendthrift children, replenishing them as they are emptied.

Thanks to Kirchhoff and Wien, by the end of the 19th century the veil of inscrutability that had relegated the stars to the background of astronomers' research projects began to lift. Indeed, in the wake of a valid blackbody model, the tools for studying the sidereal world multiplied visibly. In 1879, a physicist born in the Slovenian territories of the Habsburg Empire, Josef Stefan, had discovered that the total radiation of a blackbody is proportional to the fourth power of temperature. An empirical

---

[19] Absolute temperature is measured on a centigrade scale whose zero corresponds to the minimum value theoretically allowed to this quantity (for macroscopic bodies) by the laws of physics, 273.15 degrees below zero of the Celsius scale. Since temperature measures the average value of the kinetic energy possessed by the elements of the system, the existence of an absolute zero corresponds to the (fundamental) state of minimum possible energy.

[20] *Wilhelm Wien Biographical*, in "Nobel Lectures, Physics, 1901–1921", Elsevier, Amsterdam 1967.

[21] I. Müller, *A History of Thermodynamics: The Doctrine of Energy and Entropy*, Springer, Berlin 2007. For further information see also E. Fermi, *Thermodynamics*, Dover, New York 1964.

result of such importance that the Viennese Ludwig Boltzmann,[22] sublime mind in the body of an insecure and fragile man, immediately wanted to prove it with the methods of thermodynamics. Together with Wien's law, Stefan-Boltzmann's law, as it came to be called, acquired an enormous interest in studies of the nature of light, to which two other giants would devote themselves at the beginning of the 20th century: Max Planck, as Kirchhoff's successor at the chair of theoretical physics in Berlin, and the young Albert Einstein, while still earning his living as a clerk at the patent office in Bern. In astronomy, this law also allowed the total luminosity of a star to be related to its radius and photospheric temperature (the former giving the area of the radiating surface, the latter the flux per unit area).

In short, the measurements of color and brightness, combined with those of distance, gave astronomers an idea of the size and thermal conditions of distant and mysterious sources. This alone would have been enough to make Comte eat his hat, but it was not all. Light had another enormous bonus in store. It was discovered by an obscure lecturer at the University of Prague, Christian Johann Doppler.[23] To introduce the matter, let us use the words of Leonardo da Vinci: «I say: if you throw two small stones at the same time on a sheet of motionless water at some distance from each other, you will observe that around the two percussions numerous separate circles are formed; these will meet as they increase in size and then penetrate and intersect one another, all the while maintaining as their respective centers the spots struck by the stones».[24] Elsewhere the Florentine genius adds: «If a stone is flung into motionless water, its circles will be equidistant from their centre. But if the stream is moving, these circles will be elongated, egg-shaped, and will travel with their centre away from the spot where they were created.»[25]

Inspired by observations of simple natural phenomena such as those described by Leonardo, Doppler proposed for sound that radial motion changes the pitch of a pure tone, making it higher for the listener as the

---

[22] C. Cercignani, *Ludwig Boltzmann: The Man Who Trusted Atoms*, Oxford University Press, Oxford 2006

[23] A. Eden, *The search for Christian Doppler*, Springer-Verlag, Berlin 1992; E.N. da Costa Andrade, *Doppler and the Doppler Effect*, in "Endeavor", 18, 1959, pp. 14–19.

[24] *Manuscript A, 61r*, Institut de France, Paris, quoted in S. Chopra, *Leonardo was a geophysicist. An interview with José M. Carcione*, in "Recorder", Vol. 38, No. 10, 2013, pp. 19–21.

[25] *Manuscript I, 87*, cit.

source is coming closer to him, and lower as it is moving farther away.[26] This is a phenomenon we are familiar with today. It happens, for example, when we are standing on the sidewalk and an ambulance passes us and drives away with its siren wailing. We have all experienced the change in pitch of the sound that quickly drifts from higher (approaching phase) to lower (receding), compared with the frequency of the siren when the ambulance is stopped. Doppler also brought ether waves, or light,[27] into the picture. Actually, the only evidence he was able to devise to prove the reality of his effect related precisely to the colors of stars in binary systems, as we shall see.

But common experience does not confirm the existence of such an effect for light. In fact, the ambulance in our example does not vary its color when the siren changes its pitch. If it is yellow, it continues to look that way, whereas the analogy with sound would make it appear bluish when the car is coming toward us and reddish when it is running away. The solution to this riddle lies precisely in the law that Christian Doppler was able to derive by applying simple kinematic considerations to the hypothesis that sound and light are wave phenomena[28]: the relative frequency

---

[26]G. Wolfschmidt, *Christian Doppler (1803–1853) and the Impact of Doppler Effect in Astronomy*, in "Acta Universitatis Carolinae, Mathematica et Physica", 46, Suppl., 2005, pp. 199–211; K. Hujer, *Christian Doppler in Prague*, in "Journal of the Royal Astronomical Society of Canada", 57, 1963, pp. 177–180.

[27]Recall that until the pioneering experiments of Albert Abraham Michelson and Edward Williams Morley in the late 19th century, light was thought to propagate in a mysterious medium called the ether (see also p. 482).

[28]In deriving his famous formula, Doppler had also treated light as a longitudinal wave. He was aware of the transverse wave hypothesis formulated for light by the Frenchman Augustin-Jean Fresnel (cf. M. Suhail Zubairy, *A Very Brief History of Light*, in *Optics in Our Time*, Springer, Berlin 2016, pp. 3–24) in 1821, but he was unwilling to accept it. Fortunately, his error did not affect the result, as was later proved by his friend Bernard Bolzano, a Bohemian political agitator and mathematician of Italian descent, well known for his rigorous approach to mathematical limits. Bolzano, who held Doppler's qualities as a scientist in enormous esteem, acted as a true promoter of his weak, sickly, and difficult colleague. See, for example, this passage from an 1839 letter of recommendation for the publication of a mathematical paper: «Mr. Doppler has already demonstrated his very promising abilities to the scientific community through his numerous published works in mathematics and physics. The expectations raised by his hitherto published works would multiply when one enters into personal acquaintance with him. You are not only struck by how many highly interesting and fruitful ideas, in many areas of knowledge, that so young a scientist is able to produce, but you also convince yourself with the greatest pleasure that this exceptional spiritual power combines with an amiable character, genuine unaffected determination and with that pure love of science and truth» (I. Stoll, *Christian Doppler — Man, work and message*, in "The Phenomenon of Doppler", Poligrafia Eds. Branch 6, Prometheus, Prague 1992, pp. 13–29).

shift increases in proportion to the radial velocity of the source (positive if receding and negative if approaching), measured in units of the speed of the waves carrying the signal.

Now everything is crystal clear. Since the speed of light is nearly a million times that of sound, the source must be moving very fast relative to the observer to produce a noticeable effect, not far from the speed of light. Thus, at the insignificant speeds of motor vehicles, the color change of the ambulance is out of the question. The Austrian physicist was aware of this fact, and as an astronomical verification of his formula he proposed to consider the «colored light of binary stars».[29] This was not a correct choice, however. Even the orbital velocities in binary systems are still too modest — at most a hundred kilometers per second, or ~0.03% of the speed of light — to produce significant effects on the stellar colors. The different colors of star pairs in binary systems are due to mainly different intrisic properties to the objects, such as initial mass, and, in prospective binaries, different evolutionary phases.

It is a different matter if, instead of a generic color, you are looking at a distinct spectral feature such as a line.[30] Doppler, though not at all familiar with spectra, had a hutch: «It is almost to be accepted with certainty that this will in the not too distant future offer astronomers a welcome means to determine the movements and distances of such stars which, because of their unmeasurable distances from us and the consequent smallness of the parallactic angles, until this moment hardly presented the hope of such measurements and determinations».[31] Words that anticipate the great revolutions in astrophysics and cosmology of the 20th century, spoken by a scientist of whom we might have known nothing had he not had this authentic and unique flash of genius.

Christian was born in Salzburg in 1803, where he would have grown up to be a stonemason like his parents, prosperous but rough people. Instead, he was directed to study because his precarious health was unsuitable for manual labor. The plan was to make him an accountant, but his bright

---

[29] From the title of Christian Doppler's article on his effect, *Über das farbige Licht der Doppelsterne*, published in the "Proceedings of the Bohemian Society of Sciences" in 1843.

[30] The same consideration would apply to the sound of the ambulance. If, instead of the pure sound of a siren, the ambulance were to announce its presence by playing Beethoven's 9th Symphony, the shift in frequency would be equal but spread over the entire spectrum, and only a refined ear would notice it.

[31] Eden, *The search for Christian Doppler*, cit., p. 133.

intelligence was noticed by a local mathematician, so he was sent to high school and then to university. After graduating in Vienna, he worked for a few years as an assistant to his mathematics professor. Then he was sent on his way, as happens to the precarious without patrons even today. Flooding the Habsburg Empire with his curriculum vitae proved useless. To make a living, he was forced to take a job as an accountant in a textile factory in Salzburg. Highly frustrated in his ambitions, he thought of emigrating to America, but just as he was packing his bags, he was called to Prague to teach mathematics at a secondary technical school. A humiliation for an already unhappy and sickly young man, but safer than an uncertain future across the ocean. After much effort, in 1841, now married to a fellow villager with whom he would have five children, he finally obtained a position as professor of applied geometry and elementary mathematics at the Prague Polytechnic. Lacking a clear scientific goal, he wandered among various topics until his interest in the aberration phenomenon discovered by Bradley led him to devise the effect that now bears his name. The following year, at the meeting of the Bohemian Royal Society, he was admitted to illustrate it in the natural science session.

It was not exactly a success. Colleagues greeted the communication with warmth and skepticism. They questioned the general idea and in particular Doppler's claim that the colors of stars in binary systems depend on orbital velocities (which is a fair criticism). In any case, there was still no experimental evidence for the phenomenon, which was difficult to reproduce even for sound waves alone because of the modest speeds attainable at the time; and, as we know, since Francis Bacon and Galileo Galilei, physicists behave like St. Thomas. They want to "touch" in order to believe. Only three years later, however, a 28-year-old professor of chemistry and meteorology at the University of Utrecht, Christophorus Buys Ballot, also known for discovering the direction of low-pressure vortices in the northern hemisphere, produced a first positive test. His intention was *destruens*: he wanted to disbelieve the reality of the Doppler effect and had devised a way to do it for sound.

His strategy was to use the railroad line that had recently been built between Utrecht and Amsterdam. On the straight and level stretch between Utrecht and Maarssen, the British locomotive could reach the spectacular speed of 40 kilometers per hour, or about 3 percent of the speed of sound in air. Ballot arranged things so that during the passage of the convoy, some witnesses standing on both sides of the track could verify the phenomenon of sound jumping on a musical note produced by some trumpeters on board an open car. The experiment was at the limit of feasibility. To reduce the uncertainty, Ballot repeated the test several times (in particular, he had to redo it in warm weather after a first attempt in a very cold February had to be abandoned because of snow and wind). He even changed roles, putting the musicians on the ground and the witnesses on the train. Eventually he became convinced that the relative movement actually changed the pitch of the sound, and with great intellectual honesty he told the world about it. In any case, he remained one of the most ardent critics of the Doppler effect applied to light. He was wrong and would have to change his mind again.

During these same years, Doppler risked his hard-earned job when his students accused him of being too strict in examinations. He was saved by the growing fame his discovery brought him: just another episode in the hard life of a misunderstood genius with a bad temper! When the revolutionary upheavals that swept Europe in 1848 broke out in Prague, the stern professor fled back to Vienna. In the Habsburg capital, he finally found some academic recognition and was even appointed first director of the new Institute of Physics established by the young emperor, Franz Joseph. One of his students was the Augustinian monk Gregor Johann Mendel, the founder of modern genetics. Doppler died in Venice five years later, consumed by a chest disease (probably, as with Fraunhofer, contracted in his youth in his parents' shop, as his father and brother did) and griefed over the continuing bitter criticism of his discovery; especially from his academic colleague Joseph Petzval, with an endless controversy[32] that would eventually involve the young Ernst Mach. He bequeathed to mankind an extraordinary idea about one of the most fruitful light phenomena for the knowledge of the physical world, with innumerable applications even in daily life,[33] but still lacking experimental verification. But how to make a source move so fast that it could compete with light?

---

[32]The dispute with Petzval was to be submitted to arbitration by the Austrian Academy of Sciences, but Doppler was unable to attend the session because he was leaving for Venice, where he hoped to recover some of his health, and where he died instead 4 months later. His absence was seen by his detractors as an attempt to avoid a negative verdict. Cf. D.D. Nolte, *The fall and rise of the Doppler effect*, in "Physics Today", 73, 3, 2020, pp. 30–35.

[33]From medicine to high way patrol, from automatic doors to GPS, to the motions of stars, galaxies, and the universe as a whole, the Doppler effect is ever-present.

The answer had already been given by Hippolyte Fizeau in a communication presented to the Philomatic Society of Paris in 1848.[34] The French physicist, perhaps unaware of his Austrian colleague's work, had theorized on his own that radial motion of a source would cause the spectral lines to shift toward the red side (redshift) if the source was moving away relatively to the observer, or toward the blue side (blueshift) otherwise.[35] It was precisely the Doppler effect applied to spectra, noticeable even at modest speeds and much easier to perceive and measure than a qualitative change in color. Twelve years after Fizeau, a very young Ernst Mach, then Boltzmann's black beast[36] but also Einstein's inspiration,[37] had even indicated the strategy for measuring wavelengths using as references the lines imposed on the spectrum of a star by the Earth's atmosphere, which do not move because they are extraneous to any radial motion of the celestial source.

The experimental proof did not take long. It was an astronomical test, not surprisingly, because only in the sky one could hope to find suitable conditions for producing measurable signals. It was May 1868. In the austere setting of the Royal Society of London, before an intrigued and bewildered audience, William Huggins claimed to have spectroscopically found[38] that the star Sirius was moving away from the Sun at 47 kilometers per second.

---

[34] J. Lequeux, *Hippolyte Fizeau: Physicien de la lumière*, EDP Sciences, Les Ulis 2014.

[35] Perhaps it is not useless to point out that when we consider the entire electromagnetic spectrum, from γ rays to radio waves, the terms "redshift" and "blueshift" lose their original meaning (associated with visible light, which is confined between blue and red) and acquire that of "in the direction of the longer wavelengths" or "of the shorter wavelengths", respectively.

[36] It is possible that the bitter scientific competition between the Moravian Mach and the Austrian Boltzmann, both professors in Vienna, one of philosophy and the other of theoretical physics, was one of the causes that led the unhappy Ludwig to commit suicide in 1906. A brilliant and innovative physicist, Boltzmann was loved by no one. Conservatives blamed him for watering down causalism with statistics, progressives for his mechanistic faith; everyone accused him of being an algebraic "terrorist". See Cercignani, *Ludwig Boltzmann*, cit.

[37] Generally attributed to Mach, who, however, never wrote about it, is a statement, fully sponsored by Einstein, that the inertia of any system is the result of the system's interaction with the rest of the universe: every particle in the cosmos has an influence on every other particle. This principle would preclude the existence of an empty universe, in which, for example, the centrifugal reaction loses meaning because there is nothing to anchor a reference system to.

[38] W. Huggins, *Further Observations on the Spectra of some of the Stars and Nebulae, with an Attempt to determine therefrom whether these Bodies are moving towards or from the Earth* . . . , in "Philosophical Transactions of the Royal Society of London", 158, 1868, pp. 529–564.

The estimate was made by applying the Doppler formula to the measurement of the shift of the Fraunhofer F line[39] with respect to the so-called "laboratory position" (that corresponding to a null radial velocity) and correcting the result for the orbital motion of the Earth. It was a gamble that perhaps only an amateur could afford. Not many people still believed that light actually behaved as the Austrian physicist had predicted. With great audacity, Huggins had opened the way to the use of a formidable new scientific instrument capable of transforming the static image of the deep sky into another in motion: an indispensable prerequisite for the generalized use of the laws of Newtonian mechanics and thus for the evolution of astronomy into astrophysics (and celestial mechanics). A few years earlier, the Jesuit Angelo Secchi had already ruled out the existence of sidereal velocities much greater than that of Sirius. Two observations that, by confirming the reality and usefulness of the Doppler effect, immortalized the Austrian physicist and simultaneously demonstrated the ridiculousness of his interpretation of the colors of binary stars.

The two pioneers of astronomical spectroscopy, Huggins and Secchi, were men of very different social background, education, and motivation. The first[40] was born in London in 1824, in an elegant building of the shopping district, to parents of Calvinist confession, wealthy silk cloth merchants. The shop opened on the ground floor of their house. It was frequented by rich bourgeois ladies who wallowed among the colored fabrics in a desperate attempt to reconcile female vanity with that austerity that Queen Victoria would have made a rule of life. William, an only child, received a good education. He attended Cambridge University for a while, but did not complete his studies. The well-off young man did not need a degree to pursue his hobbies, which included stargazing and music. A violinist, «always rather an intellectual than a perfervid player», as his wife Margaret would later define him, and «Christian unattached»,[41] he found the scope of existence in astronomy.

---

[39] In modern spectroscopic terminology, the Fraunhofer F line is called $H\beta$ and is part of the hydrogen series named after the Swiss Johann Jakob Balmer. In 1885, this unknown professor of geometry at the University of Basel found the famous empirical formula that gives the wavelengths of all the lines in the homonimous series (in the optical range). The Balmer formula had the same role in validating the nascent quantum theory of the atom as Kepler's laws had played in Newtonian gravity.

[40] B.J. Becker, *Unraveling Starlight: William and Margaret Huggins and the Rise of the New Astronomy*, Cambridge University Press, Cambridge 2017.

[41] I.S. Glass, *Revolutionaries of the Cosmos: The Astro-Physicists*, Oxford University Press, Oxford 2006, pp. 118 and 140.

Six years older than the English gentleman scientist, Secchi[42] came from Reggio Emilia, a city in the Duchy of Modena, which had just returned to the House of Habsburg-Este after a twenty-year experience of republican government under the wings of the newborn tricolor flag and the Napoleonic eagles. The youngest of many children, he was born in 1818 into a family of modest and God-fearing people, a social ghetto from which it was difficult to escape. The father was a poor, uneducated artisan who died when Angelo was a child, leaving a void of affection and resources to make ends meet. The mother, who took special care of her youngest son, managed to enroll him in the Jesuit high school after elementary school. There the boy immediately showed unusual intelligence, will, and dedication to study. Prudent and grounded, he obeyed meekly, but his sharp, ironic eyes revealed a universe of curiosity. To satisfy it, at the age of fifteen he asked to join the Company. This choice of life guaranteed him a first-class university education at the Roman College. Mainly classical subjects, but also mathematics and natural sciences. A *summa* of knowledge that made him a good physics teacher. But that was not his destiny. To throw him into the arms of Urania was a dramatic event that involved the entire Society of Jesus. In 1848, just one year after his ordination, Don Angelo had to share the fate of the confreres whom Pope Pius IX had asked to leave Rome and Italy because he could no longer guarantee their safety. What happened?

During the eighteenth century, the Jesuits had gradually acquired a leading role in the Church and distinguished themselves as a cultural elite.[43] They had thus attracted the daggers of the Masonic and Enlightenment intelligentsia, the "party of the philosophers", and the currents within the Catholic world that favored a limitation of papal power in the interests of the nation-states. «*Sint ut sunt aut non sint*» (Let them be as they are or let them not be), the Superior General of the Order, Lorenzo Ricci, had proclaimed facing the request to take a few steps back to satisfy the opponents. An entrenchment that first cost a series of expulsions, from Portugal, France, Spain, and the Kingdom of Naples, and then led to the temporary end of the Society, dissolved by the Franciscan Pope Clemente XIV Garganelli in 1773. Of the two options of Ricci had won the «*aut non*

---

[42]I. Chinnici, *Decoding the Stars: A Biography of Angelo Secchi, Jesuit and Scientist*, Brill, Leiden 2019.
[43]W.V. Bangert, *A History of the Society of Jesus*, Institute of Jesuit Sources, Boston (MA) 1986.

sint». The Jesuit communities were dispersed, the libraries expropriated, the property plundered. In 1814, shortly after the fall of Napoleon, the Order was reconstituted just in time to allow the opening of the school in Reggio Emilia, then frequented by Secchi. But gradually the music resumed as before. The discontent against the Jesuits exploded in Switzerland in 1847 and spread like wildfire to Rome. While the Mazzinians raised their banners on the Capitol and temporarily replaced papal sovereignty with a republican government,[44] the followers of Ignatius of Loyola were urged to leave Italy as a precaution. Among them was Angelo Secchi, who after a short stay in England, where he learned the language and improved his mathematical knowledge, reached the College of the Order in Georgetown, in Washington, D.C., the oldest Catholic institution of higher education in the United States. He was only thirty years old. The American experience would deeply influence him and lead him to astronomy.

Thirty-two was also the bearded and elegant Huggins when, in 1856, after taking over the family business when his parents voluntarily retired, he decided to close it down, sell his house and shop, and move to the South London hills to devote himself entirely to his passion, astronomy. He was an amateur in the strictest sense of the word, with no obligations to anyone; even freer than William Herschel, who had nevertheless been in the pay of a king. He began his observations casually, with no real project in mind. Eclectic and opportunistic, he always tried to direct his efforts where his nose told him there was a possibility of a scoop. But he also gradually acquired a well-deserved reputation as a meticulous observer, cautious in his conclusions. When he took possession of the new observatory house at Tulse Hill, he was still an unknown novice. It was his attendance at the Royal Astronomical Society that made his fortune. Joining this circle of top amateur astronomers as a Fellow, while seeking ideas and inspiration from the more experienced, he found a friend and mentor in the Reverend William Rutter Dawes. A keen observer of binary stars and an influential scientist, Dawes took such a liking to him that he gave him a very valuable

---

[44]The Roman Republic, the second after that of 1799, was a kind of prolongation of that "spring of the people" that had vainly inflamed Europe in 1848. It was proclaimed in 1849, after Pope Pius IX had fled to Gaeta, under the protection of the King of the Two Sicilies. It lasted only six months and was defeated by a coalition led by Napoleon III, Emperor of France; the same one who, 17 years later, would favor the victory of the Piedmontese of the Kingdom of Sardinia over the Austrians in the so-called Second War of Independence.

8 inch objective lens purchased in America from Alvan Clark.[45] The latter was a painter from a family of whalers who, shortly before the middle of the 19th century, had founded a laboratory of astronomical optics in Cambridge (MA) with his two sons, and had rapidly gained fame. So it happened in the New World!

The sale of the lens was not a trivial commercial transaction, but a tangible sign of esteem and friendship. High-quality optics were a rare and valuable commodity, especially in England, where the community of gentlemen scientists flourished and demand was as high as the high taxes on optical glass imposed by Her Majesty's government. In search of technological excellence, Huggins used Dawes' lens as the objective of a telescope with an equatorial mount. Manufactured by Thomas Cooke, a renowned astronomical instrument maker in York, the instrument rotated around the polar axis by means of a clockwork motor, so that the astronomer only had to make small corrections for diurnal motion to keep it pointed during observations. It was a device that allowed him to concentrate on his scientific goal, leaving the bulk of the work of tracking the daily carousel of stars to his mechanical assistant. Dawes,[46] nicknamed "the eagle's eye", also taught Huggins the profession of astronomy and the full meaning of observing the sky. An education which, if it did not fill all the gaps of a lacking scientific education, certainly instilled in this brilliant amateur the ambition to participate with the purpose of winning, in complete contradiction to Coubertin's motto.[47] This explains why, in 1858, while the world was going crazy over Donati's comet, the second brightest of the 19th century that even the man in the street could observe, William instead devoted himself to studying binary stars and the surface of Jupiter, which required a great deal of specialization, and waited for the "big one" to come along. That arrived in 1860, when the first reports of Kirchhoff's and Bunsen's discoveries landed in England from Berlin.

---

[45] *Alvan Clark*, in "Proceedings of the American Academy of Arts and Sciences", 23, 2, 1887–1888, pp. 315–317.

[46] N. English, *The Eagle-Eyed Reverend William Rutter Dawes: A History of Visual Observing from Harriot to Moore*, Springer, Berlin 2018, pp. 200–215.

[47] The Olympic creed, shaped by the ideals of the French educator and historian Pierre de Coubertin, says: «*L'important dans la vie ce n'est pas le triomphe mais le combat; l'essentiel ce n'est pas d'avoir vaincu mais de s'être bien battu*» (The important thing in life is not the triumph but the struggle; the essential is not to have won but to have fought well); https://web.archive.org/web/20090324234949/http://multimedia.olympic. org/pdf/en_report_1303.pdf.

The claim that the chemical and physical nature of the Sun was written in the spectra aroused great interest in the British scientific community, but it was not entirely surprising to a community that had been playing with prisms since Newton. There were also some authoritative voices of dissent. Michael Faraday himself, who enjoyed an immense reputation for his contributions to electromagnetism and electrochemistry, as well as for his many inventions, openly declared that he did not understand how light could carry the connotations of matter, temporarily giving Comte reason. In a short time, however, the idea was metabolized and became the subject of research both on Earth, in university and private laboratories, and in the sky, mostly by amateur astronomers. Huggins, too, decided to quench his thirst at this «fountain of water in a dry and thirsty land».[48] It was just the kind of desert he longed to scour in search of El Dorado.[49] He wasn't unaware of the dangers that a new and unexplored path might hold for an obscure amateur, but as a good son of Albion, he knew well that if you risk nothing, you gain nothing. With an attitude more buccaneer than pioneer, he was careful not to embark on a systematic program of spectroscopy of the stars and nebulae of the boreal sky, as the methodical Herschel, by whom everyone in England was inspired, would have done. He continued to pounce where, thanks to the discoveries of others, there seemed to be an opportunity for a big score. The attack began with the Sun.

In 1861 James Nasmyth had raised the question of the texture of the solar photosphere, where he had observed filamentous willow leaf structures chaotically arranged everywhere except at the edge of the spots.[50] It was still the time when, in the absence of objective documentation, the community had to evaluate and eventually accept the reports of naked-eye explorations made by those who had powerful instruments and excellent eyesight. Observations that sometimes bordered on the impossible and were therefore not easily repeatable. Though still a neophyte when he made his announcement to the Manchester Literary and Philosophical Society, Nasmyth was believed by all but Dawes, mostly out of jealousy towards a newcomer who claimed to have made a scoop as soon as he entered the business. The combative reverend found it necessary to intervene when he realized that the

---

[48] J.L. Heilbron (ed.), *The Oxford Guide to the History of Physics and Astronomy*, Oxford University Press, Oxford 2005, p. 18.

[49] O. Struve, *Stellar Spectroscopy*, in "Science", ns, 106, 2749, 1947, p. 204.

[50] T.E. Margrave Jr., *Review of Visual Observations of Solar Granulation*, in "Journal of the Washington Academy of Sciences", 58, 2, 1968, pp. 26–31.

alleged discovery was about to have the approval of John Herschel, who was struggling with the new edition of his astronomical bible, the *Outlines of Astronomy*. Huggins sided with his friend Dawes, but without going too far. He made some observations of his own and described them by introducing the term "granules", for which he received the congratulations of De la Rue. At the time, this was a kind of safe conduct in the world of British gentlemen scientists, who were swarming with intelligence, but also with more or less long and sharp knives. A community where a few "barons" set the agenda and the many peons struggled to survive by keeping a low profile. Despite his ambition, Huggins was still one of the peons.

The turning point, sudden and unexpected, came on a warm evening in late August 1864.[51] William had decided to point the spectroscope-equipped telescope at a planetary nebula in the Dragon, now known as NGC 6543 or Cat's Eye. A ball of light framing a bluish star. William Herschel had discovered it eighty years earlier during his sky surveys. «I was fortunate in the early autumn of the [...] year, 1864, to begin some observations in a region [that of the nebulae] hitherto unexplored»,[52] would have nostalgically recalled old William writing his memoirs in 1897. Why this choice? Because no one had ever observed the spectrum of a nebula, and yet many clues suggested that this heterogeneous family of celestial objects held important secrets. Huggins had no idea what he was going to see, but he hoped it would be something that might shed some light on the matter and earn him a proper reputation. «After so many years we find ourselves without a guide in the endless wilderness of nebulae»,[53] complained the Reverend Thomas William Webb, an Anglican minister who in his spare time explored the secular sky with a 10 inch reflector armed with a silver mirror; an innovative technology about which we will speak later (p. 328). The observational inputs were numerous and often contradictory, and did not help to untie the fundamental knot of the problem, namely the nature of the scattered light. Unresolved stars or luminous fluid such as a hot gas? The corollary to the unresolved question was the distance and size of the objects in play.

---

[51] M. Beech, *The Pillars of Creation: Giant Molecular Clouds, Star Formation, and Cosmic Recycling*, Springer, Berlin 2017, pp. 38–41.
[52] M.J. Crowe, *Modern Theories of the Universe: From Herschel to Hubble*, Dover, New York 1994, p. 184.
[53] Becker, *Unraveling Starlight*, cit., p. 69.

An idea was gaining ground — also supported by William Herschel at a certain stage of his mental journey in search of a model for nebulae — that there were two families, different in nature and properties. One family was that of the magnificent spirals observed by Lord Rosse. The other was made up of intrinsically nebulous objects close enough to show us signs of change, both photometric and morphological, on short time scales.[54] There were many examples of these unique modifications. In 1837, while at the Cape of Good Hope, John Herschel had observed the variability of the star η Carinae[55] and the associated nebula called the Keyhole. And a quarter of a century later he was astonished to learn that the British educator Eyre Burton Powell had recorded a change in the shape of the object from Madras. Webb himself had reported possible variations in the brightness of the nebula around the blue-white supergiant Merope, one of the Pleiades. In 1862, John Russell Hind, a prolific English observer, had discovered the variability associated with the nebulosity of the star T Tauri.[56] Phenomena that, as we know today, accompany the newborn stars or those extra large and dying. A curiosity: as an asteroid hunter, Hind wanted to name 12 Victoria his second prey, 12th in the list of discoveries. This caused an international controversy. The rule was to give small planets names, preferably female, taken from classical mythology. Since Queen Victoria was reigning at the time, Hind was accused of flattery. But the cunning Englishman stood his ground, explaining that Victoria was referring to the Roman goddess and had won. As in, what matters is winning!

Let us return to Huggins and his account of this first observation. He would have liked to point the telescope at the nebula suggested by Webb, but it was not the right time of year, so he decided to make the best of it. The Cat's Eye was a promising target because it was particularly colorful.

---

[54] R.F. Hirsh, *The Riddle of the Gaseous Nebulae*, in "Isis", 70, 2, 1979, pp. 196–212.

[55] G. de Vaucouleurs, *The Wonder Star: Eta Carinae*, in "Astronomical Society of the Pacific Leaflets", 6 281, 1952, pp. 244–251.

[56] This star gave its name to a family of variable stars, the classical T Tauri. These are young objects (less than 10 million years old) surrounded by a protoplanetary disk of gas and dust from which the star's future planetary system may form. You can enjoy the images of some of these planetary disks produced by the Atacama Large Millimeter/submillimeter Array (ALMA), a radiointerferometer located at an altitude of 5,000 meters in the Atacama Desert in Chile, the result of a partnership between Europe, North America, and East Asia.

On the evening of the 29th of August, 1864, I directed the telescope for the first time to a planetary nebula in Draco.[57] The
reader may now be able to picture to himself to some extent
the feeling of excited suspense, mingled with a degree of awe,
with which, after a few moments of hesitation, I put my eye to
the spectroscope. Was I not about to look into a secret place of
creation? I looked into the spectroscope. No spectrum such as
I expected! A single bright line only! At first, I suspected some
displacement of the prism, and that I was looking at a reflection of the illuminated slit from one of its faces. This thought
was scarcely more than momentary; then the true interpretation flashed upon me. The light of the nebula was monochromatic, and so, unlike any other light I had as yet subjected to
prismatic examination, could not be extended out to form a
complete spectrum [...] The riddle of the nebulae was solved.
The answer, which had come to us in the light itself, read: Not
an aggregation of stars, but a luminous gas.[58]

Since, according to Kirchhoff, only luminous matter in the gaseous state
emits light «consisting of certain definite refrangibilities», Huggins cau-
tiously concluded that «we must probably regard these objects, or at least
their photo-surfaces, as enormous masses of luminous gas or vapor».[59] He
was right. The discovery, which complicated rather than solving the mystery of nebulae, brought him, among other things, the cooption into the
Royal Society, the first of many honors. Perhaps most important was a note
in the "Progress in Astronomy" section of the Royal Astronomical Society's
1865 Annual Report, which stated that «the relation between terrestrial
physics and the physics of the sidereal heavens is rapidly becoming more
intimate, the boundary line which was once supposed to divide them is
gradually disappearing, while new and unexpected fields are opening up for
the application of the results of experimental philosophy, in distant regions
of space where formally they were supposed to have little or no concern».[60]

---

[57] The nebula is NGC 6543, currently known also as the Cat's Eye.
[58] *The Scientific Papers of Sir William Huggins, Historical Statement*, W. Wesley &
Sons, London 1909, p. 106; W. Huggins, *The New Astronomy: A Personal Retrospect*,
in "Nineteenth Century", 41, 1897, pp. 916–917.
[59] Ivi, p. 114.
[60] Becker, *Unraveling Starlight*, cit., p. 77.

Two years later, Huggins brought home another rich bingo thanks to a chance discovery.[61] Around midnight on a May day in 1866, the sky over Millbrook, some fifty miles northwest of Dublin, was unusually clear and the air barely crisp; and a nearly new moon left the clover fields in darkness. John Birmingham, an eclectic Irish landowner who dabbled in astronomy, was returning home from a visit to some friends. While admiring the star-studded sky, he noticed that a bright new star had appeared in the Corona Borealis. Only two hours before Birmingham's discovery — as was later ascertained — the German astronomer and geophysicist Julius Schmidt, then director of the Athens Observatory, had observed the same field with the naked eye without noticing anything abnormal: a sign that the new star had suddenly emerged from the cosmic darkness.

The sudden appearance of a "new" celestial source was not an unknown event, but certainly rare. Today we understand that the event originates from a controlled explosion by which a star pathologically compressed to reach very high densities (white dwarf) periodically frees itself of the excess matter spilled on its surface by a companion in an expanding phase (red giant). This powerful sneeze is accompanied by a burst of light that then gradually fades over time. The phenomenon of "nova" is different in intensity and outcome from that of the supernova, which instead involves the destruction of the star and leaves at most a remnant in the form of a neutron star or a black hole, in addition to the damage done to the surrounding interstellar medium (supernova remnant). The glow produced is also very different in the two cases, as is the frequency of the events. It is estimated that 30 to 60 novae explode each year in the Milky Way, while only three galactic supernovae are known in a millennium, the "Chinese" of 1054 A.D. and the Tycho and Kepler stars of 1572 and 1604, respectively.

The first authentic nova of the modern age[62] had been sighted in 1670 in the Vulpecula (Little Fox) by Father Anthelme Voituret, a French monk who was passionate about celestial changes, and independently by Johannes Hevelius. Only two were discovered, in Sagitta (Arrow) and Ophiuchus, bright enough to be seen with the naked eye. That's it. Therefore, the object of the Corona Borealis deserved some attention. Birmingham hurried to collect the spectrum and at the same time sent a letter to the "Times" in London to guarantee the authorship of the discovery. However,

---

[61] Ivi, pp. 86–91.
[62] M.A. Hoskin, *Novae and Variables from Tycho to Bullialdus*, in "Sudhoffs Archiv", 61, 2, 1977, pp. 195–204.

fearing that the newspaper would not publish it, which it did, he also sent a copy to Huggins. The latter had a very good reason to get to work immediately. In the meantime the nova had been sighted by others, so much so that within two hours a second report had arrived at Tulse Hill from a gentleman astronomer in Manchester.[63] Everyone was looking for Huggins because his interest in this kind of phenomenon was well known. Some time earlier, Father Secchi had informed the Royal Astronomical Society that he had found discrepancies with the positions (wavelengths) of the spectral lines of Betelgeuse published by Huggins. They could be — the Jesuit surmised — manifestations associated with the same phenomenon that had caused the capricious jumps in the star's brightness noticed by John Herschel thirty years earlier. What more could one ask for than a nova to test the hypothesis? Especially since the Franco-German astronomer and painter Hermann Goldschmidt, the luckiest of asteroid hunters, had speculated in 1826 that the phenomena might be signs of a recurring explosive mechanism.[64] A brilliant insight!

A good privateer, Huggins sensed an opportunity for a big score. So he invited a neighbor, William Allen Miller, also a gentleman scientist and a pioneer in spectroscopy, to join him in observing the celestial novelty. After all, it was a matter of getting visual impressions and, as the saying goes, two are better than one. It turned out to be a wise precaution, because what the two friends saw was hard to believe: the spectrum of the nova manifested a classical stellar continuum, furrowed by absorptions (the lines produced by the passage of radiation through a gas, according to Kirchhoff's bible), overlaid by a series of bright lines. A symbiosis, then, between the ordinary stellar spectrum and that of nebulae such as the Cat's Eye. Abandoning traditional caution, Huggins ventured to formulate the hypothesis that the emissions were caused by a hydrogen cloud around the star.[65] To confirm this, he pointed to a vague haze that seemed to encircle the nova. Enthusiastic, he and his friend Miller rushed to report the observations and interpretation to the Royal Astronomical Society. At the same time, like a good bloodhound, William did not let go of his quarry, following the object even as it faded. Eventually he convinced himself that he had discovered

---

[63]S.F. Whiting, *The Tulse Hill Observatory Diaries*, in "Popular Astronomy. A Review of Astronomy and Allied Sciences", 25, 1917, p. 160.

[64]Becker, *Unraveling Starlight*, cit., p. 88.

[65]Huggins used sodium to create a reference spectrum with which he calibrated those of celestial objects. He also used a spark to produce the hydrogen spectrum.

a burning star; an object that had exploded for some reason and expelled large amounts of hydrogen. This hypothesis accounted for the rapid rise in brightness and subsequent decline due to the gradual consumption of the ejected material. Not a bad hypothesis at the time! Eventually the nova became too faint to observe, partly due to the luminosity of the sky background brightened by a rising moon.

But Huggins pondered the discovery, wondering what the nova's spectrum would have looked like before the explosion. He was intrigued by the stars whose spectra showed emission lines. These too, he speculated, must have been cataclysmic in nature. He became convinced that spectroscopy could help to understand not only the chemical composition, but also the structure of stars, and especially what caused the variations in brightness. He hoped that a systematic study of stars with emission spectra (albeit non-variable) might reveal the possible connection, if any, between nebulae, novae, and ordinary stars. Meanwhile, «the observatory became a meeting place where terrestrial chemistry was brought into direct contact with celestial chemistry».[66]

Another two years passed, and in 1868 Huggins produced a third brilliant result, showing how to use the Doppler effect to determine the radial velocity of celestial bodies. This exploit baptized perhaps the most productive astronomical instrument of all time. It might have been enough for a self-made man who, thanks to a clever and courageous use of the spectroscope and the interpretative tools of the new physics just born in Germany, had been able to climb the highest spires of the temple of Urania. But in 1875, now 51 years old, William took a wife. Perhaps it would be more accurate to say that he was chosen by an Irish girl, 24 years his junior, who was infatuated with astronomy and the man who, in her opinion, embodied it. With her vivid eyes, Margaret Lindsay Murray[67] brought new lymph into the gentleman scientist's life, suggesting further lines of research as well as the systematic and skillful use of photographic emulsions. The partnership produced interesting results, as well as some, albeit excusable, scientific gaffes, such as when the couple tried to interpret a magnificent photographic image of the M31 nebula in Andromeda. It had been taken in late 1888 by a certain Isaac Roberts, a wealthy Welsh builder with a

---

[66] E. Harrison, *Cosmology: The Science of the Universe*, Cambridge University Press, Cambridge (MA) 2000, p. 82.

[67] P. Aldrich Kidwell, *Women Astronomers in Britain*, 1780–1930, in "Isis", 75, 3, 1984, pp. 534–546.

passion for astronomy, with a 4 hour exposure on a 20 inch reflector made for him by Howard Grubb of Dublin. The image confirmed the existence of the spiral structure that Lord Rosse had already seen and drawn in M33, the Triangulum Galaxy, and M101, the Pinwheel Spiral in Ursa Major, and which Roberts read as the signature of an accreting planetary disk around a newborn star.

Margaret Huggins believed this too, as we can realize from this passage in her notebook:

> For another, Mr. Roberts of Liverpool has just sent us a photo-graph of it which shows clearly & unmistakably a solar system in progress of making. Mr. Roberts' photograph of this object is certainly the most interesting photograph of a celestial object as an object, ever taken. How Laplace would rejoice over it ![68]

In a letter to the Irish mathematician and physicist George Gabriel Stokes, secretary of the Royal Astronomical Society, her husband added the dose:

> [Mr. Roberts' photograph] reveals for the first time to the eye of man its true nature. A solar system in the course of evolu-tion from a nebulous mass! It might be a diagram to illustrate the Nebular hypothesis! I never expected to see such a thing [...] There are some 6 or 7 rings of nebulous matter already thrown off, & in some of them we see the beginning of planetary condensation & one exterior planet fully condensed [...] The central mass is still larger, to compare it with the solar system, say as large as the orbit of Mercury. The rings are all in one plane & the position is such that we see it obliquely.[69]

Huggins's prestige was such that his ideas, even the wrong ones, made opin-ion. When he died in 1910 at the age of eighty-six in London, where he had always lived, all of England, which had awarded him four honorary degrees from Cambridge, Oxford, Edinburgh, and Dublin,[70] wept. A practice that does not belong to countries that reward imagination more than thought.

Much more measured and cautious was the work of the other great pio-neer of spectroscopy. In 1849, with the fall of the Second Roman Republic

---

[68]See Whiting, *The Tulse Hill Observatory Diaries*, cit., p. 162.
[69]L. Belkora, *Minding the Heavens: The Story of our Discovery of the Milky Way*, CRC, New York 2002, p. 200.
[70]*Sketch of William Huggins*, in "The Popular Science Monthly, June 1982, pp. 260–266.

and the restoration of temporal power to the Pope, Angelo Secchi, like so many of his brothers, was able to return to Rome. America had profoundly changed him, rekindling his passion for astronomy. So he gladly accepted the appointment as Director of the Observatory of the Roman College — but he would have done it anyway, out of obedience. Don Angelo was a leader. Frugal and hasty, devoted and respectful of his superiors, he knew how to think big. He took an observatory in really bad shape and turned it upside down. With the money of a legacy generously given to him by a brother, he managed to acquire a Merz 22 cm refractor; an instrument of small size compared to the American giants, but sufficient for a modern astronomy. With the approval of Pope Pius IX, he moved the observatory to the flat roof of the Church of Sant'Ignazio, next to the College, using the four massive pillars designed to support a dome that had never been built, replaced by the Jesuits with a magnificent *trompe-l'œil* (French for "deceiving the eye"), and took the field, immediately becoming a reference point for the astronomers of rising Italy; a nation, however, to which he never felt he fully belonged.[71]

In addition to his observatory, for fifteen years he worked on various theoretical and practical problems, not only in physics and astronomy (Father Secchi was a renowned geologist), but also on the tasks that the Pope of the "great illusion"[72] gradually assigned to him. In 1863 he discovered spectroscopy, an observational technique that would occupy him for the rest of his scientific life.

> Just when it seemed that the field of astronomical research was about to come to a standstill, and that we would only have to pick up where others had gathered riches; behold, a new discovery comes to open an infinite horizon that will one day end by revealing to us the physical nature of the stars and by showing us the quality of the matter that composes them. This is the spectroscopy and its applications made by Kirchhoff and Bunsen.[73]

---

[71]Cf. R. Buonanno, *The sky above Rome. The places of astronomy*, Springer, Milan 2007, pp. 123–146 (in Italian).

[72]In 1848, Pope Pius IX, after giving the illusion that he was supporting the cause of Italian independence by joining Piedmont's war against Austria, had backed down for fear of creating a schism within the German Church.

[73]A. Secchi, *Le scoperte spettroscopiche in ordine alla ricerca della nature dei corpi celesti*, Tipografia delle Belle Arti, Rome 1865, p. 4 (in Italian).

Unlike Huggins, who was attracted by nebulae and variable phenomena, Secchi turned to ordinary stars, trying to classify them in an effort to understand these mysterious machines. «You ask what is the use of classification, arrangement, systematization», Settembrini explained to Hans Castorp in Thomas Mann's *The Magic Mountain*. «I answer you; order and simplification are the first steps toward mastery of a subject — the actual enemy is the unknown».[74] In the same vein, Secchi had stated in plain language:

> The most salient fact verified in these researches has been this: that although the stars are very numerous, their spectra are nevertheless reduced to a few well-defined and distinct forms, which for brevity we have called Types.[75]

Persistent and tireless, Secchi was able to observe the spectra of over 4,000 stars in a few years, thanks to the use of a prism placed in front of the objective lens, which enabled him to view more than one source at a time. The impressive collection was divided into four classes based on color and other characteristics such as lines and bands.[76] The physical concepts needed to interpret this picture were still lacking, proving the peculiarity of a science that often finds something new before it is ready to receive it. But the path of sidereal astrophysics was now mapped out. From the rooftops of ancient Rome, the Jesuit priest, a fervent and orthodox believer, with few means and immense self-denial, had united the secular heavens with the Earth, forever rejecting Aristotelian dualism. However, he could not fully enjoy the earthly glory that came with it. His ordeal began in 1870, when the King of Savoy's *Bersaglieri* (the infantry corps of the Italian army) entered the Eternal City through Porta Pia. Torn between the need to collaborate with the "Italians" and the obligation of fidelity to the deposed papal king, Don Angelo displeased both, until a stomach cancer took him away before he was sixty. His death deprived the newborn Society of Italian Spectroscopists, the first professional organization in the world specialized in "physical astronomy", of its founder and president. Only three weeks had

---

[74] Quoted by A. Sandage in *The Classification of Galaxies: Early History and Ongoing Developments*, in "Annual Review of Astronomy and Astrophysics", 43 (1), 2005, pp. 581–624.

[75] Id., *Le stelle. Saggio di astronomia siderale*, F.lli Dumolard, Milan 1877, p. 45 (in Italian)

[76] Secchi's classes are 1. α *Lyrae* stars, blue-white, with intense hydrogen lines; 2. stars of the α *Herculis* type, with large absorption bands; 3. yellow solar-type stars with thin lines of hydrogen and metals; 4. reddish stars dominated by carbon lines.

passed since the death of Giovanni Maria Mastai Ferretti, Bishop of Rome, exile, executioner, and deposed ruler of a temporal reign that had lasted about two millennia. Almost a sign to underline that the new science of the heaven wanted for itself what belonged to Caesar, unhindered by what belonged to God.

Father Secchi's scientific legacy was collected overseas by the director of the Harvard University Observatory, Edward Charles Pickering, who in 1885 began a spectroscopic survey of the stars using, like Secchi, a lens prism.[77] «Observation is the first step of philosophy»,[78] was the Baconian approach to science of this brilliant young physicist, former professor at MIT. The gigantic project required a close-knit team to carry out the observations and the subsequent analysis. The necessary funding was provided by Mary Ann Palmer, Henry Draper's widow. With a considerable personal fortune, she agreed to finance the cataloging and not only, as long as the work carried her beloved husband's name. When it was completed, in 1890, the Henry Draper Catalog contained the spectra of more than 225,000 stars. The systematic survey had also revealed several novae, hundreds of variables, and a long list of peculiar stars.

To classify the spectra, Pickering used a team of women, who were cheaper and more reliable than men for an activity that required patience and self-denial. This "harem", as the women's collective was jokingly called,[79] included a Scot, Williamina Fleming, former housekeeper at Pickering's, Antonia Maury, granddaughter of Draper who carried in her features a quarter of his Portuguese blood, Henrietta Swan Leavitt, the Cepheid Muse, of whom we shall speak of again in the next chapter, and Annie Jump Cannon, rich and intelligent woman, passionate about astronomy. It was to Annie that the "Sultan" entrusted the task of classifying spectra, after he had removed Antonia Maury from her post because she had been carried away by the game of multiplying classes to the detriment of synthesis, to the great sorrow — we must believe — of William of Occam. This turned out to be one of Pickering's many happy intuitions. The thirty-three-year-old daughter of a Delaware shipbuilder and state senator, Annie

---

[77] E.E. Barnard, *Development of Photography in Astronomy (II)*, in "Science", ns, 8, 195, 1898, pp. 386–395.

[78] D.A. Kronick, *"Devant Le Deluge" and Other Essays on Early Modern Scientific Communication*, The Scarecrow Press, Lanham (MD) 2004, p. 226.

[79] D. Sobel, *The Glass Universe: How the Ladies of the Harvard Observatory Took the Measure of the Universe*, Penguin, New York 2017; M.W. Rossiter, *"Women's Work" in Science, 1880–1910*, in "Isis", 71, 3, 1980, pp. 381–398.

had an iron will and a unique ability to concentrate, partly because of a deafness contracted at an early age by scarlet fever. In a short time she was able to identify and rearrange the original Harvard classes, borrowed from Secchi's and indicated by the letters of the alphabet, into the sequence still in use today, O, B, A, F, G, K, M, based on the intensity of the hydrogen absorption lines. Miss Cannon also dictated the following witty mnemonic rule to help remember the strange list of alfabetic characters: "Oh, be a good girl, kiss me". But she went no further. No one knew that her empirical sequence was directly related to the temperature of the stellar photosphere. The Indian physicist Meghnad Saha, a pariah destined to enter the good salon of science, would have discovered it in the Charleston years.

At a time when women everywhere were denied the right to vote and their access to work was restricted, especially to those activities considered the exclusive prerogative of men, the "Harvard calculators", though underpaid, were an exception; all the more remarkable when compared to the male-dominated ostracism exercised in the same years at the University of Göttingen against Emmy Noether,[80] perhaps the greatest female talent in the field of mathematics. She was marginalized because of her gender, on the pretext of a lack of women's restrooms on the university campus, despite the support of two giants such as David Hilbert and Felix Klein. A sign of the (relatively) greater openness of the New World compared to a Europe still incapable of renewal. Pickering's harem gave birth to another fundamental discovery, a tool for measuring the distances of stars. It was an alternative to parallax and more powerful than the classical geometric method; a lockpick to unhinge the boundaries of the Milky Way and reveal a cosmos populated by scattered island universes in the depths of space and time. We will now see how it happened.

---

[80]M.B.W. Tent, *Emmy Noether: The Mother of Modern Algebra*, AK Peters, Wellesley (MA) 2008.

Chapter 12

The meter and the measure: Distances of celestial bodies, from Aristarchus to Miss Leavitt

> If we knew what we were doing
> it would not be called research,
> would it?
> Albert Einstein, attributed

> There are two possible outcomes:
> if the result confirms the hypothesis,
> then you've made a measurement.
> If the result is contrary to the hypothesis,
> then you've made a discovery.
> Enrico Fermi, attributed

On the evening of August 20, 1885, the thirty-four-year-old Ernst Hartwig was on duty at the achromatic 9 inch refractor of the Imperial University Observatory in Dorpat, Estonia. Everything in the Baltic city was German: the name itself[1] of Germanic origin, the customs, traditions, and even the language used at the university, despite the long domination of first the Swedish kings and then the tsars. Even the Observatory, founded in 1802 by the will of Alexander I Romanov, the tsar defeated by Napoleon at Austerlitz and then victorious in the French Emperor's ill-considered Russian campaign, had a distinct Germanic character. Germans were the first director, the famous Friedrich Georg Wilhelm von Struve, and the first prestigious instrument, purchased in 1824 from the Bavarian firm of Joseph Fraunhofer, then the largest telescope in the world; and German was Hartwig himself.

As soon as the sky looked dark enough, Ernst decided to point the now obsolete but still very powerful telescope at the brightest point of the great

---

[1]Since 1946, the city of Dorpat has been renamed Tartu, and its observatory, moved to the Tõravere hill, now shares this new denomination.

nebula in Andromeda. Over the past two weeks, he had peered into the mysterious ball of light of M31 several times. By now he knew the field by heart, including the Milky Way stars projected onto the nebula. So, despite a nearly full moon, the presence of an intruder jumped out at him: a new star almost at the center of the luminous cloud.[2] Hartwig realized immediately that he had stumbled upon a treasure! Excited, he thought he had found a clear case for the Laplacian theory of the formation of planetary systems. The source that had suddenly appeared could only be the central star of a gaseous disk pregnant with future planets. An achievement worthy of a Nobel nomination, if we were to judge it today. There was no time to lose. The news had to be communicated immediately to the scientific world in order not to be overtaken at the finish line.

But his director, also a Baltic German, who as a young man had distinguished himself as a daring explorer of the Siberian steppes, imposed an embargo on him. It is well known that old age and courage do not go well together! Unwilling to risk making a fool of himself and the Tsar who paid his salary, Peter Schwarz wanted assurances that the new source Hartwing had discovered was not just a false reflection, perhaps caused by the dazzling light of the Moon. After a week of meticulous checks, Hartwig finally received the go-ahead and was able to send the news of his discovery to the Central Office for Astronomical Telegrams; a service that the *Astronomische Gesellschaft* had established three years earlier in Kiel to disseminate *urbi et orbi*[3] news about the sky, especially those concerning transient phenomena that required real-time reactions. In a world drunk with nationalism, observational astronomy could only be a shared science; by vocation, because the sky belongs to everyone, and by necessity, because more eyes see better than one and, if located in distant places, might succeed in beating bad weather conditions.

When the information came out, it was discovered that the star had already been noticed by a Frenchman three days before Hartwig's observation, by a Hungarian baroness two days later, and also by the great

---

[2]L.A. Marschall, *The Supernova Story*, Springer, Berlin 1988, pp. 94–101; P. Hodge, *The Andromeda Galaxy*, Springer, Berlin 1992, pp. 5–7; D.E. Beesley, *Isaac Ward and S Andromedae*, in "Irish Astronomical Journal", 17, 2, 1885, pp. 98–102.

[3]Latin phrase meaning "to the city [Rome] and to the world", used by the pope as bishop of Rome (*urbi*) and head of Christianity (*orbi*), and more broadly as a synonym for "everywhere in the world".

Maximilian Wolf[4] at the Heidelberg Observatory on the Königstuhl. However, for one reason or another, none of the three had paid any attention to the phenomenon. Lucky for Hartwig. In the months that followed, S And, as the strange star was called, was steadily tracked by several observers until, in February 1886, it became too faint even for the most powerful instruments,[5] and the darkness of the cosmic night swallowed it before daylight would in any case take over the faint glow of M31. But what the hell was it? Astronomers had no doubt about the nature of the phenomenon: not the whim of a fledgling star, as Hartwig, whose discovery had already earned him a promotion to director of the Bamberg Observatory in Upper Bavaria, had imagined. Rather, it was the roar of one of those rare stars that from time to time suddenly appear in some corner of the sky and then slowly disappear into the nothingness from which they came: the mysterious novae stars we met in the previous chapter. As being associated with M31 — no one considered the hypothesis that S And could be a source interposed between the Earth and the nebula — it could therefore be used to estimate the distance of its nebula, as long as there was another nova of known distance with which to make the comparison; assuming, of course, that at the maximum of the light curve all the novae had the same intrinsic brightness. Too bad that at the moment there was no trace of such a touchstone (or template, to put it in astronomical jargon).

After only 16 years, the object of desire suddenly materialized near Algol, in the constellation Perseus.[6] It was noticed by a Scottish priest who was returning home at night after visiting some friends. Raising his eyes to the sky by chance, the man of God had seen a star shining in a place where, according to his memory, there should not be one; a situation very similar to that which in 1866 had led the Irishman Birmingham to the discovery of the nova in the Corona Borealis that we mentioned in the previous chapter (p. 290). In less than 24 hours, the brightness of the intruder increased by 120,000 times, making the new star visible to the naked eye. For a couple of days the rise continued, albeit slowly, to the point where the object became one of the brightest sources in the night sky. But the party did

---

[4]See Knill, *Wolf, Maximilian Franz Joseph Cornelius*, http://people.math.harvard.edu/~knill/history/wolf/bea_proofs_wolf.pdf.

[5]G. de Vaucouleurs, H.G. Corwin, *Andromedae 1885: A Centennial Review*, in "Astrophysical Journal", 295, 1985, pp. 287–304; S. van den Bergh, *The Enigmatic Supernova S Andromedae*, ivi, 424, 1, 1994, pp. 345–346.

[6]*GK Persei, The New Star of the 20th Century*, in "American Association of Variable Star Observers", 2010, https://www.aavso.org/vsots_gkper.

not last long. Immediately after this outburst, the star began to fade, and after a few months of slow decline, GK Persei, also known as Nova Persei 1901, returned invisible to those not equipped with a telescope. Yes, but its distance?

The problem was solved by the detection of a faint nebula around the star, as had been seen in an earlier case by Huggins. It was noticed within a few days by Camille Flammarion and Eugéne Michel Antoniadi in France, Max Wolf in Heidelberg, and George Ritchey at the Yerkes Observatory in the United States: a series of "big shots" to confirm with independent observations a discovery that would have important consequences.[7] In fact, in 1903 the Argentinean-American astronomer Charles Dillon Perrine communicated the result of a spectroscopic study of the nebula with the 36 inch reflector recently donated to the Lick Observatory in California by the English politician and gentleman scientist Edward Crossley.[8] He in turn had purchased the instrument from the eccentric millionaire Andrew Ainslie Common.[9] At the age of forty, this engineer had chosen astronomy, throwing himself into the fray with his great intelligence, his powerful economic resources, and a genuine passion for innovation. Without concealing the remaining doubts, Perrine had made the plausible assumption that the light shown by a knot of the luminous envelope was the reflection of that «emitted by the nova at the time of its greatest brightness».[10]

The idea of a light echo was taken up by Jacobus Kapteyn, whom we have already met in the census of the Milky Way, and the Austrian Hugo von Seeliger, professor of astronomy and director of the University Observatory in Munich. They hypothesized that the milky cloud marked the front of the radiation produced at the time of the nova's appearance and then reflected by pre-existing circumstellar matter. The two scientists were able to estimate the star's distance by comparing the angular dimension of the cloud with the distance the light had traveled in the time since the burst appeared. The calculation, though very uncertain, showed that Nova Persei

---

[7]C.D. Perrine, *Photographs of the Nebulosity Surrounding Nova Persei*, in "Publications of the Astronomical Society of the Pacific", 14, 86, 1902, pp. 150–153.

[8]J.E. Keeler, *The Crossley Reflector of the Lick Observatory*, in "Publications of the Astronomical Society of the Pacific", 12, 75, 1900, pp. 146–167.

[9]J. Lankford, *Amateurs and Astrophysics: A Neglected Aspect in the Development of a Scientific Specialty*, in "Social Studies of Science", 11, 3, 1981, pp. 275–303.

[10]C.D. Perrine, *The Spectrum of the Nebulosity Surrounding Nova Persei*, "Astrophysical Journal", 17, 1903, pp. 310–314.

could be 300 light-years from Earth.[11] The peculiar meter had been boldly calibrated[12] and could be applied to the case of S And by comparing the apparent magnitudes of the two stars, assuming that the two objects had the same nature and therefore the same intrinsic brightness. Since the maximum of Nova Persei's light curve appeared to be 250 times brighter than that of S And, the distance of the latter had to be $300 \times \sqrt{250} \simeq 4,700$ light-years, provided that the assumptions on which the rickety costruction stood were justified; and in fact not all of them were, with serious consequences for cosmology,[13] as we will see in a moment.

It must be said that the measurement of celestial distances has always been a headache for astronomers, a source of real scientific misdeeds, but also of astonishing surprises, and even today an arena for the best minds. In the good old days, when the cosmological problem was confined to the sphere of the fixed stars and conditioned by the inability to reveal the parallactic oscillations of the stars, two tightrope walkers of natural philosophy, Eratosthenes of Cyrene and Aristarchus of Samos, first devised the way to estimate the terrestrial radius and the distances of the Moon and the Sun. A Greek-speaking Libyan, Eratosthenes lived in the third century B.C., between the glorious exploits of the Greek Alexander and the pounding advance of the Roman legions. The chronicles remember him, not without a touch of resentment and perhaps envy, as a kind of spare hand: an eternal runner-up. Instead, he had prestigious duties. He was tutor to the pharaoh's heir and third rector of the Library of Alexandria, which, along with the adjacent Temple of the Muses or Museion, was the center of culture and memory in the Mediterranean world. Even the method he devised for measuring the radius of the Earth may seem like a clever gimmick. Instead, it was an authentic piece of scientific skill, made by drawing on diverse knowledge.[14]

Like Pythagoras long before him, Eratosthenes was convinced that the Earth was spherical. So, when he learned that in Siene (modern Aswan), a

---

[11] Corresponding to a parallax of about $0.''01$; P. Couderc, *Les auréoles lumineuses des Novæ*, in "Annales d'Astrophysique", Vol. 2, pp. 271–302; see also P.C. van der Kruit, *Jacobus Cornelius Kapteyn: Born Investigator of the Heavens*, Springer, Berlin 2015, pp. 311–315.

[12] Subsequent measurements would have put the distance of GK Persei at 1,500 light-years, 5 times greater than that estimated by Kapteyn and von Seelinger.

[13] Cf. M. Capaccioli, *L'espansione dell'Universo*, Corriere delle Sera, Milan 2019, series "Viaggio nell'Universo", A. Ferrara (ed.) (in Italian).

[14] J. Dutka, *Eratosthenes' Measurement of the Earth Reconsidered*, in "Archive for History of Exact Sciences", 46, 1, 1993, pp. 55–66.

city in Lower Egypt just founded by Ptolemy III, the midday Sun reached its zenith on the occasion of the summer solstice, he thought to take advantage of the fact that in the Nile Delta, in Alexandria where he lived, the star still cast a small shadow at the same moment. The geographer constructed a geometric model of the phenomenon, assuming that, for an observer on Earth, the rays coming from the Sun could be treated as a bundle of parallel lines. Under these far from obvious hypotheses, the length of the residual shadow was sufficient to calculate, in the context of Euclidean geometry, the angle between the two locations with respect to the center of the terrestrial sphere. After the direct determination of the distance between Alexandria and Siene, all that remained was the application of a simple proportion. The resulting value for the length of the Earth's radius still amazes us: 40,000 Greek stadia, which, with due uncertainty about the true value of this unit,[15] agrees with the modern estimate[16] by better than 15%.

A few decades before Eratosthenes' stroke of genius put the practice of the Earth on hold, Aristarchus had set his sights on the Moon and the Sun.[17] From the duration of lunar eclipses, he had determined that the distance of the celestial body was equivalent to 60 terrestrial radii, which is incredibly close to the modern average.[18] Again, the model implied bold hypotheses about the shape of bodies and the behavior of light. Then, by measuring the angle of elongation between the Moon and the Sun at the exact moment of the first or last quarter when the illuminated and dark parts of the Earth's satellite are equal, he had fixed the ratio between the distances of the two celestial bodies at a number between 18 and 20: this time blatantly wrong with respect to the value accepted today of about 400. A sensational hoax? Not a chance! The cause of the error was not in the method itself, but in the extreme difficulty of applying it. A small inaccuracy in the value of the angle of elongation that the astronomer had to measure was enough to break the bank. The elongation in quadrature is $89°51$. Instead, Aristarchus estimated «a quadrant minus one thirtieth

---

[15]One stadium was equal to 600 Greek feet. Unfortunately, the length of a foot varied from place to place. So we can only take an average value for the stadium, which is about 160 m.

[16]C.C. Carman, J. Evans, *The Two Earths of Eratosthenes*, in "Isis", 106, 1, 2015, pp. 1–16.

[17]J.L. Berggren, N. Sidoli, *Aristarchus's On the Sizes and Distances of the Sun and the Moon: Greek and Arabic Texts*, in "Archive for the History of Exact Sciences", 61, 3, 2007, pp. 213–254.

[18]Note that due to the very elongated orbit of our satellite, there is a difference $\Delta d/d \sim$ 10% between the maximum and minimum distance of the Moon from the Earth.

of a quadrant,»[19] which is 87°. This may seem like a small difference, and it is; but the angle becomes the argument of a trigonometric function that diverges near the right angle. Quirks of mathematics that a good experimenter should always be aware of. Perhaps Aristarchus was too, but he certainly did not worry that his estimate might be seriously flawed. The geocentrists were satisfied with a meager universe, with an extension of no more than 10,000 terrestrial radii, as Ptolemy would later adopt;[20] just enough to contain the orbit of Saturn.

A century has passed. In the city of Nicaea, in that Anatolian peninsula on whose shores, blessed by a warm sea and enriched by the traffic of ships and caravans, the philosophy of nature had flourished with the Milesian Thales, Anaximander, and Anaximenes, Hipparchus wanted to reconsider the question of the distance of the Moon, appealing to the phenomenon of diurnal parallax.[21] The small apparent displacements of the celestial bodies due to the changing point of view of observers at different locations on the Earth's surface had been known for some time, but to take advantage of this fact it was necessary to find a way to make simultaneous observations. While today a couple of cell phones would be enough to solve the problem (at least until the minute propagation time of the electromagnetic signal becomes critical), two centuries B.C. there really seemed to be no way to connect the clocks of two distant stations. Instead, Hipparchus found a loophole in a total solar eclipse in 187 B.C. that was visible in the Hellespont, where he lived. In the Egyptian Alexandria, astronomers measured a maximum coverage of the disk of only 4/5 due to the different angle of view. Once the base, i.e. the distance between the two locations, had been determined using an elementary geometric construction for a mathematician of his caliber, the philosopher of Nicaea calculated that a distance to the Moon should be between 70 and 80 terrestrial radii: too much. Later, by revising Aristarchus' method based on lunar eclipses, he found a smaller value, adding to the confusion until, almost three centuries later, Ptolemy

---

[19] A. Mancini, *Aristarchus' Book on the Sizes and Distances of the Sun and the Moon*, sd, p. 31, http://www.aristarchus.it/doc/ARISTARCHUS_BOOK_1_sect.pdf.

[20] Cf. R. Fitzpatrick, *A Modern Almagest. An Updated Version of Ptolemy's Model of the Solar System*, 2010, https://farside.ph.utexas.edu/books/Syntaxis/Almagest/node3.html; see also T. Lu, H. Aslaksen, *The Copernican Revolution and the Size of the Universe*, https://citeseerx.ist.psu.edu/viewdoc/download?doi=10.1.1.175.3206&rep=rep1&type=pdf.

[21] G.J. Toomer, *Hipparchus on the Distances of the Sun and Moon*, in "Archive for the History of Exact Sciences", 14, 1974, pp. 126–142.

used his authority to freeze the scales of his geocentric system. Hipparchus retains the great merit of having opened the way to the parallax method, which in time would become a powerful meter within the boundaries of the Solar System, and in the sidereal world the black beast of the heliocentrists.

Whatever it is, the solar parallax[22] remained essentially a mystery for 15 centuries. The values adopted by astronomers were hypothetical, since no one could measure this quantity directly, for the excellent reason that it is really small, less than 9 seconds of arc; an angle that no human eye can resolve. Halley finally entered the game, in his old age, with all the vigor that a scientist of his fame could afford. Dissatisfied with the accuracy of the planetary distances, he urged his young colleagues to use the method of the transits of Venus on the disk of the Sun to improve the unit of measure that makes dimensional Kepler's third law.[23] This is how it works.[24] As mentioned above, it happens from time to time that Venus, rather than the Moon, is positioned in a line between the Earth and the Sun. This would occur about once every year and a half if the planet's orbit were perfectly coplanar with that of the Earth. Instead, the transits occur in pairs, separated by an interval of 8 years, alternating between 121.5 and 105.5 years. Miracles and coincidences of the complex planetary clock. The phenomenon is infinitely less spectacular than a canonical solar eclipse, because the disk of Venus, seen from the Earth, is a diaphragm more than a thousand times smaller than the Moon. Without the predictions of astronomers, its effects would certainly not be noticed. In fact, all that can be seen through a telescope, with adequate eye protection, is a black dot that, in a few hours, crosses the solar disk from side to side along a line more or less parallel to the ecliptic. Really nothing special! But this sky walk can be used to measure a fundamental quantity, the Astronomical Unit (AU). Here is the procedure to perform the exercise.

Try to imagine watching the transit from two locations at different latitudes. It is intuitive that the corresponding trajectories of Venus on the solar disk will be parallel, but not coincident. From their angular separation, it is possible to calculate the distance to the planet, provided we

---

[22] The angle subtended by the Earth's equatorial radius at the average distance of our planet from the Sun.

[23] E. Halley, *Methodus singularis qua Solis Parallaxis sive distantia a Terra, ope Veneris intra Sol conspiciende, tuto determinari poterit*, in "Philosophical Transactions of the Royal Society of London", 29, 348, 1717, pp. 454–464.

[24] For more information, see L.W.B. Browne, *Halley's Method for Calculating the Earth-Sun Distance*, in "Archive for History of Exact Sciences", 59, 2005, pp. 251–266.

know the length of the meridian arc connecting the two observers on Earth. Then, using Kepler's third law, which gives us the distance to Venus in AU for free, a simple proportion will finally give us the coveted radius of the Earth's orbit. This is easier said than done, not only because of the rarity of the event, which occurs at most twice in a human lifetime and can only be observed when the sky is clear. The biggest pitfall is the small separation between the two orbits, which is difficult to estimate with sufficient accuracy. Here is where Halley's trick comes in. The visionary English genius understood that the angular distance between the two orbits could be reconstructed from the difference in the time taken for the planet to travel from edge to edge of the solar disk,[25] typically in the order of a few minutes and therefore easy to determine. It was the winning strategy, already anticipated by James Gregory in his *Optica promota* of 1663, that was desperately needed.

In 1716, knowing that he was out of the game because of the timing of the event — he would have to live more than 104 years to enjoy the next transit — Halley left his moral legacy to posterity:

> I recommend it therefore again and again to those curious astronomers who, when I am dead, will have an opportunity of observing these things, that they remember my admonition, and diligently apply themselves with all imaginable success; in the first place, that they may not by the unreasonable obscurity of a cloudy sky be deprived of this most desirable sight, and then, that having ascertained with more exactness the magnitudes of the planetary orbits, it may redound to their immortal glory.[26]

What a strange destiny! To make surprising predictions, such as that of the return of his comet, knowing that he would not live enough to see the phenomenon. In any case, his admonishment was taken seriously. The proposal was reasonable, the proponent authoritative, and the problem worthy of attention. From that time to the present, albeit with different motivations, there has always been a rush to observe the solar transit of the

---

[25] The speed being equal, the time taken to cross a chord of a circumference is obviously proportional to the length of the chord. Then, knowing the lengths of two chords, it is trivial to calculate the distance between them.

[26] In J.D. Fernie, *Marginalia: Transits, Travels and Tribulations, I*, in "American Scientist", 85, 2, 1997, pp. 120–122.

blue planet on the rare occasions when it occurs. In the 18th century, however, this meant costly expeditions, often to remote and exotic lands, which some men fearlessly faced, driven by the fever of knowledge. Another heroic epic of knowledge that had its pathetic standard-bearer in the Frenchman Guillaume-Joseph-Hyacinthe Legentil de la Galaisière. His misadventures are the paradigm of what men are willing to suffer in order not «to live like brutes, but to follow virtue and knowledge».[27]

Legentil[28] had been quietly working at the Paris Observatory, hunting nebulae for César-François Cassini, son of Jacques and grandson of Giovan Domenico, when, at the invitation of the Académie Royale des Sciences, he agreed to go to the French colony of Pondicherry, in southeast India, to observe the transit of Venus across the Sun. Exactly the one of 1761 that Halley had referred to in his admonition. After taking care of business and saying goodbye to his young wife, who was certainly not happy to be left alone for so long, he embarked for the Far East with a mountain of instrument cases. After a long and uncomfortable sea voyage, now in sight of Malaysia, "Guillaume the Unlucky" learned that his destination had just been conquered by the army of His Majesty George III. The British interpreted the Seven Years' War, in which the young William Herschel had been involved, as an excellent opportunity to increase their imperial dowry at the expense of the French. We can only imagine the surprise and the bitter disappointment of the transalpine astronomer. It is likely, however, that he did not use the picturesque jargon that would later make General Pierre Cambronne famous on the battlefield of Waterloo to vent his anger. A noble background and theological studies had made him a well-rounded and reasonable man.

Lacking an alternative, he built one on the spot, changing his destination within the Indian peninsula. But each time he landed, he was forced to retrace his steps. The fatal day came while he was still sailing. To make matters worse, the sky looked perfectly clear that day. Guillaume tried to observe the astronomical phenomenon from the deck of the ship, but the waves made it impossible. At that point, he could have gone home safe and sound, albeit with his tail between his legs. Instead, he decided to stay in the area and wait for the next transit, which would take place

---

[27] D. Alighieri, *Divine Comedy*, Hell, XXVI, vv. 118–119.
[28] B. Foix, *Une présence française. Le voyage savant de l'astronome Le Gentil (1725–1792)*, Edilivre, Saint-Denis 2019 (in French).

in eight years. He invested this time in drawing maps of the sites and studying ancient Indian astronomy. As the new mini-eclipse approached, he moved to Manila, Philippines, where he was turned away by the Spanish. Meanwhile, with the Peace of Paris, King George III, now mad but far from foolish, had returned Pondicherry to the French, keeping the much juicier Canada for himself (shortly thereafter, the French would retaliate by coming to the aid of the North American colonies during the Revolutionary War). Taking this as a sign of destiny, Legentil sailed for India, as desired by the Paris Academy that sponsored him. Arriving in Pondicherry, he worked hard to set up his small observing station and be ready to make the most of the event. It seemed that nothing could go wrong this time, and instead, at the moment of truth, the sky, which had remained clear for a whole month, clouded over. It was enough to drive a man crazy. This time, Legentil had to give up and decided to return home. Along the way, he was plagued by dysentery, pirates, and a stormy sea that forced him to land in Réunion, in the Indian Ocean, where a Spanish ship charitably took him in. Arriving in Paris in the fall of 1771 — Good knows how — poor Guillaume discovered that he had been left for dead. He had lost his wife, who had remarried thinking she was a widow, and all his possessions, which had been divided among his heirs. Incredible but true, he even had to pay the legal fees of a useless lawsuit to regain his property. His chair at the Académie had also been recycled, but he managed to get it back thanks to the personal intervention of the sovereign, Louis XV the Beloved. At least that! The only positive result of this painful experience were two volumes in which he told of his *Voyage dans les mers d'Inde*. Who knows how many astronomers, sailing by sea or passing through jungles and deserts to reach remote places to observe an eclipse, have thought about this story with concern. Surely all of them will have crossed their fingers, hoping that the many efforts of the journey would be rewarded by good weather and a less problematic return home.

With the solar parallax problem solved, at least for the time being, another problem remained, that of the distance to the stars. After the collapse of the Aristotelian-Ptolemaic cosmology, no one believed in a celestial sphere with a radius of about 80 million kilometers, as the Alexandrian astronomer had postulated. Thanks to the telescope, Galileo had given the sidereal universe a new, three-dimensional image that assigned each star its own distance. But which one? «To try to determine the distances of the stars from us is a difficult matter and one that requires a subtle mind»,

Brahe had already commented in 1573, in his *De stella nova* (On a New Star), «because they are so incredibly far from the Earth».[29] Geocentric to the bone, the Dane ruled out the possibility of using the annual parallax, which he had tried in vain to measure with scientific scruples. Unlike him, the Copernicans relied on this direct geometric determination. But despite their efforts, the phenomenon continued to elude them, leaving the conservatives with the satisfaction of seeing the heliocentric hypothesis invalidated and the progressives with the worry of explaining why the Earth's rotation was not registered by the stars.

Pending a solution, which the rise of the Copernican system made only a matter of time and technological progress, astronomers had worked to invent fanciful methodological shortcuts to map the cosmos in depth, giving the stars a third radial coordinate in addition to the two that identify their projections on the sphere. The lynchpin of the various strategies was always the same, the attenuation of the luminous flux in proportion to the square of the distance, as Kepler had already proposed.[30] With this idea in mind, the ingenious Huygens tried to compare the light of Sirius with the effect of the Sun when viewed through a small hole made in a disk placed at such a distance as to block the star exactly. It was an elaborate procedure that implied an intrinsic similarity between the Sun and the star that dominates Canis Major, and assumed that the astronomer was able to remember at night what he had seen during the day. The result, astonishing at the time, assigned to Sirius a distance 28,000 times greater than that of the Sun: an enormous value, and yet 20 times less than the figure we accept today (much of the error is due to the assumption that the two stars have the same intrinsic brightness).

Newton went one step further than Huygens by seeking a nightly comparison, while still maintaining the assumption that all stars are equal. Following Gregory's suggestion, he chose Saturn as his touchstone, whose relative brightness he calculated by requiring the planet to reflect a quarter of the incident sunlight. Then, by measuring the brightness of Sirius relative to the planet, he obtained a distance of one million AU for the star. It is not surprising that under these conditions the sidereal distances came

---

[29] M. Marett-Crosby, *Twenty-Five Astronomical Observations That Changed the World: And How to Make Them Yourself*, Springer, New York 2013, p. 110.

[30] J.B. Hearnshaw, *The Measurement of Starlight: Two Centuries of Astronomical Photometry*, CUP, Cambridge 1996. In fact, this property was already known in Dante's time, as shown by the example of the three mirrors used by Beatrix in Canto II of *Paradise* (vv. 97–105) to explain to the poet that lunar spots cannot be deep wells.

out resoundingly wrong. In fact, the few measurements that agree with today's must be considered mere coincidences. The authors of these uncertain determinations were themselves aware of their limitations. "I want parallax", sidereal astronomy seemed to cry out. Yet, as we have already seen, for two centuries after the introduction of the telescope there was no way to flush out the phenomenon. The frustration of researchers was effectively expressed by John Herschel in a speech to the Royal Astronomical Society on February 12, 1841. Acknowledging that distance measurement had been the dream of every astronomer, the son of the great William called the path «hopeless, endless, and exhausting»,[31] and recalled the mocking fate of many who felt close to the result but saw it placed just beyond their technological capabilities. In fact, Herschel's remark was only a rhetorical device to emphatically celebrate the first victory, achieved *in extremis*[32] by the German Friedrich Wilhelm Bessel at the expense of the tsar's astronomer, Friedrich Georg Wilhelm von Struve, and a third competitor, the Scotsman Thomas Henderson Friedrich.

Bessel[33] joined the Urania club almost by accident. Born in 1784 in the western plains of Germany to a modest family of civil servants, he left school early to join an import-export firm in Bremen as an unpaid apprentice. On the docks of the commercial port, he was exposed to land and sea geography, foreign languages, and the various aspects of navigation, which led to his interest in astronomy. Promoted to bookkeeper, he served the company during the day, which finally paid him a salary, and at night he exercised his talent for mathematics by studying and then calculating, purely for practice, the orbits of comets. Heinrich Olbers noted him for a paper in which Bessel had refined the orbital elements of Halley's Comet by introducing some old observations into the data set. Certain that he had discovered a talent, the influential physician and amateur astronomer persuaded the young man to accept an underpaid temporary position at the Lilienthal Observatory near Bremen, the very place where the *Himmelspolizei* had been formed as an association seven years earlier (see p. 188). A leap in the dark for a 21-year-old without even a diploma. But, as Leonida Rosino, a master of Italian science of the twentieth century, liked to tell

---

[31] A.W. Hirshfeld, *Parallax: The Race to Measure the Cosmos*, Dover, New York 2013, p. xiv.

[32] Latin locution for "at the point of death", which in a broad sense means "at the very last minute".

[33] J.F.W. Herschel, *A Brief Notice of the Life, Researches, and Discoveries of Friedrich Wilhelm Bessel*, in "Annual report of the Royal Astronomical Society", 9, 1847, pp. 4–16.

his students and collaborators: "astronomers are paid little for work they would do for free".

At Lilienthal, under the direct control of Johann Schröter, Bessel was able to fill the gaps in his education and soon became a refined scholar. His story could have ended here, like that of so many dark talents to whom fate denied a chance. His chance came when the Prussian king, Frederick William III, although committed to containing the avalanche that was igniting Napoleonic Europe and Germany in particular, decided to equip the Albertina University in Königsberg with an astronomical observatory and chose him, a twenty-six-year-old without a university degree, to be its first director and holder of the chair of astronomy. The formal obstacles to this appointment, authentic taboos for the Prussian mentality, were pragmatically overcome thanks to an honorary doctorate awarded to Bessel by the University of Göttingen on the recommendation of Gauss, who had met the young fellow for a few hours in Bremen in 1807 and had gained a very good impression of him;[34] clearly the grumpy genius knew how to judge people. What followed were great satisfactions, honors, co-options in famous academies, scientific successes, and prestigious job offers, accompanied by a decidedly unhappy private life. Bessel lost two of his four children, and he himself died young after a painful illness. However, he managed to achieve immortal fame for having brought astrometric measurements to a very high level of precision through innovative data reduction techniques and, above all, for having made the first reliable estimate of the parallax of a star.

When Bessel took up the problem,[35] he had already had the heliometer built by Fraunhofer for his observatory: a telescope designed for accurate angular measurements of the properties of the solar photosphere.[36] The objective lens was divided into two halves, mounted in such a way as to allow controlled and measurable sliding of one half relative to the other. Each source in the field provided two images, the angular separation of which could be known with great precision by means of a micrometer connected to the gear for translating one of the half lenses relative to the other. From the hands of an optical magician, the heliometer became an almost perfect instrument in the hands of a master of precision. Bessel's strategy

---

[34]G. Shaviv, *The Life of the Stars: The Controversial Inception and Emergence of the Theory of Stellar Structure*, Springer, Berlin 2009, p. 140.

[35]F.W. Dyson, *Measurements of the Distances of the Stars*, in "Science", ns, 42, 1070, 1915, pp. 13–22.

[36]E.H. Geyer, *The Heliometer Principle and Some Modern Applications*, in "Astrophysics and Space Science", 110, 1, 1985, pp. 183–192.

to reveal the parallactic oscillation of a given star A was to map its angular separation over time with a very close star B, particularly faint and therefore presumably fixed (typically faintness implies large distance and consequently negligible parallax and proper motion). Each time the distance between the two stars was measured, the astronomer rotated the objective to align the section with the pair, then shifted half a lens until the image of A coincided with that of B. This method avoided the drawbacks of atmospheric refraction, which is irrelevant in pairs of very close stars, and limited those of atmospheric turbulence.

As a target, Bessel chose 61 Cygni, the star that Piazzi had called "flying" because of its very high proper motion: a reddish star, barely visible to the naked eye, which promised to be particularly close to the Sun and therefore vulnerable to the German's refined method of investigation. However, its extreme weakness made it unusable in less than perfect weather conditions. Discouraged by the difficulties, Bessel was about to give up when he learned that Struve, one of his competitors in this scientific race, seemed to be close to the goal. He too had a jewel of Fraunhofer's, the large refractor with an achromatic objective of 9.6 inches from the Dorpat Observatory, with which he had been following Vega, the brightest star in the Lyra, for years. The third party was Henderson, who at that time presided over the southern hemisphere. He was in fact at the Cape of Good Hope Observatory in South Africa, although reluctantly because of the difficult environmental conditions. He had learned from colleagues working on the island of Saint Helena — the same island where the Emperor of France had sadly completed his glorious parabola some fifteen years earlier, and where Halley had begun his brilliant career — that the dominant star of the constellation Centaurus manifested a particularly high proper motion. He had therefore decided to compete with this dowry, in the illusion that he would give his country another primacy and perhaps earn the reunification with civilization.

Motivated by the competition, Bessel gritted his teeth and managed to complete the cycle of observations, winning the race at the finish line. In December 1838, he published his determination of the distance to the 61st star in the constellation Cygnus in the "Astronomische Nachrichten" (Astronomical News), the world's oldest astronomical journal. A parallax of 0.314 arcseconds placed the star 10 light-years from the Sun. It was the revenge of Aristarchus, the end of a nightmare for the heliocentrists, and the beginning of a new era for mapping the cosmos. Within a few months, the other two competitors also reached the finish line, and their approximate results were quickly packaged to close the gap. The whole world applauded Prussia's accuracy, not knowing that this quality would soon become a dull and murderous obsession. But the beautiful coin also had a bitter downside. Bessel's measurement showed how small the range of the annual parallax was in relation to the sidereal distances; totally insufficient to satisfy the needs of the new science called astrophysics, and especially of cosmology. It could only be used as a solid starting point for the construction of a scale of cosmic distances based, for example, on the law of attenuation of the flux.

We will now try to understand how,[37] by imagining that the parallax enables us to measure the distances of all the stars within a sphere of radius D_1. Let us further assume that within this horizon there are a few objects from a homogeneous family of stars, bright enough to be visible beyond D_1 (e.g. up to $D_2 > D_1$), with characteristics that allow their unambiguous identification, and all having on average the same intrinsic brightness. Let us call them standard candles for convenience, and let us also assume that their light is not absorbed by dust or gas on its way to us, or at least that we are able to correct for such effects. Under these conditions, we can first calibrate the intrinsic brightness of the family by the parallax applied to the standard candles contained in D_1, and then use it to measure the distances of the others that are visible and usable up to the limit D_2. This could be the first in a series of similar steps, relying on the probability, increasing with distance, of finding members of other families of stars with the properties of the standard candles, perhaps less frequent (per unit volume) but brighter, and thus able to lead us further out.

An example will help us understand better. In the hunt for variable stars, Pickering had collected a large set of photographic plates with the two major instruments of the Harvard Observatory. One of these was the large 38 cm Cambridge (MA) refractor, built in Germany in 1847 by Merz and Mahler, for some years the largest telescope in North America and a true pioneer of astronomical photography. The other instrument was the Bruce Astrograph, named for its donor. In 1889, Miss Catherine W. Bruce, a New York philanthropist with a latent interest in astronomy, had donated the sum of $50,000 to the Harvard Observatory for the construction of the instrument and its dome on a site in southern Peru at an altitude

---

[37]For more information, see Capaccioli, *L'espansione dell'Universo*, cit.

of 2,400 m near the city of Arequipa, on the slopes of a towering active volcano. Manufactured by Alvan Clark of Cambridgeport, Massachusetts, the Peruvian telescope was equipped with a 60 cm achromatic doublet and had an impressive field of view, 25 square degrees, recorded on 14 × 17 inch plates, larger than an A3 sheet. Its mission was to make a photographic map of the southern sky, which was inaccessible from Harvard and generally little known. In fact, it had been explored sporadically by the French and somewhat better by the British with limited expeditions to South Africa and the island of Saint Helena.

The task of searching for variable stars on the long exposure plates (2 to 5 hours) taken every few days with Bruce's astrograph was entrusted to Mrs. Henrietta Swan Leavitt in the early 20th century.[38] It was a tedious job, like all those in Pickering's harem, for which Henrietta, hired as a "human computer", was underpaid but did not mind,[39] and which would consign her to the history of science. She used a special microscope to alternate in the eyepiece the image of the same area of the sky taken in two different exposures in order to detect any differences. With infinite patience and commendable meticulousness, she detected and catalogued 2,400 variables of all kinds. About 2 percent of these stars, identified on images of the two southern nebulae called the Large and Small Magellanic Clouds (later recognized as satellites of the Milky Way), enabled her to discover the period-luminosity relation of the Cepheids, one of the most productive empirical laws in the entire history of astronomy.

But what are (classical) Cepheids? They are stars whose brightness changes cyclically over a period of 1 to 60 days,[40] named after the prototype of the class, the fourth brightest star in the constellation Cepheus. The fickle stars had already been studied in the late 18th century by a sickly and brilliant young Englishman, John Goodricke.[41] For the next hundred years and beyond, these pulsating stars did not arouse much interest, so much so that Pickering, in a profoundly male chauvinist world, set a woman to work on them: at the beginning of the 20th century, science, like war, was

---

[38] G. Johnson, *Miss Leavitt's Stars: The Untold Story of the Woman Who Discovered How to Measure the Universe*, W.W. Norton & Co., New York 2005.

[39] She had enough money to live on and a progressive deafness that prevented her from socializing.

[40] J.J. Kolata, *Elementary Cosmology: From Aristotle's Universe to the Big Bang and Beyond*, Morgan & Claypool Publishers, San Rafael (CA) 2015, pp. 44–45.

[41] C. Gilman, *John Goodricke and His Variable Stars*, in "Sky and Telescope", 56, 11, 1978, pp. 400–403.

considered a man's business, and night observation even more so, because of the fatigue and physical and moral dangers and the isolation of the sites.

Mrs. Leavitt recognized the Cepheids of the Clouds by their peculiar behavior.[42] After measuring the apparent magnitude of each star at different epochs, she was able to estimate the period and construct what is known as the "light curve", that is, the trend of brightness over a complete oscillation. It was what her boss had asked her to do. But Henrietta went further. Ignoring the distance to the Clouds, she understood that it should be much greater than their sizes. She therefore felt justified in assuming that the objects of each nebula were all at the same distance from us: a bit like saying that all the inhabitants of New York are at the same distance from an observer in San Francisco. This consideration allowed her to compare the luminosities of the different stars in the same nebula. In fact, the San Francisco observer in our example cannot know the heights of the New Yorkers if he ignores the distance between the two cities, but he is sure that the apparently taller individuals are really taller. So the shy "calculator" realized that the brighter stars make their oscillations over a longer period of time. Using a logarithmic scale, the correlation between average luminosity ϕ and period P became particularly easy to describe: $\log \phi = a + b \log P$. Of the two coefficients, the slope b was the same for both Clouds, proving an identical behavior of the Cepheid populations in both nebulae, while the zero point a took different values due to the different distances of the two stellar systems from the Sun.

The report that an excited Henrietta presented to her gruff employer was something of a miracle, for it opened a powerful new avenue for sampling distances in the deep sky where parallax could not venture.[43] The Cepheids are in fact intrinsically bright stars, easily recognizable as such by the regular variation of their luminosity (and not only),[44] and are accessible everywhere in the Milky Way and beyond — although this property was not entirely clear in the early 20th century. To see how the meter works, suppose we have identified a Cepheid in the Milky Way and estimated its

---

[42] S. Webb, *Measuring the Universe: The Cosmological Distance Ladder*, Springer, Berlin, 1999, pp. 142–146.

[43] J.D. Fernie, *The Period-Luminosity Relation: A Historical Review*, in "Publications of the Astronomical Society of the Pacific", 81, 483, 1969, pp. 707–730. See also H.S. Leavitt, E.C. Pickering, *Periods of 25 Variable Stars in the Small Magellanic Cloud*, in "Harvard College Observatory Circular", 173, 1912, pp. 1–3.

[44] The Cepheids have peculiar trends of luminosity, color, and photospheric radial velocity over time.

average brightness and period. By plugging this last information into the Leavitt relation that applies to, say, the Large Cloud, we can deduce what the luminosity should be if the star were instead in that nebula. The comparison of the two luminosities, measured and calculated, gives the distance of the star relative to the Large Cloud. Conversely, if the galactic Cepheid we use has a known distance, we derive the distance of the Large Cloud in parsecs (or any other linear unit). Today, we are well aware that to effectively use this formidable tool for mapping the deep cosmos, we need eagle eyes, that is, modern telescopes of high quality and large aperture. Fortunately, a visionary and determined giant, George Ellery Hale, had long since taken on the problem.

Chapter 13

Symphony from the New World: George Hale and the big science

> Try to turn every disaster
> into an opportunity.
>
> John Davison Rockefeller

> There are few earthly things
> more beautiful than a university
> where those who hate ignorance
> may strive to know,
> where those who perceive truth
> may strive to make others see.
>
> John Masefield,
> poem for the installation of the rector
> of Sheffield University (1946)

After the resounding successes of Bessel and Huggins, and the technological prowess of William Herschel and Lord Rosse, European astronomy was catching its breath in terms of large instruments for observing the heavens. Fueled by coal and steel, the Industrial Revolution had decisively shifted interests and investments toward the sciences applied to peace and war;[1] two sides of the same coin. Profit, together with power, was the key word of a productive bourgeoisie that was not willing to go along with the pure drive of Odysseus. Paradoxically, it was this same bourgeoisie, perhaps even more crude and cruel, that instead made the fortune of science and higher education on the North American continent. Here, as early as the second half of the nineteenth century, some "new men",[2] who quickly became immensely wealthy, decided to redeem their humble beginnings, or to wash away the

---

[1] P.N. Stearns, *The Industrial Revolution in World History*, Routledge, New York 2020; M. Meriggi, *L'Europa dall'Otto al Novecento*, Carocci, Roma 2006 (in Italian).

[2] Among the ancient Romans, a "*homo novus*" was the first person in his family to reach high positions in the state organization without belonging to the noble class.

blood they had shed for their rise, or simply to secure immortality, by offering scientists opportunities unimaginable in the Old World.[3] Almost everywhere, from the Atlantic to the Pacific coasts, from the industrial North to the plantations of the slave-owning South, projects for large observatories equipped with ever more powerful telescopes proliferated, as did new universities. And while in Europe astrophysics and cosmology favored the interpretation and mathematical formalization of phenomena, in the New World, pragmatic, daring, and still imbued with the spirit of the frontier, the "great book of nature" was scrutinized in search of facts, often without the guidance of *a priori* ideas. Among the pioneers of this new epic of celestial observation was the legendary figure of George Ellery Hale,[4] a scientist whose story is intertwined with that of great American capitalism, ruthless and visionary,[5] deregulated and philanthropist.

George was born in Chicago in 1868, three years after the end of the bloody war that had brought the secessionist Confederate states into the Union, and a few months before the Great Fire destroyed and looted ten square miles of the Lake Michigan city. This disaster, which left more than 100,000 people homeless, was the source of the Hale family's economic fortune. William Hale, George's father, played an active role in rebuilding Chicago's skyline. He manufactured the hydraulic elevators, for which he held the patent, for the soaring new iron and concrete skyscrapers that would, within a few decades, be the stage for the criminal raids of famous gangsters amid rivers of bourbon and women in Charleston gowns: Al Capone, Dion O'Banion, and George "Bugs" Moran. He possessed «a boundless energy and tenacity his son would inherit».[6] George's mother was quite the opposite: a sickly, depressed woman who cultivated a gray streak of Calvinist austerity. Mr. and Mrs. Hale had been traumatized by the loss of their first two children, but they were wealthy enough to accommodate the wishes of their fragile and emotionally unstable third child. So when he asked them for a telescope to watch the transit of Venus across the

---

[3] A. Greenspan, *Capitalism in America: A History*, Penguin, New York 2018.

[4] W.S. Adams, *Biographical Memoir of George Ellery Hale*, in "National Academy of Sciences, Biographical Memories", XXI – Fifth Memoir, http://nasonline.org/ publications/biographical-memoirs/memoir-pdfs/hale-george-ellery.pdf. For a rich bibliography on George Hale, cf. R. Berendzen, *Pioneer of American Astronomy*, in "Journal for the History of Astronomy", 5, 1974, p. 63, note 1.

[5] C.R. Morris, *The Tycoons: How Andrew Carnegie, John D. Rockefeller, Jay Gould, and J.P. Morgan Invented the American Supereconomy*, H. Holt & Co., New York 2005.

[6] W. Sheehan, D.E. Osterbrock, *Hale's "Little Elf": The Mental Breakdowns of George Ellery Hale*, in "Journal for the History of Astronomy", 31, 2000, pp. 93–113.

disk of the Sun, they readily complied. Their gift could have remained the expensive whim of a rich, spoiled 14-year-old. Instead, for George, it was the prelude to an extraordinary career as a scientist and science manager, strongly supported by his parents.

The parents' attention peaked when they gave their son a real observatory, built of bricks in the courtyard of the family home. There, the young man, now a college student, could indulge his passion for the Sun as the archetype of the stars. It turned out to be a good investment, for George was indeed a very gifted person. He alternated long periods of hyperactivity with moments of deep depression accompanied by severe headaches. His immense energy, iron will, exuberance, logorrhea, reduced need for sleep, sharp intelligence combined with exhausting work rhythms were all signs of a mood disorder known to psychiatrists as hypomania: certainly a bad companion in life, but also a formidable tool for excellence. Young Hale would make the most of this dual talent, combining it with the advantages that his personal wealth and social connections could bring him. The same path had been successfully traveled three centuries earlier by the Danish aristocrat Tycho Brahe.

The United States was a vast construction site, with some things completed and functioning, but most still under development. Since there was no university in Chicago appropriate to their son's social status, the Hales enrolled him at the Massachusetts Institute of Technology in Boston, a 30-year-old engineering school that already had an excellent reputation.[7] The project was to make him an engineer, but old William did not object to his son following his own scientific inclinations. So, in 1890, George defended his B.S. thesis with the design of an innovative telescope capable of producing monochromatic images of the Sun: a brilliant idea that would later be realized in the family's private observatory and serve as a model for future generations of solar telescopes. It was the first sign of unparalleled ingenuity and creativity. The story could have ended here, as is often the case with the whims of the rich that last «*l'espace d'un matin*».[8] Instead,

---

[7] J.R. Cole, *The Great American University: its rise to preeminence, its indispensable national role, why it must be protected*, PublicAffairs, New York 2009.

[8] «*Mais elle était du monde, où les plus belles choses // Ont le pire destin; // Et Rose, elle a vécu ce que vivent les roses, // L'espace d'un matin*» (But she bloomed on earth, where the most beautiful things have the saddest destiny; And Rose, she lived as the roses live, for the space of a morning); from the "*Consolation à M. du Périer*" for the loss of his daughter, by the French poet and writer François de Malherbe, lived at the turn of the 16th and 17th centuries.

fate had much more in store for the scion of one of Illinois' most prominent families. The opportunity came in 1891, when oil tycoon John Davison Rockefeller agreed to fund an American Baptist Church initiative to endow Chicago with its own university.

The family and childhood of this giant of capitalism were of the kind that built the Big Apple's extravagant reputation as the melting pot of all good and all evil.[9] Rockefeller was born in New York, the gateway to the New World for a flood of immigrants of all ethnicities who came from Europe seeking their fortunes overseas or simply fleeing hunger, constant war, and religious persecution. His father had come to the city from the woods of western New York State, where he had been a lumberjack, and had transformed himself into a witty and chaotic charlatan. In his shadow, the boy grew up with the cynical belief that in business one should «trade dishes for platters».[10] Having entered the oil business as a true pioneer, thanks to the black gold extracted and distributed by his Standard Oil, John quickly became the richest man in America; a fortune that in the 20th century, with the boom of the automobile, would continue to grow with every fill-up at an EXON, Mobil, or Chevron pump, to name just a few of his companies. Then, in the same spirit as Enrico degli Scrovegni — the banker who, in the early 14th century, in order to save his soul and that of his usurer father, had donated to Padua and to the world the chapel of the same name, frescoed by Giotto — the ruthless financier, the same one who would inspire Walt Disney to create the cartoon character of John D. Rockerduck, Uncle Scrooge's cynical and unscrupulous rival, wore the mask of the philanthropist.

It must be said that for Rockefeller the founding of the University of Chicago was not an operation of farsighted patronage, but simply another affair, the best of his life, as he himself later declared. Contrary to the gerontocratic canons of the European tradition that prevailed even in the New World academies, he entrusted the creation and then the direction of the university to the 35-year-old William Rainey Harper,[11] a former child prodigy and historian of religion, with a mandate to maintain a high

---

[9] R. Chernow, *Titan: The Life of John D. Rockefeller, Sr.*, Vintage Books, New York 2004.

[10] Cf. J.P. Hunter III, *Money for Power*, Library of Congress: Washington (DC) 2013, p. 541. Bill Rockefeller said of himself that he often cheated his children in order to prepare them for life.

[11] Cf. H. Holborn Gray, *Searching for Utopia: Universities and Their Histories*, University of California Press, Berkeley (CA) 2012.

standard. The young president was a down-to-earth utopian. His program foresaw «bran splinter new, yet as solid as the ancient hills»,[12] that is, a complete revolution in teaching methods and programs aimed at building the productive backbone of the country. Among other initiatives, he decided to involve George Hale in the operation. He hoped that in this way he could take over the family observatory, including the beautiful 12 inch reflector with which it was equipped. But his intention was discovered and the proposal was scornfully rejected by the Hales. The following year Harper came again. This time George's father led the dance. With the disarming bluntness of a businessman, old William said he was willing to donate the observatory to the university on the condition that his son be associated with it as a professor of astrophysics. This was no small request, since George was not even a graduate student. He also asked for a formal commitment of a large sum, $250,000 (equivalent to about $25 million today), to be used for the construction of a big observatory, of which his son should be the director. Once the agreement was reached, George was given a professorship, but without a salary. He clearly did not need it to make a living. In the meantime, while still a student, he had married and used his honeymoon to visit Lick Observatory in California. What else could one expect from a man addicted to astronomy?

Pacta servanda sunt.[13] In order to obtain funding, Harper turned to Charles Tyson Yerkes,[14] an urban rail transport entrepreneur. Yerkes urgently needed to clean up his image in the eyes of his new young wife and the good society of Chicago, the city where he had moved after the Great Fire to rebuild his fortune. His name, in fact, had been tarnished by the systematic use of bribery and blackmail to save himself from prison after a financial crash, and then by the scramble to gain a monopoly on public transportation in Chicago. As a perfect tycoon, the Quaker Yerkes was willing to pay whatever he was asked, provided the observatory that bore his name was the largest in the world: «I do not care what it costs, send me the bill»,[15] he said, taking leave of Harper and Hale. Great times!

---

[12]Ivi, p. 46.

[13]"Agreements must be kept"; legal maxim attributed to Roman politician and jurist Domitius Ulpianus, 2nd–3rd century.

[14]J. Franch, *Robber Baron: The Life of Charles Tyson Yerkes*, University of Illinois Press, Urbana (IL) 2006.

[15]R. Evans, *Microwave Background: How It Changed our Understanding of the Universe*, Springer, Dordrecht 2015, p. 47.

George Hale, who inherited his father's attitude of thinking big, was the right man in the right place at the right time. He had vision, charisma, a good reputation, the right connections, and solid protection. Exempted from teaching, which Harper probably thought he was incapable of doing, he was given the responsibility of carrying out the great project in the village of Williams Bay, Wisconsin, 150 kilometers from Chicago. The site, a plot of land a few yards from Lake Geneva, had been donated to the university by a private benefactor. It was suitable for both nighttime and daytime astronomical observations.

The assignment was an open invitation to the new professor, who immediately rolled up his sleeves to define the project. But the decision to build a large 40 inch refractor that would make Yerkes famous was the result of chance rather than Hale's design. George had heard of the financial difficulties that plagued Pickering's project to install a giant telescope on Mount Wilson, a peak in the San Gabriel Mountains near the city of Pasadena, north of Los Angeles, and decided to seize the opportunity to procure a critical component for his venture, the objective lens. This is the story.

With no resources of his own, the creative director of the Harvard College Observatory had managed to bring the University of Southern California into the enterprise, harnessing the ambition of West Coast pioneers to compete with the snooty society of the East. In this marriage of interests, the dowry consisted of a substantial financial contribution promised by a banker, a former mayor of Los Angeles, who was one of the founders of the university itself. But the philanthropist had died unexpectedly in 1892, leaving no written record of his pledge. Relying on word of mouth, the Provost had already hired the firm of Alvan Clark & Sons[16] in Cambridgeport (MA) to fabricate a huge achromatic doublet from glass blanks cast by the Parisian firm of Mantois. He wanted an objective lens with a diameter of 1 meter, a weight of 225 kilograms, and a focal length of 62 feet for a monster refractor 18 meters long. The loss of funding was a blow to the Californians, who had set out to improve the aperture record of the world's largest telescope already in their possession.

For five years, in fact, the leader had been a refractor on Mount Hamilton[17] the highest peak of the Devil's Range, a few hundred kilometers

---

[16] D.J. Warner, R.B. Ariail, *Alvan Clark and Sons, Artists in Optics*, Willmann-Bell, Richmond (VA) 1995.
[17] D.E. Osterbrock, J.R. Gustafson, W.J.S. Unruh, *Eye on the Sky: Lick Observatory's First Century*, University of California Press, Berkeley (CA) 2010.

north of Mount Wilson and overlooking San Francisco Bay. The University of California at Santa Cruz had begun the astronomical colonization of this site in 1876, the same year that Sitting Bull's warriors in the state of Montana defeated the 7th U.S. Cavalry at Little Big Horn: a battle of modest size compared to the massacres that were taking place in Europe, but pregnant with meaning and for that reason entered into legend. For the first time, the need for clear skies outweighed the practical difficulties of managing a resource far from population centers. What might have discouraged a European, certainly did not frighten men accustomed to deserts and prairies.[18] The funds for the enterprise came in 1876 from the will of a carpenter who had become a millionaire during the Gold Rush. By funding the observatory, the bearded James Lick, real estate baron, intended to secure a pharaonic abode for his mortal remains.[19] It took 12 years to build the roads, erect the buildings, and assemble the instruments. In 1888, Lick Observatory, the Northern Hemisphere's first astronomical garrison on a high mountain, was inaugurated with the commissioning of the giant refractor, a telescope with a 91 cm diameter lens made by the Clark firm on glass from France. The majestic instrument, a tube 17.5 m long and weighing dozens of tons, had been produced by a company in Cleveland (OH) founded a decade earlier by Worcester Warner and Ambrose Swasey: two technicians who had escaped from the prestigious Pratt & Whitney, today the queen of the aerospace industry.

Let us return to Williams Bay, on the shores of Lake Geneva. After securing Clark's commitment to complete the difficult optical work, Hale used Yerkes's money to purchase ownership of the giant lens that the University of Southern California could no longer afford. *Mors tua, vita mea!*[20] Meanwhile, a renowned Chicago architect, Henry Ives Cobb,

---

[18] «The possibility that a complete astronomical establishment might one day be planted on the summit seemed more a fairy tale than a sober fact». This 1887 statement by Captain Richard S. Floyd, president of the Lick Trust, gives a good idea of the difficulties encountered in the astronomical colonization of Mount Hamilton; http://collections. ucolick.org/archives_on_line/bldg_the_obs.html.

[19] The original idea of Lick, an eccentric philanthropist, was to erect a giant statue of himself and his parents or to build a pyramid in San Francisco to rival those at Giza. Later, thanks to the intervention of the president of the California Academy of Sciences, his good friend, he changed his mind and opted for a unique tomb, a great astronomical observatory that would connect him to the stars. In James Lick's deed of trust we read: «A telescope superior to or more powerful than any telescope yet made [...] and also a suitable observatory connected therewith».

[20] "Your death, my life" is a Latin locution of medieval origin.

designed and built a majestic Victorian Gothic structure.[21] A dome 23 m in diameter would house the telescope built by Warner and Swasey, who had already proven themselves at Lick Observatory. Hale wanted to play it safe. The long tube, mounted on a tall metal column so that it could be pointed even near the zenith, had and has an imposing and improbable appearance: 20 ton of iron, reminiscent of a fascinating Metropolis machine. Despite its size, the instrument was so well balanced that it could be swung by hand with minimal force and in great safety. The observer could reach the focus via a hydraulic platform almost as wide as the shadow of the dome; probably a solution inspired by the family business of hydraulic elevators. The building was designed with an eye to the future — as was always Hale's custom — to accommodate other telescopes as well as physics, optics, and chemistry laboratories.

Within a few years everything was ready for use. In October 1897, the largest astronomical reflector of all time[22] was inaugurated with pomp and ceremony in the presence of the elite of American astronomy. Hale hoped to make great and spectacular discoveries. To publicize them, he had already founded "The Astrophysical Journal",[23] a specialized journal published by the University of Chicago Press. For a century it would be a reference point in the celestial sciences, with the highest impact factor among astronomy publications. An indefatigable and generous organizer, he also promoted in 1899, along with the mathematician Simon Newcomb, director of the U.S. Naval Observatory, and Charles Pickering of Harvard, the creation of the Astronomical and Astrophysical Society of America, which became the American Astronomical Society in 1915. Five years later it was the turn of an International Union for Cooperation in Solar Research, which by the end of World War I would become the International Astronomical Union (IAU), the United Nations of astronomers worldwide.[24] From the outset, the staff of the Yerkes Observatory could boast men of the caliber of James Edward Keeler,[25] planetary astronomer and close associate of Hale who

---

[21] D.E. Osterbrock, *Yerkes Observatory, 1892–1950: The Birth, Near Death, and Resurrection of a Scientific Research Institution*, University of Chicago Press, Chicago (IL) 1999.

[22] An even larger refractor, built for demonstration, was shown at the Paris World's Fair in 1900, but it proved to be of poor quality and was discarded.

[23] G.E. Hale, *The Astrophysical Journal*, in "Astrophysical Journal", 1, 1895, pp. 80–84.

[24] W.S. Adams, *The History of the International Astronomical Union*, in "Publications of the Astronomical Society of the Pacific", 61, 358, 1949, pp. 5–11.

[25] G.E. Hale, *James Edward Keeler*, in "Science", 12, 297, 1900, pp. 353–357.

would later have a brief but splendid career in California. Intending to explore photographically the 79 spiral nebulae surveyed by Lord Rosse, in 1898 Keeler discovered to his surprise that objects of this kind were far more numerous. Extrapolating his counts across the sky, he predicted the existence of 100,000 spirals,[26] an impressive number for those times.

Hale also engaged Edward Emerson Barnard,[27] whose name is associated with important discoveries in both the planetary and galactic fields. Born in 1857 to a very poor family in the secessionist and slave-owning South, Barnard had begun to scan the skies for amusement. His opportunity came when a businessman offered him a large cash prize for each comet he discovered. In a short time, Edward found eight of them, raking in a fair amount of money and making a name for himself as a skilled observer. His fellow citizens then decided to tax themselves to send him to college, a form of patronage unheard of in the Old World, where the advancement of deserving young people without financial means was traditionally the province of munificent nobles or the Church. But Barnard was not a bookish man. He quickly left the university, which would later award him an honorary degree anyway, and went to work as an astronomer in California at Lick Observatory and then with Hale at the University of Chicago (where a degree was apparently not a requirement for selection of faculty members, starting with the patron himself). At Williams Bay, observing with the 40 inch telescope, he was able to use the skills he had acquired as a boy practicing in a photographer's studio. He created the first catalog of images of the dark nebulae scattered across the Milky Way's stellar mantle, and convinced himself that they were giant clouds of dust, large reservoirs of material for new star formation. He also discovered a star with remarkable proper motion, which turned out to be the second closest star to the Sun after the α Centauri system. Edwin Powell Hubble also received his Ph.D. from Chicago. We will talk more about this giant later in the chapter 15.

Meanwhile, the 19th century turned over a new leaf with incredible celebration. After a long period of wars and popular uprisings to assert the political, economic, and commercial supremacy of the great historical powers, and to redistribute overseas lands and spheres of influence in Asia, it

---

[26] R.W. Smith, *Beyond the Galaxy: The Development of Extragalactic Astronomy, 1885–1965,* Part I, in "Journal for the History of Astronomy", XXIX, 2008, p. 100.
[27] J.T. McGill, *Edward Emerson Barnard Commemoration,* in "Popular Astronomy", Vol. 36, 1928, pp. 336–339.

seemed that the world was about to enjoy the pleasures of a *Belle Époque* in peace. It was only a grand illusion, fueled by the marvels of the technological spillovers of scientific discovery, the rising wealth of the bourgeoisie born of the second industrial revolution, and the dull selfishness of the old powers. A slumber of political common sense and social attention that would soon lead to the tragic awakening of the Great War.

After two and a half centuries of slow advancements, telescope technology had also taken a real leap forward with the introduction of glass mirrors made reflective by a silver coating. This was the egg of Columbus. Refractive optics had reached its limits because of the weight of the huge lenses, the stratospheric cost of machining, and the objective difficulty of producing pure blanks in large sizes. Mirrors, on the other hand, supported from the back instead of the edges, could theoretically support any size and required only one optical surface to be polished. However, it was essential to overcome the serious drawback of the historical speculum, with the rapid aging of its optical surface. The solution was found in 1835 by Justus von Liebig, a German scientist who pioneered organic chemistry and biological agriculture:[28] the very man whose name can be found on the packaging of a well-known brand of bouillon cubes for having invented the process of extracting and preserving volatile meat proteins in the mid-19th century. Liebig was in his early thirties and, thanks to the foresight of Alexander von Humboldt, had been teaching at the old Lutheran University of Giessen in central Germany for more than a decade, despite having no formal academic training, when he found a way to deposit thin films of silver obtained by reducing a salt of the precious metal. The trick was simple and could be used to make any smooth surface reflective. In fact, silver has the ability to reflect 96% of visible light, with the exception of blue and violet, for which it is almost completely opaque.

Liebig's silver plating process, first shown to the public at the Great Exhibition in London in 1851, revolutionized the manufacture of mirrors, which until then had been based on the use of mercury according to the recipe of 16th-century Venetian glassmakers. The craftsman would spread the liquid metal on a sheet of tin and work it with his hands or a spatula, causing serious damage to his health. Once dried, the amalgam had a delicate reflective surface that was protected by a layer of glass. Thus were born the magnificent mirrors that still adorn the walls of the ballrooms of

---

[28]W.H. Brock, *Justus von Liebig: The Chemical Gatekeeper*, CUP, Cambridge 2002.

the Grand Canal palaces: a much more expensive and risky process than spreading a layer of silver nitrate on a smooth substrate and then reducing it with an agent. The German chemist reaped great economic benefits from his invention. In 1856, silvering attracted the interest of an Alsatian astronomer, Carl August von Steinheil.[29]

At the time, Liebig and Steinheil were both living in Munich, the one having been appointed to the local university by the King of Bavaria, the other having taught at the same university for decades and then founded an optical-astronomical company there, the C.A. Steinheil & Söhne, specialized in the manufacture of lenses, mirrors, and pioneering instruments for measuring light. Steinheil had been a student of great abstract minds such as Gauss and Bessel, and yet he cultivated a real vocation for experimenting with new technologies. He had long wondered, out of scientific curiosity and commercial interest, how to overcome the difficulties of using the speculum in astronomical mirrors. Before meeting Liebig, he had unsuccessfully tried a method of electrically depositing copper on optically machined mirror surfaces. Silvering solved almost all of his problems, allowing him to use glass as a rigid, recyclable optical substrate on which to deposit the reflective film.

The initial part of the manufacturing process for silver-coated glass mirrors, such as the pioneering 36 inch made for Common in 1879 and then given to Lick Observatory, was conceptually similar to the classical speculum optical processing. Instead of the copper-tin alloy, the starting element was a homogeneous, bubble-free, stress-free disk of glass paste with a typical thickness-to-diameter ratio of 1 to 6. One of the two faces was hollowed out by rubbing a watery solution of abrasive powders (iron oxides) with a tool made of the same glass material until a surface with the desired curvature was obtained. Ingenious tests guided the optician in his grinding and confirmed when the mirror was ready for the next phase. León Foucault had devised one such method, based on studying the structure of the image produced in the focal plane of the mirror by a point source at infinity. The trick was to study this image by intercepting part of it with a sliding blade perpendicular to the optical axis. By projecting the result of the "partial eclipse" onto a screen, manufacturing defects were detected.[30]

---

[29] R. Cohn, J. Russell, *Carl August von Steinheil*, Bookvika, Memphis (TN) 2012.
[30] For further information, see S.C.B. Gascoigne, *The Theory of the Foucault Test*, in "Monthly Notices of the Royal Astronomical Society", 104, 1944, pp. 326–334.

Once polished, the mirror was ready for the first of many silver coatings. Silvering was an operation for specialists, made up of many little secrets acquired through experience. In principle, it was as simple as cooking a fried egg; but few know how to turn this poor dish into a masterpiece for the palate. After being thoroughly washed and degreased, the shiny surface of the mirror was dipped in a solution of silver nitrate, which, when treated with glucose or formaldehyde, released the metal. As the silver precipitated, it fixed to the surface, forming a thin, uniform, and highly reflective film. The advantages were a reduced weight compared to speculum mirrors, with a consequent gain in the manufacturing of the supporting structure; a 50 percent increase in reflectivity, which more than compensated for the need for a second mirror in Newtonian and Cassegrain telescope mounts; a lower cost of raw material, since only a surface free of impurities and bubbles was required; less sensitivity to thermal variations, which were a major cause of degradation in optical performance; and, most importantly, an end to the nightmare of having to repolish the reflective surface every six months. When the silver coating became too dark and opaque (silver tends to oxidize under the influence of sulfur in the air), it was sufficient to wash it off with an acid and replace the old layer with a fresh one, without having to touch up the optical surface of the substrate. Normally, each observatory had both the equipment and the technicians to carry out this delicate operation, which involved the manipulation of the fragile glass disk.[31] Since it took only a few days to complete the job, it was no longer necessary to have a second spare mirror. The only disadvantage, not entirely negligible, was the opacity of the silver to the blue and violet radiation. The problem did not concern the eye, which has its highest sensitivity in the yellow color, but the photographic plates, which were then making their entry into astronomy. The first emulsions were in fact particularly sensitive to precisely that blue/UV light that silver refused to reflect. For this reason, too, refracting telescopes still experienced a period of glory in the second half of the nineteenth century, appreciated especially by those planetary astronomers who frowned upon the diffraction figures (crosses) created by the secondary mirror mounts of reflectors (so-called "spiders").

However, making large lenses was enormously expensive, and making them very large was technologically daunting and, in any case, futile because of their extreme thickness (causing strong absorption) and weight (causing undesirable deformation under the effect of gravity). So when astronomers'

---

[31] M.G. Raymond, *Silvering of Telescope Mirrors*, in "Popular Astronomy", Vol. 26, 1918, pp. 385–386.

bulimia for mega-optics exploded in the 20th century, mirrors took over. The limitations imposed by the use of silver as a reflective material were mitigated by the invention of photographic emulsions and other detectors sensitive to yellow-green light, and then eliminated by the replacement of the precious metal with aluminum: a slightly less performant material in terms of reflectance ($\sim$ 90%), but much less selective for light colors. The aluminum coating process, invented between the two world wars by a young professor at the California Institute of Technology, John Donovan Strong, involves sublimating rings of very pure metal by means of an electric discharge in a low-pressure environment (billions of times lower than atmospheric pressure) and letting the aluminum vapor settle on the previously cleaned glass surface. All you need is an airtight container, a set of vacuum pumps, and a battery of electrical resistors to produce a rapid rise in the temperature of the aluminum, up to about $1,000°C$. Under these conditions, the metal goes directly from a solid to a vapor state. Free to move, the aluminum atoms form a cloud that expands, finding very little resistance from the remaining air molecules and eventually sticking to the obstacles it encounters. A proper geometry in the distribution of the aluminum rings ensures a uniform atomization over the entire optical surface to be treated. The advantage of aluminum over silver is not only that it reflects blue and violet light well, but also that it oxidizes less quickly than the precious metal, which, as every housewife knows, tends to quickly form a blackish film. The price is a loss of $\sim$ 10% of reflectance in the yellow-red region: a small sacrifice compared to the benefits.[32]

We must now go back to Hale. In 1894, he had convinced his father to buy him a blank of good glass from the French company Saint-Gobain, a historic company that Jean-Baptiste Colbert, First Minister of State at the time of the Sun King, had wanted to specialize in the production of wine bottles. It was an investment of 300,000 euros in today's value that the old William had willingly made in order to enhance the career of his visionary scion. George planned to make a mirror 1.5 m in diameter, which would equip a large and modern reflector and thus consolidate the record of the Yerkes Observatory. But President Harper turned a deaf ear and would not loosen the purse strings to fund the project. He apparently had other priorities for his university. Meanwhile, George's father had died, and

---

[32]In the near-infrared range (0.7–10 μm), there is a way to optimize signal losses due to the poor performance of aluminum by resorting to gold, which reflects as much as 98–99%. This is the solution chosen for the 18 mirror segments of NASA's James Webb Space Telescope (JWST), which is operational since 2022.

with him the promise of funding for optical processing had disappeared. So, in an ideal response to the exhortation to American society attributed to New-York Daily Tribune journalist Horace Greeley, "Go West, young man, go West and grow up with the country", Hale decided to try his luck in California, and he went to raise funds at the Carnegie Institution for Science, an organization founded in 1900 in Washington, D.C., by the multimillionaire Andrew Carnegie.[33] Cleverly, he made sure to be preceded by a letter of recommendation from Sir William Huggins, in which the highly decorated English scientist praised the reflector, calling it «telescope of discovery of the future».[34]

Once again, a Scrooge McDuck broke into American astronomy, bringing with him the power of an immense fortune, one of the largest of all time. A poor Scottish émigré, Carnegie had amassed an immense fortune in the 1860s thanks to his business genius, quickly becoming the king of the steel industry and making Pittsburgh his capital. Then, after selling his company to the Wall Street gurus, the financier John Pierpont Morgan, he had decided to use a significant part of the proceeds, $350 million (equivalent to 35 billion today), for the good of culture and humanity.

For the site of the new observatory on the Pacific coast, Hale chose Mt. Wilson, which years before had been selected to house the large glass eye of the phantom 40 inch telescope now in his hands. A singular nemesis, accomplished with the intervention of Pickering, who had courted this site for two decades, without both money and luck. George first visited the mountain in 1903, climbing a steep mule trail, a classic icon of the West, told by Hollywood and even through the evocative notes of a Ferde Grofé composition. He was immediately convinced that the place was exactly what he was looking for. The summit enjoyed the favorable climatic conditions of Southern California and, unlike today, did not suffer from the pollution of the lights of Los Angeles.[35] On this occasion, however, Hale

---

[33] Morris, *The Tycoons*, cit.

[34] Cf. M. Simmons, *Building the 60-inch Telescope*, https://www.mtwilson.edu/building-the-60-inch-telescope/.

[35] Already in his *Optiks*, Newton had written: «If the Theory of making Telescopes could at length be fully brought into practice, yet there would be certain Bounds beyond which Telescopes could not perform. For the Air through which we look upon the Stars, is in a perpetual Tremor; as may be seen by the tremulous Motion of Shadows cast from high Towers, and by the twinkling of the fix'd stars. The only remedy is a most serene and quiet Air, such as perhaps be found on the tops of the highest Mountains above the grosser Clouds» (I. Newton, Sir, *Opticks: or, A Treatise of the Reflections, Refractions, Inflections and Colours of Light*, 4th ed., William Innys, London 1730).

lacked foresight. Soon after, thanks to "black gold", the frontier town would explode demographically and economically, becoming the second largest metropolis in North America by population and the first largest, but also the nightmare of Mount Wilson astronomers.

In 1904, with Carnegie's money, Hale took a 99-year lease on 40 acres, built the observatory edifice, and arranged for the grinding of his own mirror: an unprecedented feat that became the paradigm for adventures to come. He also commissioned the construction of the telescope's supporting structure in San Francisco, which was miraculously saved when the great earthquake of 1906 devastated the city. Paradoxically, this part of the work was the easiest for the energetic George, who was also a very capable and lucky general. The real difficulties and risks lay in the transportation of the instruments at high altitudes, by arms and on the backs of mules, over steep trails that in places were no more than half a meter wide, through centuries-old forests, dodging sheer cliffs and wild animals. At every corner there might be a hungry puma ready to pounce on man and beast, or a vagabond ready to slaughter the unfortunate for half a silver dollar or a flask of mescal. A true epic, worthy of a John Ford movie, perhaps with Antonín Dvořák's symphony. *From the New World* as the soundtrack. After four years of feverish work and a few accidents — the mirror, entrusted to the expert hands of George Ritchey,[36] was almost finished when, for unknown reasons, it was irreparably scratched and had to be reworked — the adventure ended in 1908 with the inauguration of the telescope, then the largest in the world.[37] Because of its modernity and power, the 60 inch aperture instrument would open new windows on the fledgling field of nebular astrophysics, fulfilling Huggins' promise. When Carnegie visited the mountain settlement and the headquarters in the village of Pasadena on Mount Wilson, he was so impressed that he doubled Hale's funding out of the blue. There were smiles then, and there are tears now.

Hale understood that to unlock the secrets of the life of stars and nebulae, astronomers would have to equip themselves with ever more powerful instruments. But his true passion remained the Sun. While still in Chicago, he had designed a large horizontal solar telescope, capable of returning an image of the star almost the size of a football, and equipped

---

[36]D.J. Mills, *George Willis Ritchey and the Development of Celestial Photography*, in "American Scientist", 54, 1, 1966, pp. 64–93.

[37]D.E. Osterbrock, *Pauper & Prince: Ritchey, Hale, & Big American Telescopes*, University of Arizona Press, Tucson (AZ) 1993.

for spectroscopy.[38] The time had come to make it. As usual, he turned
to a patron to finance the venture, this time one of the fairer sex. Miss
Helen Snow agreed to sponsor the instrument on the condition that it bear
her father's surname. The Snow telescope was transported from Chicago
to Mount Wilson, where it was placed within walking distance of the 60
inch reflector. A coelostate[39] with an aperture of 75 cm reflected the image
of the Sun on a flat mirror that sent it at a distance of 250 m, on the other
side of a rudimentary tunnel cooled to minimize the turbulence induced by
the strong heat. There, a third mirror focused the beam onto the slit of
a spectrograph. So it was that in 1908, between one project and another,
the indomitable George found the time to prove, three hundred years after
their discovery, that sunspots were vortices of gas moved by intense magnetic fields. To do so, he relied on a physical effect discovered 12 years
earlier by the Dutchman Pieter Zeeman: the doubling of spectral lines in
the presence of a strong magnetic field. It was the proof of the power of the
new science, astrophysics, born from the equal union of two ancient disciplines, astronomy and physics (ideally bringing together in a single book
the two great *philosophiae naturalis'* works of Aristotle, the *Physics* and
the *On the Heavens*). Curiously, the birth took place in Italy — a nation
grappling with the serious problems caused by the recent reunification of
the peninsula into a single kingdom — thanks to a handful of solar and sidereal astronomers: Giuseppe Lorenzoni, Arminio Nobile, Lorenzo Respighi,
Angelo Secchi, and Pietro Tacchini. In 1871 they founded the *Società degli
Spettroscopisti Italiani* (Society of Italian Spectroscopists) and endowed it
with a journal, the "*Memorie*" (Memories), which for several decades played
a leading role in the scientific world.[40]

While the 60 inch reflector covered itself in glory with pioneering
research in quantitative spectroscopy, stellar parallaxes, photometric measurements, and the first photographs of nebulae, Hale was already thinking
about a larger telescope. As always, he needed a sponsor to match the
Carnegie Institution's contribution. He found that donor in John Daggett
Hooker, a Los Angeles industrialist eager to make history. As soon as

---

[38] F.H. Seares, *The Mount Wilson Solar Observatory*, in "Publications of the Astronomical Society of the Pacific", 29, 170, 1917, pp. 155–170.

[39] Optical system consisting of two mirrors used to track the Sun in its apparent motion.
It is designed to return the image of the star on a fixed path defined by the position of
the observing instrument.

[40] I. Chinnici, *The "Society of Italian Spectroscopists: Birth and Evolution*, in "Annals
of Science", 65, 3, 2008, pp. 393–438.

George had the money in hand, he ordered from Saint-Gobain, which knew how to make glass, a blank for a 100 inch mirror: a monster made from over two tons of molten glass and cooled very slowly to avoid the onset of cracks. The challenge was at the limit of technology and possibly beyond. When the blank arrived in America, Hale was unpleasantly surprised to find it filled with bubbles. His mental health, already compromised, collapsed under the enormous stress, so much so that he was forced to stay in a nursing home for some time. His patron Carnegie was very concerned: «But pray show your good sense by keeping in check your passion for work, so that you may be spared to put the capstone upon your career, which should be one of the most remarkable ever lived».[41] Not a bad tribute indeed. He, like Hale, knew from direct experience the truth of Confucius' saying: "Choose a job you love and you will never have to work a day in your life".

In desperation, Hale ordered a second blank, but it proved worse than the first. Meanwhile, Gavrilo Princip, a 19-year-old Bosnian nationalist, shot and killed Archduke Franz Ferdinand, heir to the throne of Austria-Hungary, and his beautiful wife, Sophie, Duchess of Hohenberg, while Their Highnesses were visiting the city of Serajevo. This assassination was the bloody and romantic prelude to one of the largest and most senseless massacres in history. Saint-Gobain ceased production of astronomical glass to concentrate on the war effort. Hale, with the courage of despair, had already decided to try his luck with the first blank. He had entrusted it to the expert hands of George Ritchey,[42] a stubborn Irish-born optician whom he had hired when the Yerkes Observatory was founded, using his father's money to pay his salary, and who had already successfully polished the family's 60 inch mirror.

In five years of hard work, Ritchey and his team got to the bottom of the problem and, like Michelangelo with his marbles, transformed a shapeless mass of pitted glass into a magnificent paraboloid. It was not exactly the kind of mirror the engineer wanted to make. Ritchey had innovative proposals for reducing coma aberration that he would later develop in France, working with the astronomer and inventor Henri Chrétien in his laboratory

---

[41]Quoted in M.S. Longair, *A Brief History of Cosmology*, in W. Freedman (ed.), *Measuring and Modeling the Universe*, CUP, Cambridge 2004, Vol. 2, p. 4.

[42]D.E. Osterbrock, *The Canada-France Telescope and Ritchey, George-Willis: Great Telescopes of the Future*, in "Journal of the Royal Astronomical Society of Canada", 87, 1, 1993, pp. 51–63.

in Paris. But Hale demanded a specific result and kept his man's fantasies in check until the work was completed. Then, fed up with his constant criticism and complaints, he fired Ritchey and forced him to emigrate. Perhaps the only bad action of a generous man who, this time paying more attention to his stomach than his brain, was unable to appreciate the ideas of his brilliant collaborator.

In November 1917 — while on the Italian front the breach opened by the Germans at Caporetto was turning into a terrible defeat, later redeemed by a victorious counterattack that put things back to the way they were but at the cost of hundreds of thousands of lives — the Hooker telescope, designed by Francis Gladheim Pease around Ritchey's polished mirror, a monster of 100 tons moved electrically on a cushion of mercury, opened its great silver eye for the first time.[43] Hale himself tried to point the instrument at Jupiter. The effect was shocking. The image of the planet appeared blurred and flickering. The blow was promptly overcome when it was realized that the cause was a trivial thermal problem. The instrument had been exposed to the Sun all day and had not yet reached thermal equilibrium. A few hours later, pointed at Vega, it returned an almost perfect image of the star.

Hale had once again hit the bull's eye, overcoming all difficulties, from economic to technical and logistical. As in a long march of science, the faith of a visionary and the muscles of the pack animals used to carry the pieces to the top of the mountain had prevailed. One of the mule drivers was named Milton Lasell Humason:[44] another "man without letters". Having joined the staff of Mount Wilson astronomers thanks to Hale's fine nose, he would then, as we will see (p. 371), play a crucial role in the discovery of the law of the expansion of the universe. This, too, could happen in the America of big dreams.

With its extraordinary capacity to explore the depths of the cosmos, the Hooker 2.5 m reflector, larger and much more sophisticated than Lord Rosse's Leviathan, had an effect similar to that of Galileo's telescope. The leather and glass tube of the Pisan pioneer had effectively taken astronomy from the narrow horizon of the Solar System to the vast sea of stars. Within

---

[43]L.O. Cooper, *The New 100-inch Reflector for Mount Wilson*, in "Journal: Publication of the Pomona College Astronomical Society", 4, 1915, pp. 141–142; H. Spencer Jones, *The 100-inch Telescope of the Mount Wilson Observatory*, in "The Observatory", 43, 1920, pp. 122–124.

[44]L. Yount, *Modern Astronomy: Expanding the Universe*, Chelsea House, New York 2006, p. 29.

a few years, the iron and silvered-glass giant on Mount Wilson would open to astronomers the "realm of the nebulae", relegating the Milky Way to the status of an isolated universe. A new paradigm that Ritchey would celebrate ten years later:

> This series of telescopes, by revealing to all men, graphically, by means of exquisite photographs, a Universe of which the Earth, the Sun and the Milky Way are but an infinitesimal part, will bring to the world a greater Renaissance, a better Reformation, a broader science, a more inspiring education, a nobler civiliza-tion. This is the Great Adventure. These telescopes will reveal such mysteries and such riches of the Universe as it has not entered the hearth of man to conceive. The heavens will declare *anew* the Glory of God.[45]

Since the rest of this story depends largely on the quality of photomet-ric measurements[46] of stars and nebulae, a brief digression into the slow progress of this observational technology is in order.[47]

The advent of the telescope had multiplied the number of visible stars, but for nearly 170 years there had been no significant improvement in the methods of sidereal photometry. Stellar magnitudes were still far from having the deep physical meaning and, more importantly, the precision they have today. They served mainly to confirm the celestial geography entrusted to the fanciful figures of asterisms. It is true that the usual Halley had noted that a star of the first magnitude taken to ten times its distance would appear to be the sixth, thus alluding to the law of flux attenuation with the inverse of the square of the distance and the logarithmic nature of the magnitude scale; but this was a sporadic case (see also the note at p. 310).

---

[45] G. Ritchey, *The Modern Reflecting Telescope and the New Astronomical Photogra-phy*, in "Transactions of the Optical Society", 29, 1928, p. 197; F. Hargreaves, *Obituary Notices: Richey, George Willis*, "Monthly Notices of the Royal Astronomical Society", Vol. 107, 1947, p. 38.

[46] The term "photometry" was coined by the Swiss-German polymath Johann Heinrich Lambert, who used it as the title of his book *Photometria* in 1760; M.K. Shepard, *Introduction to Planetary Photometry*, CUP, Cambridge 2017, p. 11.

[47] Cf. R. Miles, *A Light History of Photometry: From Hipparchus to the Hubble Space Telescope*, in "Journal of the British Astronomical Association", 117, 4, 2007, pp. 172–186; E. Budding, *Introduction to Astronomical Photometry*, CUP, Cambridge 1993, pp. 12–21.

Not surprisingly, the music changed with William Herschel. Busy constructing the light curves of the variable stars he was discovering in large numbers, the astronomer and musician decided to measure the differential fluxes of his sources against a sequence of nearby stars of similar brightness and photometric stability. The criterion had been introduced by Brahe and further developed in the first half of the 18th century by the Frenchman Pierre Bouguer,[48] former child prodigy, expert on ships and hydrography, who could afford to beat Leonhard Euler in hydraulics and dabbled in photometric measurements in his spare time. Bouguer was convinced that the human eye was not capable of estimating the quantitative difference between two sources of light, but was good at determining when they appeared to be of equal brightness. He was absolutely right. In fact, the eye operates in an involuntary mode, without giving any indication to the consciousness, by means of an adjustable diaphragm, the iris, which serves to regulate the average incoming luminous flux. In short, the Frenchman had homologated the astronomer's eye as a precise comparator, and even suggested the way to take full advantage of this faculty for estimating stellar magnitudes. This involved simply dimming the brightness of an assumed standard star until it matched that of the source under study, thereby obtaining an intensity ratio. The practical example of the photometer built a century later by von Steinheil, the professor of mathematics and physics at the University of Munich whom we met earlier in this chapter (p. 329), will help us to understand the concept better.

Steinheil was an excellent inventor. In 1836, he created an instrument that made it possible to compare two stars of different magnitudes by focusing them in the same eyepiece through an objective lens divided in two along one diameter. A system of prisms made each of the two stars impact only one half of the objective, while the instrument's mechanics allowed the two half lenses to move parallel to the optical axis. With this trick, each source could be blurred independently of the other and enough to produce two identical patches of light. After calibration, the relative sliding (de-focusing) of the two half-lenses provided an objective measure of the intensity ratio between the two stars. By this ingenious stratagem Steinheil was able to analyze, among other things, the extinction that starlight undergoes due to the Earth's atmosphere as the zenith distances change

---

during the night and thus the mass of air opposing the passage of radiation varies.

The concept was taken up in 1867 by another German, Karl Friedrich Zöllner,[49] whose interests included optical illusions and spiritualism. Zöllner had the idea to compare real stars with a reference source made with a kerosene lamp. The brightness of the artificial star was dimmed by an exotic physical phenomenon, the cross effect of two light polarizers. The approach was promising, but in practice the device worked poorly because of the difference between the real stars and the artificial source. The former was sharp and trembling; the latter was pale, diffuse, and stable. All this made comparisons difficult, if not impossible. Pickering found a way to overcome this obstacle with a trick. He built an optical system capable of observing the North Star simultaneously with the object whose magnitude he wanted to determine. Everything else was the same as in Zöllner's photometer. By rotating the two Nicol prisms[50] relative to each other, the light from the reference star was attenuated until the two sources appeared equal. In this way, the flux measurement was reduced to that of an angle. But why the Polaris? Because this star never sets and does not change its height above the horizon, so the amount of atmospheric absorption it undergoes remains constant (except for meteorological variations).

The advent of photometric comparators created the need to establish a relationship between intensity ratios and the magnitude scale inherited from Hipparchus and Ptolemy and never abandoned. By the 19th century it had become clear that the human eye, like the ear, is a logarithmic device; its response does not change in proportion to the absolute variation Δs of the stimulus s but to the relative variation, $\Delta s/s$. This is why the light of the same match produces very different responses in bright sunlight and in a dark room. Once the functional relationship was identified, the problem was to find the scale factor. It won the proposal of the British astronomer Norman Pogson,[51] that a difference of 5 magnitudes was equal to 100 in the flux ratio (as already anticipated by the usual Halley). It was an excellent compromise between adherence to the scales of the ancient astronomers and the need for a numerically simple recipe for modern measurements.

---

[49]D.B. Herrmann, *Karl Friedrich Zöllner*, B.G. Teubner, Leipzig 1982.

[50]Rhombohedral crystal of Iceland spar, a variety of calcite that produces polarized light.

[51]V. Reddy, K. Snedegar, R.K. Balasubramanian, *Scaling the Magnitude: The Fall and Rise of N.R. Pogson*, in "Journal of the British Astronomical Association", 117, 5, 2007, pp. 237–245.

It happened in 1856. In that year, the bloody Crimean War, which had flared up to maintain the balance of power in the Old World, ended with the defeat of the Russian Empire by the alliance of the Ottoman Empire, France, the United Kingdom, and Sardinia-Piedmont. And while in the United States premonitory signs of the will to secede appeared, in far-away China the Western powers unleashed the shameful II Opium War in response to the Taiping revolt, crystalline example of the most blatant imperialism.

Then came the age of photography, a technique that for decades remained primarily in the hands of gentlemen scientists. It was actually an amateur astronomer, the Anglican priest Thomas Espin,[52] who first experimented in 1883 with estimating magnitudes from the thickness of stripes left by stars on plates exposed with a transit telescope.[53] The method worked quite well, but was affected by the different rate of diurnal motion of stars at different polar distances. It was replaced by direct measurement of the diameters of well guided star images; a technique pioneered by Pickering and his "human computer" Henrietta Swan Leavitt at Harvard (p. 315). The benefits of photography went far beyond a simple improvement in measurement accuracy, which was pushed down to two hundredths of a magnitude (equivalent to an uncertainty of 2%). The plates could be used both as an acquisition tool and as an archival document. They permitted the observation of very faint stars because of their ability to integrate information during long exposures. And they could be easily analyzed during the day in a heated laboratory, sitting comfortably in front of instruments that did not require the use of the eye as a light comparator. A new kind of indoor professionalism began to be demanded of the astronomer, who for millennia had specialized in observing the night sky.

As always, however, not all that glitters is gold. In the midst of so many advantages, photographic plates posed a serious problem. Emulsions were sensitive to blue light, while the eye is known to perceive all iris colors but with a particular preference for yellow (which, by the way, is the color of the

---

[52]T.E.R. Phillips, *Obituary Notices. Fellows: Espin, Thomas Henry Espinell Compton*, in "Monthly Notices of the Royal Astronomical Society", 95, 1935, pp. 319–322.

[53]A transit telescope can only be moved in the plane of the meridian. Since it cannot follow the celestial sphere, it produces images in which stars leave longer streaks the longer the exposure time and the greater the polar distance of the field of view. However, there are modern versions of these instruments, such as the 9.2 m aperture Hobby-Eberly telescope installed at McDonald Observatory in West Texas, which allow limited tracking around the meridian by moving the secondary mirror.

Sun as seen through the atmosphere). This difference made it difficult to compare historical measurements by eye with those obtained with the new technology. A good reason to welcome with open arms the arrival on the scene, at the end of the 19th century, of the so-called "photocells". Their history begins with the discovery of the photosensitivity of certain materials. In 1839, the very young Alexandre-Edmond Becquerel had noticed how electric current is generated during some chemical reactions induced by light:[54] first step on this path towards the recognition of electrons and their way of escaping from the embrace of atoms with the intervention of energy projectiles called photons. Children of Becquerel's intuition are also the panels that today cover the roofs of some houses and that, by stealing light from the Sun and transforming it into electricity, help to reduce the pollution of the planet and the burden of electricity bills.

In 1892 George Minchin, an Irish professor working in London, decided to test on the stars the photometric capabilities of a photovoltaic cell of his own invention: a kind of voltaic cell in which a selenium plate struck by light develops a potential difference. He found the necessary telescope in Ireland, where William Monck, professor of moral philosophy and his friend, let him use a refractor with a 7.5" aperture made in America by Clark. It was a christening and a flop. Although the telescope's large aperture focused an intense beam of light on selenium, the two friends could barely detect the signal from Venus and Jupiter. With shameless foresight, Minchin declared that there would be «little difficulty in obtaining fairly accurate measurements of the light of the stars of the first and second magnitudes».[55] He was right, but it was an American who did it, Joel Stebbins. We will mention this and other stories of modern electronic detectors in the last chapter of this book. For now, we will take a stroll through the wonders of 20th century physics, at least the part that is mainly related to the celestial sciences.

---

[54] In J.E. Buerger, *French Daguerreotypes*, The University of Chicago Press, Chicago (IL) 1989, p. 88.
[55] Miles, *A Light History of Photometry*, cit., p. 178.

Chapter 14

Fundamental bricks:
Two new sciences, relativity and
quantum mechanics

All science is either physics
or stamp collecting.
Ernst Rutherford, attributed

The universe may have a purpose,
but nothing we know suggests that,
if so, this purpose has any similarity to ours.
Bertrand Russell, *Why I am not a Christian*

The dawn of the 20th century was accompanied by great expectations for the future. How many good wishes in New Year's toasts, and how much awe at the wonders of technology, the child of the new science, the hope for a renewed social order, and the worry about a rapidly expanding market, increasingly omnivorous, ruthless and global, seemingly unbreakable. For thirty years, the guns had been silent in Europe after the Prussian army's dazzling victories over its Austro-Hungarian brethren and then over its lifelong enemies, the French. Dress rehearsals for the maelstrom of fire and death into which the delirium of a Germany *über alles* (above all)[1] would drag the world twice in the first half of the new century, and about which the nations of the *Entente cordiale*, indolently and vainly entrenched in anachronistic fences, were not paying due attention. In this atmosphere of unreal peace, the guns were already thundering on the borders of the Old Continent, crumbling the foundations of ancient empires with "feet of clay". In the *Belle Époque*, the strongest tensions were rooted in the accentuated imbalances in the social structure of the great nations. The complex transformation brought about by capitalism, with famine and hunger,

---

[1]From *Das Lied der Deutschen*, the German national anthem since 1922.

degradation, but also a growing awareness of human and civil rights, conspired to awaken the masses, urbanized by the Industrial Revolution, and summarily made less rough. A demand for redemption that the epigones of the Congress of Vienna and the new rich vainly believed they could control and, if necessary, extinguish in blood. In North America, after the attempted secession of the Southern states by the Confederates, only the rifles of the cavalry crackled to rid the land and its natural resources of the inconvenient presence of the natives. A genocide like so many others in Australia, Africa and Asia, told like an epic novel and opposed by few.

The Western model, armed with technology and pushed to its limits by European imperialism and American liberalism, was winning everywhere: in India, in China, and especially in Japan. The mysterious Empire of the Rising Sun was engaged in a deliberate restructuring of the millennia-old medieval order in an effort to adapt to competition with the West: one could summarize it by saying that everything had to change so that nothing would change.[2] In the Russian Empire, peasants and factory workers were preparing for the great revolution that would break out in November[3] 1917, while World War I was still raging, leading to the construction of the first socialist state in history; an event that would decisively shape the entire 20th century. Romanticism was now at its lowest ebb, undermined by a neo-positivist model that lacked the ideal impulses of the Enlightenment. And artists, bewildered by machines and the reversal of values, sought new expressive means to read souls and tell the world, sometimes going so far as to forge incestuous alliances with rampant national and racial selfishness, proclaimed as always in the name of God.

In science, the 20th century was preparing to reap what the previous century had sown. It began with a tremendous insight by Max Planck.[4] The forty-year-old austere professor at the Friedrich-Wilhelms University in Berlin was struggling with the vagaries of the blackbody invented by his predecessor Kirchhoff: an omnivorous ideal body whose spectrum no

---

[2]Sentence made famous by *The Leopard* by Giuseppe Tomasi di Lampedusa, a novel published in English by Collins and Harvill Press, London 1960.

[3]The October Revolution broke out in Petrograd on the night of November 6-7 in the modern Gregorian calendar, corresponding to October 24 and 25 in the Julian calendar. See p. 131.

[4]The literature on Max Planck is vast. For example, see, B.R. Brown, *Planck: Driven by Vision, Broken by War*, Oxford University Press, Oxford 2015.

one could shape with the tools of classical physics[5] was reproduced in the laboratory using a large insulated cavity (similar to a well-protected oven) that could be inspected through a tiny hole. According to classical thermodynamics, instead of falling to zero, the radiation in equilibrium with this strange object should surge into the ultraviolet, causing an energy catastrophe totally at odds with the evidence.

Although he no longer had the freshness and boldness of youth that usually accompanies revolutions in scientific thought, Planck found a way to set things right: a solution to which he confessed to having been led by «an act of desperation».[6] He suggested that the electromagnetic radiation described and interpreted as waves by Maxwell and Heinrich Hertz was produced and absorbed in discrete, indivisible units, the energy quanta (*Lichtquant*); true and distinct atoms of light in the Greek sense of that word, then baptized "photons". In 1905, Einstein would identify the photoelectric effect as the *experimentum crucis* to validate this quantum model of light already promoted by Newton, competing (apparently) with the continuous model advocated by Huygens:[7] a discovery that would earn the physicist *par excellence* the Nobel Prize 17 years later. Interestingly, Einstein's model was not readily accepted by scientists of the caliber of Hendrik Lorentz, Robert Millikan, and Max Planck himself, all Nobel laureates. The inventor of the mathematical quantum, a "reluctant revolutionary", was deeply skeptical of the physical reality of photons, given their inability to reproduce the complex phenomenology of interference, as he noted in the 1913 letter recommending the cooptation of his friend Albert Einstein into the Prussian Academy of Sciences: «That he may sometimes have missed the target in his speculations, as, for example, in his hypothesis of light quanta, cannot really be held too much against him, for it is

---

[5] It should be emphasized again that the blackbody owes its name to its (theoretical) ability to absorb all radiation that falls on it, regardless of frequency and angle of incidence. This does not mean that a blackbody cannot emit. On the contrary, it must emit, like any warm body. Its spectrum has no lines: just a pure continuous, with a maximum at a temperature inversely proportional to the wavelength (Wien's law) and tending to zero on either side, at very low and very high frequencies.

[6] In A. Montwill, A. Breslin, *The Quantum Adventure: Does God Play Dice?*, Imperial College Press, Dublin 2012, p. 2.

[7] The continuous model consisted of energy waves propagating in an ether and oscillating transversely to the direction of propagation.

not possible to introduce really new ideas even in the most exact sciences without sometimes taking a risk».[8]

That same year, 1905, the 26-year-old Einstein, then an obscure clerk at the Bern Patent Office, had also reformulated Newtonian mechanics in order to generalize Galilean relativity to all physical phenomena.[9] His purpose was to ensure that the laws of electromagnetism and optics, as well as those of mechanics, were the same in all inertial reference systems.[10] He had been led to believe that in a vacuum light travels at a constant speed regardless of the motion of the source and the observer, as Maxwell's equations claimed. Another curious property with embarrassing consequences. In fact, it implied the relativity of the concept of simultaneity and thus the existence of an intrinsic link between space and time, the inseparable coordinates of an event. It also implied the equivalence of matter and energy, capable, under certain conditions, of transforming one into the other. Even the most fanatical followers of Isaac Newton had to accept this new *Welthanschauung* (worldview), although it was so far from common sense.

After another handful of years, still a young man, the Danish Niels Bohr, devised an unorthodox way to stabilize the model of the atom proposed by his mentor, the New Zealand physicist Ernest Rutherford, and to reproduce the phenomena in the interactions between matter and energy:[11] culmination of the fulminating epic of the intimate exploration of matter and baptism of a new science. The story begins in 1904 in the Cavendish Laboratory at Cambridge University, founded thirty years earlier by Maxwell. Working there at the time was the Scotsman Joseph John Thomson, who had discovered the electron in 1897. Faced with the need to accommodate these new particles in an atomic structure, he had imagined positively charged clouds in which negative particles (electrons) were embedded like candied fruit in a plum cake. However, this model, which for the first time explained the

---

[8] A. Pais, *Subtle is the Lord: The Science and the Life of Albert Einstein*, Oxford University Press, Oxford 2005, p. 382.

[9] H. Günther, *Elementary Approach to Special Relativity*, Springer, Singapore 2020.

[10] Inertial reference systems are all those in which the first principle of dynamics holds: in the absence of forces, a body remains at rest or moves with rectilinear and uniform motion. Given an inertial system (for example, to a good approximation, that associated with the so-called fixed stars), so are all others that move with respect to it in a rectilinear (that is, they do not rotate) and uniform (they do not accelerate) motion.

[11] J.L. Heilbron, *Niels Bohr: A Very Short Introduction*, Oxford University Press, Oxford 2020; see also E.N. da Costa Andrade, *The Birth of the Nuclear Atom*, in "Scientific American", 195, 5, 1956, pp. 93–107.

composite nature of the "uncutables", had a short life. Rutherford, then a student of Thomson, soon demolished it, showing in a famous experiment in 1909 that the volume in which matter is mainly concentrated is tiny compared to the size of atoms: a compact nucleus of heavy positive electric charges (protons) surrounded by a crowd of negative electrons. But how does it work?

This formidable question was answered by the Danish physicist Niels Bohr, who was working under Rutherford in Manchester at the time. He hypothesized that electrons orbited the atomic nucleus exclusively on privileged stationary orbits, in which accelerated charged particles did not have to pay the toll of emitting and thus losing energy. In order to isolate the right "orbits", Bohr devised a Planck-like quantization rule capable of interpreting the selectivity of atoms in absorbing and emitting photons. In fact, he suggested that the exchange of energy with the outside world occurred through the electron jumps from one stationary configuration to another. «At all events there is something going on which is inexplicable by the older mechanics»,[12] his master would comment laconically, with perfect reason. Much later, recalling those days, Einstein would say: «That this insecure and contradictory foundation [of physics] was sufficient to enable a man of Bohr's unique instinct and tact to discover the major laws of spectral lines and of electron shells of the atoms together with their significance for chemistry appeared to me like a miracle — and appears to me as a miracle even today. This is the highest form of musicality in the sphere of thought».[13]

Structurally, Bohr's model appeared to be the microscopic analog of the Solar System: a central nucleus that, like the Sun, dominates a swarm of "planets", the electrons, placed on orbits of "well-tempered" sizes. But the ingredients soon lost the identity of material particles. It happened in 1925 when the Austrian physicist Erwin Schrödinger, a man of the most diverse interests, eccentric and contradictory in his behavior, proposed to model electrons as clouds of probability of being in a certain place with a certain speed. This idea followed a year later the proposal made by Louis de Broglie, a French nobleman with improbable hair, in his doctoral thesis: to attribute to material particles the same dual nature as that of light, and to recognize in them also a wave connotation. It would fall to Werner Heisenberg, a 26-year-old German physicist with a light-hearted gaze and

---

[12]R. Spangenburg, D.K. Moser, *Niels Bohr, Atomic Theorist*, rev. ed., Chelsea House, New York 2008, p. 50.
[13]Ibid., p. 51.

an exasperated patriotism, to close the circle by revealing, in 1927, one of the most astonishing behaviors of the complicated world of the infinitely small:[14] the existence of a limit to the knowledge of physical quantities that creates a barrier between the real world, where $2 + 2 = 4$, and a virtual world, where $2+2$ can take any value as long as it happens over a sufficiently short period of time.

These were discoveries and hypotheses that led to unimaginable advances and spectacular applications. But because of their scope and, let's face it, their seeming extravagance, they had to be tested, in the laboratories of physicists on the ground and also in the sky, where there are conditions unavailable on Earth[15] (exploiting the now victorious idea of a unified cosmos). Such was the case with the test of general relativity made on the occasion of the total solar eclipse of 1919. Only four years earlier, at the end of 1915, while millions of soldiers were marching and dying in the trenches dug after the sudden end of the mirage of a quick war, Einstein had published his revolutionary conception of gravity as the effect of the interaction between the properties of matter (mass, momentum, and impulse) and the geometry of space-time. In short, «space-time tells matter how to move; matter tells space-time how to curve».[16] The new theory was based on the postulate that the effects of constant acceleration are equivalent to a uniform gravitational field.[17]

Faced with such a radical paradigm shift, the scientific community remained quite skeptical, including Einstein's good friend Max Planck: "As an older friend, I must advise you against it [...] You will not succeed, and even if you succeed, no one will believe you", seems to have been his uninspiring comment on what Albert had called a "theory of incomparable beauty".[18] In order to abandon all reservations and leave the old, safe path laid down by the great Newton, a convincing experimental proof was in order. This was identified with the verification of a strong conjecture of Einstein's theory, the deviation of the otherwise straight path of light rays

---

[14]Cf. D.C. Cassidy, *Uncertainty: The Life and Science of Werner Heisenberg*, W.H. Freeman & Co., New York 1993; C. Rovelli, *Elgoland*, Adelphi, Milan 2020 (in Italian).

[15]D. Lindley, *Uncertainty: Einstein, Heisenberg, Bohr, and the Struggle for the Soul of Science*, Anchor Books, New York 2008.

[16]J.A. Wheeler, *Geons, Black Holes, and Quantum Foam: A Life in Physics*, W.W. Norton & Co., New York 1998, p. 235.

[17]R. Stannard, *Relativity: A Very Short Introduction*, Oxford University Press, Oxford 2008.

[18]Brown, *Planck: Driven by Vision, Broken by War*, cit.

as a consequence of the distortion of space-time around a massive body. In practice, it was to test whether, at the time of the Sun's transit over a star field, the celestial sources around our star appeared shifted from their usual position, as if the plane of the sky had been temporarily expanded locally. To simultaneously have the Sun and the darkness needed to see the stars, the first good opportunity of a total solar eclipse was taken to perform the test.

It was the year 1916. At this point in the game, as the fiercest battles raged between the "Saxon" armies on either side of the English Channel, the director of the Cambridge Observatory, Arthur Stanley Eddington, entered the scene: an Englishman of a fervent and determined character who, for the sake of truth, had no problem coming to the aid of a German Jew. As secretary of the Royal Society, Eddington had been able to read Einstein's three papers on general relativity almost immediately after their publication, despite the war embargo. Perhaps any scientist imbued with patriotism would have treated them with less curiosity or even contempt,[19] but not Eddington, who instead devoured them and fell in love with the ideas they contained. He had the cultural tools to understand the power and beauty of a physical theory capable of accounting for a tiny orbital quirk of the planet Mercury that could not be explained by the classical methods of celestial mechanics. In short, this is the story.

As early as the 19th century, observations had shown that the major axis of Mercury's orbit retrogrades by 574 arc seconds in the time interval of a century.[20] By appealing to planetary perturbations and even to the imperfect sphericity of the Sun, classical celestial mechanics could justify a large part of this systematic backward slip.[21] A crumb was left out, just 43 seconds per century, but enough to make Le Verrier say in 1859 that this was a big problem worthy of astronomers' attention. Well, even this crumb found its full justification in the context of general relativity.

---

[19]The clash between intellectuals on opposite sides intensified when, in October 1914, 93 eminent German scientists, including Planck, signed a manifesto addressed to the educated world (*Aufruf an die Kulturwelt*) to deny the atrocities committed by the Kaiser's army in Belgium. An episode that was considered outrageous in France and especially in England.

[20]It is common practice in the literature to add to this value another 5,021 seconds due to lunisolar precession, for a total of 5,600 seconds. But the effect accounted for by general relativity concerns only the first number. Precession is just a change in the orientation of the reference system.

[21]N.T. Roseveare, *Mercury's Perihelion from Le Verrier to Einstein*, Oxford University Press, Oxford 1982.

Einstein himself announced this at the end of 1915, proposing two further tests of his new theory: the curvature of light rays and a shift of spectral lines towards the red (redshift), both induced by the presence of a strong gravitational field.

The Dutchman Willem de Sitter was the bridge between Cambridge and the Berlin Academy. Because of his country's neutrality, he was able to talk to all sides in the war. A friend of Einstein, he taught in Leiden, in a fairy-tale setting worthy of an Andersen fable, with some of Europe's finest scientific minds. These included Hendrik Antoon Lorentz, famous for his research in electromagnetism and electrodynamics and for anticipating some aspects of special relativity, and Paul Ehrenfest, an Austrian Jew with a tragic fate, a student of Ludwig Boltzmann and also a friend of Einstein. It would be these scientists, with their enthusiasm and personal relationships, who would sow the seeds of relativity in an Old World deafened by bombs and maddened by fear and hunger. But it was Eddington, above all, who freed the theory from the pincers of disbelief. An infatuation that somehow also saved his skin.

A strict Quaker, Arthur was a tenacious pacifist.[22] In 1916, when conscription was introduced in England and he, too, at the age of 35, was called to wear the uniform, he declared himself willing to serve the country as an ambulance driver or a farmer, but not to take up arms to kill. A conscientious objection that could have cost him dearly. Instead, despite the harsh laws of war, he emerged unscathed. At his repeated request, thanks to the intercession of the powerful Astronomer Royal Andrew Crommelin, who lobbied the War Office on his behalf, and Frank Watson Dyson, who was organizing the expedition for the total solar eclipse of May 29, 1919, Eddington was temporarily exempted from military service. He was allowed to remain in Cambridge to take care of the scientific enterprise, the first useful opportunity to test the theory of a quaint German who claimed to take the place of Newton.[23]

At the eleventh hour, on the eleventh day, in the eleventh month of 1918, the Great War ended with the material and moral defeat of Imperial Germany. Just in time for the Cambridge professor to set sail for the archipelago of São Tomé and Príncipe without fear of U-boat ambushes. At the same time, Crommelin set sail for Sobral, Brazil. In the small

---

[22]M. Colson, *Albert Einstein and Sir Arthur Eddington*, Gareth Stevens Publishing, New York 2015.

[23]Cf. A. Bettini, *The 1919 Total Solar Eclipse*, in "Il Nuovo Saggiatore", 35, 3–4, 2019.

independent state off the coast of west-central Africa, the Sun would go out for a few seconds, giving Eddington time to photograph a stellar field subjected to gravitational interference from the Sun itself. He hoped to find the change predicted by relativity, and indeed he claimed to have seen it. Reduced with modern techniques, the material collected by the passionate British astrophysicist seems to prove nothing. Perhaps he made a mistake in his search for an answer, or perhaps, blinded by the desire to prove the truth of a theory that had completely seduced him, he forced his hand on the data. Good for Einstein, who suddenly became a real star. «Dear mother! Today a joyful notice. H.A. Lorentz has telegraphed me that the English expeditions have really proven the deflection of light at the Sun».[24]

The verification of the general theory confirmed the model of a closed universe[25] that the German physicist had elaborated as early as 1917 as a direct application of the principles of relativity to cosmology. It succeeded in keeping under control the consequences of gravity, the dominant force on a large scale,[26] while at the same time satisfying certain metaphysical-theological preconceptions. Although a believer, Einstein was a *sui generis* Jew. A great admirer of Baruch Spinoza, he had drawn from the 17th-century Sephardic philosopher, exiled from his community for reckless reflections on religion, the belief that God and nature were to some extent synonymous, and that therefore the cosmos must reflect the same attribute of divinity: an immovable and timeless eternity. But how can this property be satisfied in the absence of a lever to compensate for the tendency of gravity to produce a collapse? Without losing heart and without renouncing his faith in the Spinozian *"Deus sive Natura"*, Einstein drew from his hat a purely repulsive force proportional to distance (through a constant called "cosmological" and denoted by the Greek letter Λ), capable of exactly balancing the gravitational attraction on matter represented as a homogeneous and isotropic fluid. This extension of the spectrum of the forces at play may seem arbitrary to us, almost the desperate bluff of those who, with their backs against the wall, no longer know which way to go; but this is not so. In formulating his hypothesis, Einstein had faithfully followed the Galilean method, masterfully exemplified by Newton's

---

[24] K.C. Fox, A. Keck, *Einstein A to Z*, John Wiley & Sons, Hoboken (NJ) 2004, p. 81.

[25] For a detailed study, see A. Liddle, *An Introduction to Modern Cosmology*, John Wiley & Sons, Chichester 2003.

[26] The only other force then considered was the electromagnetic, which wins on small scales but has no effect on scales where matter is electrically neutral.

solution of the famous apple problem. However, once a way was found to save appearances, the cosmological model had to be put to the test through a meticulous process of falsifying its predictions.

Shortly thereafter, far from the moral and material misery of war, Willem de Sitter had studied the behavior of a static model of a spatially flat universe with zero density, that is, without matter, but subject to Einstein's force of repulsion. It was a mathematical exercise to explore a state that did not seem so far from the truth (here and now, we should add). In the absence of the actor of gravity, the universe becomes a prisoner of the only repulsive force that, contrary to the premises, induces the expansion of the metric substrate. An alarm bell for the static hypothesis, no less resounding than the other about the fragility of Einstein's creature, incapable of withstanding even the slightest disturbance and therefore highly precarious.

Five years have passed. The defeat of the Black Eagles had shattered the geopolitical framework of Old Europe, redrawing borders and proposing new, ephemeral balances. Germany was busy licking its deep wounds and hatching sinister plans for revenge, while the Habsburg Empire and Tsarist Russia no longer existed. Sovietization, which had begun in 1917, was consolidating. In Petrograd, as the beautiful imperial capital overlooking the North Sea was now called, a veteran of the Great War had resumed his studies. Born in 1888 to entertainers, Alexander Friedmann cultivated many scientific interests in addition to a passion for mathematics and ballooning. He was working primarily on theoretical meteorology when, thanks to Ehrenfest, who had been his teacher and also maintained personal ties to Russia,[27] became aware of Einstein's work, now confirmed by Eddington. The year was 1922. After the bloody civil war, the peasant and worker uprisings triggered by the requisition of grain and the militarization of labor, which were bloodily suppressed, and a devastating famine due to an unusually severe drought, a new era seemed to be opening up, full of hope with the creation of the Soviet Union, the first socialist state in history, a clear symbol of the final victory of Bolshevik communism. Free from prejudice and fully able to tame the harsh formalism of relativity, Friedmann sought to find a solution to Einstein's equations for a universe as homogeneous and isotropic as the previous one, but relieved of the constraint of eternal paralysis. Dynamic equilibrium was the egg of Columbus. The cosmic repulsion imagined by

---

[27] A. Belenkiy, *"The Waters I am Entering No One Yet Has Crossed": Alexander Friedman and the Origins of Modern Cosmology*, 2013, https://arxiv.org/abs/1302.1498.

Einstein was no longer necessary, replaced, to Occam's delight, by ordinary inertia, invoked to compensate for gravity, as in the common experience of a stone thrown upward by an initial impulse. But beware! The example is as revealing as it is misleading. The stone is in fact moving in the pre-existing and completely autonomous space of Newtonian mechanics. Instead, what changes in general relativity is the structure of spacetime, which can shrink or grow, thus changing its "gravitational effect" on matter.

The solutions found by Friedmann produced an infinite number of models characterized by an origin in time and by a decelerated expanding metric[28] triggered by a mysterious initial impulse, later mockingly christened the Big Bang.[29] What made the difference was the value of the matter-energy densities. If it was higher than a certain critical limit, it implied a geometry of space similar to that of the sphere described by Gauss, where the total volume is finite and the parallel lines meet sooner or later. It entailed an expansion that, after some time from the beginning, reversed into a symmetric contraction. This behavior is similar to that of a stone thrown upwards which, having reached the top of its trajectory, stops going up and falls back. A density below the critical threshold, instead, is incapable of arresting the expansion, which, although the velocity continues to decrease, does not stop growing in an infinite space with a hyperbolic metric, where the parallel lines diverge at infinity.

And our beautiful everyday space, the one with zero curvature, where parallel lines do what Euclid preached, that is, remain equidistant without ifs or buts? Among Friedmann's solutions this is but one. It corresponds exactly to the value of the critical density that divides open (hyperbolic) from closed (spherical) models. It is a statistically improbable universe because it is unique among infinite possibilities. Ask yourself what probability you have of picking a given number, say 0, from an ideal (infinite) bag containing the entire set of integers. This universe expands indefinitely and never stops, tending asymptotically to zero velocity.

When Einstein read the paper that the Russian had sent for publication in the journal "Zeitschrift für Physik",[30] and which contained only the spherical model, he was very upset and reacted immediately in the same journal.

---

[28] Only for the first half of the (finite) life of closed models, which actually ends with a "Big Crunch" (from which a second identical cycle can start).

[29] H. Kragh, *Naming the Big Bang*, in "Historical Studies in the Natural Sciences", 44, 1, 2014, pp. 3–36.

[30] "Journal for Physics", a German peer-reviewed scientific journal founded in 1920 by Springer and published until 1997.

«The work cited contains a result concerning a non-stationary world», he sentenced, «which seems suspect to me. Indeed, those solutions do not appear compatible with the field equations».[31] A slap on the wrist given too hastily! Since Friedmann could not leave the Soviet Union, he first wrote a letter and then asked for the help from one of his friends, Yuri K. Krutkov, a theoretical physicist, who was able to corner the famous scientist and challenge his objections. The two met in Berlin in May 1923 on the occasion of Lorentz's farewell lecture. Put on the ropes, Einstein had to grudgingly admit that he had committed «a calculation error» and that «the results produced by "Herr Friedmann" were both correct and clarifying».[32] Quite a satisfaction. In a letter to his sister, Krutkov wrote «On Monday, May 7, 1923, I was reading, together with Einstein, Friedmann's article in the "Zeitschrift für Physik". [...] I defeated Einstein in the argument about Friedmann. Petrograd's honor is saved!».[33] But the unknown meteorologist with the aura of a Bolshevik intellectual could not enjoy this success for long. Two years later, in September 1925, typhoid fever took his life. He was only 37 years old.

Shortly thereafter, and quite independently, a Belgian priest rediscovered Friedmann's spherical solution. Georges Lemaître had come to astronomy by the high road.[34] Thanks to a scholarship he won after entering the Malines-Brussels seminary for adult vocations, he was able to move to Cambridge and attend Eddington's lectures: «a brilliant student [...] of great mathematical ability»[35] who soon became well acquainted with general relativity. He then landed at the Harvard Observatory and at MIT as a Ph.D. student in physics. In America he met great astronomers and heard directly from Henry Norris Russell, president of the American Astronomical Society, the enthusiastic account of the discoveries made by a certain Hubble, the man from the Far West who had succeeded in proving the extragalactic nature of some nebulae (cf. chapter 15). But not only that. He learned that in Flagstaff, on the south rim of the Grand Canyon, another unknown

---

[31] H. Gutfreund, J. Renn, *The Formative Years of Relativity: The History and Meaning of Einstein's Princeton Lectures*, Princeton University Press, Princeton (NJ) 2017, p. 81.
[32] Ivi, p. 82.
[33] Quoted in E.A. Tropp, V.Ya. Frenkel, A.D. Chernin, *Alexander A. Friedmann: the Man who Made the Universe Expand*, CUP, Cambridge 2006.
[34] D. Lambert, *The Atom of the Universe: The Life and Work of Georges Lemaître*, Copernicus Center, Kraków 2016.
[35] Words of Eddington, quoted by S. Mitton, *The Expanding Universe of Georges Lemaître*, "Astronomy and Geophysics", 58 (2), 2017, pp. 28–31.

scientist, Vesto Slipher, claimed to have measured the radial velocities of the same nebulae, finding them to be remarkably large and essentially all positive (i.e. recessive).

In 1925, Father Georges returned to Europe with a wealth of knowledge and observational suggestions, and was appointed to teach at the Catholic University of Leuven in Flanders. He would serve there for the rest of his life. Two years later, at the age of 33, he was ready to launch his attack on the cosmos with a solution of Einstein's equations consistent with Slipher's measurements. Thanks to the support of the cosmological constant, his time-dependent model of a homogeneous and isotropic universe with positive curvature emerged from an initial singularity. From there, it grew at a decreasing rate, then stabilized, and finally resumed an exponential expansion, in the manner of de Sitter, as a result of the dilution of the cosmic fluid. No real novelty compared to Friedmann — about whom Lemaître knew nothing until Einstein told him of this Russian maverick during a Solvay Conference[36] in 1927 —, apart from the fact that the Belgian was moving in the right circle because of his close connection with Cambridge. His solution was submitted to the father of relativity, now treated as a physics guru, who hastily commented: «your calculations are correct, but your physics is abominable».[37] Lemaître was not discouraged by the uninspiring response and deeply flawed judgment. Instead, he worked out two remarkable aspects of his model, also present in Friedmann's solutions, but ignored by the Russian mathematician (who had little time to work on them before his death). First, the question of the origin of time, which his friend and teacher Eddington disliked so much that he defined it «repugnant».[38] Let us follow Fr. George's reasoning. If the cosmos is expanding as time progresses, then if we turn back the hands of the clock, it will appear to be contracting, becoming smaller and smaller, until the length of the cosmic scale reaches zero. This is the moment when the "cosmic egg" — a term

---

[36] Organized by the International Solvay Institutes for Physics and Chemistry, founded by the Belgian chemist, industrialist, and philanthropist Ernest Solvay, the conferences, held in Brussels since the first one in 1911, deal with outstanding problems in physics and chemistry.

[37] A. Zee, *Einstein's Universe: Gravity at Work and Play*, Oxford University Press, Oxford 2001, p. 69.

[38] Cf. S. Appolloni, *"Repugnant", "Not Repugnant at All": How the Respective Epistemic Attitudes of Georges Lemaître and Sir Arthur Eddington Influenced How Each Approached the Idea of a Beginning of the Universe*, in "IBSU Scientific Journal", 5 (1), 2011, pp. 19–44.

coined to mock this idea — unfolds and gives birth to its creature, the "baby universe".

The scientific community greeted the proposal with caution. The introduction of an act of creation into secular cosmology, especially by a Catholic priest, could not fail to arouse some suspicion, not entirely unjustified given the strong interest in the subject shown by Pope Pius XII.[39] With great intellectual honesty, Lemaître tried to bring the debate back to the level of physics, freeing it from ideological and theological reservations. He clearly stated that his theory was in no way a mathematization of *Genesis* and that it should remain «completely outside any metaphysical or religious context».[40] Not for nothing had he been educated by the Jesuits! Eventually, Lemaître was believed by the former advocate of a static universe. In 1933, when they were both on a lecture tour in California, Einstein dared to tell him that his model was «the most beautiful and satisfying explanation of creation» he had ever heard.[41] A 180-degree turn by a great "natural philosopher" who also knew how to make amends for his mistakes without flinching.

The second key aspect of Lemaître's elaboration of his model concerns the discovery of a direct proportionality between the distance of a source from the observer and the speed at which it appears to be receding as a result of cosmic expansion. A legacy of the lectures he had heard in America, this law could have been immediately taken as a test of the theory. The Belgian priest presented it in a paper written in French, which, for this reason too, had little circulation. Science now spoke German and English. And when the restless Eddington finally had the paper translated into English (with the crucial passage deleted by Lemaître himself, no one knows why),[42] it was already too late. By now, Hubble had hit the jackpot,

---

[39] A controversial figure, Pope Pacelli, scion of a noble and very conservative family, led the Catholic Church through a stormy period for the world by taking strong positions only on theological issues.

[40] J. McCann, *Georges Lemaître: the Priest Who Proposed the 'Big Bang'*, in "An Irish Quarterly Review", Vol. 105, No. 418, 2016, pp. 212–224.

[41] D. Topper, *How Einstein Created Relativity out of Physics and Astronomy*, Springer, New York 2013, p. 175.

[42] In the original paper (G. Lemaître, *Un univers homogène de masse constante et de rayon croissant, rendant compte de la vitesse radiale des nébuleuses extragalactiques*, in "Annales de la Société Scientifique de Bruxelles", 47, 1927, pp. 49–59), the indicted sentence read: «*Utilisant les 42 nébuleuses extra-galactiques figurant dans les listes de Hubble et de Strömberg, et tenant compte de la vitesse propre du Soleil, on trouve une distance moyenne de 0,95 millions de parsecs et une vitesse radiale de 600 km/s, soit 625 km/s à 10^6 parsecs. Nous adopterons donc $R'/R = v/rc = 0,68 \times 10^{27}$ cm^{-1} (Eq. 24)*»

thanks to the firepower of the Mount Wilson telescopes and the ability to promote his own image and research. Focused more on understanding the sky than exploring it, Europe had missed its chance.

In fact, the idea of a relationship between distance and recession velocity has been circulating for some time. It was promoted by the first spectroscopic observations of spiral nebulae, mostly from the United States. As early as the second decade of the twentieth century, Slipher had collected the spectra of 25 objects, noting, as we have said, the peculiar excess of positive velocities in his sample and their vague correlation with the distances of the sources as inferred from the apparent sizes and brightness of the nebulae. Slipher worked at the observatory that Percival Lowell, scion of a wealthy and cultured Boston family, had built for himself in 1894 on Mars Hill on the outskirts of Flagstaff, Arizona. It was a magnificent and wild place, perched on the Colorado Plateau at 8,000 feet, nestled in an endless coniferous forest bounded to the north by the Grand Canyon and beyond by the towering white crests of the San Francisco Peaks. Another true icon of the Old West, where even today it is not difficult to find, mingled with the "palefaces", the enigmatic faces of the Native Americans.

Born in 1855, Lowell showed a real passion for astronomy from an early age. As an adult, after graduating from Harvard in 1876, he spent half his time tending to the family investments in Boston and the other half, between 1983 and 1894, traveling in the Far East, where he began to test his skills as a keen observer and writer.[43] Then, on the threshold of forty, he decided to give his soul, body, and purse to Urania. His conversion had begun with reading the English translation of the book *Astronomie populaire* by Camille Flammarion, the seductive French popularizer. But the real impetus came when Lowell accidentally came across some fascinating observations of Mars made by the director of the Brera Observatory, Milan, during the Great Opposition of 1877. These are the facts.

A few years after the first unification of the peninsula under the Savoy crown, in 1866, Giovanni Virginio Schiaparelli had received a generous grant

---

(Using the 42 extragalactic nebulae listed by Hubble and Strömberg, and taking into account the proper velocity of the Sun, we find an average distance of 0.95 million parsecs and a radial velocity of 600 km/s, that is 625 km/s at 10^6 parsecs. We will therefore adopt $R'/R = v/rc = 0.68 \times 10^{27}$ cm^{-1} (Eq.24)). In the English translation published in the "Monthly Notices of the Royal Astronomical Society" in 1931, it became: «From a discussion of available data, we adopt $R'/R = 0,68 \times 10^{27}$ cm^{-1} (Eq. 24)», where the reference to the recession velocity had disappeared.

[43] A.L. Lowell, *Biography of Percival Lowell*, MacMillan, New York 1935.

from the Italian Parliament to purchase a refractor telescope with an objective of 22 cm aperture and 3 m focal length from the trusted firm of Georg Merz in Munich.[44] An excellent medium-sized instrument that arrived just in time to look at the Red Planet in the best conditions, when Mars was in great opposition, at the shortest distance from the Earth. This event occurs about every 15 years, because, in addition to the fact that the Sun and the planet are in diametrically opposite positions (it happens every synodic period of about 780 days), a favorable placement of the two celestial bodies in their respective elongated orbits is also required (Earth must be close to aphelion and the planet to its perihelion). Aware that he was playing a game with many famous and well-equipped competitors, Schiaparelli had literally lived attached to the eyepiece, filling countless sheets of paper with drawings of what he saw. This promising research continued for years. Eventually, the astronomer reported that he had detected «depressions in the ground [of Mars] not very deep, extending in a straight direction for thousands of kilometers, over a width of 100, 200 kilometers or even more».[45] The features would later prove to be optical illusions.

Not wanting to go too far, Schiaparelli only commented on the alleged scoop by referring to geological structures modified by seasonal climatic variations: «these channels probably represent the main mechanism, through which water (and with it organic life) can spread over the dry surface of the planet».[46] But the issue was too big and greedy not to stir controversy and at the same time stimulate fanciful speculation, at a time when the inauguration of the Suez Canal in 1869 had popularized hydraulic megastructures; not least because of a naïve English translation of the Italian text in which the use of the term "canal" instead of "channel", as Schiaparelli intended, implied the intervention of an alien intelligence. Martians became more fashionable than ever, loved and feared by a science fiction audience; and Lowell set about to root them out.

This was no mere expensive whim, but a true missionary obsession, of which the Arizona observatory was the most successful and fruitful product. While vainly peering at Mars through the formidable 61 cm refractor built for him by Alvan Clark & Sons and writing books about the little green men,

---

[44]L. Botta (ed.), *Giovanni Virginio Schiaparelli. L'uomo, lo scienziato*, Cristoforo Beggiami, Savigliano (CN) 1983 (in Italian).
[45]G.V. Schiaparelli, *La vita sul pianeta Marte*, in "Natura e Arte", IV, fasc. 11, 1895, p. 3 (in Italian).
[46]Ibid.

Lowell recruited to his team a mild-mannered young man from Indiana who could think with his head without ever raising it too high. Vesto Slipher was born in 1875 to poor farmers: a hardscrabble childhood that would give him a strong, healthy body. This is how two younger colleagues described him when he was now over 60 years old: «V.M. Slipher, thirty-five years my senior, was always ahead of us "boys" climbing the mountain — we puffing and panting and he, disgusted, waiting for us to catch up», and «V.M. at age sixty-five could chop wood with the best of them».[47] He was already feeling lucky about his new life as a scientist.

Not seeking glory, he meekly did what the director-master told him to do, and so he obeyed without a murmur when, in 1909, he was assigned to observe nebulae. It had happened that Lowell, distressed by repeated failures in his hunt for Martians, had decided to temporarily suspend the project in order to catch his breath and raise some money to launch another crusade. His goal now was to find Planet X, which he thought should sail slowly and coldly beyond the known boundaries of the Solar System, torturing Neptune's orbit a bit: as if to say, what's done is rendered! Lowell dreamed of a quick success that would enhance his scientific reputation and attract important collaborators to Flagstaff with whom he could revive the Mars crusade. Although he planned to circumscribe the area to be surveyed by using Neptune's orbital perturbations in the manner of Le Verrier and Adams (see chapter 8), he still needed a telescope with a large field of view to explore the vast haystack of the sky in search of the tiny needle. This abrupt turn in the observatory's historic mission freed up much of the observing time on the Clark refractor, which was converted to an astrophysics instrument, equipped with a spectrograph, and given to Slipher with a program focused on nebulae.

At the time, capturing the spectra of these diffuse objects was no joke, even for those with access to large-aperture telescopes. To see any trace of a signal required grueling exposures of many hours, often protracted over several nights. And not infrequently, at dawn, after a sleepless night in the cold, astronomers would realize with despair that they had worked for nothing. The plate, properly developed, fixed, and washed, showed nothing useful. But sometimes it went better. The spectrum showed one or two emission lines or even a Fraunhofer line in absorption on a faint continuum trace. In this case, it became possible to measure the radial velocity of

---

[47]W.G. Hoy, *Vesto Melvin Slipher (1875–1969). A Biographical Memoir*, National Academy of Sciences, Washington (DC) 1980.

the source by invoking the Doppler effect and assuming that the shifts of the lines from the laboratory positions (provided by a standard spectrum) resulted from the radial motion of the source relative to the observer. A bold conjecture[48] at the time which did not fail to raise doubts, and which only future would confirm definitively, as the case of quasars, which we will introduce in the last chapter, masterfully demonstrates.

Lacking any previous experience, Slipher chose his targets more or less randomly, learning from mistakes. He also tried to interpret the wealth of data he was accumulating. He discovered, for example, that some nebulae had internal motions,[49] which he later attributed to rotation: a first step toward mass measurements.[50] In particular, he noticed the considerable asymmetry in the distribution of systemic radial velocities[51] in his sample, with a significant excess of receding motions compared to approaching ones. But when he tried to draw conclusions, he had to deal with the pitfalls of statistics based on small numbers. The apparent correlation between the directions of the velocities and the positions of the sources in the sky led him to believe that he had detected the apical motion of the Sun in the Milky Way: a plausible conclusion, already successfully proposed by Herschel using stars, but in this case wrong. Nevertheless, this work earned him a standing ovation at the summer meeting of the American Astronomical Society in 1914, just as the Austro-German troops were testing the war plan prepared by strategist Alfred von Schlieffen on the French Maginot Line, thus initiating a senseless slaughter.

As for the intriguing inverse correlation between recessive velocities and angular sizes of sources, Slipher saw it but did not do much with it. The observation anticipated, if only slightly, the birth of modern cosmology, and left the astronomer with no interpretive suggestions to support the

---

[48]The Doppler hypothesis was supported by the fact that the shift was not constant, but proportional to the wavelength; a property that could be verified when there were multiple lines in the spectrum.

[49]Their spectral lines were not straight, as the shape of the spectrograph slit would suggest, but showed different shifts (radial velocities, in Doppler terms) at different positions along the major axis of the image of the nebula.

[50]When comparing gravitational attraction and centrifugal reaction under the assumption that spectral shifts track the rotational velocity of nebulae, mass is the only unknown. The rest (apparent size and distance) can be provided by observations.

[51]The systemic velocity of a set of gravitationally bound objects is that relative to the center of gravity. In the case of regular and symmetric nebulae, the center of gravity was thought to fall at the point of maximum light density, usually coinciding with the geometric center of the system.

empirical law that a handful of data seemed to indicate. It is true that Slipher missed the boat, but it was not for lack of courage or imagination on the part of this shy and reserved man, who really found himself in the right place at the wrong time. Carl Wilhelm Wirtz, much better educated than Slipher, behaved in the same way; then, after Hubble's discovery of the expansion of the universe, in 1936 the German scientist weakly tried to assert his role as a pioneer.[52] Some other astronomers around the world did the same. They saw the phenomenon, commented on it, but did not know what to make of it.

How could they be blamed? Nobody knew yet the rank and role of spiral nebulae. In 1890, an authoritative historian of astronomy, the Irishwoman Agnes Clerke, had pronounced a peremptory verdict, which she repeated in 1905: «The question whether nebulae are external galaxies hardly any longer needs discussion. It has been answered by the progress of research. No competent thinker, with the whole of the available evidence before him, can now, it is safe to say, maintain any single nebula to be a star system of co-ordinate rank with the Milky Way».[53] Instead, in 1911, the British astrophysicist Arthur Eddington, not yet in his thirties, commented on the hypothesis that the spiral galaxies were instead "island universes": «If confirmed, the hypothesis opens to our imagination a truly magnificent vision of system beyond system [...] in which the great stellar system of hundreds of millions of stars (our galaxy) [...] would be an insignificant unit».[54] The only established data concerned the mixed nature of these

---

[52]C.W. Wirtz, *Ein literarischer Hinweis zur Radialbewegung der Spiralnebel*, in "Zeitschrift für Astrophysik", 11, 1936, p. 261 (in German). Wirtz should also be remembered for introducing the concept of the "K-correction" into cosmology. The idea is as follows: due to displacement, a source moving radially shows a spectrum that is both stretched and shifted with respect to laboratory conditions. These two phenomena must be taken into account when estimating monochromatic magnitudes, i.e. fluxes inside a finite interval of wavelengths (in jargon, the bandwidth).

[53]D.W. Sciama, *The Unity of the Universe*, Dover Publications, Mineola (NY) 2009, p. 48. In this context, it seems interesting to report the remark of the American astronomer Franck W. Very at the end of his article on *The nebula about Nova Persei 1901*, published in 1902 in "Astronomische Nachrichten" (Vol. 158, 3, pp. 33–42). It shows how two parties coexisted, with real supporters of one thesis and the other: «In conclusion, I desire to subscribe the opinion of Professor Cleveland Abbe, as expressed by him in the Monthly Notices of the Royal Astronomical Society for 1867 (vol. 27, p. 257), that the true nebulae are galaxies, or complex aggregations like our own Milky Way, and are composed of stars (either simple, multiple, or in clusters) and of gaseous bodies of both regular and irregular outline».

[54]Cited in J.S. Tenn, *Arthur Stanley Eddington: A Centennial Tribute*, in "Mercury", Vol. XI, No. 6, 1982, p. 179.

objects, containing both stars and gas, and the peculiar tendency to move away from the Sun at velocities much greater than those characteristic of individual stars. The matter had to be clarified as soon as possible, for the benefit of cosmologists and also of the general public, which, especially in America, was becoming more and more informed about the progress of research and more and more eager to know what world they were living in. So much so that in 1920 the scientific controversy over spiral nebulae became the subject of a famous and notable public debate, although its notoriety may have been exaggerated compared to its merits. Rather, it responded to the desire to link the discovery of galaxies to the burgeoning development of American society and to the need of the organizer, George Hale, to better understand how to translate the question into an observing program.

In 1913, George, along with his brother William and his sister Martha, had donated the conspicuous sum of $1,000 per year (equivalent to 20,000 euro today) to the Washington Academy of Sciences for the purpose of organizing a cycle of five yearly conferences in memory of their father William, who had passed away 15 years earlier. The intention was «to give clear and comprehensive outline of the broad features of inorganic and organic evolution in the light of modern research».[55] For the last lecture of the series, which had been opened by a prestigious lecture by Sir Ernest Rutherford, George had in mind to discuss the extravagant theory which a German of Jewish faith had published during the war years. However, wishing to please the secretary of the Academy, who considered Einstein a visionary,[56] he dropped the idea and proposed that his father be celebrated with a debate on the size of the universe. But the secretary had to object to this suggestion as well. «I have a sort of fear, however, that the people care so

---

[55] *General Notes*, in "Publications of the Astronomical Society of the Pacific", 26, 154, 1914, p. 164.
[56] The general secretary of the Academy, Charles Greeley Abbot, who was an astrophysicist, expressed the following opinion: «As to relativity, I must confess that I would rather have a subject in which there would be a half dozen members of the Academy competent enough to understand at least a few words of what the speakers were saying if we had a symposium upon it. I pray to God that the progress of science will send relativity to some region of space beyond the fourth dimension, from whence it may never return to plague us»; D.H. Clark, M.D.H. Clark, *Measuring the Cosmos: How Scientists Discovered the dimensions of the Universe*, Rutgers University Press, New Brunswick 2004, p. 69.

little about island universes, notwithstanding their vast extent»,[57] he had commented ironically. He recommended that they choose a topic in geology or biology, which he thought would be more appealing to the general public. This time Hale resisted. After all, he was footing the bill. If not Einstein, it had to be his beloved astronomy to take the stage for an afternoon of debate between a proponent of the reductive thesis that all nebulae are components of a single star system, the Milky Way, and a supporter of Kant's thesis of a vast archipelago of countless galaxy-like islands. Two astronomers from the great California mountain observatories were chosen as the champions of this unique contest: Harlow Shapley for Mount Wilson and Heber Curtis for Lick (the latter joined the joust after Lick's director, William Campbell, withdrew).

The 35-year-old Shapley certainly seemed the more folkloric of the two contenders.[58] His round face, like that of an eternal child, and his strong accent betrayed a peasant background. In fact, he was born on a farm in the rural and deeply conservative heart of the Midwest, a land of "redskins" and farmers, ranchers, and boatmen on the two great rivers, the Mississippi and the Missouri. His childhood reminds us of the pages full of Sun, dust and puddles of *Tom Sawyer*. After dropping out of elementary school in rural Nashville, Harlow gained experience as a reporter for a local newspaper. Returning to the classroom and completing his secondary education, he had decided to take a college course in astronomy while waiting for enrollment in journalism school to open: only because astronomy was the second of the disciplines listed alphabetically, the first being archaeology, a name — he later joked[59] — he couldn't even pronounce. A temporary parking space that would instead become a lifelong love.

After graduating, Shapley was awarded a scholarship to Princeton University, an old and prestigious colonial college in New Jersey. The head of the small but prestigious astronomy department was Henry Norris

---

[57]M.A. Hoskin (ed.), *The "Great Debate": What Really Happened*, in "Journal for the History of Astronomy", 7, 1976, pp. 169–182.

[58]B.J. Bok, *Harlow Shapley (1885–1972): A Biographical Memoir*, National Academy of Sciences, Washington (DC) 1978; D. West, *Harlow Shapley. Biography of an Astronomer: The Man Who Measured the Universe*, CreateSpace Independent Publishing Platform, Scotts Valley (CA) 2015.

[59]In fact, Shapley had always been a lover and devotee of the humanities; cf. K.F. Mather, *Harlow Shapley, Man of the World*, in "The American Scholar", 40, 3, 1971, pp. 475–481.

Russell, who, along with the Danish astronomer Ejnar Hertzsprung, had just discovered the diagram linking the intrinsic luminosities of stars to their colors: a kind of Rosetta stone, essential for deciphering stellar evolution. For this work, carried out independently, both astronomers had used the distance indicator newly discovered by Henrietta Snow Leavitt: the relationship between the pulsation period and the average luminosity of the Cepheid variables (p. 315). It was the aristocratic, Presbyterian Russell who introduced Shapley, a Missouri farm boy, to this magical instrument that would soon make him famous and, paradoxically, be the architect of his ultimate defeat. The story is a bit of a mystery, but full of fascination and consequence. It began in 1914, when Shapley, having completed his apprenticeship at Princeton, obtained a position on Hale's staff in Pasadena.[60]

He brought with him to California a project suggested by Solon Bailey, the man who had built and run the Arequipa Observatory (p. 314): to study the family of those dense, compact clusters of many tens of thousands of stars called "globulars" because of their appearance. Harlow was already familiar with a few dozen of them, a third of which were projected onto the constellation of Sagittarius, in the dusty heart of the Milky Way. Like all the young men at the observatory, he was assigned some duties, for which Hubble did not discount anyone. But he was also given enough freedom for his own research, to which he could devote his share of observing time at Mount Wilson's large refractors. He planned to complete his survey of globular clusters and to observe RR Lyrae, short-period pulsating stars also known as "cluster variables" because they are endemic to these star systems. He had chosen them as standard candles in the belief that their intrinsic luminosity was some kind of universal constant. He was wrong, but not by much.

It was still necessary to calibrate these stars, and that's when the devil decided to put his tail on this thing. Shapley knew that RR Lyrae might be related to some rare long-period Cepheids also found in globular clusters, and devised the following strategy of attack. First, he managed to calibrate some of these stars close enough to estimate their distances using geometric methods. Then he assumed that this small sample was representative of the cluster variables, given the similarity of the light curves, and so the game was played. The procedure was plausible *a priori*, but in practice it proved

---

[60]B.J. Bok, *Harlow Shapley, Cosmographer*, in "The American Scholar", 40, 3, 1971, pp. 470–474.

to be incorrect. The cluster Cepheids, while morphologically similar to the classical Cepheids, are somewhat fainter. They are low-mass stars going through the same phase of instability as the other pulsating variables, but at a later stage in their lives and with a very different internal structure. Older ladies that still look girlish in some features, but have lost some of their luster due to age.

Unaware of the devil's trap, Shapley made his first serious mistake by trading candles for lanterns, which greatly lengthened his distance scale. Fortunately for him, however, and in spite of the devil who had concocted the Cepheid trap, he made a second mistake that somewhat compensated for the first. In order to estimate the distances of 11 classical Cepheids, he used the so-called "statistical parallax", a geometric indicator which, unlike the apparent brightness associated with it, is not affected by any possible extinction of light. Convinced that space was perfectly transparent, Harlow neglected interstellar absorption. The result was a short calibration which acted on his distance scale in the opposite direction to the error made with the Cepheid type, shortening it and partly putting things back in place. The matter of the Cepheids would be clarified a quarter of a century later, in the years of the Second World War, by the German Walter Baade, then working in the United States, with spectacular observations of the M31 nebula, which we will discuss in chapter 17. As for extinction, a problem well known to astronomers since the 18th century, in 1930 Robert Trumpler, a Swiss-born American astronomer, explained the reddening of stars by the presence of interstellar dust within the galactic disk. Light struggles to pass through and some of it, especially in the blue region of the spectrum, is lost along the way, making the celestial sources appear fainter (and redder) and thus more distant.

Using the distances provided by his somewhat tortuous construction, Shapley was able to work out the distribution of globular clusters in space. The result astonished him! These compact star systems were arranged to form a spheroidal cloud centered in Sagittarius, at a point he estimated to be 50,000 light-years away. With the kind of high note that distinguishes a great tenor from a mere chorister, the 33-year-old Harlow then ventured the hypothesis that the center of the globular cluster distribution coincided with that of the galactic disk. But what about the Sun? While leaving it in the plane of the disk where Herschel and Kapteyn had placed it, Shapley assigned the star a huge distance from the heart of the vast stellar system, too far even by modern measurements (26,000 light-years). It was Copernicus' revenge on those who, having lost the game of the Solar System, had tried to restore man to a pivotal position in the cosmos. It was a great and courageous insight that, with the proper corrections, still forms the paradigm of the local universe today. But Shapley did not stop there. Dazzled by some false clues and by his own prejudices, he thought he could win big with the proposal that his Milky Way, a disk with an extension of 300,000 light years and a thickness of 30,000, represented the cosmos itself since it included all other celestial objects: stars, stellar clusters, and nebulae of all kinds. This was the thesis that he enthusiastically accepted to defend[61] at the "Great Debate" scheduled for 1920 at the National Academy of Sciences, proud that Hale, his boss, had chosen him for the duel, and hopeful that this showcase would help him advance his career. The death of Pickering, the father-master of the Harvard Observatory, had opened up a prestigious position and Harlow, certainly not lacking in ambition, had given it some thought.

The other contestant, instead, had accepted the invitation to confrontation rather coldly. In contrast to Shapley's unyielding certainties about his own abilities and achievements, Heber Doust Curtis[62] maintained a healthy doubt, fueled by some seemingly inexplicable facts. For about ten years he had been photographing nebulae with the Crossley telescope, trying to get something out of them.[63] It was the discovery of some novae in spirals such as NGC 6946, the Fireworks Galaxy between the northern constellations of Cepheus and Cygnus, that made him choose the field of supporters of the island universe concept. Looking back at the M31 plates in the Mount Wilson archive, he had found 6 other novae, but none as bright as S And (see p. 301). These stars were 100,000 times fainter than Nova Persei; just enough, he thought, to conclude that the host spirals must be well outside our stellar system. But if M31 is so far away, Shapley replied, turning the argument on its head, it must be 100 times brighter than a typical galactic nova, and that fact would be astonishing, to say the least! The vaguely sarcastic remark did not bother Curtis as much as the result of

---

[61] Shapley model was received positively by the astronomical community. Eddington, for instance, said: «I think it is not too much to say that this marks an epoch in the history of astronomy, when the boundary of our knowledge of the Universe is rolled back to hundred times its former limit». Ibid.

[62] R.R. McMath, *Heber Doust Curtis, 1872–1942*, in "Publications of the Astronomical Society of the Pacific", 54, 138, 1942, pp. 69–71.

[63] The plates collected at Lick Observatory between 1895 and 1913 showed 700,000 nebulae; a number to be compared with the 8,000 objects listed in the NGC catalog, which was compiled without the aid of photography.

the astrometric work done in Pasadena by a detached Dutch astronomer, Adriaan van Maanen.[64]

Adriaan passed for a real ace when it came to proper motions. Hale had hired him on the recommendation of Kapteyn, who now held the keys to his heart. A godsend for George Ritchey, who, busy polishing the mirrors of Mount Wilson Observatory, had neither the time nor the expertise to put into practice an idea that had been rattling around in his head for some time: to compare two plates of the M101 spiral taken five years apart and see if the internal radial motions discovered spectroscopically by Slipher and Wolf could also be visualized as angular displacements. He then asked the Dutch to look into the matter. Van Maanen set to work immediately. Soon after, having completed the measurements of the positions of about twenty details in the image of the nebula on the pairs of plates, he announced that he had found the sign of a clear rotation of M101 on the plane of the sky. This was only the beginning of the saga. Within a few years, the pieces of evidence he produced multiplied, crossed with the spectroscopic detection of rotational velocity components along the line of sight, thus putting the proponents of the island universes in serious trouble. The velocities required to produce the observed angular drifts would even have to exceed the speed of light if the nebulae were located far beyond the boundaries of the Milky Way. An absurdity! For the survival of the Kantian model, one could only hope that van Maanen had made something wrong.

On Monday, April 26, 1920, at the Academy's annual meeting, the great confrontation between the two cosmological theses took place;[65] a kind of modern, democratic, live version of the Galilean *Dialogue concerning the two chief world systems*. Interviewed half a century later, Shapley — who, unlike Curtis, did not want to discuss the subject of nebulae — recalled that afternoon as follows:

> I think won the "debate" I was right and Curtis was wrong on the scale of the universe. It is a big universe, and he had it as a small thing [...] Curtis won as to the spirals [...] I had the misfortune of finding some blue stars in the center of the Milky Way [...] On that too-flimsy basis, I assumed that there was no absorption in space, or very little [...]. I call that my

[64]F. Seares, *Adriaan van Maanen, 1884–1946*, in "Publications of the Astronomical Society of the Pacific", 58, 1946, pp. 89–103.

[65]R.W. Smith, *The Expanding Universe: Astronomy's "Great Debate"*, 1900–1931, CUP, Cambridge 1982.

blunder because I faithfully went along with my friend, Van Maanen. He was wrong on the proper motions of galaxies, that is, on their cross-motions. Therefore I was off-beat for a while, whereas Curtis and Hubble and some others discredited the Van Maanen measures and questioned his results. I stood by Van Maanen. That was the blunder — if you call it that way. I mean that I followed a false trail for a while. Now I've been complimented more than once on the fact that when I found it was false, I switched immediately to the truth. I was the leading supporter of Hubble and Curtis after we got onto what was what — namely, that there's enough absorption of light in space to dim down my clusters. They were not as far away as my observations had indicated.[66]

A somewhat disjointed story, as are all speeches reported in their entirety, without any reworking of the original text, and especially tamed by hindsight. In fact, the debate ended in a draw, with neither winner nor loser. Each of the contenders remained of his own mind, though not for long. For his part, Hale had fully achieved his hidden purpose: to clear his head about the scheduling of observations at the 60 and 100 inch telescopes. «We are planning an extensive attack on spirals, with special reference to internal motion, proper motion, spectra of various regions, novae, etc.», he had written to Campbell, director of Lick Observatory, late in 1919, «and here again *I should be glad to know what Curtis has in hand, so that our work may fit in with it to advantage*»[67] (emphases supplied). We shall soon see how this great patron of American astronomy would put into practice the lesson he had just learned for the benefit of knowledge of the deep cosmos.

The Great Debate gave Shapley the coveted promotion.[68] His brilliant performance was reported to the President of Harvard University, who the following year offered him the chair of astronomy at the prestigious institution and shortly thereafter the directorship of the observatory. This goal was achieved despite some handicaps: being too young and competing with his teacher, the powerful Russell, without the support of Hale, who feared that the new assignment might compromise the scientific efficiency of his

---

[66]Cf. *Oral History Interviews: Harlow Shapley* (1966), https://www.aip.org/history-programs/niels-bohr-library/oral-histories/4888-1.

[67]Hoskin, The "Great Debate", cit., p. 169.

[68]O. Gingerich, *How Shapley came to Harvard: Snatching the Prize from the Jaws of Debate*, in "Journal for the History of Astronomy", 19, 3, 1988, pp. 201–207.

thoroughbred. But it was also a Pyrrhic victory. Sitting at his predecessor's octagonal table, which rotated like the drum of a Colt to allow a quick transition from one project to the next, Shapley would soon see his world system dismantled by his greatest rival, Edwin Powell Hubble, whom he himself had paradoxically helped to arm.

Four years his junior, Hubble was also from Missouri, but had grown up in a big city.[69] Shortly after his birth in 1889, his father, a stern and domineering lawyer in the fire insurance business, had in fact moved his large family to Chicago, in a very different environment from the Midwestern ranch where the teenage Harlow had spent his younger years. As he grew up, Edwin became a tall, well-built, handsome boy who, despite his brilliant grades in all subjects, was especially noted in school for his athletic prowess. He practiced high jump, track, basketball, and boxing, and he must have been really good if, as legend has it, he was offered to fight for the world heavyweight title with the then-champion, black Jack Johnson. It was his excellence in sports that earned him first the scholarship with which he graduated from high school, and then the prestigious Rodhes Scholarship to attend Queen's College in Oxford. Here, to fulfill a promise to his dying father, he chose to study law, relegating his precocious passion for astronomy to spare time.

After graduating from law school, he returned to America in 1913, now completely British in manner, accent, and dress. To support himself, he taught high school physics and Spanish and coached basketball. He also practiced law for about a year after passing the bar exam, but without much enthusiasm. Then, at the age of 25, he decided to follow his first love, Urania. He enrolled in college again, this time at his home in Chicago, and in three years of hard work, thanks to diligent attendance at the Yerkes Observatory, earned the degree of Doctor of Philosophy with a thesis on a photographic study of faint nebulae. It was a kind of premonition. He wanted to work as an astronomer, but there was no place for him at Yerkes. Not much of a problem, though. He had already been spotted by Hale, who continued to visit Chicago. Being the extraordinary connoisseur of men that he was, the patron offered to bring him to California. The new giant 2.5 m telescope was about to open its eye of Polyphemus on the summit of Mount Wilson, and George needed young and healthy astronomers to make it work at full speed. Meanwhile, the Great War had broken out in

---

[69] G.E. Christianson, *Edwin Hubble: Mariner of the Nebulae*, Farrar, Straus & Giroux, New York 1995.

Europe. In deference to Monroe's doctrine, which encouraged Americans to mind their own business, the United States had watched from afar the bloody stalemate of modern European armies, perfectly equipped to sow death and nailed to rot in opposing trenches. But by early 1917, in response to indiscriminate attacks by German U-boats and to defend America's own economic and financial interests overseas, Congress had decided to abandon neutrality and take the field in support of the *Entente*.

Eager to enlist as a volunteer, Hubble hurried to graduate. After signing up for the infantry, he sent a laconic telegram to Hale, who was waiting for him with impatience: «Regret cannot accept your invitation. Am off to the war».[70] Knowing the character, it is likely that Edwin saw the path of war as more direct and faster than that of science to make a name for himself. His enormous ego could benefit from a lucid and lively intelligence, a good dose of opportunism, a peculiar skill in self-promotion, and a genuine disposition to fight, reinforced by sports.[71] But you can't win them all. The training at home took too long, and when the brilliant reserve major was finally sent to France, it was too late to fight. The conflict was drawing to a close, and with the impending peace, the dreams of glory of the young American warrior also vanished. No honors on the field. Even a minor shrapnel wound, if there ever was one, was caused by an accident in the barracks, not by the fury of battle. In any case, Hubble remained in the service for a year, shuttling between Germany and England on legal assignments, but also frequenting scientific circles. During this time he refined his character and developed the British officer's style that would characterize him for the rest of his life: high boots, walking stick, and cardigan, cleverly combined with the somewhat snobbish accent he had acquired during his Oxonian stage, with a taste for pipes, dogs, and fly-fishing.

After a farewell to arms and an internship at Cambridge, he returned to the United States and in 1919 finally accepted the position on Mount Wilson that Hale had kept warm with his incomparable nose. For the ambitious man that he was, the sacrifice of living on an isolated and wild peak was well rewarded by the extraordinary opportunity to use the largest telescope in the world. A fortune that the Major — as everyone now called him — never let go of. Nor did he miss the beautiful Grace Burke-Lieb, daughter of an influential Southern California millionaire and young widow, whom Edwin

---

[70] S. Hughes, *Catchers of the Light: The Forgotten Lives of the Men and Women Who First Photographed the Heavens*, Independent Publisher, Chicago (IL) 2013, p. 901.
[71] Christianson, *Edwin Hubble*, cit., pp. 1–420.

married in 1924. It was a perfect partnership, destined to fabricate and consolidate the myth of the brilliant and reclusive explorer of the skies in scientific circles and public opinion, in the media and in the salons of nearby Hollywood. Grace, an emancipated and vivacious woman, would take care of public relations, even volunteering to escort distinguished guests up and down the mountain to visit her husband.

In addition to the scientific facilities, the logistics of the observatory included the so-called "Monastery":[72] a well-organized system of separate living quarters for those who worked during the day (clerks, technicians, and solar astronomers) and at night (sidereal astronomers and their assistants), a few offices, a library, a kitchen, and the famous dining room, where the ritual of meals was consumed with a strictly hierarchical liturgy. At the head of the table sat the astronomer who was to observe that night with the 100 inch; a privilege of role rather than rank, blatantly reinforced by the right to scan the rhythm of the courses through a bell. What a difference from the carefree and irreverent style of American society depicted by novelists and the nascent film industry. For pragmatic training at the Mount Wilson facilities, Hubble was entrusted to Milton Humason,[73] a kindly man who had arrived on the mountain at the age of fifteen, driving mules to carry the 60 inch pieces. Then, through marriage to the daughter of the site's chief engineer, he had obtained a job as a janitor. Willing and curious, he had started to visit the domes at night and help the astronomers. Noticing his natural talent as an observer, Hale had not hesitated to promote him to the scientific staff, despite the protests of his more gallant colleagues. Humason fell in love with the Major at first sight, and Edwin quickly realized that he had found in "Milt" the perfect collaborator: an intelligent and effective arm to complement his fertile and unprejudiced mind.

Hubble chose spiral nebulae as the subject of his research at the suggestion of Hale himself, who had been impressed by the inconclusive results of the Great Debate. A field with room for anyone with imagination and dedication, an El Dorado in perfect harmony with a lone, unblemished, and fearless knight of the kind of celluloid heroes Hollywood was beginning to promote. He started photographing larger objects with the Hooker telescope, looking for the end of a red thread to which he could attach himself to turn a simple observation into a project. Finally, in 1923, the

---

[72] A brief history of "the Monastery, accompanied by numerous photographs, is available in the form of video clips at https://www.mtwilson.edu/vt-the-monastery/.

[73] R.L. Voller, *The Muleskinner and the Stars: The Life and Times of Milton La Salle Humason, Astronomer*, Springer, New York 2016.

fish fell into the net. A star peeked out of a disk of the Great Nebula in Andromeda. Without hiding his excitement, as befits a British gentleman, Hubble marked the object's position by hand and wrote a dry "Var" (for variable) on the side. It was not the first variable he had serendipitously seen in M31, but this time the source was not a simple nova, a category of stars disqualified by the S And anomaly. The light curve clearly pointed to a Cepheid. A few years earlier, some Cepheid variables had been observed in spiral M22 by John Charles Duncan on Mount Wilson, but the astronomer had not realized the significance of his discovery. Similar observations had been made in Germany by Wolf and Baade. But none of them had had the legacy of Shapley's work.

With the Cepheids as cosmic gauges, Hubble obtained an approximate distance to the nebula through the method developed by Shapley while this latter was still working in Pasadena. The value that came out was much larger than the admittedly enormous size of the new Harvard professor's Milky Way. Now the Major knew what to look for. Within a year, he discovered 11 more Cepheids in M31 and 22 in the Triangle Nebula, which gave the same distance of 900,000 light-years for both objects. The news leaked out among insiders and even reached the editorial offices of leading newspapers. The Mount Wilson astronomer had produced solid evidence that at least these two spirals should be considered of the same nature as the galaxies imagined by Kant.[74] From the clinic where he was hospitalized for aggravation of his chronic illnesses, Hale gloated.

In fact, the Estonian astronomer Ernst Julius Öpik had reached the same conclusion for M31 two years earlier. However, his subtle dynamical argument[75] was not taken too seriously, even though the paper had appeared in the "Astrophysical Journal", the publication founded by Hale. But by then, the extra-galactic hypothesis was knocking hard at the door, and fewer and fewer people remained in doubt. Among them, paradoxically, was Hubble himself. Hanging over his work like a sword of Damocles[76] were

---

[74] J. Gribbin, *Galaxies: A Very Short Introduction*, Oxford University Press, Oxford 2008.

[75] For a detailed analysis, see E. Opik, *An Estimate of the Distance of the Andromeda Nebula*, in "Astrophysical Journal", 55, 1922, pp. 406–410.

[76] Anecdote about a certain Damocles who, envying the power of a tyrant, was invited to take his place. During the big investiture party, Damocles noticed a sharp sword hanging above the throne, suspended by a thin horsehair, and understood the message: power brings with it great dangers that are always lurking. Hence the phrase "Sword of Damocles" to indicate a looming danger of which there is no known time when it might materialize.

the assumptions that he himself had listed as possible sources of error: the true association of the variables with the nebulae, the absence of any significant and anomalous absorption of light, and, above all, the universality of the nature and properties of the Cepheids. He also feared that Shapley's calibration might be wrong — or at least he was pleased to point out this possibility, given his bitter rivalry with the Harvard director. But what kept him awake at night was the unexplained result of van Maanen's work: those proper motions that contradicted the extragalactic scenario.

Hubble thought it wise to wait a little longer before coming out of the closet officially. But in late 1924, he was urged to announce his discovery by the director of the Princeton University Observatory, Henry Norris Russell. Gripped by a patriotic fervor, and perhaps also by antipathy toward Shapley, who had wrested the Harvard professorship from him, Russell was determined to publicize and reward Hubble's discovery at the Society's annual meeting in Washington, D.C., later that year. The Major, unable to resist such an invitation, thought it best not to appear in person. He mailed the manuscript at the last moment. This behavior upset Russell, who, in a letter to Joel Stebbins, secretary of the Society, accused Hubble of being an "ass".[77] Despite these uncertainties, the announcement, read by Stebbins to the assembly, earned him a thunderous applause and a share of a nice $1,000 check, the equivalent of six months' salary at Mount Wilson! In his formal letter of thanks to Russell, Hubble wanted to clear up his hesitation: «The real reason for my reluctance in hurrying to press was, as you may have guessed, the flat contradiction to van Maanen's rotations».[78] It was only ten years later that he would finally dispel his fears by repeating the Dutchman's measurements and proving that the proper motions were the result of a mistake or perhaps a desire to give reality to the nonexistent: a fairly common and entirely human attitude, but one that rarely at a distance pays off.

Curtis and Shapley were both in the room when Stebbins, in his capacity as secretary of the Society, read Hubble's paper. No one knows what they said to each other afterwards. In any case, Shapley was already aware of the matter. «Here is the letter that destroyed my universe»,[79] he had

---

[77] Christianson, *Edwin Hubble*, cit., p. 160.

[78] A.S. Sharov, I.D. Novikov, *Edwin Hubble: The Discoverer of the Big Bang Universe*, CUP, Cambridge 1993, p. 39.

[79] W.L. Freedman, L.C. Ho, *Measuring and Modeling the Universe*, CUP, Cambridge 2004, p. 24.

commented laconically with his student Cecilia Payne — future variable star specialist and icon of feminism in science — waving a sheet of paper that accompanied the plot of the light curve of the first Cepheid found by the Major. He preferred to treasure it, happy to have saved the part of his work that he felt most compared him to Copernicus': having relegated the Sun to a peripheral star in the Milky Way. In fact, after congratulating himself on the discovery — «I do not know whether I am sorry or glad to see this break in the nebular problem. Perhaps both», he would later write to Hubble[80] — he almost immediately began to study this new class of objects, which he renamed galaxies. Hubble, instead, continued to call them extragalactic nebulae because, in his opinion, «the term *nebulae* offered the values of tradition while the term *galaxies* the glamour of romance».[81] Two parallel lives worthy of the pen of Plutarch. Hubble lived mostly in contemplation of himself, using his brilliant legal talent to place his products. In his memoirs, Shapley portrayed the rival as one who «just didn't like people. He didn't associate with them, didn't care to work with them».[82] Remaining a "country boy" from Missouri in manners and appearance, Harlow was annoyed by the Major's distinctly Oxonian expressions such as "Bah Jove" or "to come a cropper", while reluctantly admitting that Hubble's affected style exerted a strong fascination on the ladies.

On the contrary, Shapley showed a deep concern for the world around him. A brilliant orator and popularizer of science, a demanding but thoughtful professor, self-confident but always plagued by doubt, a staunch pacifist, after World War II he ended up in Senator Joseph McCarthy's net because of some openings to the Soviet Union. Born in an enclave of true conservatives, as only the truly poor can be, he was accused of communist sympathies. Although both Harlow and Edwin were well known and acclaimed during their lifetimes, neither one of them succeeded in winning the Nobel Prize, in part because it was not customary at the time to award it to followers of Urania. The coveted prize would be conferred in 2012 on one of Shapley's sons, Lloyd Stowell, for his mathematical contributions to

---

[80]Christianson, *Edwin Hubble*, cit., p. 159.

[81]E. Hubble, *The Realm of the Nebulae*, Dover Publications, New York 1958, p. 18. The Swedish astronomer Knut Lundmark, a third actor in the saga of galaxy discovery, liked to call them anagalactic nebulae, preferring Greek negation (alpha privative) to pure neologism.

[82]S. van den Bergh, *Hubble and Shapley: Two Giants of Early Observational Cosmology*, in "The Journal of the Royal Astronomical Society of Canada", 105, 6, 2011, pp. 245–246.

economics, demonstrating the nurturing qualities of Harlow and his wife Martha Betz.

We return to the story of the fundamental building blocks of the cosmos. The greatest merit of Hubble's work was that it finally convinced the skeptical community of the existence of some nebulae of equal dignity to the Milky Way. But besides spirals, of which no more than 40 examples were known at the turn of the century thanks to Lord Rosse's observations and Isaac Roberts' legendary photographs, what else could be called galaxies in the infinite and heterogeneous zoo of diffuse objects? There was an urgent need to learn how to separate the wheat from the chaff, possibly on the basis of the appearance of the objects, and to bring some order to the complex morphology of the two main classes of objects, galactic and extragalactic, for the benefit of language and the search for ideas for subsequent exploration.[83]

As early as 1908, the German Max Wolf had attempted to classify nebulae into some 20 classes using photographs taken with the 30 inch refractor, but without establishing any connection between them and without taking into account the physical nature of the objects themselves. Dissatisfied with the result, he concluded that no two nebulae in the sky are alike. In 1925 an amateur astronomer, the Englishman John Henry Reynolds,[84] became interested in the matter, which seemed hopeless. At the beginning of the twentieth century, this rich and generous heir to a prosperous industrial business had donated to the Egyptian government[85] a reflector with a 30 inch mirror purchased from Andrew Common. The instrument went to equip an observatory at Helwan, 30 km south of Cairo, capable of taking advantage of the good atmospheric conditions in the desert.[86] At home in Birmingham, Reynolds used another telescope of the same aperture, which he had built with his own hands and which, when the city sky became prohibitive because of light pollution, he donated to the Mount Stromlo

---

[83]A. Sandage, *The Classification of Galaxies: Early History and Ongoing Developments*, in "Annual Review of Astronomy and Astrophysics", 43, 2005, pp. 581–624.

[84]M. Johnson, *Obituary Notices: Reynolds, John Henry*, in "Monthly Notices of the Royal Astronomical Society", 110, 1950, pp. 131–133.

[85]It may be recalled that Britain had invaded Egypt in 1882 as a preventive measure to protect the Suez Canal. Formally, Egypt continued to be ruled by a Chedivè, a title equivalent to viceroy, but power was entirely in the hands of the British High Commissioner.

[86]V.G. Fesenkov, *The Scientific Work of the Helwan Observatory*, "Soviet Astronomy", Vol. 2, 1958, pp. 256–259.

Observatory in Australia, again with the aim of furthering the exploration of the Southern Hemisphere.

Reynolds personally observed the heavens whenever his responsibilities as a businessman, educated patron, and music lover allowed. He also obtained a wealth of photographic material from Helwan, thanks to close contacts made during his travels in Egypt and the collaboration of the English astronomer Harold Knox-Shaw. Summarizing what he had been able to observe by studying photographs of nebulae, in 1920 he proposed a classification of spirals representative of the different degrees of condensation: five classes arranged in a sequence not very different from the one that would later make Hubble famous. But in 1927, as the data at his disposal grew, he too fell into the trap of wanting to perfect his scheme by introducing many classes and turning a taxonomy into a mere description.

Hubble, for his part, had already addressed the problem in his Ph.D. thesis, but using the low-resolution material collected at Yerkes with a 24 inch aperture and a short focal length reflector. After settling in Pasadena, he was able to resume his research using the photographic surveys of Keeler and Perrine, of Curtis with the Crossley reflector at Lick Observatory, and of Ritchey and Pease with the powerful 60 inch at Mount Wilson. The reference frame had changed from that of a few years earlier. Knox-Shaw and Reynolds had introduced the concept of amorphous nebulae: armless objects we now call elliptical galaxies. And in 1918, Curtis had identified spirals with prominent central bars. To this harvest of images and speculation, Hubble could add observations made by himself, his assistant Humason, and his colleague John Charles Duncan with the 60 and 100 inch reflectors. By 1923, he finally felt ready to present his own morphological scheme[87] to the American Section of the Commission on Nebulae established by the International Astronomical Union.[88] Although the question of the nature of these objects was still open, knowing that he had a trick up his sleeve, he ventured a division of all nebulae into two main types of objects, those within the Galaxy and those extragalactic. He then proposed a further subdivision of the latter into the three major classes of elliptical (E), spiral (S), and irregular (I) nebulae, arranged in a sequence that he thought reflected an evolutionary path. In fact, he was inspired by the

---

[87] Cf. A.R. Sandage, *The Hubble Atlas of Galaxies*, Carnegie Institution of Washington, Washington (DC) 1961.

[88] R. Hart, R. Berendzen, *Hubble's Classification of Non-Galactic Nebulae, 1922–1926*, in "Journal for the History of Astronomy", 2, 1971, pp. 109–119.

model of galaxy formation proposed by the British theorist James Jeans, a professor at Trinity College, Cambridge, according to which nebulae should evolve from spherical to ellipsoidal shapes, then flatten out into disks and develop spiral arms. The proposal would completely lose its appeal when it was discovered a few decades later that ellipticals, unlike spirals, are composed mostly of old stars; but at the time it seemed to confirm, on an interpretive level, the morphological scheme Hubble had devised.

At this point, the story takes a mystery-movie turn. While the scheme proposed by Hubble was being studied in the context of the still unresolved problem of the nature of spirals, in 1926 the 37-year-old Swedish astronomer Knut Emil Lundmark published a classification model of extragalactic nebulae that in its main lines seemed to be a photocopy of that of his Mount Wilson colleague. In fact, he considered the three main families of amorphous ellipticals, spirals, and Magellanic irregular types. These were further subdivided according to the light gradient toward the center. Hubble instead distinguished spirals by the morphology of their arms. Plagiarism? That was how the sanguine Major understood it, and he became furious with the colleague with whom he had even shared an earlier paper on nova Z Cen 1922. When he read the communication that Lundmark had presented at a conference in Cambridge, he lost his patience and wrote very angrily to Vesto Slipher, President of the "Commission on Nebulae" of IAU:

> I see that Lundmark has published a "Preliminary Classification of Nebulae" which is practically identical with my own, except for the nomenclature. He calmly ignored my existence and claims it as his exclusive idea. I am calling this to your official attention because I do not propose to let him borrow the results of hard labor in this casual manner[89]

Shapley, who hated the Major in his heart, sided with Lundmark and confided to Russell that he trusted the Swedish classification, which was superior to Hubble's, because it was based on better evidence. But as had happened to Leibnitz, Lundmark had to contend with a giant of science and communication who counterattacked with the extraordinary clarity of a great lawyer and the skill of a consummate storyteller.[90] He promptly published his revised version of the nebula classification, complete with a

---

[89]Ivi, p. 116.
[90]But see how things went in P. Teerikorpi, *Lundmark's Unpublished 1922 Nebula Classification*, in "Journal for the History of Astronomy", Vol. 20, No. 3, 1989, pp. 165–170.

footnote on the first page bluntly accusing Lundmark of plagiarism.[91] Entitled *Extra-Galactic Nebulae*, the paper began with a peremptory statement:

> Galactic nebulae are clouds of dust and gas mingled with the
> star of a particular stellar system; extra-galactic nebulae, at
> least the most conspicuous of them, are now recognized as sys-
> tems complete in them selves, and often incorporate clouds of
> galactic nebulosity as component parts of their organization.[92]

A firm stance whose generality clashed with the paucity of scientific evidence on distances and sizes, limited to a total of 6 objects, including the Magellanic Clouds. Too little evidence to derive a rule. Hubble, who still felt the ground slipping beneath his feet and wanted to secure a parachute, appealed to a «principle of uniformity» that should be applied to the countless fainter nebulae. In short, by staking his reputation on an idea of nature that does not jump, the Major presented the first lucid and comprehensive description of galaxies, ideally marking the beginning of the era of extragalactic astrophysics. The work included the proposal of a morphological sequence consisting of elliptical, spiral, and irregular shapes. Two families of spirals were distinguished, normal and barred, and three subclasses according to the decreasing dimensions of the central bulge and the thickness and winding of the arms. In 1936, by now certain of his own facts, Hubble presented, as a result of a Sillman Memorial Lecture given at Yale,[93] the famous tuning fork representation of the complete scheme, which later became the icon of extragalactic taxonomy. He also introduced a new intermediate class between ellipticals and spirals to make the transition between the two classes less abrupt. His prestige was such that no one dared to criticize this "invention" which assigned to the so-called S0 galaxies[94] the role of *trait d'union* between two massive galaxy families, although they themselves were of rather low mass. Hubble was keen to preserve the continuity of shape suggested by Jeans' evolutionary model, so much so that he named the spirals along his sequence "early" and "late" (adjectives of time).

---

[91] E. Hubble, *Extra-Galactic Nebulae*, in "The Astrophysical Journal", 64, 1926, pp. 321–369.
[92] Ivi, pp. 321–322.
[93] The lecture was published under the title *The Realm of the Nebulae*.
[94] De Vaucouleurs would later call these objects "lenticular galaxies".

In addition to the construction of the morphological classification scheme of galaxies, on the modification and improvement of which astronomers would focus their efforts for many decades to come, Hubble's paper contained numerous physical considerations. For example, statistics on the different types of objects, estimates of the mass-to-brightness ratios of extragalactic nebulae, a first evaluation of the average mass density of the universe of galaxies, and even the curvature of the spherical geometry of Einstein's static universe. No comment! The Major was indeed a self-centered opportunist, but also a scientist of genius, with an extraordinary flair, an uncommon capacity for synthesis, and an ambition capable of overcoming hesitation and making him unscrupulous. Qualities and flaws that come to the fore when one examines the events of his second great discovery, the expansion of the universe.[95]

The idea that spiral nebulae participate in a general receding motion that increases with distance from the observer had been in the air for over a decade, thanks largely to the obscure and valuable work of Slipher. In 1925, Lundmark had noticed that the redshifts of angularly smaller, and therefore presumably more distant, spirals were greater than those of nearby and larger galaxies.[96] Two years later, Lemaître had developed an interpretative model of the phenomenon, including comparisons with observational evidence, but, as we already know, he had not bothered to adequately publicize it (see p. 356). In 1928, Hubble traveled to Leiden, Holland, to attend the third IAU General Assembly. Many years later, in an interview for the American Institute of Physics, his assistant Milton Humason recalled how «Dr. Hubble came home rather excited about the fact that two or three scientists over there, astronomers, had suggested that the fainter the nebulae were, the more distant they were and the larger the red shifts would be. And he talked to me and asked if I would try and check that out».[97] Sniffing out the prey, the great hunter had unleashed his weapons: the largest telescope in the world, practically at his disposal, and an exceptional hound, good "Milt", to squeeze to the bone. While waiting for the first results, he thought it a good idea to throw his hat in the ring by publishing in

---

[95] N.A. Bahcall, *Hubble's Law and the Expanding Universe*, in "Proceedings of the National Academy of Sciences of the United States of America", 112, 11, 2015, pp. 3173–3175.

[96] V. Trimble, *Anybody but Hubble!*, 2013, https://www.google.com/search?client=firefox-bd&q=lundamark+hubble+law.

[97] *M. Humanson's interview* (circa 1965), https://www.aip.org/history-programs/niels-bohr-library/oral-histories/4686.

1929 an article with the same data and results as Lemaître, without even mentioning the Belgian scientist. To justify this omission, it was later said that Edwin could not read Lemaître's paper since he did not know French. Instead, he quoted Lundmark, with whom he had a certain grudge, but only to demolish the arguments of his unfortunate Swedish colleague.

In a few years, the two Mount Wilson astronomers collected data on 40 new objects, bringing the redshift measurement to the distance of the Leo cluster, where the recession velocity is close to 20,000 km/s. As with the search for the nature of nebulae, Hubble had made his case in detail, proving what was already known, namely the existence of a universal law of direct proportionality between the distances of galaxies and their redshifts. A law that today bears his name, thanks to the propaganda of his friend Humason. To leave no room for doubt, he himself later declared «I consider the velocity-distance relation, its formulation, testing and confirmation, as a Mount Wilson contribution and I am deeply concerned in its recognition as such».[98] Lemaître did not take it too well, but, as we know, history is written by the victors, and by now American astronomy was winning over European astronomy. The absurdity is that Hubble remained critical of the true nature of redshift throughout his life. He strongly doubted that it represented a real velocity,[99] so much so that he preferred to temper the use of the term by adding the adjective "apparent". «Although the discovery of the expansion is often attributed to Hubble with his 1929 paper», his favorite student Allan Sandage reportedly admitted in 2009, «he never believed in its reality».[100]

The Major was not alone in his doubts. In 1934, Walter Baade and Frit Zwicky wrote bluntly in an otherwise seminal paper: «We ourselves are by no means convinced that the universe is expanding».[101] Their skeptical position was the consequence of a hypothetical alternative redshift mechanism to the purely Doppler one, proposed as early as 1929 by Zwicky

---

[98] Letter to De Sitter quoted in Sharov, Novikov, *Edwin Hubble*, cit., p. 64.

[99] In a letter to Willem De Sitter, Hubble wrote: «Mr. Humason and I are both deeply sensible of your gracious appreciation of the papers on velocities and distances of nebulae. We use the term 'apparent' velocities to emphasize the empirical features of the correlation. The interpretation, we feel, should be left to you and the few others who are competent to discuss the matter with authority». (Ivi, p. 67).

[100] H. Nussbaumer, L. Bieri, *Discovering the Expanding Universe*, CUP, Cambridge 2009, p. XV.

[101] W. Baade, F. Zwicky, *Cosmic Rays from Super-Novae*, in "Proceedings of the National Academy of Sciences of America", 20, 5, 1934, p. 261.

himself and called the "tired light" model.[102] Photons should lose energy due to interactions with other particles along their path to the observer, the more the longer the path. Repeatedly proposed, this theory, which restored stillness to the cosmos, has never found experimental confirmation.

In 1953, the great astronomer died of a stroke, full of glory and yet disconsolate. He did not have time to see the results of the cosmological program he had initiated. His untimely death — he was only 64 years old — contributed to the magnification of the myths that he, his wife, the enterprising Grace, and his friends had built around his figure as a lonely explorer of the cosmos.[103] Perhaps the most sensational of his achievements is that associated with the visit Einstein made to Mount Wilson Observatory in 1931 as he was preparing to leave Nazi Germany for good. It is said that during an excursion to the 100 inch telescope, Hubble had the opportunity to reveal to the illustrious guest his recent discovery of the expansion of the cosmos, with the result that Einstein immediately rejected his static model, claiming that it was the «biggest blunder» of his life. Or so said the Russian George Gamow, a great scientist but also an unrepentant prankster, in his 1956 article in "Scientific American"[104]. In any case, the claim was corroborated by the English physicist John Archibald Wheeler, who was at Princeton at the time. He said that he heard Einstein say to Gamow, in reference to the cosmological constant, "that was my biggest blunder of my life"».[105] Instead, some historians think it more likely that it was the usual Eddington who spoke to Einstein about this matter, and that the famous sentence was never uttered. This would explain why Herr Albert wrote the American astronomer's name in his diary as it is pronounced in German: Hubbel instead of Hubble. The two great scientists were undoubtedly introduced and probably exchanged a few words, but Einstein may not have even read any of the American's papers.

Whatever happened, Einstein explicitly repudiated the repulsive force he had invented to balance gravity: «Since I have introduced this Λ [cosmological] term, I had always a bad conscience. But at that time I could

---

[102] F. Zwicky, *On the redshift of spectral lines through interstellar space*, in "Proceedings of the National Academy of Sciences of America", 15, 1929, pp. 773–779.

[103] Cf. D. Overbye, *Lonely Hearts of the Cosmos: The Story of the Scientific Quest for the Secret of the Universe*, Picador, London 1991.

[104] G. Gamow, *The Evolutionary Universe*, in "Scientific American", 195, 1956, pp. 136–154

[105] E.F. Taylor, J.A. Wheeler, *Exploring Black Holes: Introduction to General Relativity*, Addison Wesley, Boston (MA) 2000, p. G-1.

see no other possibility to deal with the fact of the existence of a finite mean density of matter. I found it very ugly indeed that the field law of gravitation should be composed of two logically independent terms, which are connected by addition. About the justification of such feelings concerning logical simplicity, it is difficult to argue. I cannot help to feel it strongly and I am unable to believe that such an ugly thing should be realized in nature».[106] An admission of guilt that makes this great man even greater, for he was well aware that «only death can save one from making mistakes».[107]

---

[106] G. Weinstein, *George Gamow and Albert Einstein: Did Einstein Say the Cosmological Constant Was the "Biggest Blunder" He ever Made in His Life?*, arXiv: History and Philosophy of Physics, 2013, https://arxiv.org/ftp/arxiv/papers/1310/1310.1033.pdf.
[107] Ibid.

Chapter 15

The eighth wonder:
The Palomar giant telescopes
and the new ideas

My, what big eyes you have!

Charles Perrault,
Little Red Riding Hood

The immense distances to the stars
and the galaxies mean that
we see everything in space in the past,
some as they were before
the Earth came to be.
Telescopes are time machines.

Carl Sagan, *Pale Blue Dot*

With his record-breaking telescopes and the astronomers he selected and cultivated, Hale had turned the exploration of the heavens into a major field of scientific research. In the process, he created the conditions that would settle once and for all the question of the nature of nebulae and usher in a new era of our knowledge of the cosmos. By the 1920s, curiosity had pushed the horizon of knowledge far beyond the boundaries of the Milky Way, the classic Hercules columns of 19th-century astronomers. But even the big calibers mounted on the mountains of California were no longer adequate to sail the endless ocean of galaxies. Hale was well aware of this. So while Europe was distracted licking its wounds from a cruel war and nurturing the subtle and poisonous germs of nationalism, in America of the Great Illusion the irrepressible George was already weaving the plot to provide his champions with an even more powerful weapon than the already formidable Hooker telescope. In 1923, the worsening of his illness forced him to relinquish the management of Mount Wilson to Walter Adams, solar astronomer and granite manager. This resignation, however, did not in the least affect his extraordinary faculty for thinking big. In between bouts of mental illness, he had begun to fantasize about an instrument with an aperture of 200, if not 300 inch, and to explore its physical feasibility with the technical assistance of Francis Pease, astronomer and engineer, Ritchey's former assistant at Yerkes.[1]

The construction of the Hooker telescope had been a valuable experience. experience.[2] Although Ritchey was a pain in the neck, and at one point Hale ended up resenting him, he had a unique ability in the field of optics. The rosary of his criticisms of St. Gobain's ill-fated blank cast, which the technician ruthlessly rattled into the patron's ears, was accompanied by a list of valuable suggestions for the future. He continued to insist, like Marcus Porcius Cato with his "*Carthago delenda est*" (Carthage must be destroyed), on the imperative need to select, before any other action, a material for the main optics that would be insensitive to the inevitable temperature variations during the observing session. The mirror of the Hooker telescope was made of ordinary glass, and its large deformations with temperature variations gave room to grunts from the most demanding astronomers, such as Hubble, because every brave knight pretends to ride a good warhorse. A second proposal concerned the geometry of the primary mirror.[3] Ritchey proposed the innovative design of a thin disk on a supporting honeycomb structure to lighten the weight of the glass monolith while making its structure more rigid. This stratagem would also help reduce the cooling time in the casting process and thus control the gas accumulation responsible for the bubbles that had plagued both 2.5 m blanks — a serious problem that we already mentioned in chapter 13. Last but not least, the reduction in glass mass enabled a faster response to temperature changes (an important feature for getting the telescope up and running immediately after sunset).

The requirements for the primary optics were clear, but no one knew how realistic they were in terms of feasibility, time, and cost. In short, the main mirror represented the greatest uncertainty of the entire project. In fact, despite the stringent performance requirements set by Hale, who wanted to end his career with a true technological high C, the telescope's structure and envelope did not seem to present unbridgeable difficulties. The Great War had improved mechanical technologies by developing know-how

---

[1] R. Florence, *The Perfect Machine: Building the Palomar Telescope*, Harper Perennial, New York 2011.

[2] J.H. Tiner, *Exploring the World of Astronomy: From the Center of the Sun to the Edge of the Universe*, Master Books, Green Forest (AR) 2013.

[3] H.J. Abrahams, *The Ritchey-Chrétien Aplanatic Telescope: Letters from George Willis Ritchey to Elihu Thomson*, in "Proceedings of the American Philosophical Society", 116, 6, 1972, pp. 486–501.

that could be usefully redirected to peacetime activities. But the mother of all problems, as is often the case in the world of science and beyond, was the mountain of money needed to get the project done. Extrapolating the cost of the 60 and 100 inches, Hale estimated that the budget for a modern reflector would increase in proportion to the cube of the aperture. For a 200 inch telescope, this formula led to a minimum commitment of $5 million. This was a huge sum, equivalent in purchasing power to about $350 million today, and equal to 2/3 of the total investment in the European Southern Observatory's Very Large Telescope, the record-breaking terrestrial optical instrument in contemporary optical astronomy. Hale's estimate was essentially correct.[4]

Unable to milk the overly generous Carnegie Institution, and not knowing where else to turn, the untamed George relied on the generic «lure of the uncharted seas of space».[5] He trusted in the seduction that the latest astronomical discoveries could exert on the vast public of a young and exuberant nation, where no goal seemed impossible, provided the will was strong. A pragmatic enlightenment that had matured on Wall Street, in the factories of the Northeast, and on the prairies of the Midwest, and was celebrated on celluloid by Hollywood. Hale hoped that by listening to the good stories of science, some big fish would take the bait.[6] It had always happened in his previous experiences as an astronomy tycoon, and it happened again this time. After reading the article on *The Possibility of Large Telescopes* that the astronomer had delivered to the pages of the historic New York monthly "Harper's Magazine", the powerful Wickliffe Rose, a senior executive of the Rockefeller Foundation, contacted him for some information on the project.[7]

---

[4]Experts estimate that the cost of a terrestrial optical telescope of aperture D grew as $D^{2.77}$ before 1980, and as $D^{2.45}$ if built after that date, due to new technologies. For instruments in orbit, the trend is also a power law, but with a much less steep slope, of the order of $D^{2.0}$. Cf. R. Giacconi, *Telescope Costs*, in "Science", 281, 1998, p. 1961. For extremely large telescopes, see L. Steppa, L. Daggert, P. Gilletta, *Estimating the costs of extremely large telescopes*, http://www.gsmt.noao.edu/documentation/SPIE_Papers/Stepp.pdf.

[5]G.E. Hale, *The Possibilities of Large Telescopes*, in "Harper's Magazine", 156, 1928, pp. 639–646; Id., *The 200-Inch Telescope*, in "Scientific American", 154, 5, 1936, pp. 237–240.

[6]*A History of Palomar Observatory*, Palomar Observatory, https://sites.astro.caltech.edu/palomar/about/history.html.

[7]W.S. Adams, *Biographical Memoir of George Ellery Hale (1868–1938)*, in "National Academy of Sciences of the United States of America Biographical Memoirs, XXI, 5, 1939, pp. 179–241.

It was the spring of 1928, and the great country was dozing off, enjoying the battle for the White House. Herbert Hoover, a stone-hearted Republican, would win it in a few months. «We in America today are nearer to the final triumph over poverty than ever before in the history of any land [...] We shall soon, with the help of God, be in sight of the day when poverty will be banished from this nation»,[8] he had shamelessly proclaimed upon accepting his party's nomination. Unfortunately, he was dead wrong. In the first year of his presidency, the most serious and dramatic systemic crisis of all time would erupt in the United States and promptly be exported abroad: a kind of economic and financial pandemic capable of infecting the whole world, with few exceptions, initiated by the Wall Street stock market crash of October 24, 1929, the famous Black Thursday, with the largest sell-off of stocks in the history of the United States. By a singular coincidence, even the 5 m telescope, like Lord Rosse's Great Leviathan, was born at the dawn of a period of immense suffering, decadence, and famine.

Sensing an opportunity to score, Hale responded to Rose with a letter of strategic rationale and justification:

> The 100 inch Hooker telescope of the Mount Wilson Observatory has solved many fundamental astronomical and physical problems beyond the reach of our 60 inch reflector, and prepare the way for an attack on still more important problems that demand greater light-gathering power for their solution. Among these outstanding questions are: (a) The structure of the universe [...], calling for a more intensive study of the Galaxy, of which our solar system is a minute part, and especially of the vast region of spiral nebulae ("island universes") beyond the Milky Way [...] (b) The structure of spiral nebulae [...] The evolution of spiral nebulae [...] The constitution of matter, since the enormously greater range of mass, temperature, pressure, and density of the heavenly bodies presents opportunities for discovery far beyond the possibilities of laboratory experiments.[9]

The wily fisherman's net was closing in on the big fish with a seductive offer of a tasty lure. The 66-year-old Rose, a historian by training, manager and

---

[8] C. Reef, *Poverty in America*, Infobase Publishing, New York 2007, p. 133.

[9] O. Struve, V. Zebergs, *Astronomy of the 20th Century*, Macmillan, New York 1962, p. 133.

philanthropist by vocation, was indeed a big shot. At the time, he was chairman of the International Education Board for «the promotion and/or advancement of education, whether institutional or otherwise, throughout the world»,[10] established under the auspices and with the capital of the Rockefeller Foundation;[11] and he had a real passion for big science, as he demonstrated on this occasion. In fact, while Hale was trying to squeeze some money out of him to initiate a feasibility study, which he felt was urgently needed given the extraordinary nature of the project he had in mind, Rose astonished him by volunteering to finance the entire project with $6 million, more than enough to make a 200 inch reflector. Not even six months had passed since the first contact!

Once the general agreement was in place, it was necessary to identify the party to whom the funding would be allocated, since the Mount Wilson Observatory was already under Carnegie's control. The choice was the California Institute of Technology (Caltech), which Hale himself had founded a few years earlier in the city of Pasadena with the physicist Robert Millikan, Nobel laureate in 1923.[12] Caltech and the Carnegie Foundation then formed a partnership called the Mount Wilson and Palomar Observatories: a research institute destined to be the beacon of world astronomy for half a century. But to complete the mosaic and leave, one important and not insignificant piece was missing. A shrewd and realistic manager, Rose was concerned that the new observatory had the financial resources to ensure its long-term operation: endowment funds that the Rockefeller Foundation was not prepared to provide. The money had to be found elsewhere, and quickly, or the support for the project would be lost. Once again, the magician Hale was able to pull a patron out of the hat. Henry Robinson, a wealthy banker from Los Angeles, promised to donate the necessary funds for the observatory's operation every year. Thus, at the end of 1928, Rose's $6 million arrived, enabling the first major stone to be laid for the "Cathedral of Astronomy".

---

[10] *The Rockefeller Foundation: A Digital History*, https://rockfound.rockarch.org/international-education-board.

[11] Shortly after the IEB was established, Rose wrote these words: «In a democracy non-governmental institutions serve an important purpose; they have greater freedom for initiative and for experiment; they thus serve to stimulate and guide governmental effort» (ibid.).

[12] J.R. Goodstein, *Millikan's School: A History of the California Institute of Technology*, Norton, New York 1991.

While on the West Coast of the United States the astronomers were rubbing their hands with impatience to open the dances, but also with the anxiety typical of those who know they have embarked on a daring adventure, on the East Coast Shapley was gritting his teeth in anger and disappointment.[13] The Harvard director had hoped that instead of subsidizing another telescope for the California sky, the Rockefeller Foundation would choose to partner with the University of Michigan to build the Bloemfontein Southern Station in South Africa. Certainly, he had good reasons for wanting to do so, chief among them the exploration of the southern hemisphere, richer in stars and yet little visited by astronomers. But fortunately for science, Rose did not listen to him.

The construction of the giant reflector,[14] which now bears the name of its godfather, began with the establishment of a pyramidal organization and division of labor. Hale headed a small board of directors, flanked by an advisory committee of eminent scientists and technical experts. Then there were a number of committees, each responsible for a major subsystem such as optics, mechanical assembly, handling, site, and civil works. In short, a small, fierce, and well-organized army of professionals to plan, decide, design, supervise, manage, manufacture, and test the telescope of telescopes, a "perfect machine" worthy of the wonderful sky it would go to study. Moreover, no astronomer in the observatory was allowed to withdraw from the project or defect, perhaps citing the priority of his own research or career. Adams, for example, when confronted with the request to appoint his thoroughbred as director of Yerkes, had replied curtly that Hubble should stay on to complete the 200 inch.[15]

First of all, Hale had to make sure that a glass disk of the desired size and quality could actually be cast. Few doubted that it could be done, but no one knew for sure how long it would take, how much it would cost, and most importantly, in which way and with what materials. Hale opted for fused silica, a glass made by liquefying a silica sand at a very high temperature and then cooling it slowly enough to prevent it from recrystallizing. The resulting amorphous material has extraordinary properties. In particular,

---

[13]D.H. DeVorkin, *Henry Norris Russell: Dean of American Astronomers*, Princeton University Press, Princeton (NJ) 2000, p. 313.

[14]R. Preston, *First Light: The Search for the Edge of the Universe*, Penguin, London 1996.

[15]D.E. Osterbrock, *Yerkes Observatory, 1892–1950: The Birth, Near Death, and Resurrection of a Scientific Research Institute*, The University of Chicago Press, Chicago (IL) 1997, p. 116.

it is ten times less sensitive to temperature changes than ordinary glass, as Hubble bluntly claimed, intolerant of the vagaries of the 100 inch "bottom of the bottle". There was one firm in the United States with an established track record of producing fused silica. The General Electric Company, heir to the industrial enterprise begun in the second half of the 19th century by the prolific inventor and entrepreneur Thomas Alva Edison, designed and manufactured everything from small appliances to impressive one-of-a-kind pieces of high-tech electromechanical technology. One of its flagship products was the incandescent light bulb, for which GE had developed special expertise in thermal glass processing. The company's owner, the Anglo-American engineer Elihu Thomson, accepted Hale's challenge. For four years, beginning in 1927, a handful of technicians at the company's plant in Lynn, Massachusetts, worked hard to produce satisfactory blanks of molten quartz of gradually increasing size, hoping to reach the fateful 5 m diameter. But results were lagging, largely because of the difficulty of controlling the microbubbles trapped in the final product. A huge loss of time, confidence, and money. Faced with the production of a measly 60 inch blank, the operation had already absorbed 10% of the total funds available for the project.

Hale realized it was time to change course or the project would fail. So he terminated his contract with General Electric and, in late 1931, turned to another company in New York that had specialized in the production and processing of borosilicate glass.[16] Since 1915, Corning Glass Work marketed a product with excellent thermal properties called Pyrex, which was used for fire dishes and laboratory test tubes. Borosilicate glass had been invented in 1893 by Otto Schott, the chemist who, along with Karl Zeiss and Ernst Abbe, was responsible for Germany's dominance in the world of optics, and was sold under the name Duran. The technical information was brought to the United States by Corning's director of research, Eugene Sullivan, who had met Schott while working on his doctoral thesis in Leipzig. Once the contract was signed, the project was placed in the hands of George McCauley, a physicist in the company's research laboratory: a quiet, methodical man who was undeterred by the enormous weight of responsibility and expectation placed on his shoulders. «It will be no different than making a bean pot, except in the methods employed»,[17]

---

[16] R.N. Wilson, *Reflecting Telescope Optics, I. Basic Design Theory and its Historical Development*, Springer, Berlin 2007.

[17] Preston, *First Light*, cit., p. 43.

he would reply laconically to those who asked him what strategy he had in mind for producing the monstrous Pyrex mirror. Clint Eastwood couldn't have said it better.

The preparatory phase lasted over two years, but yielded flattering results. Gradually larger and larger borosilicate disks were successfully produced, up to a diameter of 120 inches, and distributed almost everywhere on the American continent to equip new instruments. Lacking any previous experience, McCauley tried step by step to understand where the weak points of the process lurked and how to provide for them as the diameter of the glass disk grew. By March 1934, as America celebrated the end of Prohibition with rivers of legal whiskey, everything was ready at the Corning plant for the final act, the casting of the 5 m diameter, 75 cm thick blank (the rule of thumb used by the manufacturers of classical astronomical mirrors to ensure the necessary rigidity was to fix the thickness of the glass disc to one-sixth of the diameter). The molten glass was poured onto a steel mold with more than a hundred firebricks of various sizes bolted to the bottom with the purpose of creating the famous honeycomb pattern dreamed by Ritchey. With this strategy, McCauley intended to halve the 40-ton weight of the simple solid disk while reducing the cooling time, estimated at nine years for a Pyrex monolith, to reasonable levels. After casting, the mold would be sealed in an electric furnace to control the temperature drop until the blank solidified.

In order to transfer the vitreous magma, heated above the melting point, so that it would survive the short journey from the furnace to the mold and prevent a dangerous increase in viscosity on contact with the refractory bricks, McCauley had organized a cordon of workers armed with giant ladles moved on overhead tracks. The final act began in the early morning of Saturday, March 25, 1934, under the watchful eyes of a crowd of onlookers and press and radio reporters ready to celebrate the feat: a hellish event comparable in size and audacity to the casting of the statue of Perseus described by Benvenuto Cellini in his autobiography and which we can still relive thanks to a historical film available on the Web.[18] Everything happened within half a day. The most serious and decisive accident occurred when three firebricks broke loose from the bottom of the tank. The enormous heat emitted by 20 tons of glassy mass had melted the steel bolts holding

---

[18]E. Jacobs, *Old, Weird Tech: How to Cast a 20-Ton Piece of Glass for a Telescope*, in "The Atlantic", 2011, https://www.theatlantic.com/technology/archive/2011/02/old-weird-tech-how-to-cast-a-20-ton-piece-of-glass-for-a-telescope/70942/.

them in place. Desperate attempts to control the rebellious bricks by pushing them with steel rods, exposing the workers to a veritable inferno of
heat, were in vain. In the end, to save what could be saved, it was decided
to crush the mutineers, knowing that this brutal therapy would irreparably
damage the geometry of the system.

McCauley did not flinch, and with great lucidity, after declaring that he
would still attempt a second cast,[19] he decided to proceed to the next step,
the cooling of the ill-fated first cast, which was accelerated because it had
now become only a test. This was a wise decision, for three months later,
when he reopened the sarcophagus containing the glowing glass mass, he
found a defective but perhaps workable blank, albeit with much patience
and a little more time and money. Much later, McCauley would recall with
these words[20] the dilemmas of those difficult days when it came to deciding
whether to settle for the result or to start the process all over again:

> While we would have wished to salvage the present one, the
> work of grinding away the great quantity of glass required to
> re-establish its intended rib structure seemed very questionable
> as a cost-saving measure. Besides, no one could predict with
> certainty that the grinding would be completed without that
> one single fracture to a rib that would be sufficient to doom the
> disk to the scrap, or cullet, pile.[21]

In the end, the contractor and the client agreed to try their luck again with a
second casting, in part because they now knew how to approach the critical
elements of the process. The mold was redesigned to allow air to circulate
inside the refractory bricks, facilitating heat dissipation. The anchors were
strengthened by replacing the steel in the bolts with a chrome-nickel alloy,
and the hot glass transport crew was increased. Thus, on December 2, 1934,
in just 6 hours, the largest glass monolith ever built took shape in its mold
without a hitch. This time, MacCauley did not want any curious onlookers
around. The giant Pyrex cake was promptly placed in the thermostatic
vault for the long annealing process. Two months at a constant temperature
to set, then another eight months of controlled cooling at just one degree

---

[19]Preston, *First Light*, cit., p. 43.
[20]G.V. McCauley, *Making the Glass Disk for a 200 Inch Reflecting Telescope*, in "The
Scientific Monthly", 39, 1, 1934, pp. 79–86.
[21]Id., *Corning Glass Works and Astronomical Telescopes*, unpublished manuscript,
1965, https://www.cmog.org/article/glass-giant.

per day. Like a perfect soufflé, the oven had to remain sealed throughout the heat treatment period, with no possibility of inspecting the product. All the "McCauley's chefs" had to do was wait and hope for a good result.

This feat was well received by the press. Emerging from the deepest recession, America lived with curiosity and pride its primacies, useful also to compensate for an inferiority complex towards the Old Continent and its glorious roots: the same roots that in Europe instead fueled selfishness and revanchism, disguising them in the garb of rhetoric and *grandeur*. In the same year, 1934, Adolf Hitler, in imitation of the Italian dictator Benito Mussolini, seized absolute power in Germany with the title of Führer, to the dull indifference of the victorious great powers of the Great War. In faraway China, Mao Zedong led the Red Army of the Communist Party in a long march to escape the encirclement of Chiang Kai-shek's Kuomintang troops in a historic retreat, laying the foundation for building a pillar of the contemporary world.

There was ten months of trepidation in Corning. The fear peaked when a storm that flooded the Chemung River cut power to the annealing furnace for three long days, leaving the mass of glass to dissipate its heat uncontrollably. Finally, the sarcophagus was opened, and McCauley could see for himself that this time the process had produced a perfect Pyrex disk, supported by an elegant honeycomb texture. It was a great success for Corning and a solid step forward for the telescope project.[22] The rest of the way, the engineer told himself, would be all downhill. Eight years had passed since the start of the realization phase, too long for Hale to see his ultimate dream come true. In fact, he would die shortly thereafter, in 1938, at only seventy years of age.

Freed from all obstacles, the majestic blank was exposed to the press, which made a media case out of it. Lowell Thomas, the journalist and radio commentator whose reports from the Middle East gave Lawrence of Arabia planetary fame, described the Pyrex disk meltdown as «the greatest item of interest to the civilized world in 25 years, not excluding the World War».[23] A clear exaggeration, but one that testifies to the participation of the entire nation in the scientific enterprise and its spectacular technological impact. The blank was then packaged, boxed, and placed vertically on a special

---

[22] An interesting collection of images related to the casting of the 200 inch mirror can be found at https://www.cmog.org/collection/exhibitions/mirror-to-discovery.

[23] D. Dyer, D. Gross, *The Generations of Corning: The Life and Times of a Global Corporation*, Oxford University Press, Oxford 2001, p. 14.

flat rail car for shipment to Pasadena. On the other side of the continent, Caltech opticians were waiting to turn it into a true astronomical mirror. The train journey took almost a month. A slow pilgrimage across the dusty plains of the Midwest still scarred by the Depression, the stony expanses of Arizona, and the Rocky Mountains. The convoy traveled by day, at moderate speeds to limit vibrations. At night, it stopped at an isolated station platform, illuminated by spotlights and guarded by armed men whose job was to protect the precious cargo from the crowds of onlookers who flocked to peek at the "monster at every good opportunity". On Easter morning of 1936, the blank reached the Caltech laboratory created *ad hoc* to care for the giant: a shed 50 m long and 15 m high, with no windows to limit contamination and equipped with a powerful overhead crane with a capacity of 50 tons. The blank would remain in this shelter for more than 11 years to undergo the grueling treatment that would transform it into the limpid eye of the world's largest telescope, an icon of 20th-century astrophysics and American greatness.

The first treatment was to grind both sides of the blank to make them parallel. This process took an entire year. Then the Pyrex block was fed into a machine designed to mechanically repeat the same actions that optical masters had performed for centuries to grind and polish telescope mirrors, using tools made of the same material as the blank, lots of water, and a mountain of abrasive powders. Dripping with the milky water of machining, the metal mammoth was mechanically programmed to couple translations with asynchronous circular motions. The blank, placed face up on the same load supports it would later find aboard the telescope, was held in slow rotation while a Pyrex tool, also rotating, provided grinding along a predefined path: the so-called "Lissajous figures," obtained by orthogonally composing two harmonics of slightly different frequencies. Movement on the surface of the blank was ensured by a bridge structure: all together, a more robust than versatile robot, conceived and designed by Caltech engineers. The tools, metal disks from 12 to 200 inches in diameter, had their working surfaces paved with a mosaic of Pyrex blocks. The grinding was slow and methodical, punctuated by long pauses to allow the glass to dissipate the heat generated by the rubbing. The opticians, like croupiers with chips on the green table, used long spatulas to help spread the watered metal oxides evenly. The goal was to produce a perfectly smooth spherical surface that would later be retouched into a rotation paraboloid with a relatively short focus: only 16.5 m, a little more than three times the diameter of the glass disk.

Despite its somewhat casual appearance due to the large amount of water dripping from the blank, the workshop was a maniacally clean place. Extreme care had to be taken to prevent even the smallest piece of solid material from slipping under the sanding tool. A scratch was enough to destroy months of work. For this reason, no one could enter the plant without first wearing the proper clothing, headgear, and footwear: a comical baker's uniform that had earned the technicians and engineers the nickname "men in white. Marcus Brown, an optician who came out of nowhere — he was a truck driver from Mount Wilson and, like many others, had been promoted by Hale for his skill and dedication — led the team with calm and determination. «Glass won't ever do what you expect. It has as many moods as a movie star», he used to tell his young staff, then adding: «If you don't know what you're doing, don't do anything until you find out».[24]

For six long years, the grinding went on monotonously, without major stumbles, and just when the goal was in sight, it had to be interrupted by the entry of the United States into the war, in 1942. "*Ubi major, minor cessat*" (where the major is, the minor becomes negligible). The large Pyrex dish was carefully packed up, and the laboratory staff returned to making prisms, lenses, and other optical trinkets for the U.S. Army, or simply shipped off to one of the various fronts, East or West. The work of polishing and testing could only be resumed after the surrender of Japan, which had been struck by two atomic bombs, and was completed in October 1947, when the mirror was finally accepted by the customer. Optical tests certified an accuracy of one millionth of a centimeter; each point on the optical surface deviated on average less than one-twentieth of the wavelength of visible light from the ideal parabola.[25] Beautiful and thin, you might say! In fact, after treatment at Caltech's Beauty Farm, the mirror weighed 15 tons, 25% less than when it entered the optical shop.

All other activities related to the construction of the telescope and infrastructure had long since been completed. The white dome loomed imposingly over the summit of Palomar Mountain, the site Hale had chosen for the observatory: a flat, very dark peak 5,000 feet (1,700 meters) above

---

[24] M. Pendergrast, *Mirror, Mirror: A History Of The Human Love Affair With Reflection*, Basic Books, New York 2003.

[25] An impressive level of precision, best perceived by saying that the difference between the machined and theoretical surface area would remain on the order of a millimeter if the mirror were rescaled to the distance between Rome and Florence.

sea level in San Diego County, about 90 miles (150 kilometers) southeast of the city of Los Angeles. In 1934, Caltech had purchased 120 acres of land on the mountain from the Beech family for $12,000, and the County of San Diego undertook to build a road to the site, popularly known as the "Highway to the Stars". The "dovecote", as the Spanish colonizers had called the place because of its alleged abundance of pigeons (*palomas* in Spanish), had been under consideration ever since Hale had sought a home in California for his 60 inch reflector. But even though it was the darkest spot in the area, it had been discarded due to the complete lack of infrastructure (especially roads). Then things changed. The same thirty years that had been enough for the little El Pueblo de Nuestra Señora de los Ángeles del Río de la Porciúncula to grow so large that it dazzled Mount Wilson, had brought a bit of civilization to the big nothing on the border with Mexico. It must be said that Hale, haunted by the grumbling of his own astronomers, was fully convinced that he had to consider the choice of site carefully. Thus, before deciding, he had examined other solutions besides the California mountains, such as the plateau of the Grand Canyon, where Lowell had placed his observatory. For this reason, he had sent Hubble to Arizona in 1928 with a small refractor for a hasty siting campaign. The mission proved inconclusive, but it did allow the Major to spend a few days in Flagstaff in the company of Slipher, whose data he was then using to build the empirical case for the expansion of the cosmos. It would have been an excellent opportunity to discuss with his colleague and author of the observations. This is exactly what the egocentric Edwin was careful not to do.[26]

We now come to the giant telescope mount. From the beginning of the enterprise, this had proved to be the least problematic segment, thanks to the formidable legacy of technological and industrial know-how left by World War I.[27] With his usual managerial flair, Hale had assembled a team of skilled engineers around the mechanical project. He had entrusted them with solving the problems posed by the oversized optics and responding to astronomers' specific requests, such as being able to point the instrument north, something the 100 inch could not do because of structural obstacles. Turning design ideas into parts of the instrument required someone who

---

[26] D. Lago, *Edwin Hubble's Silence*, in "Origins of the Expanding Universe", in *ASP Conference Proceedings*, 471, 2013, pp. 269–278.

[27] B. Rule, *Engineering Aspects of the 200 Inch Hale Telescope*, in "Publications of the Astronomical Society of the Pacific", 60, 355, 1948, pp. 225–229.

knew the market for engineering firms in a country the size of a continent. In 1934, Hale hired Captain Clyde McDowell, an engineer and sailor with experience in shipbuilding. He also enlisted as draftsman and designer Russell Porter, a singular figure of multifaceted genius: architect, Arctic explorer, builder, inventor, cartographer, photographer, and writer. His fascinating and realistic representations of the instrument and the dome[28] served to guide the project and its realization, together with the 1:10 scale model that McDowell wanted to test the main technical solutions. The most important one was the geometry of the coupling of the telescope tube with the polar axis of the equatorial mount. It could not be a fork, because that would involve long lever arms and cause intolerable bending. But astronomers also did not want the yoke solution adopted for the 100 inch, which limited the movements of the telescope and put the north polar cap out of reach. The compromise solution was a horseshoe yoke that retained the advantages of the fork without suffering from its drawbacks.

For the mechanical realization, McDowell chose in 1936 the Westinghouse Electric Corporation, a rival of General Electric that had entered the business with the first, unsuccessful production of the mirror. The giant horseshoe of the polar axis took shape in their Philadelphia factories, supported by hydrostatic pads to reduce friction during the tracking of the celestial sphere. The thin layer of oil made it possible, among other things, to motorize only one end of the axis without introducing excessive torsion, and to move the 530-ton instrument smoothly and with high precision. The telescope tube was designed by Mark Serrourier, a young and brilliant Caltech engineer, with the aim of minimizing the effects of unavoidable mechanical deflections on the optical alignment. Instead of focusing on stiffness, which would have increased the overall weight of the imposing structure, Serrourier had chosen the strategy of compensating for the deformation of one end of the telescope tube with respect to the support, i.e. the elevation axis, by a similar passive bending of the other end, thus transforming two mechanical failures into a parallel translation. A Columbus' egg, a bit like the saying: "If you can't win it, go along with it". This is a strategy now widely adopted in the design of the latest generation of telescopes.[29]

---

[28]Cf. P. Ré, *Russel Porter (1871–1949) and the glass giant of Palomar*, http://www.astrosurf.com.
[29]Wilson, Wilson, *Reflecting Telescope Optics, I. Basic Design Theory*, cited above, p. 449.

The undertaking required overcoming many practical difficulties, as well as the construction of oversized facilities in which some of the larger parts could be reshaped, milled, or simply reheated. The instrument was finally preassembled at the Westinghouse plant in Philadelphia, where a first inauguration was held on April 30, 1937, with a ceremony attended also by Einstein.[30] This was the only known interaction between the famous physicist and the large telescope, two major players in 20th century cosmology. The iron giant was then duly dismantled, its parts packed and shipped by sea through the Panama Canal to the Port of San Diego, and from there along the "Highway to the Stars" to Mount Palomar. The great enclosure awaited them;[31] a dome formed by a hemisphere 42 m in diameter and weighing 1,000 tons, which rested on a 20 m high masonry cylinder by means of dozens of wheels that allowed smooth rotation without friction or vibration. Although the era of servo control was still in its infancy, no effort was spared to automate as many operations as possible. The dome, for example, followed the clockwise rotation of the telescope thanks to a robot whose naivety would make us smile today. Also automatic was the unrolling of the curtain that protected from the wind the opening made on the surface of the dome to give access to the night sky: a huge observation slit that was closed at the end of the night's operation by sliding two curved doors in front of it. The observatory was completely self-sufficient in terms of energy. A modern diesel engine plant provided all the energy required for the instruments and the 18 outbuildings where dozens of technicians and astronomers lived and worked.

The dome and telescope were ready by April 1939, but the mirror was still missing. Originally expected in 1940, it would not reach the summit until seven years later in an adventurous two-day transfer. The operation to move the delicate and cumbersome jewel to the pristine and wild summit of Mount Palomar began at 3:30 a.m. on November 18, 1947, when the San Diego Highway Patrol gave permission to proceed. The mirror had been loaded on a special flatbed trailer equipped with multiple axles to increase safety at critical points along the route, with the 35 ton weight properly distributed. It was pulled by a single lead truck and flanked by two

---

[30] C.A. Chant, *Completion of the Tube of the 200 inch Telescope (with plates VI-VII)*, in "Journal of the Royal Astronomical Society of Canada", 31, 1937, pp. 241–243; *A History of Palomar Observatory*, cit.

[31] C.S. McDowell, *As The 200 Inch Telescope Develops*, in "Scientific American", 155, 5, 1936, pp. 253–257.

others equipped to help push it from the rear when needed. Another truck carried spare parts and accessories. A few cars followed, packed with journalists and cameramen recruited to capture the feat. The roads involved in the slow march had previously been blocked by police, bridges reinforced and, where necessary, the deepest potholes filled. The 160-mile journey was made almost at a walking pace, with one stop planned for the night. The operation was hampered by a heavy downpour that turned to snow in the mountains. Upon arrival at the observatory, the mirror was placed in the aluminization chamber, a monumental chapel where Sleeping Beauty received the kiss of life in the form of a surface layer of pure aluminum. Less than a millionth of a millimeter thick, it gave the glass the sheen of polished silver. A final, fundamental tweak to transform the Pyrex block into the most sophisticated trap ever built by mankind to capture the faintest messages from the faintest celestial sources.

Six months later, on June 3, 1948, the telescope, though not yet fully operational, was solemnly inaugurated in the presence of a thousand people, all seated in the Pharaonic dome to welcome this "eighth wonder". Hale was not among them. For the past decade, death had kept him from his dream. «It is a beautiful day. The Sun is shining and they are working on Palomar»,[32] he had said one of the last days, looking at the blue sky from the window of the clinic where he was hospitalized. He disappeared too soon, but he was not forgotten in the moment of triumph. In fact, the "New York Times" proposed to name the large mirror and the telescope after him. It was a posthumous tribute to the man who had done more than any other to arm the hand of twentieth-century astronomers, and a medal for the well-known ability of American journalism to influence history; a remarkably different behavior from the servility and acquiescence of the press in some countries of Old Europe.

Not even Hubble, the noblest of Hale's warriors, could enjoy the giant's favor for long. He had a clear vision of the new machine's potential. When a BBC reporter asked him in 1948 what discoveries he thought he would make with the new instrument, he replied brilliantly: «We hope to find something we didn't expect».[33] So the honor of officially opening the Great Eye, a kind of technical inauguration that astronomers romantically call first light, was reserved for him, the observatory's scientific and media star. It happened on a clear, cold night in January 1949. Crouched in the cage suspended above

---

[32] Florence, *The Perfect Machine*, cit., p. 313.
[33] V. Rubin, *Bright Galaxies, Dark Matters*, AIP Press, Melville (NY) 1996, p. 76.

the mirror — where the light reflected from the aluminum patina met some large lenses to wash out the aberrations induced by the Pyrex paraboloid and to regain a sufficiently large corrected field before hitting the photographic plate — the Major ordered to point the telescope on an old friend of his, the equatorial nebula NGC 2261.[34] The result was both astonishing and disappointing. The machine worked like a Swiss watch. The image of the delicate gas fan associated with the variable R Monocerotis stood out in all its glory in the eyepiece. But the stars did not appear as pinpoints as they should be. The cause of the aberration was quickly discovered. The edges of the mirror were not shaped as required by the optical design. There was also some residual astigmatism and even mechanical problems in the tracking system.

The director of the observatory was faced with a Hamletic dilemma: begin observations with a long-awaited, potentially perfect, but still somewhat sluggish instrument, or postpone its commissioning yet again and run the risk of having to rework the optics. Resisting pressure from those willing to settle for a start, Adams ordered, with his heart in his throat, that the mirror be disassembled and shipped to Caltech for the necessary improvements. The prestige of the institution he represented was at stake; a reputation based on scientific and technological records that the media, which had been very attentive to the project and valuable partners up to that point, would not hesitate to destroy, out of intellectual honesty or simply for the sake of news, if the Cyclops still proved inadequate. In short, it took several months of work and heartbreak before Adams could finally declare the hunt for the stars open.

While technicians scrambled to fix the optical defects, Hubble decided to take a vacation with his wife to practice his much-loved and very British fly-fishing. He was in Colorado, in the middle of the great nothing of the red rocks, when he had a heart attack. He came out alive, despite logistical difficulties, but not the same man. The proud lion had lost his claws just as the hunt was getting rich; a genuine torture of Tantalus, the mythological figure who, after death, was condemned by the gods for his sins to a hunger and thirst that could not be quenched forever. The Major held out for four years, delegating the management of the observations to two promising students, Allan Rex Sandage and Halton Christian ("Chip") Arp. In the fall of 1953, when even the guns of faraway Korea fell silent, making the Cold War

---

[34]G.E. Christianson, *Edwin Hubble: Mariner of the Nebulae*, Farrar, Straus & Giroux, New York 1995, p. 316.

a little less icy, his heart finally gave out. «For several years he had suffered from a heart ailment, but his health had improved considerably within the past year, and his death comes as a severe shock to all his friends and associates. It is difficult to estimate the loss to astronomy», Adams would have written in his *Obituary*.[35] Instead, Edwin felt that he was nearing the end. A few days before he died, he was walking with his wife by the 200 inch dome. Perhaps in a moment of tenderness, Grace reminded him of the portrait a colleague had painted of him before they met: «[Edwin] is a hard worker, he wants to discover the universe; this shows how young he is».[36] He must have shaken his head, sadly aware that he would not live to see the end of the great cosmological program he had begun.

At the time of this great man's death there was no poet to sing his praises. His refusal to adapt to the changes in the style of astronomers, now so easy and so different from his Oxonian ideal of hierarchical protocols and old manor stereotypes, had slowly crumbled the immense popularity of his character, carefully constructed together with Grace. But even he, the "Lone Star" of astronomy, would not have been forgotten by the compatriots who had so enjoyed the tales of his discoveries.[37]

With the commissioning of the Hale telescope, Mount Palomar became the navel of the astronomical world, the Mecca for every ambitious observer of the sky, and it would remain so for almost half a century thanks to its giant. But another instrument, unique of its kind, had already been operating at the observatory for a decade: a telescope with an aperture of only 18 inches, but with a corrected field of view of $4° \times 4°$, equivalent to an area of sky covered by 64 full moons arranged like the squares of a chessboard. Perfect for extensive sky surveys, also because of its short focal length, which enhanced its luminosity and reduced exposure times. The project of this formidable explorer came from Germany. A soft-spoken young scientist, destined to become a kind of Darwin of the stars, had brought it to Pasadena from Hamburg.

Walter Baade was born in 1893 in a small Westphalian hamlet, the first of four children of an elementary school teacher.[38] A congenital deformity

---

[35] W.S. Adams, *Obituary*, in "The Observatory", Vol. 74, 1954, p. 32.

[36] N.U. Mayall, *Edwin Powell Hubble*, in "Bibliographical Memoirs NAS 41", https://apod.nasa.gov/diamond_jubilee/1996/hubble_nas.html.

[37] A. Sandage, *Edwin Hubble (1889–1953)*, in "The Journal of the Royal Astronomical Society of Canada", 83, 6, 1989, pp. 351–362.

[38] D.E. Osterbrock, *Walter Baade: A Life in Astrophysics*, Princeton University Press, Princeton (NJ) 2001.

in his hip had from the start endangered his integration into high-ranking Prussian society, so much so that his father hypothesized an ecclesiastical career for him. But evil does not always do harm. The physical defect would have spared the young man the slaughterhouse experience of the First World War. Like every bourgeois scion of the Kaiser's perfect hive, Walter received an excellent classical education. Upon entering the United States a few years later, he would declare that he spoke German and English and could read French, Latin, Greek, and Hebrew. After graduating from high school, he enrolled in a math and physics program at the University of Göttingen, Gauss's alma mater and a mathematician's paradise ever since. At the time, the Institute of Astronomy was headed by a fine scientist, Johannes Franz Hartmann, known for the optical test of the same name and for having discovered the spectroscopic signs of interstellar gas.[39] But the real big guns sat in the mathematics department; people of the caliber of David Hilbert and Felix Klein, two giants of 20th century science.

Baade loved practical astronomy. From Hartmann he learned spectroscopy and the art of photometric measurement, which he would later put to excellent use. However, conditioned by the Jesuit academic logic that favored opportunity over inclination, he ended up working for three years as an assistant to the elder Klein, overwhelmed by the maieutic enthusiasm of a teacher who thrived on incomprehensible abstractions. Nevertheless, this was an important period of his life and education, spent in a melting pot of ancient knowledge and revolutionary new ideas. While his exhausted and disbelieving homeland contemplated its own moral and material ruins, harshly punished and unnecessarily humiliated by the victorious powers,[40] he managed to forge a lasting friendship with some young scientists who were to become pillars of twentieth-century physics: first and foremost among them Wolfgang Pauli. The story is this.

In 1919, Klein had urged the German physics guru, Arnold Sommerfeld, to write an essay on relativity for his monumental *Enzyklopädie der*

---

[39]In 1904, while serving as an astronomer at the Potsdam Observatory, Hartmann observed in the spectrum of the binary star δ Orionis the presence of ionized calcium lines that did not participate in the Doppler oscillation of the other lines. From these steady-state characteristics he deduced that somewhere along the line of sight between the Sun and δ Orionis there was a gas cloud causing this absorption (S.J. Dick, *Discovery and Classification in Astronomy: Controversy and Consensus*, CUP, Cambridge 2013, p. 84).

[40]The memorial to the students of the University of Berlin who died in the Great War bore the motto: "*Victis invictis victuri*" (to the defeated the undefeated who will win): a sign of the overwhelming thirst for revenge of a nation humiliated by the defeat.

mathematischen Wissenschaften (Encyclopedia of Mathematical Sciences), but received a polite refusal along with the suggestion that the task be assigned to one of his students,[41] Wolfgang Pauli, a 19-year-old Austrian boy prodigy and godson of Ernst Mach. Klein invited the young man to Göttingen to work on the article and instructed Baade to meet him at the train station and help him settle in. Although profoundly different, the two became very good friends.[42]

As we have said, Baade preferred practical astronomy to Klein's abstruse mathematics. Nearing graduation, he applied for a position at the Hamburg Observatory and in 1920 got it on the strength of excellent references from his teachers. As a novice, he was assigned a boring and routine task: measuring the positions of comets and asteroids (he would discover 10 of them in his lifetime). But he was also allowed to conduct his own research with the one-meter Zeiss reflector, the largest telescope on German soil at the time. A voracious and omnivorous reader of scientific literature, Walter knew and appreciated the work being done on nebular variable stars across the ocean. He therefore became passionate about the study of RR Lyrae. He also tried to contact Shapley, the man who had built his scientific fortune on these variables, but without much success. The American leader frowned upon intrusions into a field he considered his own, and he also feared that the young German's discovery of a few halo variables, faint and far from globular clusters, might disrupt the cosmological picture that had made him famous.

Walter did not give up. He combined the necessary ambition with a good-natured character and a genuine intellectual honesty, a rare commodity among great explorers. Wanting to see America's legendary astronomers up close, he openly asked Shapley to host him on a year-long tour of the United States and Canada, with the goal of visiting the major observatories, including Mount Wilson. With great foresight, his director, whose esteem he had earned, offered to pay half the cost of the trip. A great effort for those lean times in the Weimar Republic. The Rockefeller Foundation paid

---

[41]Sommerfeld's other bright students were Werner Heisenberg, Peter Debye, and Hans Bethe, all three Nobel laureates.

[42]They remained friends for life, despite Pauli's difficult temperament. They met for the last time in 1958 at the Solvay conference on *The Structure and Evolution of the Universe*, where Walter enjoyed the great popularity he had gained with the discovery of stellar populations (see later). Both Pauli and Heisenberg attended, and William Lawrence Bragg chaired the meeting. An all-star cast to celebrate the queen of the 20th century sciences!

the rest. Thus, in 1926, Baade sailed from Hamburg to the New World, where he met those who were writing the history of modern cosmology and became known and appreciated for his scientific and human qualities. Back home, he was commissioned to participate in the 1927 expedition to the Arctic Circle and the 1929 expedition to the Philippines, both to observe a total solar eclipse. The phenomenon on which astrophysics had been built in the second half of the 19th century was back in vogue after Eddington promoted it as a litmus test for falsifying general relativity. On both missions, Baade was accompanied by Barnhard Voldemar Schmidt, an Estonian optician of German father and Swedish mother, who worked as a freelancer at the Hamburg Observatory.[43]

Schmidt was a reserved and introverted man who communicated with difficulty and reluctance with others, with a few exceptions, including the quiet Baade. He had been born in 1878 on a rural Baltic island, a flat strip of land a stone's throw from Tallinn, within the borders of the Russian Empire. An inquisitive and creative boy, he had lost his right hand and forearm in an accident while experimenting with gunpowder.[44] So had gradually turned to optics, working the bottoms of beer bottles to make photographic lenses until he became a very skilled professional. By the time he moved to Germany in 1901, he had turned his passion into a successful business. Using only his left hand, he ground lenses and mirrors of the highest quality and sold them through Zeiss-Jena. His customers were scientists of the rank of Hermann Vogel, director of the Potsdam Observatory, and Karl Schwarzschild. In short, he was an optical tightrope walker with a limited education but an instinctive craftsman's sensibility. Walter Baade later said of him: «A highly unusual man, this Bernhard Schmidt. He always worked in a claw-hammer cutaway coat and striped trousers of formal attire. He rebelled at any regular working hours. Money meant nothing to him. He liked his schnapps, and chain-smoked good cigars [...] His friends were few for he was shy and retiring. He prized his independence above everything else».[45]

Schmidt was at the height of his popularity and prosperity when fortune turned against him with the outbreak of the Great War. Considered an

---

[43] G. Wolfschmidt, *Bernhard Schmidt and the Schmidt Telescope for Mapping the Sky*, in "Baltic Astronomy", 20, 2011, pp. 187–194.

[44] E. Schmidt, *Notes on the Childhood and Youth of Bernard Schmidt*, in "Irish Astronomical Journal", Vol. 3, 1955, pp. 240–245.

[45] F. Dyson, *Astronomy in a Private Sphere*, in "The American Scholar", Vol. 53, No. 2, 1984, p. 178.

enemy because he was born in the shadow of the Romanov double-headed eagle, he was first interned by the Germans and then placed under close police surveillance. He lost everything and was never able to rise again. He even tried to flee to his native Estonia to escape the post-war inflation that ravaged Germany. No one seemed to want the services of a one-handed optician. In 1926, he finally found peace at the Hamburg Observatory in the Bergedorf district. The director, Richard Schorr, offered him hospitality as a "voluntary colleague",[46] with a laboratory but no duties, no hours, and no salary. It was there that the lonely Barnhard met Walter Baade, a kindred spirit. The two formed a kind of friendship that was cemented over the long days of sailing the Indian Ocean in search of the Black Sun: an enforced idleness spent on the upper deck loungers, downwind, with a blanket over their laps. During one of the interminable conversations, Schmidt let slip that he had designed an optical scheme for a telescope that could quickly cover large areas of the sky and return aberration-free images. Music to the ears of a man who wanted to hunt for variable phenomena and needed an instrument to detect them.

Back in Hamburg, Baade suggested that Schmidt build a prototype of the wide-angle camera he had talked about on the ship. The Estonian demurred, explaining that he still had to find a way to make the refracting component that was an integral part of the design. But it did not take him long. «By the summer of 1930 [Schmidt] had completed his first 14 inch [telescope]», Baade would later recall. «He called me one sultry Sunday afternoon to say it was ready. From an attic window of the observatory he trained it on a cemetery. "Can you read the names on the tombstones?" he asked. "Yes", I replied, "but I can see only one thing: the optics are absolutely marvelous"».[47] Small and compact, the *Schmidt-Kamera* consisted of a single fast spherical mirror with a short focal ratio (typically $f/2$), stopped by an annular diaphragm at the center of curvature to homogenize its optical properties with respect to direction. In fact, a spherical mirror is axisymmetric and therefore appears with different cross-sections at parallel beams with different inclinations. The Schmidt design eliminates (almost) all vignetting and restores isotropy to the mirror over a wide solid angle. Each ray uses virtually the same fraction of the surface of the spherical surface (that of the shadow of the aperture).

---

[46] E.J. Öpik, *Bernhard Schmidt*, in "Irish Astronomical Journal", Vol. 3, 1955, pp. 237–240.
[47] Ibid.

The absence of an axis of symmetry magically made primary aberrations disappear, giving the mirror a much wider field of view than in conventional telescopes, which are limited by coma (see also p. 155). Unfortunately, solving one problem created another, since parallel rays reflected from a spherical surface do not converge at a single point. The outermost rays have a shorter focus than the innermost. But the ubiquity of spherical aberration, which does not exempt even on-axis sources, was a blessing in disguise in this case. The defect could be corrected over the entire field by a single optical element, a thin objective lens, properly shaped, placed at the center of curvature, and designed to keep chromatic aberration within tolerable limits.

However, as we have often said, optics is like a short blanket: if you pull on one side, you inevitably expose the other. A first problem is that a spherical mirror returns a focal surface that is also spherical. Therefore, the detector, be it a celluloid film or a glass plate, must be brought into the right shape by means of a holder with a convex bottom, taking advantage of the plasticity of the material. Alternatively, the field can be flattened by an optical system which, in its simplest form, consists of a convex lens placed in contact with the detector. A second complication is that the focus formed inside the instrument is difficult to access, which is why the term "Schmid chamber" is often used instead of telescope.[48] This makes direct viewing access cumbersome and impractical, so Schmidt telescopes must be accompanied by finders of sufficient brightness. From a manufacturing standpoint, the greatest difficulty is in machining the thin aspherical corrector. Schmidt devised an ingenious method of making it, based on the plasticity of glass. He placed a plane-parallel plate on a hollow metal cylinder of appropriate diameter and thickness and sealed it to create a moderate vacuum inside. He then ground the top surface of the pressure-deformed glass sheet until it was flat, so that, left free to stretch, it would assume the desired geometry.

The Schmidt telescope made it possible, if not easy, to undertake a project that had previously been considered beyond human capability: to photograph the entire sky in a reasonable interval of time. Suffice it to say that the instruments used for the *Carte du Ciel* proceeded in shots of a few plates per night, and with a rather small field of view ($2° \times 2°$). Baade

---

[48]This impediment is eliminated in the Schmidt-Cassegrain configuration, in which a second mirror reflects the beam back to the holed primary so that light can be collected behind it.

was well aware of this. He managed to convince the sad-eyed Estonian genius to publish his invention with a three-page article, which is all he left us.[49] Fourteen years his senior, Schmidt did not live long after this exploit. A staunch pacifist, he did not recognize himself in Hitler's Germany, but he felt too old to get back into the game. Discouraged but still active, he fell ill and died of a lung infection in 1935. Within a few years, his name would become very popular in the astronomical world.

Things turned out quite differently for Baade. On his return from the Philippines, he learned from Pauli that the University of Jena wanted to ask him to head the local observatory: «I don't know what telescope they have there, but the weather will certainly be no worse than in Hamburg», was the biting encouragement of the friend.[50] Walter tried to negotiate. He asked for a one-meter reflector with a good spectrograph, an increase in the staff of astronomers and technicians, and an expansion of the library, a landmark for an avid reader like himself. He counted on financial support from Zeiss, which did not arrive because the company was still struggling with post-war reconstruction. Eventually Baade gave up. But he passed off his resignation from the prestigeous post as a sacrifice not to leave the observatory, simultaneously asking for a pay raise. Tough and pure, but not idealistic to the core!

Once the Jena deal was closed, Baade concentrated on the study of galaxies, which Hubble had just elevated to the status of fundamental building blocks of the cosmos. In 1931 he received a second job offer, this time from an American institution, the Mount Wilson Observatory. Knowing his talents, Director Adams had personally chosen him to strengthen the team of stellar photometrists in preparation for the commissioning of the 200 inch, which he hoped would be closer than it was. This time Baade did not hesitate. He enthusiastically accepted the invitation and promptly moved to California with his family. He would not return to Germany until after his retirement, in another post-war period, once again all uphill for his homeland. In Pasadena, he found another young German-speaking astronomer, Fritz Zwicky, with whom he would collaborate very profitably and then argue to the point of brawling for the rest of his life.

---

[49] B. Schmidt, *Ein lichtstarkes komafreis Spigelsystem*, in "Mitt. Hamburger Sternwarte in Bergedorf", 7, 36, 1938, pp. 15–17 (in German).
[50] Osterbrock, *Walter Baade*, cited above, p. 30. The source gives the concept but not the exact words spoken by Pauli.

Zwicky[51] had come to America in 1925 at the age of 27. A Swiss citizen, he had been born in Varna, a city and seaside resort in the Principality of Bulgaria on the Black Sea, the first of three children of a prominent Swiss industrialist with diplomatic ambitions and a Czech woman born in the shadow of the Habsburg black eagle. To ensure a successor in the family business, his father had sent his young son to his native village in the canton of Glarus, an Alpine valley on the border with Lichtenstein, for a German education and trade training. But Fritz was much more drawn to the sciences, as well as to his mountains, climbing and daredevil skiing. Many years later, the gruff astronomer would soften as he recalled all this: «Mountains were once my big adventure but it is over since a long time; I still dream from the wonderful days sometimes, read also a few pages from a mountain book. But the thought of doing again active mountain climbing has faded».[52]

When it came time for college, Fritz chose a course in mathematics and experimental physics at the *Eidgenössische Technische Hochschule*, the Swiss Federal Institute of Technology in Zurich. Einstein taught there, drawn by the reputation of the place where he had failed the entrance exam 20 years earlier. Herman Weyl, one of the greatest thinkers of the 20th century, supervised Zwicky's thesis at the end of the course. From 1928, Wolfgang Pauli would also hold the chair of theoretical physics at this prestigious university, protected from the racist claws of the Führer. After graduating with a thesis on ionic crystals under the supervision of the Dutchman Peter Debye, a future Nobel laureate in chemistry, and Paul Scherrer, co-founder of the European Organization for Nuclear Research (CERN), Zwicky came to the United States in 1926 on the usual Rockefeller Foundation scholarship to join the research group of experimental physicist Robert Millikan at Caltech. It is said that his choice of California was motivated by the presence of high mountains for hiking, climbing, and skiing; and it was there, a stone's throw from Mount Wilson, that this massive man with a heavy German accent, a bad temper, and a boundless imagination discovered the sky.

---

[51] J. Johnson Jr., *Zwicky: The Outcast Genius Who Unmasked the Universe*, Harvard University Press, Cambridge (MA) 2019; A. Stöckli, R. Müller, *Fritz Zwicky: An Extraordinary Astrophysicist*, Cambridge Scientific Publishers, Cambridge 2011.
[52] Many of the anecdotes quoted here can be found at https://www.dynamical-systems. org/zwicky/Zwicky-e.html.

Millikan was a big shot. Three years before Zwicky arrived at Caltech, he had won the Nobel Prize "for his work on the elementary charge of electricity and on the photoelectric effect".[53] The brilliant Swiss post-doc, who had a cult of genius and liked to call himself Einstein's disciple, was expected to contribute to quantum mechanics, the new, exquisitely European science of the infinitely small. But the marriage between the two phenomena was not a happy one. It is said that in one of his moments of rage, Zwicky even accused his titular boss of never having a good idea. Not bad for a rookie! «Well, good, young man, and what about you?», a piqued Millikan is said to have replied, only to receive an arrogant but truthful reply: «Every two years I have a good idea. Give me a topic, I will give you the idea!».[54] Without losing his aplomb, Millikan invited his young assistant to get out of there and go next door to measure himself with astrophysics.[55] Given how things turned out later, mankind should be grateful to Millikan for this outburst. Fritz did not allow it to be repeated twice, and without flinching he tackled head-on the problem of the emerging modern cosmology, the science of the infinitely large.

Although they were two very different characters, one gentle but not submissive, the other aggressive and domineering, Fritz and Walter bonded immediately. In addition to language and a passion for *sauerkraut* (sour cabbage), they shared a common interest in the galaxies and an impressive scientific and humanistic education. Baade had brought to Pasadena some photographs taken with Barnhard Schmidt's telescope. When Zwicky and Hubble saw them, they were thrilled and thought about getting this terrific fish-eye camera as soon as possible. The Major, accustomed to thinking big, aspired to an instrument of considerable aperture, suitable for flushing out his feeble targets. More realistically, the young Swiss declared that he would be satisfied with something small, as long as it could be procured quickly. Zwicky won, simply because at the observatory there were neither funds nor manpower for another challenging undertaking parallel to the 200 inch. If the aperture was kept sufficiently small, said patron Hale without undue

---

[53] *The Nobel Prize in Physics 1923. Robert A. Millikan. Facts*, https://www.nobelprize.org/prizes/physics/1923/millikan/facts/.

[54] K. Winkler, *Fritz Zwicky and the Search for Dark Matter*, in "Swiss American Historical Society Review", 50, 2, 2014, p. 27.

[55] In addition to physics and astrophysics, Zwicky exercised his talents in various practical activities, collecting more than 50 patents, mainly in the field of jet propulsion. In 1929 he also developed the tired light theory as an alternative to Lemaître's kinematic interpretation of cosmological redshift (see p. 381).

enthusiasm when he was involved in the deal, perhaps something could have been nibbled away from the giant telescope's budget to disguise the retreat as functional for the scientific management of the larger instrument. No one would have minded.

So it was. Once the decision was made, with a fistful of dollars and an extra effort on the part of Caltech engineers, in a short time the 18 inch aperture Schmidt camera was designed by Russel Porter, built at Caltech by John Anderson,[56] Sinclair Smith,[57] and Albert Brower, and installed in its dome on the mountain. The mirror had a diameter of 24 inches, reduced to an effective 18 inches by the diaphragm. With its short focal length (only twice the aperture) it promised deep images with relatively short exposure times over a field with a diameter of almost 9 degrees. Monstrously wide images, with an angular surface equivalent to a mosaic of 250 full moons, were collected on circular films with a radius of 3 inches, jokingly called "cookies". Hubble snubbed the little dwarf. Instead, from the first light in September 1936, Zwicky settled into the small dome on Palomar Mountain, intending to explore the cosmos and find out whether, as the Major claimed, galaxies were randomly distributed, or rather, as he correctly thought, island universes were organized into archipelagos. From a theoretical physicist accustomed to wallowing in the most abstruse mathematics, he gradually became a prophet of observational evidence as the primary source of discovery and understanding of phenomena. In his hands, the small Schmidt telescope became a powerful tool for celestial cartography and the bloodhound with which to hunt down those hypothetical violent and transient stellar phenomena that he himself, in collaboration with Baade, had named "supernova" and about which we will tell more in the next chapter.

The success of the instrument was enormous, far exceeding that of its main user. Zwicky's aggression, his verbal abuse, a manifest arrogance — he judged humility to be a lie and said he felt like a «lone wolf»[58] — along with his contempt for the work of others, especially those who did not think as he did, quickly earned him the enmity of his colleagues. He, «the big mouth», retaliated by calling them "spherical bastards",[59] that is, bastards from any

---

[56] J.S. Bowen, *J.A. Anderson, Astronomer and Physicist*, in "Science", 131, 1960, pp. 649–659.

[57] J.A. Anderson, *Sinclair Smith*, in "Publications of the Astronomical Society of the Pacific", 50, 296, 1938, pp. 232–233.

[58] Johnson Jr., *Zwicky*, cit., p. 16.

[59] R. Kirshner, *The Extravagant Universe: Exploding Stars, Dark Energy, and the Accelerating Cosmos*, Princeton University Press, Princeton (NJ) 2002, p. 144.

point of view: an obvious jargon legacy of a mathematized *forma mentis* (way of thinking). The expression had become common in the Zwicky family lexicon. The story goes that when his first wife once opened the door of their home in Austin, Texas, to a group of graduate students invited for dinner, she called out to him from the doorway: «Fritz, the bastards are here!». Dorothy was the daughter of Senator Egbert Gates, whose vast fortune had helped keep Palomar Observatory alive during the Great Depression. His marriage to her in 1932 had made Fritz a brother-in-law of Nicholas Roosevelt, cousin of Theodore, 26th President of the United States. The two divorced amicably, and in 1947 Zwicky remarried in Switzerland to Anna Margaritha Zurcher, with whom he had three daughters.

It took only one year of operation of the 18 inch telescope to convince the management and astronomers at the Mount Wilson and Palomar Observatories of the opportunity to quickly build a more powerful twin, a 48 inch aperture Schmidt telescope.[60] Once again, funds and experience were drawn from the great lung of resources and expertise accumulated for the 200 inch giant. The 72 inch diameter spherical primary — considerably larger than the aperture of the instrument, whose size is dictated by that of the diaphragm placed at the center of curvature — was commissioned from Corning. Instead, the correcting plate was manufactured using the Schmidt method at the Pasadena Observatory laboratory, a few blocks from Caltech, which was also responsible for the design and construction of the mount. The dome that would house the new wide-angle instrument on top of Palomar was completed when the war interrupted this project as well. Work resumed with the surrender of Japan and was completed in 1949. For 13 years, the little 18 inch remained the only glass eye looking at the sky on Mt. Pigeon!

When the project was completed, scientists at Mount Wilson lined up to use the newcomer; a medium-power telescope that, with exposures of only a few minutes, was nevertheless capable of imaging areas of the sky equivalent to a mosaic of 150 full moons, recording them on square glass plates 35.5 cm on each side. They were true mines of objects and phenomena, some of them variable, transient, or even extemporaneous, scattered across the vast sky.[61] But most of the astronomers were disappointed. The

---

[60]R.G. Harrington, *The 48 inch Schmidt-type Telescope at Palomar Observatory*, in "Publications of the Astronomical Society of the Pacific", 64, 381, 1952, pp. 275–281.
[61]Cf. M. Capaccioli (ed.), *Astronomy with Schmidt-Type Telescopes*, in "Proceedings of the 78th Colloquium of the International Astronomical Union", D. Reidel Publishing Co., Dordrecht 1984.

management of the observatory had decided to use a large fraction of the time of the "Big Schmidt" — so called to distinguish it from the 18 inch — to carry out a photographic survey of the entire sky accessible from Mount Palomar, corresponding to about 3/4 of the entire celestial vault (about 30,000 square degrees). This feat would have taken several years: a mere trifle compared to the 100 centuries required for a conventional telescope to cover the same area.[62]

As usual, a sponsor was needed, and one was found in the National Geographic Society. Famous for its support of terrestrial expeditions to learn about planet Earth, this global non-profit organization agreed to fund the most comprehensive sky survey ever conceived. The program was launched as the National Geographic Society-Palomar Observatory Sky Survey, then contracted to POSS, with all due respect to geographers.[63] The vault of the sky, from the Pole to beyond the Tropic of Capricorn, was divided into 879 fields, each of which was photographed in rapid temporal succession with two different Kodak emulsions, one sensitive to blue light (coded 103aO) and one to red (103aF), in order to extract some information about the colors of the stars. The letter "a" in the plate code indicated that these emulsions were specifically designed for astronomical use, reflecting Kodak's interest in science. The plan was to push the exposure to the maximum depth compatible with the brightness of the night sky. This requirement meant that the dome had to be closed for some nights around the full moon, as the diffuse lunar glow — the same light that, according to Oscar Wilde, helps dreamers find their way[64] — raised the celestial background to intolerable levels. The work lasted almost a decade, ending in 1958. To collect the 2 × 936 plates of the complete mosaic in two colors, according to strict quality standards, more than 4,000 exposures had to be taken: an optical glass column as tall as a three-story building!

The enterprise was an all-American one, and it was rightly left to the Palomar astronomers to sift through the formidable archive. But Pandora's

---

[62] «Although certain larger instruments, such as the great reflectors on Palomar, Mount Wilson, and the Lick Observatory, could probe deeper into space, they had very small fields of view, less than the size of the Moon. A complete sky survey with one of these telescopes would have required at least ten thousand years»; G.O. Abell, *The National Geographic Society-Palomar Observatory Sky Survey*, in "Astronomical Society of the Pacific Leaflets", 8, 366, 1957, pp. 121–126.

[63] J.M. Lund, R.S. Dixon, *A User's Guide to the Palomar Sky Survey*, in "Publications of the Astronomical Society of the Pacific", 85, 504, 1973, pp. 230–240.

[64] «Yes: I am a dreamer. For a dreamer is one who can only find his way by moonlight, and his punishment is that he sees the dawn before the rest of the world»; Oscar Wilde, *The Critic as Artist*.

Box did not remain the exclusive province of those who had conceived and carried out the project. It was decided that the survey would be distributed to anyone who requested it, in the form of more expensive and rare negative film copies or cheaper prints on photographic quality paper. It was the first step toward the public domain policy of scientific databases to which materials collected by powerful ground- and space-based astronomical instruments are now usually subjected after a short proprietary period. Typically, a year after acquisition, the data, in the original and/or in the form of lists of astrophysical parameters of the cataloged objects, are uploaded to online archives, usually accessible to all without much formality.

POSS pioneered other surveys entrusted to Schmidt telescopes. In the 1970s, the mapping of the southern hemisphere was completed by the British wide-field telescope at Siding Spring Observatory in New South Wales, Australia: an instrument virtually identical to the 48 inch Palomar Schmidt, but with an achromatic corrector to overcome the defect of simple lenses in focusing light of different colors. Kodak's new fast fine-grained emulsions, called IIIaJ and IIIaF, were used for observations. They gave a better response but did not match well with the material previously used by POSS. The spectral sensitivity of a plate, combined with the other characteristics of the telescope and of the Earth's atmosphere, and with the filter chosen by the astronomer, determines the "passband" of the photons permitted to contribute to the image. Ideally, all those with wavelengths outside the passband are excluded. It is obvious that, under the same conditions, two different emulsions give rise to different passbands, i.e. images with different average colors. For this reason, in the 1980s, Palomar astronomers planned a new survey, POSS II, based on the same emulsions used by their British colleagues, to which they added a third one sensitive to the near infrared. For this occasion, the telescope, now obsolete and in danger of being scrapped, was updated with an achromatic corrector and an autoguide. The latter is an optoelectronic device capable of replacing the astronomer in keeping the instrument constantly pointed at the target during tracking (this operation requires a certain skill, either natural or artificial, as random seeing[65] complicates the perception of the true sky drift).

---

[65] Astronomers call "seeing" the set of effects (blurring and twinkling) that images of celestial objects undergo due to turbulence in the Earth's atmosphere. It is produced by rapid changes in the index of refraction along the line of sight, especially in the region immediately above the observing instrument. Seeing, whose effects increase with frequency, is the primary source of degradation of optical and infrared images, which would otherwise be limited only by the aperture of the telescope.

A new look for the old Cyclops, and a new name, too. Thanks to a clever marketing campaign, the 48 inch was in fact renamed after Samuel Oskin, a manufacturing industrialist, banker, philanthropist, and courageous California explorer.[66] This was the result of a generous donation to the Palomar Observatory. The specific funds to upgrade the "big Schmidt" and carry out the survey came instead from the Alfred P. Sloan Foundation, established by the patron of General Motors, by the National Geographic Society, and by the Federal Government of the United States. Kodak donated the photographic material and some cash. In 1999, when the work was completed, the glorious Oschin telescope was further modified for robotic operation. A year earlier, the one-meter Schmidt, built by the consortium of European countries called the European Southern Observatory (ESO)[67] and inaugurated in 1971 at La Silla Observatory in the Chilean Andes, has also been decommissioned and reduced to a digital planet hunter.[68] An improved copy of the Hamburg Schmidt, built in 1954 with a 120 cm primary and an 80 cm correcting lens, and transferred in 1976 to the Spanish-German observatory of Calar Alto in Andalusia, the ESO telescope was able to compete worthily with its two older brothers, given the spectacular weather conditions of the South American *Cordillera*.

In the second half of the twentieth century, a handful of Schmidt telescopes distributed between the two hemispheres was enough to overturn the topography of the Milky Way and the cosmos itself, and to transform from an extemporaneous fact into a fruitful systematic research the census of rapid and transient phenomena such as novae and supernovae, and even of the small celestial bodies swarming rapidly through the Solar System. Not only with images, but also with spectra obtained through objective prisms, large and slightly wedge-shaped transparent glass plates placed at the primary focus of the telescope. For example, over a 15-year period beginning in 1959, the 48-inch Palomar Mountain Explorer detected 178 supernovae,[69] and in just 12 months, in 1960, it discovered over 2,400 new asteroids (for three-quarters of which orbital parameters could also be

---

[66] Among his many exploits, Samuel Oskin visited the North Pole, traversed the Amazon in a canoe, and crossed the Alps on the back of an elephant, just as Hannibal had done when the Carthaginian general descended into Italy during the Second Punic War.

[67] G. Schilling, L.L. Christensen, *Europe to the Stars: ESO's First 50 Years of Exploring the Southern Sky*, Wiley-VCH, Weinheim 2012.

[68] C. Madsen, *The Jewel on the Mountaintop: The European Southern Observatory through Fifty Years*, Wiley-VCH, Weinheim 2012.

[69] P. Murdin, L. Murdin, *Supernovae*, CUP, Cambridge 1985, pp. 43–45.

inferred; an important loot for that sky patrol on which humanity's very survival may depend). In just three years, the instrument made it possible to determine the positions, luminosity, and in some cases even the proper motions of millions of stars in the Milky Way, extending knowledge to the distant halo surrounding the Galactic disk, made up of very old stars, as well as to the faintest stars in the solar neighborhood. An indispensable database for, among other things, NASA's effective management of Hubble Space Telescope (HST) targets. Just as Big Bertha in World War I needed observers to know where to fire its formidable projectiles, modern giant telescopes require sensing instruments to identify targets to focus on. This is in some ways a peculiarity of astronomy as a science whose immense and extremely rich laboratory already exists and cannot be manipulated, but only observed.

The Schmidt telescopes also revolutionized the technology of examining and measuring photographic material: a transformation similar to what it would take for someone to go from tending his grandmother's vegetable garden to managing a vast plantation of a wide variety of products. The heroic era of patient astronomers hunched over their microscopes in the manner of Henrietta Leavitt, inspecting backlit glass plates millimeter by millimeter, or determining the positions and fluxes of a handful of objects, now seemed almost anachronistic, given the amount of information contained in each image. Too much work for humans, and too inefficient and unreliable. The solution was to enlist the help of computers, which were incomparably faster and, despite their lack of imagination and creativity, far more accurate. To feed their insatiable appetite for data, however, it became necessary to digitize the photographic material, converting each plate into a matrix of numbers. This is how it works. Ideally, we divide the photographic image into a number of equal squares (pixels) and sample the average degree of blackening within each of them.[70] By arranging the results in a grid, whose rows and columns of numbers reproduce the position of the pixels in the image, we obtain the digital matrix that electronic processors like to work with. Of course, the size of the pixels is not random. We choose it according to our needs, always keeping in mind that

---

[70]In optics, transparency is the property that allows light to pass more or less well through a medium. It is measured by the amount of transmitted light relative to the amount of incident light, and thus can take values between 0 (total opacity) and 1 (perfect transparency). Photographic density, instead, is the inverse of the logarithm of transparency.

it should not be less than the spatial resolution of the photographic material, nor more than the minimum detail we do not want to lose. In fact, oversampling would lead to a deterioration of the signal-to-noise ratio.[71]

The first modern machines for scanning and digitizing plates were built in the 1960s for military purposes. The Cold War required aerial surveillance of the enemy's defensive and industrial installations. For the national security of each of the two blocs that ruled the planet, it was vital to extract as much information as possible from the images collected by spy planes. Fortunately for astronomers, new solid-state digital detectors allowed the high classification of photographic plate measuring instruments to be quickly lifted. In the 1970s, digital microdensitometers such as those from Photometric Data Systems (PDS) Corporation, marketed by California-based Boller & Chivens, a division of Perkin Elmer Corporation, hit the market. A real boon to research institutions that could afford to buy them. But there were also those, like the Scots and the Dutch, who preferred to design specific instruments from scratch[72] for fast and accurate scanning of large Schmidt plates, for targeted research programs, and for complete digitization of sky surveys such as POSS II. For example, the 897 plates for each of the three colors of the second *Palomar Observatory Sky Survey* were the raw material for the *Digital Sky Survey* (DSS) produced at the Space Telescope Science Institute (STScI), the second guide star catalog of HST, and the astrometric and photometric catalogs of the United States Naval Observatory, which is a kind of National Bureau of Standards for astronomy. Vast online archives containing over one million stars and 50 million galaxies, which have allowed the identification of tens of thousands of galaxy clusters, hundreds of supernovae and quasi-stellar objects, and dozens of comets and asteroids.

Obviously, digital images would be of little use if the computer to which they are fed were not properly trained to analyze them. The need, not only in astronomy, coupled with the availability of ever faster processors and ever larger memories (which have grown by a factor of more than 10^6 in half a century), has spawned a new science with a robust theoretical apparatus and countless practical applications. Image processing uses mathematical

---

[71] Noise decreases as the sampled area increases, as suggested by the fact that, to improve a noisy data set, one takes the average. It is therefore an error to reduce the sampling area below the resolution: you lose on the noise front and gain no additional information.

[72] The abbreviations of these prototypes are, for example, COSMOS, APM, SUPERCOSMOS, ASTROSCAN, and PMM.

operations, including very sophisticated ones, to manipulate, correct, and visualize images reduced to numerical matrices, to identify sources, catalog them, and extract their parameters (e.g., relative positions, size, and brightness) and even features such as stellar vs. extragalactic nature or, if applicable, the geometry of the light distribution. To detect possible transients, it will then be sufficient (in principle) to compare the object catalogs of two images of the same field taken under the same conditions but at different epochs, looking for sources that appear different (in brightness or position) in the two exposures or that show up in only one of them: a modern version of the ancient practice of variabilists who superimposed two plates of the same field exposed at different epochs, one of which was positive, to look for intruders or wanderers.

When the amount of data is huge, as in the case of large digital surveys, other needs and more daring user demands come into play. To meet them, a new discipline of astroinformatics called "data mining" has emerged.[73] It is a methodology of inquiry that straddles artificial intelligence, machine learning, statistics, and database manipulation systems. The ultimate horizon of this formidable tool is to assist the researcher in the discovery of phenomena by exploring multidimensional spaces of parameters impractical for humans, looking for correlations and peculiarities. In short, the ambition of the data miners is to elevate the computer from the rank of servant to that of scientist *sui generis*: a colleague rather than a servant to the astronomer, although, at least for the time being, man has the last word. God knows what will happen in the future if, as some fear, artificial intelligence takes over from human intelligence.

More recently, the advent of solid-state digital detector mosaics, which we will discuss in the final chapter, has allowed direct acquisition of large-format images ready for computer analysis. For example, in 2008 the Samuel Oschin telescope, upgraded with the installation of a $12,000 \times 8,000$ pixel detector capable of covering a field of nearly 8 square degrees, was dedicated to a Caltech project originally called the "Palomar Transient Factory" (PTF). This is a fully automated search for rare and unknown phenomena and variable sources. The novelty lies in the real-time data reduction that dramatically reduces the gap between discovery and follow-up investigation. PTF has discovered tens of thousands of new transient sources, including

---

[73]See, e.g., N.M. Ball, R.J. Brunner, *Data Mining and Machine Learning in Astronomy*, arXiv:0906.2173, https://arxiv.org/abs/0906.2173.

thousands of supernovae, novae, cataclysmic variables, and young stellar flares.

In the new millennium, Schmidt telescopes, which made many historic contributions to astronomy in the second half of the 20th century, have been replaced by innovative optical designs, which we will discuss at the end of our gallop. Thin corrective lenses, designed to compensate for spherical aberration, could not meet the demands for ever larger apertures due to the unacceptable increase in weight and thickness of a suspended element, as was the case with the objective lenses of the giant refractors of the early 20th century. But, it must be said, even in the world of science, everything ends. It comes to mind White Rabbit's answer to Alice's question: «How long is forever?», in Lewis Carroll's masterpiece: «Sometimes, just one second».

Chapter 16

The Sun and the other stars: How the stars work, how they live and for how long

> For what sort of thing,
> from among those you don't know,
> will you put forward
> as the thing you're inquiring into?
> And even if you really encounter it,
> how will you know that
> this is the thing you didn't know?
>
> Plato, *Meno*

> The stars are golden fruit
> upon a tree
> All out of reach.
>
> George Eliot, *The Spanish Gypsy*

Let us retrace our steps once more. In the early decades of the twentieth century, the offensive launched by a handful of daring astronomers led by Hale, armed with the powerful West Coast telescopes and supported financially by a patronage unparalleled since the Renaissance, had opened to human curiosity the vast and unexplored prairies of the realm of galaxies. «No one knew [what galaxies were] before 1900. In 1920, very few people knew. All astronomers knew after 1924»,[1] Allan Sandage would later write, proudly and perhaps a little prematurely, in praise of his master Hubble. The lessons imparted to the world by these pioneers of the heavens touched every aspect of the scientist's profession. Forged in personality and social sensibility by the experience and myth of the frontier, American astronomers were not impressed by the logistics bordering on adventure, nor by the physical exertions, dangers, and loneliness of mountaintop

---

[1] A.R. Sandage, *The Hubble Atlas of Galaxies*, Carnegie Institution of Washington, Washington (DC) 1961, p. 1.

outposts. They had chosen the sites with the best skies for their observing stations, not the most convenient, or the cheapest, or the closest to the centers of power, both political and academic, as was the custom in Europe. On the old and quarrelsome continent where it was born, astrophysics struggled to keep up with the "Yanks" in the field of observations, held back in its impulses by the conservatism of some glorious personalities of science and culture and by the scabs of history, penalized by the lack of competitive instruments and wealthy private protectors, and crushed by the spectacular successes of experimental and theoretical physics, which the powerful considered more important than the exploration of the heavens because of their repercussions on the works of peace and war. The Europeans, however, still held the scepter of research into the structure and evolution of the stars. But the actors in these brilliant investigations were mostly philosophers, physicists, engineers, or even mere amateurs, and only rarely professional astronomers. So we return to the Old World and go back far enough in time to find the first tentative answers to the question: what are "the Sun and the other stars"[2] made of, and how do they work?

In the mid-19th century, one of the most intriguing scientific cases on the table was the Sun itself. With an average density of 1.4 grams per cubic centimeter,[3] the star was conceived as a liquid sphere very close to that of water[4] in a slow and progressive cooling from an initial state of high temperature, reached at birth by some unknown cause. The story of passive evolution, which would soon find a coincidental support in the sequence introduced by Angelo Secchi to classify stellar spectra (see the note at p. 295), was metabolized to such an extent that in the jargon of astronomers the hot and cold extremes of the temperature scale of stars are still referred to by the adjectives "early" and "late" (which, however, have now lost their temporal meaning).

---

[2] Final words of Dante's *Divine Commedy.*

[3] To determine the mass and volume of the Sun, it is sufficient to know the radius of the Earth's orbit, i.e. the Astronomical Unit, the duration of the sidereal year, and the angular radius of the Sun, as well as the value of Newton's universal gravitational constant G. All these quantities had been measured with sufficient precision in the 19th century.

[4] The shapes assumed by the rapidly rotating liquid spheres had already been analyzed in the first half of the 18th century by the Scotsman Colin Maclaurin, a hundred years later by the German Carl Jacobi, and again, at the end of the 19th century, by the Frenchman Henri Poincaré.

A first turning point with respect to this naive model came with the proposal, which can be traced back to Kant,[5] that stars are active furnaces, «iron balls» able to burn for a long time thanks to the energy produced by an unspecified internal source. The first to speculate about the nature of the fuel was probably Julius Robert von Mayer.[6] We have already met him (p. 248) speaking of the principle of equivalence between mechanical work and heat. In the wake of Kant's suggestion, Mayer treated the Sun as a heat engine, elevating the «great secret» of its source of heat and light — in the words of William Herschel[7] — to the central problem of solar physics and astrophysics. The question to be answered was the lifetime of a star made of coal or any other traditional fuel, which was far too short compared to the age assigned to the Sun by the analysis of terrestrial rock layers and marine mollusk fossils: at least a hundred million years. Much more than 6,000 years, as the Bible scholars wanted![8] In fact, the Old Testament places the creation of the world only four thousand years before the birth of Christ. Even Kepler and Newton had fallen into the trap of understanding the Holy Book as a reliable cosmic calendar. Some went so far as to give not only the year, but the month, the day, and even the hour of the beginning of creation.[9]

---

[5] I. Kant, *Universal natural history and theory of the heavens*, Richer Resources Publications, Arlington (VA) 2008.

[6] *J.R. von Mayer*, in "Nature", 17, 1878, pp. 450–452.

[7] F. Cajori, *The History of the Conservation of Energy: The Age of the Earth and Sun*, in "Popular Science Monthly", 73, 7, 1908, p. 97; H. Kragh, *The Source of Solar Energy, ca. 1840–1910: From Meteoric Hypothesis to Radioactive Speculations*, 2016, arXiv:1609.02834; P.-M. Robitaille, *A Thermodynamic History of the Solar Constitution - I: The Journey to a Gaseous Sun*, in "Progress in Physics", 3, 2011, pp. 3–25.

[8] G.J. Whitrow, *The age of the universe*, in "The British Journal for the Philosophy of Science", 5, 19, 1954, pp. 215–225.

[9] The most famous example is that of James Ussher, the Anglican archbishop of the city of Armagh in Northern Ireland. The high prelate harbored two dominant passions: his hatred of papists and his quest for maximum precision in dating the event from which, according to Jews and Christians, everything began — by an act of God, since, according to classical philosophy, nothing in the world came from nothing. In the middle of the seventeenth century, Ussher published in the *Annals of the World* the result of his very learned reflections, according to which the Eternal Father would have put his hand to his work at 6 PM on October 22, 4004 B.C. (with reference, one must swear, to the Greenwich meridian!). The operation, which had been attempted in the Renaissance by the highly cultured Italian writer, philosopher, and physician Julius Caesar Scaliger, required immense culture and a thorough knowledge of ancient languages. Among other things, Ussher also had to tackle the daunting problem of correcting the error made by the Scythian monk Dionysius Exiguus (Dionysius the Humble; 5th-6th century) in fixing the date of Christ's birth, which the Irish had moved to the 5th year of the pagan era.

The hunting ground of priests, biblical exegetes, and the finest intellects for more than two millennia, the chronology of *Genesis* had undergone a first dramatic upheaval in 1775 with the Frenchman Georges-Louis Leclerc,[10] future Count of Buffon. Eclectic character of Enlightenment science, half globetrotter and half passionate and inventive scholar, accused of embodying the figure of "*arbiter elengantiarum*" (judge of refinement) in scientific presentations («*Le style est l'homme même*»,[11] the style is the man himself, was his motto). While still wearing the simple clothes of the *citoyen*,[12] using the resources of a foundry he owned in Burgundy, he had experienced the duration of cooling iron spheres with increasing dimensions. The extrapolation of the results to the mass of the Earth, which Buffon imagined to have been born from a splash of boiling solar matter following the passage of a comet, assigned to the planet, and therefore to the star that spawned it, a minimum life of 75,000 years. A sensational conclusion, and also a sacrilegious gamble that in other times and places would have sent the French straight to the stake of the Roman Inquisition. But in the France of Voltaire and Rousseau, this did not happen again until the Jacobins regained control of the brain and, with surgical precision, severed the heads of those who thought differently from them.

Having broken the ice with Buffon, who had exploited the gap left by the Enlightenment in the fog of the most reactionary conservatism, it would be the Scottish naturalists and geologists James Hutton and Charles Lyell — the latter a great friend of Charles Darwin — who would bring the lifetime of the Earth and its star, the Sun, into the million-year regime.[13] Numbers still very far from today's estimates, but large enough to blatantly falsify the hypothesis of energy production by pure combustion. It was Mayer who proved this, data in hand. In fact, he had the intuition to convert the measurement of the solar constant made by Pouillet in 1838 (p. 271) into an estimate of the total amount of energy released by the star every second. From there, he could calculate the fuel consumption, assuming a reasonable value for thermal power, which is the amount of energy obtained

---

[10] J.-P. Poirier, *About the Age of the Earth*, in "Comptes Rendus Geoscience", 349, 5, 2017, pp. 223–225.

[11] J. Dürrenmatt, «*Le style est l'homme même*». *Destin d'une buffonnerie à l'époque romantique*, in "Romantisme", 148, 2, 2010, pp. 63–76.

[12] He became Count of Buffon at the age of 66, but the honor did him no good because it led his son to the guillotine during the Terror.

[13] J. Repcheck, *The man who found time: James Hutton and the Discovery of Earth's Antiquity*, Basic Books, New York 2003.

by burning one kilogram of a given material. The result was discouraging. A coal Sun could have shone for a few thousand years at most before dying of exhaustion. It should be remembered that the solar constant is the amount of energy that passes through the unit area normal to the flux in one second at Earth's distance. Since it is measured at ground level, its value must be corrected for atmospheric absorption. It stands to the total energy released by the star in one second as the unit surface stands to that of the sphere with a radius equal to that of the Earth's orbit.

The whole argument suggested a universal principle: everything must change when it exhausts the energy reserve that makes it be as it is. In short, if you are broke, you must change your lifestyle! It follows that the longer a phenomenon persists in a given state, the greater must be the capital originally invested to support it, and the slower must be its consumption. To justify the longevity of the prodigal Sun, it became imperative to identify a resource adequate to the need. Perhaps inspired by the image of firemen feeding the boilers of early steamships with generous piles of coal,[14] Mayer became convinced that the solution to the solar conundrum was in a mechanism for reintegrating costs, and hypothesized that the star, born hot, could survive for a long time by raking in meteorites and comets. Much as a whale does when it scrapes the water to collect plankton. The idea is less naive than it might seem at first. In fact, it was not a simple in-flight refueling, but the simultaneous acquisition of the kinetic energy released by the bodies falling on the star. The blacksmith is able to heat the iron because each blow of his hammer converts much of its kinetic energy into heat. The work is done by the craftsman's arm. In the case of falling bodies, kinetic energy is gained at the expense of gravitational energy. «An asteroid, therefore, by its fall into the Sun», wrote Mayer, «developes [sic] from 4600 to 9200 times as much heat as would be generated by the combustion of an equal mass of coal».[15] In short, for the same mass, capture is actually a much more generous energy production mechanism than simple combustion, and it works until supplies are exhausted. Unfortunately, all

---

[14]Proving that science often finds inspiration in everyday life, in 1882 Carl Wilhelm Siemens, a famous Anglo-German engineer, inventor, and industrialist, had hypothesized an energy recovery mechanism for the Sun similar to the regeneration of blast furnaces. In the previous decade, the British climatologist James Croll, probably influenced by emerging and revolutionary economic theories, had instead alluded to "energy capital".

[15]H. Kragh, *The Source of Solar Energy, ca. 1840–1910: From Meteoric Hypothesis to Radioactive Speculations*, 2016, https://arxiv.org/ftp/arxiv/papers/1609/1609.02834.pdf.

that glitters is not gold. To balance the budget between expenses and revenues, and even to make the Sun immortal, it was necessary to assume the ingestion of 100 billion tons of minor bodies every minute. Far too much! In addition to the need for an adequate food supply, such an astonishing voracity would lead to a significant increase in the star's mass and a consequent progressive reduction of the sidereal year by 0.05 seconds per century. A phenomenon with dramatic anthropogenic implications for which there was no evidence. Mayer knew this, but hoped that the way out lay in an equivalent loss of mass through radiation, caused by some as yet unknown mechanism. He was not the first scientist to sweep an unsolved problem under the rug, and he would not be the last.

The accretion theory was independently revisited by the Scottish engineer John James Waterston, and especially by a young and rampant Irishman, William Thomson.[16] The latter, the future first Baron Kelvin, correctly believed that the dreaded increase in mass caused by the fall of meteorites had no effect on the motions of the planets where the "food" reserve was contained within the orbit of Mercury: «the added matter is drawn from a space where it acts on the planets with very nearly the same forces as when incorporated in the Sun».[17] There and then this idea enjoyed good press, but it did not convince another talented young man. Turning the tables, the Brandenburger Hermann von Helmholtz[18] proposed that the gravitational energy necessary to ensure a long life for the Sun was obtained not from the debris in the circumstellar environment, but, in a kind of autarky, from the contraction of the star, i.e. from the progressive and slow collapse of the object upon itself!

The idea was presented in a broader context in 1854. Helmholtz, who at the age of thirty-three already embodied the *homo universalis* figure (polimath) of Prussian science, had agreed to give a public lecture in praise of Kant in the city of Königsberg, where he himself taught physiology. The intention, not even too hidden, was to counter criticism of his research by

---

[16]Cf. F.D. Stacey, J.H. Hodgkinson, *The Earth as a Cradle for Life: The Origin, Evolution and Future of the Environment*, World Scientific Publishing, Singapore 2013.

[17]Lord Kelvin, *Mathematical and Physical Papers*, CUP, Cambridge 2011, Vol. 2, p. 9 (1st ed. 1882). This statement is easily understood by recalling that, by a theorem due to Newton, the external gravitational field of a spherical distribution of matter is equal to that of the total mass concentrated at the barycenter. For this and other topics of Celestial Mechanics, see E. Bannikova, M. Capaccioli, *Foundations of Celestial Mechanics*, Springer, Cham 2022.

[18]D. Cahan, *Hermann von Helmholtz and the Foundations of the Nineteenth Science*, University of California Press, Berkeley 1993.

his peer and fierce rival Rudolf Clausius, professor of Physics at the Royal Artillery and Engineering School in Berlin and *Privatdozent* at the University of Berlin (where he would become a full professor the following year). Helmholtz's agenda also included the presentation of the first two principles of thermodynamics: that of the conservation of energy and the second, still in embryonic form, on the irreversibility of physical transformations, represented by the evocative image of heat death, *der Wärmetod*, to which every closed system tends inexorably due to the impossibility of fully recovering the investments made in heat. In describing the applications of the new discipline to evolution, Helmholtz did not miss the opportunity to mention his idea of the source of solar energy. «It may be calculated that if the diameter of the Sun were diminished only the ten-thousandth part of its present length, by this act a sufficient quantity of heat would be generated to cover the total emission for 2,100 years».[19] The result was that the star, treated as a homogeneous sphere to simplify calculations, could last up to 22 million years without changing its habits, at the cost of shrinking from a semi-infinite size to its present size; far more than the paltry 32,000 years Thomson estimated for his model. Not a bad result for those times, though still very far from the modern estimate.

The Crimean War was underway: a dress rehearsal for a global conflict and a stage for the last heroics of the light cavalry, the compassion of Florence Nightingale, and the baptism of a new kind of journalism with the special correspondent of the London "Times", the Irishman William Russell. Signs of an epochal change in Western society. Marx and Engels had already infected Europe with the seeds of a revolutionary economic doctrine and a new social justice; and Charles Darwin was in the midst of working out his groundbreaking theory of the evolution of species. The great biologist and his geological friends needed a world with a long career on its shoulders to give the mechanisms of selection and natural forces enough time to do their work.

But the new physics seemed to have allied itself with the reactionary creationists, who were not at all concerned about time, which, as we know, belongs to God alone. In fact, the second principle of thermodynamics pointed to a relatively young Sun, bound, moreover, to a rapid end. A catastrophic thesis that also affected mankind, and that the great Thomson, after due consideration, would have maintained and defended

---

[19]H. Helmholtz, *Popular Lectures on Scientific Subjects*, Appleton, New York 1987, p. 190.

throughout his long career, not without fierce opponents, as the Austrian physicist Joseph Loschmidt. In 1876 he wrote that he felt obliged to fight against this pessimistic view of the world:

> Thereby, also the terroristic nimbus of the second law would be destroyed, which makes it appear to be an annihilating principle for all life in the universe; and at the same time the comforting prospect would be opened up that mankind is not only dependent on mineral coal or the Sun for transforming heat into work, but rather may have an inexhaustible supply of transformable heat available for all times.[20]

Helmholtz had elaborated his brilliant intuition in a oversimplified way because of the paucity of data, the incomplete physical basis, and the blatant falsity of some assumptions. For example, the German scientist was still thinking in terms of a homogeneous liquid Sun, practically incompressible and therefore resistant to collapse. Almost two decades should pass before the New Yorker Jonathan Homer Lane and Johann Zöllner, professor of physics in Leipzig and expert in optical illusions, would venture the hypothesis of a gaseous sphere. Moreover, a way had to be found to infer the trend of matter density from the center to the periphery of the star in order to relax the untenable homogeneous sphere assumption made when estimating gravitational heat production. However, direct observations could not provide this information. It then became urgent to construct a self-consistent model of stellar structure, capable of a mathematical formulation and free of contradictions or anomalies. The exercise seemed anything but elementary, even if it could be partially simplified by assuming equilibrium conditions, justified by the fact that if there was any evolution, it had to be very slow. Helmholtz and his British colleague Thomson — who, after some hesitation, had rejected the meteoric supply hypothesis in favor of that of the gravitational collapse — devoted themselves passionately to this task. But the first to cross the finish line were Lane and the German Georg Dietrich August Ritter.

Lane made his living in the patent business, first as an expert and consultant to the U.S. government and then on his own in Washington. This «little man, rather intellectual in appearance, who listened attentively

---

[20]W. Loschmidt, *Über den Zustand des Wärmegleichgewichtes eines Systems von Körpern mit Rucksicht auf die Schwerkraft*, in "Sitzungsberichte der Kgl. Akademie der Wissenschaften in Wien", 73, 1876, p. 135 (in German).

to what others said, but [...] never said a word himself [...] and who was quite alone in the world, having neither family nor near relative»,[21] had the insight to treat the star as a sphere of gas in which weight is everywhere balanced by pressure, invoking the same hydrostatic equilibrium that two millennia earlier had made Archimedes jump out of the bathtub to the cry of *Eureka* (I found it)! But the condition set by the Syracusan genius, while necessary, was not sufficient to solve the problem.[22] Lane then came up with a second idea. He postulated that density depended solely on pressure, according to a (power) law called "polytropic" because it could take different forms depending on the value of a free parameter. It was an arbitrariness justified by the inability to use the full form of the state equation[23] for an ideal gas, which also includes temperature as an additional independent variable. With this shortcut, the willing American was able to build a series of stellar models that, albeit with fingers crossed, provided for the first time a quantitative description of the interior of a star where the human eye could not and would never look, he told himself. Today, instead, we can "observe" the heart of the Sun thanks to the flux of neutrinos that pass undisturbed through the stellar body and tell us what is happening in the center. In fact, neutrinos interact only with the weak nuclear force and gravity, and are not affected by the strong nuclear force or the electromagnetic force. Unlike photons, whose mean free path is tiny, they can thus travel freely through the Sun, carrying information about the stellar core. But the same property that makes these messengers so pervasive poses an enormous problem when physicists try to capture them. The usual story of the short blanket![24]

It was 1870, the year of the sensational defeat of the French by the Prussians at Sedan. In Aachen, the former imperial capital on the border of Germany, Holland, and Belgium, an engineering professor at the *Technische Hochschule* (Technical University) was independently developing a model similar to Lane's. August Ritter dominated the physics of heat and fully owned the methodology and the intellectual tools required to produce

---

[21]S. Newcomb, *The Reminiscences of an Astronomer*, The Riverside Press, Boston & New York 1903, pp. 245–247.

[22]Technically, the problem is that for only one equation, that of hydrostatic equilibrium, there are two unknown functions, pressure and density.

[23]This is the thermodynamic equation relating state variables such as pressure, volume, temperature, and internal energy.

[24]G.D. Orebi Gann, K. Zuber, D. Bremmerer, A. Serenelli, *The Future of Solar Neutrinos*, in "Annual Review of Nuclear and Particle Science", 71, 2021, pp. 491–528.

results of general character.[25] The first and most important decreed that
a star is a machine with a negative heat capacity: a technical expression
to say that, when the star loses (gravitational) energy, it heats up instead
of cooling down. This would seem a physical absurdity, because normally
those who squander money should not get richer; yet this *modus vivendi*
(way of living) of stars is the key to understanding the ability of these
physical systems to generate and maintain a pressure exactly equal to that
required by hydrostatic equilibrium. The same conclusion had been reached
by Lane, and even earlier by Helmholtz using a property of mechanical
systems in equilibrium found in 1870 by Clausius — the famous virial theorem —, according to which, in the long run, the kinetic energy is just
twice the gravitational one. Ritter found also a formula for the period of
radial oscillation of a gas sphere, the first step in understanding the working
mechanisms of pulsating variables, and discovered a relation between stellar mass and absolute brightness: a fundamental pick for the construction
of an embryonic evolution theory. He was convinced that stars were born
from the collapse of diffuse clouds and in this way acquired their initial
capital of heat, with which they would then live by going through an initial
short phase of constant temperature and later a long phase of slow cooling.

Ritter's evolutionary scheme so inspired Norman Lockyer that he made
it his own, adapting it to the hypothesis that stars were formed by the
coalescence of meteoric fragments.[26] The imaginative director of the Solar
Physics Observatory in South Kensington, London, then worked out a correspondence between the characteristics of the observed sidereal spectra and
the different seasons of stellar life predicted by his conjecture. He was not
concerned with the inconsistency between the duration of his stellar model
and the age attributed to the Sun by Earth geology. The construction he
set up would not hold for long. Nevertheless, the basic idea was taken up in
1913 by the American Henry Norris Russell, whom we have already met in
his interactions with Hubble and Shapley. He tried to interpret his discovery, made simultaneously with the Dane Hertzsprung, of the existence of
two classes of stars, the giants and the dwarfs.[27] The Princeton professor
speculated that giant stars might designate the gaseous, hot state induced

---

[25] M. Meo, *Ritter, Georg August Dietrich*, in *Biographical Encyclopedia of Astronomers*,
Springer, New York 2014, pp. 1838–1839.
[26] J.N. Lockyer, *On the Causes Which Produce the Phenomena of New Stars*, in "Proceedings of the Royal Society of London", 49, 1890–1891, pp. 443–446.
[27] O. Struve, *'The Royal Road to Success': Henry Norris Russell (1877–1957)*, in "Publications of the Astronomical Society of the Pacific", 69, 408, 1946, pp. 223–226.

by formation, while dwarfs that of subsequent passive evolution. A process carried on at the expense of potential energy and limited in time — as now favored by physicists and others — because, due to the progressive cooling and increase in density, the star would have to pass from a gaseous to a liquid phase, capable of arresting the fruitful collapse.

Russell was completely off the mark. However, his model seemed to gain further support from observations. By measuring the masses and sidereal radii of eclipsing binaries — those in which the orbital plane contains (approximately) the line-of-sight, so that at each half-period of revolution one star partially or completely eclipses its companion, modulating the global luminosity of the system over time — he and Shapley had gathered evidence that the temperature of giant stars increases with mean density, and that of dwarfs decreases. Just as Lockyer's evolutionary scheme predicted, but only because once again the devil had gotten his tail into it, inducing in an error. *Déjà vu!* Mistakes along the way are physiological stages in the heroic and confusing early stages of a new line of investigation, when curiosity and imagination still prevail over facts. They are later purged by a process of falsification, with greater effort when the flaws not only lead the way, but take root in the cognitive fabric, paralyzing its immune system.

When the question of stellar structure seemed to rest on a solid foundation, thanks in part to the revision that the Swiss physicist and meteorologist Jacob Robert Emden had given to the work of Lane and Ritter, the young director of the Göttingen Observatory, Karl Schwarzschild,[28] entered the fray and posed a new and crucial problem: how does energy migrate from the hot heart of the star to the colder photospheric surface? The polytropic model that had been used up to that point considered only the convection mechanism, familiar to all of us as the one that makes the water in a pot bubble over a fire and the spaghetti swirl like pods of playful dolphins. The thermal gradient, a technical term for the degree of temperature change (in this case vertically), which is maintained by the flame under the pot in our example, creates a circulation of macroscopic fluid bubbles. In the pot, as in the star, hot bubbles rise under Archimedes' push (as long as their density remains lower than that of the surroundings), while cold bubbles take the opposite path. The result is a mixing of the fluid, which is very important in the life of stars as well as in the cooking of spaghetti.

---

[28] E. Hertzsprung, *Karl Schwarzschild*, in "Astrophysical Journal", 45, 1917, pp. 285–292.

However, as Schwarzschild pointed out, the thermal instability that causes stellar matter to boil does not occur under certain circumstances. The energy must therefore be transferred directly by photons, which, for purely geometrical reasons, try to escape from the star. Each one carries its own little legacy and is hindered in its quest for freedom by the obstruction of stellar matter particles. Just like in American football, where the running back, launched toward the touch-in-goal line, collides with the players of the opposing team, giving up part of his own momentum with each push. The example is effective, but it does not emphasize the difference in nature between the offensive player (photon) and the defensive players (material particles). The quanta of energy escaping into cosmic space pay a double toll with each collision: they give up part of their energy and change direction until, after a zigzag that can last millions of years and countless costly collisions, they are much weaker but finally free.[29] A stellar pinball machine that puts the chemical composition and physical state of the sidereal matter into play. The complication is serious and increases the difficulty of the problem. There is also a third way of transporting energy, called conduction. It is implemented by matter by transferring some of the thermal energy possessed by hotter molecules to neighboring, cooler molecules without creating macroscopic motion. It is usually very inefficient in stars because of a kind of very high resistivity.

Schwarzschild was not the first to pose the problem of radiative equilibrium, but he did so at the right time, after the fundamental discoveries of Planck and Einstein. Sadly, in May 1916, this elegant and gentle genius, the son of an Emden sister who had grown up in a muffled enclave of culture as was customary for the Central European Jewish bourgeoisie, died. He had been gravely ill for some time, stricken by the hardships he had suffered in the trenches of the Russian front, where he had been sent after voluntarily enlisting. German science had openly aligned itself with Kaiser Wilhelm and his hegemonic policies, proudly convinced of the cultural superiority of a nation heir to Goethe, Beethoven, and Kant, as the signatories of the "Manifesto of the Ninety-Three" (original title "*An die Kulturwelt!*", To the Civilized World) argued. When Karl died, he was not yet 43 years old. He left a legacy of timeless scientific contributions in various disciplines and

---

[29]In fact, the photon that leaves the photosphere is not the reduced version of the one generated in the core of the star, but rather its descendant, since very many acts of absorption and re-emission have taken place in the meantime.

a son, Martin, who was also destined for a brilliant scientific career in the United States.

In the same year, the question of opacity, i.e. the resistance of matter to the passage of radiation, was taken up across the Channel by Eddington, who did not care about the color of the flag when it came to physics, even though it had the Kaiser's black eagle drawn on it, but perhaps, — as we shall see at the end of this chapter — he did care about the color of the skin of his interlocutors. The inflexible Arthur had to grasp at straws to construct an early plausible model of a star in radiative equilibrium, because the atomic physics needed to deal with matter-radiation interaction was still in its infancy. He also had to fight hard to defend his work from criticism and ridicule, even though he was an orthodox pacifist. His colleague James Jeans had not forgiven him for the naivety in estimating the density of free electrons,[30] which are important actors because they are greedy for the photon's energy; and, as is customary among colleagues in the academy (they both taught at Cambridge), he tried to demolish the overall result. «It is right but obvious».[31] Anyway, Eddington continued undeterred, even taking some satisfaction, as in the speech he gave as president of the Royal Astronomical Society for the awarding of the Gold Medal to his arch-rival: «The Montly Notices records how we hurled at each other mathematical formulae — the most undodgeable of missiles, when they are right — and the onlooker will perhaps conclude that *someone* was badly annihilated. But it is possible that Jeans and I may still have a difference of opinion as to precisely whose corpse lies stricken on the field».[32] In the end he won.[33] Thanks to his work, in the mid-1920s it was realized how to construct an equilibrium model of a star without resorting to too many crippling tricks; and it was also understood why nature does not make stars more massive than a certain upper limit, as observations seemed to suggest.[34] The reason lies in the rapidity with which the radiation pressure increases

---

[30]These are electrons that, freed from atomic bonds by an external supply of energy, roam freely among ions (atoms with an unbalanced number of charged particles) and thus become available for interaction with photons of any energy: a breadth of vision that is denied when they are anchored to atoms and must obey quantization rules.

[31]E.A. Milne, *Sir James Jeans: A Biography*, CUP, Cambridge 1952, p. 25.

[32]R.W. Smith, *Sir James Hopwood Jeans, 1877–1946*, in "Journal of the British Astronomical Association", 88, 1977, p. 11.

[33]M. Stanley, *So Simple a Thing as a Star: The Eddington-Jeans Debate over Astrophysical Phenomenology*, in "The British Journal for the History of Science", 40, 1, 2007, pp. 53–82.

[34]Today this "Eddington limit" is set between 100 and 200 solar masses.

with increasing mass. Indeed, an XXL star could not defend itself against the powerful outward thrust of its own photons, which can overcome the cohesive force exerted by gravity.

The question of the source of energy in stars was still open. It was now clear that the gravitational reserve was woefully inadequate to power a Sun that radiometric dating assigned a minimum age of 100 million years (and a maximum age of 2 billion). «Only the inertia of tradition keeps the contraction hypothesis alive — or rather, not live, but an unburied corpse»,[35] Eddington commented as early as 1920. There had to be another "gas station" — he thought — much more generous and independent of pure gravity to fuel stars. But what? Perhaps matter itself, which could be "milked" according to the most famous equation of his hero, Einstein. «But is it possible to admit that such a transmutation is occurring? It is difficult to assert, but perhaps more difficult to deny, that this is going on and what is possible in the Cavendish Laboratory[36] may not be too difficult in the Sun. I think that the suspicion has been generally entertained that the stars are the crucibles in which the lighter atoms which abound in the nebulae are compounded into more complex elements».[37] What worried him most was the awareness of the immense power of control of the mysterious machine over the evolutionary process of a star, virtually capable of derailing the object from the path traced by pure gravitational contraction. For example, it could intervene at a late stage in the star's life, flooding it with energy in such a way that the object would expand again. In short, it no longer seemed so obvious that giants were young stars and dwarfs were the end result of evolution. Meanwhile, the estimated age of the Sun continued to grow.

The first step toward the new dating of stars had been unwittingly taken 30 years earlier, in 1895, by a German scientist, Wilhelm Konrad Röntgen, the first Nobel laureate in physics in 1901, with the accidental discovery of a «new type of invisible light» called X-rays from the standard notation of an unknown variable in mathematical analysis, since the nature of this radiation was still mysterious.[38] A year later, Antoine-Henri Becquerel

---

[35]S. Chandrasekhar, *Eddington: The Most Distinguished Astrophysicist of His Time*, CUP, Cambridge 1983, p. 17.

[36]Then the most famous atomic and nuclear physics research center in the world.

[37]A. Eddington, *The Internal Constitution of the Stars*, in "The Observatory", 43, 1920, p. 354.

[38]Today we call the physical process *Bremsstrahlung* from the German *Bremsen*, break, and *Strahlung*, radiation.

had noticed that uranium sulfate developed a very vivid phosphorescence even in the absence of any external energy input. In 1898, after discovering other "active elements", Pierre and Marie Curie became convinced that they had stumbled upon an unknown phenomenon. They called it radioactivity, in honor of radium, a metal isolated in pitchblende, a mineral known since the 15th century from silver mines in Czech mountains, and so named for its remarkable spontaneous emission of radiation. For these fundamental discoveries,[39] the three French scientists (actually Marie Curie was born and raised in Poland as Maria Skłodowskawas) shared the 1903 Nobel Prize in Physics. The spontaneous production of charged particles (α and β radiation) or energy (γ radiation) was a new and shocking fact from several points of view. Although the properties of the atomic nucleus were not yet known, it was soon understood that corpuscular emission corresponded to a metamorphosis from one element to another: a leap on Mendeleev's table that alchemists had dreamed of for many centuries.

At the University of Montreal, Canada, the New Zealander Ernest Rutherford and the Englishman Frederick Soddy quickly devised a way to use the spontaneous transformation of atomic species for chronological purposes. They discovered that the time interval required for half of the nuclei in a sample of a radioactive species to decay remained the same for the same material. A constant half-life implies an exponentially decreasing decay rate over time. It is like an unlucky investor who starts with a certain amount of money in the bank and loses half of it every day (which in this case is the half-life of the deposit). A direct numerical calculation will help you understand how the process leads to a rapid depletion of most of the capital and a long sequence of increasingly insignificant daily losses. Some isotopes have a very long half-life, others a rather short one. It was therefore to be expected that only the former would survive in the Earth's crust. This consideration suggested that the planet could be dated by comparing the relative abundances of different radioactive materials in geological samples. When the American chemist Bertram Borden Boltwood attempted this exercise in 1907, assuming that lead was the final stage of uranium decay, he found that the planet was indeed old: up to 50 times older than the 22-million-year limit that the powerful guru of British physics, Lord

---

[39]M. Malley, *The Discovery of Atomic Transmutation: Scientific Styles and Philosophies in France and Britain*, in "Isis", 70, 2, 1979, pp. 213–223.

Kelvin, had vigorously defended for nearly half a century after his conversion to gravitational collapse.[40]

For some time it was also thought that radioactivity might be the source of solar energy. The chemical composition of stars was not well known, which justified the hypothesis of an anomalous percentage of radium, although there was no observational evidence. But this proposal did not last long. Among physicists and astronomers, the idea was gaining ground that the mysterious source of the Sun's longevity was hidden in the heart of matter, whose innermost secrets were gradually being discovered: a subatomic world whose richness and diversity now contradicted the etymological meaning of the word atom, "uncuttable". When, in 1920, an English chemist with a passion for sport and music, Francis William Aston, found that a helium atom weighed slightly less than the sum of four hydrogen atoms, and was even awarded the Nobel Prize for this discovery, it seemed that the game was up. The mass defect, converted to energy using Einstein's formula ($E = mc^2$) and integrated over the total mass of the Sun, was far more than needed to sustain the star, giving it not only a long past but also a lasting future. In fact, the fusion of hydrogen into helium releases energy equal to 0.7 percent of the implied mass, which gives the Sun a nominal lifetime of about 100 billion years (if you burn out the full mass). Eddington, who had the long view, commented: «If, indeed, the sub-atomic energy in the stars is being freely used to maintain their great furnaces, it seems to bring a little nearer to fulfillment our dreams of controlling this latent power for the well-being of the human race — or for its suicide».[41] Prophetic words!

Only a year earlier, in 1919, Rutherford had succeeded in experimentally proving the existence of the proton: one of the two heavy particles that make up the atomic nucleus, two thousand times more massive than the electron and with the same electrical charge, although of opposite sign. The other particle in the nuclear partnership, called the neutron for its lack of electric charge, was discovered in 1932 by his collaborator James Chadwich, a member of the team of brilliant scientists formed in Manchester, where Rutherford had been recruited in the wake of the fame brought by the

---

[40]S.G. Brush, *The Age of the Earth in the Twentieth Century*, in G.M. Friedman (ed.), *Claude C. Albritton, Jr. (1913–1988) Memorial Issue*, "Earth Sciences History", 8, 2, 1989, pp. 170–182.

[41]J.S. Tenn, *Arthur Stanley Eddington: A Centennial Tribute*, in "Mercury", XI, 6, 1982, p. 182.

Nobel Prize. And therein lies the problem! For it seemed that in order to get four protons to overcome their mutual electrostatic repulsion and come close enough to trigger the nuclear binding forces to create and maintain a stable helium nucleus, monstrous temperatures[42] on the order of billions of degrees were required, totally implausible for stars. It was the Soviet physicist Georgy Antonovich Gamov who untied this Gordian knot, the penultimate one, as we shall see.

George Gamow,[43] as he would later rename himself in his American exile, was born in Odessa, Crimea, in 1904, when Nicholas II, the last of the Romanovs, still ruled Greater Russia. His parents, both teachers, were ethnic Slavic Jews. With the early death of his mother and the constant interruptions of public schooling caused by war events, it was his father who took charge of his primary education. It can be said that the high school student George was almost self-taught. After the war, he attended the Imperial Novorossia University in Odessa for a while and then transferred to the oldest and most prestigious academy in tsarist Russia, the University of Leningrad (a new name given to the imperial capital of Petrograd in honor of the death of the Father of the Revolution). There he was able to attend Friedmann's lectures on relativity for a short time, until the father of the dynamic universe died prematurely in 1925.

In the old imperial capital, George formed a partnership with a number of other students, including his peer Dmitri Ivanenko and a *enfant prodige*, Lev Landau, also of Jewish faith. The "Three Musketeers", as they liked to call themselves, trained their young minds by merrily wallowing in the problems of modern mathematics and the fledgling quantum mechanics founded by Werner Heisenberg[44] and Edwin Schrödinger. A training ground that would ensure all three a glorious future in science, despite the difficult conditions created by Stalin's dictatorship. After earning his doctorate, the 24-year-old Gamow was accepted as a postdoctoral fellow at the University of Göttingen. He was applying the principles of quantum mechanics to the problems of natural radioactivity when he discovered the phenomenon known as the "tunnel effect", by which even protons of relatively low energy can bypass a massive electrostatic barrier to embrace each

---

[42]Remember that temperature measures the average value of the kinetic energy of the particles of a gas, that is, it tells us how lively and aggressive they are.

[43]G. Gamow, *My World Line: An Informal Autobiography*, Viking Press, New York 1970.

[44]C. Rovelli, *Helgoland*, Allen Lane, Toronto 2021.

other in a fraternal hug secured by the nuclear force. Actually, Gamow's idea concerned the way in which the α particles (helium nuclei) could spontaneously escape the gravitational grip of an atomic nucleus, even without having the energy (in the classical sense) to do so. Listening to one of his seminars, Max Born, another German physicist who had emigrated to America to escape racial persecution by the *Deutsches Reich* (Third Reich), became convinced of the absolute generality of the phenomenon. An example will help us to understand.

Suppose we want to buy an item in a store where credit is not available. Without the necessary amount of money, we will have to forgo the purchase unless we can find someone to generously lend us what we lack. The same ironclad rule applies to classical mechanics: no one can spend energy that he or she does not have. Quantum mechanics is very different. A particle can momentarily violate the conservation of energy and cross a barrier that would otherwise be beyond its reach. But this stunt must be performed in accordance with Heisenberg's uncertainty principle: the (virtual) debt must be repaid within a time interval that is the shorter the larger the loan.[45] In the case of our example, it is as if we were granted, by an unknown sponsor and without formality, a loan equal to our needs, with the only condition that the repayment would be done faster for larger amounts. This behavior seems rather bizarre to us, and yet it really happens in nature — although there is no way to see it happen — only to experience its consequences. In fact, the "tunnel effect" is the basis of many commonly used electronic devices, such as flash memory and special scanning microscopes.

With this credit opening, even moderately fast proton pairs, such as those found at the temperatures expected in the cores of stars, became candidates for breaking through the electrostatic barrier and fusing in a tight, if temporary, embrace. Bingo? Not yet, because there was a second problem to solve. To make a helium nucleus, four protons must be joined, and this marriage must occur quickly enough to prevent the precarious intermediate pairs from divorcing. But the probability of four particles meeting at the same time is marginal. So there had to be another way to make helium and gain a lot of energy. The dilemma was solved by a physicist who, like many others with a brilliant future, had been educated at Arnold Sommerfeld's school in Munich.

---

[45]Cf. M.S. de Bianchi, *Heisenberg Uncertainty Principle and the Strange Physics of Spaghetti*, 2018, arXiv: 1806.07736.

Hans Bethe was born in 1906 in Strasbourg, a French-speaking city annexed by the Reich in 1871 after Prussia's crushing victory.[46] Although he had a Jewish mother, he was raised in the Protestant faith of his father, a university professor of physiology. Unfortunately, a simple change of religious affiliation was not enough for those who had made racial purity an obsessive and criminal goal. Hans and then his mother wisely decided to leave Nazi Germany before it was too late for them, and after a brief wandering they found a home in America. It was another very serious loss for German science, undermined by Hitler's folly, as this anecdote shows. When Planck, concerned about the fate of the Jewish chemist Fritz Haber, dared to point out to the Führer, with the force of his authority, the devastating effects of anti-Semitic policies on science, the "Bohemian corporal"[47] replied stingily: «If the dismissal of Jewish scientists means the annihilation of contemporary German science, then we shall do without science for a few years».[48] Germany is still paying the price for this criminal mistake.

Bethe possessed extraordinary abilities, which he acknowledged with unashamed self-praise, and which he was able to demonstrate on the occasion of his greatest discovery. It happened in March 1938. At the Fourth Congress of Theoretical Physicists, the best minds in the country and abroad had gathered in Washington to take stock of the burning question of the energy sources of the stars. Invited to participate in the discussion, the young German maverick quickly invented a chain of plausible reactions leading to the formation of helium from hydrogen. The first used the deuterium nucleus — an isotope of hydrogen characterized by a fragile marriage between a proton and a neutron — as a semi-stable intermediate parking lot, which avoided the need for an unlikely simultaneous meeting of four protons. In the second, better suited to more massive stars, it was the carbon nucleus that acted as matchmaker, lending itself to the development of the helium embryo. It was a great insight that earned Bethe a Nobel Prize and gave mankind the realization that stars really do act as factories of heavy elements.

Albeit through another immigrant, the United States thus made a decisive contribution to the understanding of stellar physics: a subject that, as

---

[46] S.S. Schweber, *Nuclear Forces: The Making of the Physicist Hans Bethe*, Harvard University Press, Cambridge (MA) 2012.

[47] A derogatory term used (privately) by some German generals (often from noble families), beginning with Paul von Hindenburg, to refer to Adolf Hitler.

[48] A. Beyerchen, *What We Now Know About Nazism and Science*, in "Social Research", 59, 3, 1992, p. 618.

we have said, had remained largely the preserve of Europeans between the World Wars. But in those same years, astronomers on the new continent had certainly not been idle, accumulating an impressive series of fundamental discoveries fostered by the use of their powerful telescopes. Especially those perched in the mountains of California, which, along with the lure of the "American dream",[49] had attracted the best astronomers from all over the world to the West. Thus, while scientists in the Old World were busy explaining the stars, those in the United States were interrogating the cosmos to the limits of available technology, searching for the myriad rare and unknown things just waiting to be discovered and understood.

An exciting example of this great, almost blind, chase is the discovery of the rotation of disk (typically spiral) galaxies, made by a handful of heroes who were able to expose the same spectrum for several nights in a row in order to accumulate a measurable signal.[50] Thus, in the second decade of the 20th century, Slipher (but also Francis Pease, another gem of Hale's circus) made an important discovery.[51] The emission lines of some spiral galaxies, obtained by aligning the spectrograph slit with the major axis of the objects, were not straight but showed a slight S-shape. Interpreting the observation in terms of the Doppler effect, he came to believe that these nebulae, whose true nature was still unknown, possessed a differential rotation around the central nucleus, i.e. a circular velocity that does not increase proportionally to the distance from the center, as is the case with rigid bodies. A bit like what happens to the planets as they revolve around the Sun to maintain the *status quo*[52] (state of affairs) in the Solar System: rotation produces the centrifugal force that opposes gravity.

Thus, just as the orbital velocity of the Earth allows astronomers to determine the mass of the Sun, the deformation of spectral lines could

---

[49] «The American dream, that has lured tens of millions of all nations to our shores in the past century has not been a dream of merely material plenty, though that has doubtlessly counted heavily. It has been much more than that. It has been a dream of being able to grow to fullest development as man and woman, unhampered by the barriers which had slowly been erected in the older civilizations, unrepressed by social orders which had developed for the benefit of classes rather than for the simple human being of any and every class»; J.T. Adams, *Epic of America*, Greenwood Press, Westport (CT), 1931, p. 406.

[50] V.C. Rubin, *100 Years of Rotating Galaxies*, https://ned.ipac.caltech.edu/level5/Sept04/Rubin/Rubin1.html.

[51] V.M. Slipher, *The Detection of Nebular Rotation* in "Lowell Observatory Bulletin", 62, 12, 1914, p. 1.

[52] Curiously, this Latin phrase is unknown in the classical texts. It was coined in Anglo-Saxon circles in the 18th century.

be used to infer the masses of galaxies (at least within the radial range covered by the kinematic measurements). Of course, the computational model required that these objects be in a state of dynamic equilibrium, where all forces, including inertia, are in equilibrium; and this had to be proven. Furthermore, in order to carry out the exercise — which, as we shall see, would yield amazing surprises — the distances of the objects were needed in order to convert angular measurements into metric lengths (for example, those of the radii of the galactic disks). After 1926, these determinations became possible on the assumption that spiral nebulae were in all respects equivalent systems to the Milky Way, i.e. objects much more distant than previously thought.[53] Hubble himself rolled up his sleeves to look for other "distance indicators" to complement or replace the Cepheids with which he had hit the jackpot. Meters that might be less accurate, but could be used over greater and greater distances to probe deep space.

For example, the Major speculated that the brightest stars in all galaxies have the same luminosity, and thus could serve as standard candles. The thesis was supported by Eddington's discovery of the existence of a limit to stellar masses, and thus to the brightness of stars, and, on the statistical side, by the very large number of stars in a single nebula. It was like arguing that oldest individuals in large cities all reach about the same age (a statement that may not be true for small villages). With such empirical tools, it became possible to quickly construct the topology of the cosmos down to the large cluster of bright galaxies projected onto the constellation Virgo. The new scale of distances called into question the nature of the star that appeared in 1885 near the center of the Andromeda galaxy (p. 301). More than one astronomer pondered the problem of the luminosity of the object, which was far too large for an ordinary nova: among them, certainly, Curtis. In fact, in the Great Debate with Shapley, the astronomer from Lick Observatory asserted that the division of novae into two distinct classes of magnitude «was not impossible».[54] But it was Baade and Zwicky who closed the circle. Putting together some other cases, including historical occurrences such as the 1572 nova star observed in Cassiopeia by Tycho Brahe (p. 64), after a preview of the idea at the American Physical Society meeting held at Stanford in 1933, they ventured the following year

---

[53] R.W. Smith, *Beyond the Galaxy: The Development of Extragalactic Astronomy 1885–1965*, Part I, in "Journal of History of Astronomy", 39, 2008, pp. 91–119.

[54] R.P. Kirshner, *Foundations of Supernova Cosmology*, in *Dark Energy: Observational and Theoretical Approaches*, P. Ruiz-Lapuente (ed.), CUP, Cambridge 2010, p. 151.

to propose the existence of a new category of transient stars, which they called supernovae. Stars capable of emitting immense amounts of light in an instant, obtained by converting much of their mass into energy according to the dictates of Einstein's new physics.

At that time, the two German astronomers were still good friends. The mild-mannered Walter was an avid reader with a deep knowledge of scientific literature, while Fritz, in addition to a bad temper, possessed boundless imagination and creativity. Combining their talents and knowledge, they became convinced that the explosion of supernovae should leave an extremely compact remnant composed of neutrons, the uncharged nuclear particles discovered only two years earlier by Chadwick.[55] Hyperdense stars, with a mass similar to that of the Sun and the size of an average city. In 1939 this intuition, advanced on tiptoe — «with all reserve»[56] —, became theory in the hands of Robert Oppenheimer, a New Yorker of German descent, and his collaborator, the Russian immigrant George Michael Volkoff. The two identified a mass limit, between 1.5 and 3 solar masses,[57] above which the pressure of the degenerate neutron gas was no longer able to balance gravity. A new physics that hinted at hypercondensed states of matter and did not receive all the attention it deserved. The flashes of a new great war blinded even the minds of scientists for five long years.

Baade and Zwicky went beyond proposing the existence of neutron stars. They hypothesized that the traumatic events from which the hypothetical stars were born were the sources of the flow of energetic particles from space: «We therefore feel justified in advancing tentatively the hypothesis that cosmic rays are produced in the super-nova process. It also seems reasonable to assume that a considerable part of the total radiation E_T is emitted in the form of very hard rays or energetic particles».[58] Mysterious radiation recognized at the turn of the century, which Zwicky's controversial boss, Robert Millikan, had called "cosmic rays", mistaking them for heavy photons, and which Arthur Compton — another highly decorated American physicist, discoverer of the effect bearing his name (see p. 492) — instead thought to

(see p. 492)

---

[55] A.S. Burrows, *Baade, and Zwicky: "Super-novae", Neutron Stars, and Cosmic Rays*, in "Proceedings of the National Academy of Sciences of America", 112, 5, 2015, pp. 1241–1242.

[56] W. Baade, F. Zwicky, *Cosmic Rays from Super-Novae*, in "Proceedings of the National Academy of Sciences of America", 20, 5, 1934, p. 263.

[57] The large interval is a consequence of a still uncertain understanding of the equations of state for extremely dense matter.

[58] Ivi, p. 160.

be corpuscular in nature, i.e. α and β emissions. Another goal achieved, at least in part, decades ahead of the course of established knowledge.

Although born as a theoretical physicist, Zwicky had a clear understanding of the role of exploring nature as a tool for discovery and subsequent verification and deepening: «I soon became convinced [...] that all theorizing would be empty brain exercise and therefore a waste of time unless one first ascertained what the population of the universe really consists of, how its various members interact and how they are distributed throughout cosmic space».[59] Through the years, his observational streak would definitively prevail over that of mathematical modeling of phenomena, to the point that he was led to theorize the maieutical role of the study of forms at large. The idea, presented in 1957 in a book,[60] was to attack the multidimensional and non-quantifiable problems, where causal models and simulations do not work, by identifying possible solutions and eliminating the implausible ones by comparison, without *a priori* manipulating the parameter space. To do all this, it was necessary to fill in the data matrix, by discovering and measuring, that is, exploring. This is exactly what the Swiss astronomer began to do after postulating the existence of supernovae. For a few years, from the roof of the institute and to the amusement of his colleagues, he unsuccessfully used a wretched little photographic camera that he had bought with his own money. In 1936, with the installation of the 18 inch Schmidt camera on Mount Palomar, things changed abruptly.

Sitting on the steps of a wooden staircase inside the small telescope building, a dwarf compared to the giant that rose next to it, Fritz spent his nights guiding the instrument across regions of the sky particularly rich in galaxies to maximize the chances of catching some supernova flashes. The work did not stop there. In the morning, just after breakfast, the photographic material collected during the night had to be compared with exposures of the same fields taken at another time to try to detect any stowaways: new stars near or above the galaxy images. In the first year of systematic exploration, Zwicky discovered three supernovae, which in 1941 rose to 50. Baade followed with the 100 inch Mount Wilson reflector, constructing the brightness trend over time for each supernova, the so-called "light curve", and collecting spectra. The "stalking" continued until the improvised firefly

---

[59]F. Zwicky, M.A. Zwicky, *Catalog of Selected Compact Galaxies and of Post-Eruptive Galaxies*, F. Zwicky, Guemligen 1971, p. 1.

[60]F. Zwicky, *Morphologische Astronomie*, Springer, Berlin-Heidelberg 1957.

became too faint to be observed or was swallowed up by daylight.[61] The team had gained a new member, Rudolph Minkowski, a German physicist with a fancy pedigree. His uncle was the great mathematician Hermann Minkowski, Einstein's teacher in Zurich. Like many others, he had wisely left his homeland to escape the consequences of racial laws. For him, a nation still licking the wounds of the Great Depression was far preferable to the risk of being deprived of civil rights, freedom, and even life in the educated and increasingly prosperous Germany of the Third Reich. Anti-Semitic obsession continued to bleed a country that in a few decades had become the Athens of science. A senseless massacre that Hitler completed by sending his Aryan genius to die in a suicidal war that would put the survivors on the victor's market.

The practice with galaxy clusters, i.e. with the large collections of nebulae that he explored in search of supernovae, led Zwicky to take up an idea that he had developed while still in Europe: to calculate the mass of these massive archipelagos of matter, treated as clouds of a gas of which the galaxies are particles in dynamic equilibrium under the effect of mutual gravitational attraction. He set to work, firmly convinced of the value of his research:

> For the construction of the 18 inch Schmidt telescope, its housing, a full-size objective prism, a small remuneration for my assistant, and the operational costs for the whole project during ten years, only about \$50,000 dollars were expended. This probably represents the highest efficiency, as measured in results achieved per dollar invested, of any telescope presently in use, and perhaps of any ever built, with the exception of Galilei's little refractor.[62]

He painstakingly collected the radial velocities of a thousand galaxies of the large cluster projected in the constellation Coma Berenices and their angular separations, converted into lengths thanks to the (approximate) knowledge of the distance of the cluster from the Sun. To perform the calculation, he used some indispensable simplifications and the virial theorem developed by Clausius, which we mentioned earlier in this chapter, by combining the (time-averaged) value of the kinetic energy with that of the gravitational binding, the potential energy. The result was astonishing.

---

[61] P. Murdin, L. Murdin, *Supernovae*, CUP, Cambridge 1985.
[62] In F. Dyson, *Dynamic Universe*, in "Nature", 435, 2005, p. 103.

The amount of matter calculated in this way was 200 times larger than the mass obtained by converting the luminosity of the cluster galaxies through the average mass-to-light ratio given by the rotation curves.[63] Therefore, Zwicky concluded, there must be an enormous amount of matter of an unknown nature in clusters that is not revealed by light. Amazed by his result, he not only reported the fact, but drew a series of conclusions that, if taken seriously, could have matured extragalactic astrophysics forty years in advance.[64] In addition to expressing caution about the universality of Newton's gravitational law, thus anticipating very modern speculations, he commented on the secondary evolution of galaxies (i.e., the dependence of the frequency of morphological types of galaxies on the density of the environment), the existence of intracluster light and matter, and the use of galaxies as gravitational lenses, while also stating the fundamental importance of systematic investigations. A masterpiece!

By now, the irascible Swiss had unfortunately been ostracized by the scientific community because of his hypertrophic vanity and the verbal violence with which he interacted with colleagues and even collaborators. «They considered him crazy and he considered them stupid».[65] Later, in the 23 pages of the Introduction to his *Catalog of Selected Compact Galaxies and of Post-Eruptive Galaxies*, published at his own expense in Switzerland in 1971, he would even go so far as to describe them as «scatterbrains», «sycophants and plain thieves», who «doctor their observational data to hide their shortcomings and to make the majority of the astronomers accept and believe in some of their most prejudicial and erroneous presentations and interpretations of facts», and publish «useless trash in the bulging astronomical journals». Not bad, if the intent was to make a clean sweep of friends. His personal relationship with Baade also deteriorated. Jesse Greenstein, the future powerful director of Caltech's astronomy program, reported that, during World War II, «they were a dangerous pair to put in the same room [...] Zwicky called Baade a Nazi, which he wasn't; and Baade said he was afraid that Zwicky would kill him».[66] Thus, it is not surprising that the discovery of the missing mass, a prelude to the discovery

---

[63] Note that the mass-to-light ratio only accounted for the properties of the inner regions of nebulae that could be reached by spectroscopy.

[64] G. Bertone, T.M.P. Tait, *A New Era in the Quest for Dark Matter*, in "Nature", 562, 2018, pp. 51–56.

[65] Dyson, *Dynamic Universe*, cit., p. 103.

[66] R. Prud'homme, *Interview with Jesses L. Greenstein*, in "Caltech Oral Histories", 1982; http://oralhistories.library.caltech.edu/51/1/OH_Greenstein_J.pdf, p. 26.

of dark matter that we will consider later (p. 532), did not attract the attention it deserved.

Until the mid-1970s, Zwicky's paper, which appeared in the "Astrophysical Journal" in 1937, received no more than a handful of citations.[67] But five years earlier, the Dutchman Jan Hendrik Oort had reported that in the solar region, far from the center of the Milky Way, as Shapley had shown, the census of stars, each counted by its mass, was not sufficient to explain the restoring force exerted by the Galactic disk perpendicular to its median plane. In short, even close to home there seemed to be a mass 50 percent greater than what the stars showed by their light. It was certainly not the impressive factor of 200 that Zwicky had found, but the result was in the same direction. No one paid much attention to it, even though Oort had a solid reputation as a scientist and a reliable person.

Jan was born in 1900 in West Friesland (a region of North Holland) into a progressive family with a strong tradition of church service.[68] His father was a psychiatrist with little interest in the things of God. Jan wanted to be an astronomer, and when the time came, he enrolled at the University of Gröningen, attracted by the fame of Jacobus Kapteyn. For his doctorate he went to Yale, following the American path that his teacher had wisely laid out through his friendship with Hale. Returning to Holland in 1924, he was hired as an assistant at the University of Leiden, where he completed his dissertation on high-speed stars. He was only 26 years old when, at the suggestion of Willem de Sitter (who was a good scientist and a good master), he came across a paper by the Swede Bertil Lindblad dealing with a new model of the structure of the Milky Way and its dynamics.

The problem was topical and the solution still open. On the one hand, the Gröningen school argued for a star system modeled as Kapteyn wanted: practically centered around the Sun and with a density of stars dropping to 10% even in the Galactic plane — the so-called disk, where the spiral arms that give this class of nebulae their name open out — at a distance of less than 6,000 light-years. In contrast, Shapley had shown that globular clusters appear to be distributed with spherical symmetry around a center

---

[67]See Table 1 in S. van den Bergh, *A Short History of the Missing Mass and Dark Energy Paradigms*, in "Historical Development of Modern Cosmology", ASP Conference Proceedings, V.J. Martínez, V. Trimble, M.J. Pons-Bordería (eds.), Vol. 252, in "Astronomical Society of the Pacific", 2001, p. 76.

[68]P.C. van der Kruit, *Jan Hendrik Oort, Master of the Galactic System*, Springer, Cham 2019.

projected into Sagittarius and far from the Sun. The geometry and kine-
matics of the orbits of these dense cocoons of stars linked them to high-speed
stars, so called because they appear to move at least 70 km/s faster than
the average of stars near the Sun. These misnamed objects, which actually
travel slowly (their observed high speed is due to the fact that we measure
it from the fast-moving Sun), are mostly halo objects whose orbits around
the center of the Milky Way have no preferred inclination, unlike disk stars,
which instead remain confined to one plane. Lindblad sided with Shapley.
His model of the Galaxy included several coaxial stellar subsystems of varying flattening orbiting around a common center, consistent with what the
Harvard director had found for the family of globular clusters. The greater
the flattening, the greater the average rotational velocity should be.

The idea is this. Imagine a spherical distribution of many billions of
stars obtained by putting them in circular rotation around the center on
randomly inclined orbits. On average, for every star spinning in a given
direction, there will be another star spinning in the same plane and with
roughly the same orbital radius, but moving in the opposite direction. The
sum of the two makes a zero contribution to the average velocity. In other
words, each star thinks exclusively for itself and adopts the velocity required
by the equilibrium conditions, because the system has not been assigned an
overall rotation *a priori* that would cause the stars to favor one orbital plane
over all others. Oort was deeply impressed by the Copernican simplicity
of this scenario, although hidden by a baroque mathematical formulation,
and wanted to see it more closely. It is said that he had pondered the
question for some time when, while explaining to students the model of
the galactic system developed by his masters in Gröningen, he was struck
by an idea. He interrupted the lecture and retired to his office. Two weeks
later he reappeared with a simple mathematical treatment of the problem
which, in agreement with Lindblad, showed that the Milky Way is subject
to differential rotation (different from that of a rigid body).

A stroke of genius, but also a demonstration of a strong personality,[69] all
the more so when the authority you are refuting is at your home! Published
in 1927, the work[70] immediately gave the young Oort great fame, paving

[69]R. Nityananda, *Discovering the Rotation of our own Galaxy: The Astronomer as Detective*, in "Resonance", 20, 19, 2015, pp. 869–879.

[70]J.H. Oort, *Observational Evidence Confirming Lindblad's Hypothesis of a Rotation of the Galactic System*, in "Bulletin of the Astronomical Institutes of the Netherlands", 3, 1927, pp. 275–282.

before him a highway that he was able to travel wisely, accumulating, in the course of a long and fruitful life, numerous scientific achievements (see also chapter 18). To this day, the Netherlands owes its prominent position in the world of astrophysics to this educated, elegant, and kind gentleman: a brilliant scientist and a prolific teacher, justly strict but equally generous with his students. We could easily recognize these qualities in another great scientist, Sir Arthur Eddington, were it not for his venomous and suspicious attack on Subrahmanyan Chandrasekhar: a gratuitous violence that reeks of racism.[71] The story is about the discovery of the properties of white dwarfs and takes us back in time again.

After the glorious measurement of the first stellar parallax, Friedrich Bessel, while continuing his astrometric program, had discovered that Sirius, the brightest star in Canis Major and the entire firmament, was behaving strangely. Instead of moving straight across the sky, as would be expected from a proper motion that is the projection of the star's (locally rectilinear) path in the galaxy, Sirius was following a wavy trajectory similar to that of the components in visual binary systems.[72] This feature was not unique. It was also shared by Procyon, the third brightest (γ) star in Canis Minoris. By 1844, the German astronomer had finally collected enough data to hazard a guess that each of these two objects had an unseen companion. «If we were to regard Sirius and Procyon as double stars, the change of their motions would not surprise us [...] But light is no real property of mass. The existence of numberless visible stars can prove nothing against the existence of numberless invisible ones».[73] He was right, but he never knew it. Shortly thereafter he died, sixteen years before the American telescope maker Alvan Graham Clark, son and namesake of the founder of the renowned astronomical optics company, accidentally discovered Sirius' hypothetical companion, the faint "Pup", while testing one of his instruments on the star. Although unaware of the significance of the discovery, the craftsman informed Harvard Observatory Director George Phillips Bond, who spread the news to the world. Left high and dry this time was Le Verrier, a guru of gravitational mysteries, who tried unsuccessfully to unravel the object, involving a brilliant scientist in the project,

---

[71] A.I. Miller, *S. Chandrasekhar: The student who took on the world's top astrophysicist*, in "The Guardian", October 19, 2017, https://www.theguardian.com/education/2005/mar/31/research.highereducation.

[72] N. Brosch, *Sirius Matters*, Springer, Berlin 2008.

[73] Bessel's letter to Sir John Herschel, in "The London, Edinburgh and Dublin Philosophical Magazine", XXVI, 1945, p. 261.

Léon Foucault. The only thing they got out of it was a reflector with a glass mirror 80 cm in diameter, which was then placed in Marseilles to take advantage of better weather conditions on the French Riviera than in Paris.[74]

The observation of Sirius B made it possible to follow its orbit — whose period, about half a century, was already known from the oscillations of the main star — and to calculate its mass, which turned out to be slightly less than that of the Sun. This was an unexpected and surprising result, since the intrinsic luminosity of "Pup" was not more than one hundredth of that of the Sun. How could this be explained? There would be no problem if one could prove that Sirius B was a cold star. In fact, the total luminosity of a star grows with the area of the emitting surface (i.e. the square of the radius of a spherical object) and with the luminosity per unit area, which in the ideal case (blackbody) depends on the fourth power of the temperature (Stefan-Boltzmann law). A measure was needed to solve this problem. The American Walter Adams thought about it. In 1915, while the war was already raging in Europe, he proved by a daring series of observations that Sirius B possessed the same spectrum as its majestic companion, and thus a temperature 50 percent higher than that of the Sun. An excess that may seem small, but which, according to Stefan-Boltzmann's law, corresponds to a 5-fold increase in luminosity. To cope with the total luminosity, the discovery implied for Sirius B a diameter of only 36,000 km, now reduced to 11,700 km with the latest measurements: as much as the Earth. «We have a star of mass about equal to the Sun and radius much less than Uranus»,[75] Eddington would then sentence, adding a year later: «If so, Prof. Adams has killed two birds with one stone; he has carried out a new test of Einstein's general theory of relativity and he has confirmed our suspicion that matter 2000 times denser than platinum is not only possible, but is actually present in the universe».[76] In fact, an object 40,000 times denser than water, as Sirius B appeared to be, lent itself to verifying another of the predictions of general relativity, the redshift of spectral lines caused by the need for outgoing photons to ascend a steep potential well, resulting in a toll in energy currency. Einstein predicted the phenomenon as early

---

[74] J.B. Holberg, *Sirius: Brightest Diamond in the Night Sky*, Springer, Berlin 2007.

[75] A.S. Eddington, *The Internal Constitution of the Stars*, CUP, Cambridge 1988, p. 171 (1st ed. 1926).

[76] M. Longair, *Arthur Stanley Eddington*, in P. Harman, S. Mitton (eds.), *Cambridge Scientific Minds*, CUP, Cambridge 2002, p. 232.

as 1911, well before he completed his theory, using only the principle of equivalence between gravity and inertia. In 1923, obsessed with the desire to validate relativity, Eddington wrote to his colleague Adams asking him to perform a test. He relied on the American astronomer because Adams knew how to deal with the problems of observing faint objects embedded in the diffuse light halos of bright stars, as in the case of Sirius B. Moreover, in his new capacity as director of Mount Wilson Observatory, he had privileged access to the largest telescope on the planet.

Adams did not disappoint him. Two years later, now convinced of the soundness of the difficult observations, he entrusted Hale, who was attending the annual meeting of the National Academy of Sciences in Washington, to report the first measurement of a gravitational redshift. Even the "New York Times" later broke the news to the world with an article on July 22, 1925, titled *New Test Supports Einstein's Theory*. By now, the fame of the extravagant genius of physics was such that it was no longer necessary to introduce him to the general public. Everyone knew who he was. Eddington, for his part, read Adams' result not so much as a validation of relativity, which for him was now an *a priori* truth, but rather as a confirmation that matter could reach extreme densities. No one had really believed it. However, faced with the proof presented by Adams, it was necessary to find a new model, not only for Sirius B, but also for the other two stars that at that time represented the thin family of the so-called "white dwarfs". The only way out was to imagine that in these objects the atomic nuclei had been stripped of their electronic shells, or, as physicists say in their jargon, totally ionized, in order to adapt to living in much smaller spaces.

Fortunately, the Italian Enrico Fermi and the British Paul Dirac had just developed a statistical theory that allowed physicists to treat rigorously the model of a medium composed of heavy particles (nuclei) and free electrons ("plasma"). Under conditions of extreme density, such a gas, called "degenerate", maintains itself due to the pressure of electrons that become intolerant of coexistence at the same energy level (when energy is equally divided). This behavior of identical fermions[77] is enshrined in a principle of quantum mechanics (the exclusion principle) formulated by Wolfgang Pauli in 1925. Why electrons and not others? Because they are thousands of times lighter than nuclei and therefore travel much faster. If one pretends to compress these particles into a smaller volume, they will do so, provided they are given some energy to occupy higher quantum states.

---

[77]Subatomic particles that follow Fermi-Dirac statistics, such as electrons and photons.

In the absence of this, they will resist compression, refusing even to give up their energy. This is why the pressure of a (fully) degenerate gas does not depend on temperature, as it does for an ordinary gas. The moral of this story is that, in microcosm, in certain extreme situations nothing is achieved not only by those who are broke, which is obvious, but also by those who are rich but not authorized to lay hold of their wealth.

In 1926, using the tools of new physics, the British physicist Ralph Fowler derived for white dwarfs a strange inverse proportionality relationship between mass and radius, leading to the absurdity of an object with infinite mass and zero radius. This is where Subrahmanyan Chandrasekhar[78] came in. In 1930, this child prodigy, not yet 20 years old, left his native India for England, where a doctoral fellowship at the prestigious Trinity College awaited him. The position had been sponsored by Fowler himself, who had been impressed by a publication Subrahmanyan had sent him. Chandra — as he would later be called by friends and colleagues, in keeping with the all-American practice of shortening names — came from a family of Tamil Brahmins originally from the far south of India. His father's brother, Chandrasekhara Venkata Raman, had distinguished himself in the study of light-scattering phenomena and was awarded the Nobel Prize the very year his nephew moved to Europe: an honor made all the more prestigious by the fact that it was bestowed upon a Hindu aristocrat who was worth little more than a pariah in an Anglophone society infected by colonialist racism.

The journey took several weeks. Instead of relaxing in the warmth of the Sun on the ship's passenger deck, the young Subrahmanyan remained locked in his cabin, brooding over the strange result of the theory that forced fuelless stars to evolve into faint planet-sized remnants. It was a total dedication to work that would characterize the Indian astrophysicist throughout his long scientific life.[79] Chandra, who was in total control of

---

[78]K.C. Wali, *Chandra: A Biography of S. Chandrasekhar*, The University of Chicago Press, Chicago (IL) 1991.

[79]In a lecture at Princeton in 1986, Chandrasekhar told the story of the time when he was teaching at the University of Chicago and, as a supervisor, used to verify that students were working in the evenings. He did this by checking that the windows of their rooms on campus were lit. The young Martin Schwarzschild, who probably had very good personal reasons for going out at night, had invented a trick to fool him. He had placed a cardboard figure on the platter of a gramophone and, with the help of a lamp, created the effect of someone walking around the room. But Chandrasekhar, who was a keen observer, discovered the trick after noticing, night after night, the constancy of the frequency with which light and shadow alternated. Neither he nor Martin told the public what the consequences of this were.

contemporary physics, realized that the purely quantum treatment of the problem was inadequate. In fact, as the mass increases, so does the energy of the particles in the degenerate stellar gas, to the point where the velocity becomes comparable to that of light. This discovery required a switch to a relativistic formulation of the problem.[80] Soon the young man was able to develop a theory that 53 years later would make him the second in his family to receive the most coveted prize for a physicist. His achievement surprised everyone. In fact, the mass required to reach zero dimension was not infinite. The condition was met by a degenerate star just 1.4 times more massive than our Sun (a limit now named after the Indian physicist). But what happens to stars that cross this threshold? They become unable to resist gravity and are destined to collapse into objects of enormous density. A conclusion, Chandrasekhar wrote, that opened the door to other scenarios.

When the work was published after three years of maturation, it was bitterly contested by Eddington (and also by Edward Arthur Milne, Chandrasekar's personal friend and professor at Oxford). The Quaker astrophysicist used his sharp irony and the weight of his authority to publicly ridicule his young colleague in writing.

> Chandrasekhar shows that a star of mass greater than a certain limit remains a perfect gas and can never cool down. The star has to go on radiating and radiating and contracting and contracting until, I suppose, it gets down to a few kilometers' radius when gravity becomes strong enough to hold the radiation and the star can at least find peace. I felt driven to the conclusion that this was almost a *reductio ad absurdum* of the relativistic degeneracy formula. Various accidents may intervene to save the star, but I want more protection than that. I think that there should be a law of Nature to prevent the star from behaving in this absurd way.[81]

Another anecdote concerns William Wilson Morgan, an astronomer at Yerkes, a specialist in the spectral classification of stars and galaxies, and a friend of Chandra. Shortly after moving to California, William sent a postcard to his colleague in Chicago with a message that went something like this "Great day. Went to the beach with the family". Shortly thereafter, Chandra wrote back: "It's raining and cold here. I'm at work and a day ahead of you!".

[80] Chandrasekhar's procedure would soon inspire Oppenheimer and Volkoff's work on neutron stars.

[81] S. Chandrasekhar, *The increasing role of general relativity in astronomy*, "Observatory", 92, 1972, p. 169.

Like Einstein, Eddington did not believe that Schwarzschild's solution of the field equations of general relativity, which we will consider in the next chapter, made physical sense. Therefore, as the crusader for science that he felt he was, Sir Arthur — he was knighted in 1930 — had to destroy a theory that instead offered the star a way to evolve into a gravitational singularity.

The tough stance,[82] ambiguously tinged with the colors of racism, forced Chandrasekhar to leave first England and then Europe for the United States, where he would become a famous professor at the University of Chicago, as well as the historic editor-in-chief of the "Astrophysical Journal", the most prestigious among the professional journals. But already in 1927 Eddington had to make a first *mea culpa*:

> We learn about the stars by receiving and interpreting the messages which their light brings to us. The message of the Companion of Sirius when it was decoded ran: "I am composed of material 3,000 times denser than anything you have ever come across; a ton of my material would be a little nugget that you could put in a matchbox". What reply can one make to such a message? The reply which most of us made in 1914 was — "Shut up. Don't talk nonsense."[83]

He would have been compelled to make a second retraction later.

---

[82] Cf. J.G. Mena, *Case Study: Chandrasekhar and Eddington*, 2018, http://www.mcs. st-and.ac.uk/~ras/Summer2018/Chandra/Chandra.html.
[83] Eddington, *Stars and Atoms*, cit.

Chapter 17

Pioneers at the time of the Cold War: The crisis of cosmology and the rustle of the Big Bang

Concerning a circle
beginning and end are common.
Heraclitus, fragment 103

While human ingenuity may devise
various inventions to the same ends,
it will never devise anything more beautiful,
nor more simple, nor more to the purpose
than nature does, because in her inventions
nothing is lacking and nothing is superfluous.
Leonardo Da Vinci, *Codex Atlanticus*

On September 1, 1939, the Wehrmacht (defence force),[1] supported by the Luftwaffe, which played the artillery role with its Ju-87 dive bombers, the *Sturzkampfflugzeug*, broke through the Polish border with a lightning pincer attack (*Blitzkreig*) that would bring German tanks to the outskirts of Warsaw in a single week. Suddenly awakened from their slumber, France and England rushed to declare war on Germany. Thus began the largest and bloodiest conflict in the history of the planet, heralded by the annexation (*Anschluss*) to the Third Reich of Austria and the German-speaking region of Bohemia and Moravia, the Sudetenland. Few were spared the heavy toll in lives, infrastructure, and conscience. Although genetically reluctant to get involved in the intricacies of Europe, the United States was drawn into the conflict by the treacherous attack on its naval base at Pearl Harbor on December 7, 1941. By launching Zero torpedo bombers at American battleships in an unsuccessful attempt to maim the enemy before it realized

---

[1]A hypocritical name, the result of an unjust peace at Versailles that had completely demilitarized Germany with an operation that, to paraphrase the Prince of Talleyrand, was worse than a crime: it was a mistake.

what was at stake, the strategists of the Rising Sun, the third partner in the Rome-Berlin-Tokyo Axis, had risked shaking the lethargy of the overseas giant. The Soviets had already been in the field for six months, invested by Hitler's army, which seemed to know no obstacles and did not fear the deadly "General Frost".

In Europe, it all ended in May 1945 with the material and moral defeat of the Third Reich (and, as early as 1943, of its Italian ally), and in the Far East four months later with the unconditional surrender of Japan, bowed by the Marines and the atomic horror. *Little Boy* and *Fat Man* — single bombs conceived in Los Alamos, New Mexico, by some of the brightest minds in physics — had incinerated two populous cities, killing or condemning to a slow death more than 150,000 civilians, a number of people equal to what the battle for the conquest of the Japanese island of Okinawa had cost the two contenders. This equation, used to estimate the cost in human lives for the final defeat of the unbending "Japs" with conventional weapons, made the most brutal massacre in history acceptable and even appropriate.

The guns fell silent for a time, but nothing was the same. Borders, leadership, landmarks, and customs changed. A new map of alliances and tensions took shape; new states emerged, while others lost their identity or sovereignty. The war had advanced scientific knowledge, particularly attack and defense technologies. Clad in the robes of peace, these innovations spilled over into society, renewing the style and quality of life and work in the industrialized countries and widening the gap with the vast peripheries of the world. In short, the Second World War was a turning point between two very different epochs. War, the mother of all things, had given birth to a new season in which everything had to begin anew in a seemingly reformed way, in a cyclical course where history, despite Cicero's belief in *"historia magistra vitae"*,[2] never really succeeds in becoming a teacher of life.

The global metamorphosis also reverberated in the sciences of the universe, which decisively took over the American passport thanks to the weight of the U.S. victory, the power of the dollar, and a far-sighted policy of recruiting brains all over the planet — applied with cynical opportunism, without worrying too much about the "criminal record", as for

---

[2]The full text from Cicero's dialogue *De Oratore* (On the Orator) is: «*Historia vero testis temporum, lux veritatis, vita memoriae, magistra vitae, nuntia vetustatis*» (History truly is witness of the times, light of truth, life of memory, teacher of life, messenger of antiquity).

example in the case of Wernher von Braun[3] — and thanks to superb observational instrumentation. To consolidate the leading position established at the beginning of the 20th century, largely owing to the work of George Hale, the 200 inch reflector at Palomar Mountain was finally put into operation. No one could compete with this giant, for which a slogan was coined as bold as it was potentially true: "one plate, one discovery". Walter Baade was among the first to use it, to further the significance of an observation he had made in wartime.[4]

It had happened in 1944. As a German and an "enemy alien", Baade's personal freedom had been restricted since the United States entered the war. He was allowed to move only within Los Angeles County, between the city of Pasadena, where he had his home and office, and Mount Wilson, where he went to observe. The pressure to use the 100 inch reflector had been eased considerably because most of the astronomers had been drafted to serve the country: the younger fellows sent to one of the two open fronts in the East and West, and the older scientists engaged in the war projects undertaken at the observatory. Hubble himself, despite his fame and age, had been transferred to Maryland as a senior administrator of Aberdeen's Proving Ground, a design and testing service for equipment for the U.S. Army.[5]

Under these conditions, Baade had the luxury of choosing the best nights for transparency and atmospheric stability, conditions not uncommon in the California mountains, especially in summer and fall. But that's not all. The sky at Mount Wilson was particularly dark because of the blackout imposed by the war on the metropolis of Los Angeles. The same city lights that were shining down on the observatory and dazzling the sky, to the detriment of stargazing, could have led to an attack by Japanese planes or submarines on the populated area, and were therefore turned off. As a final bonus, he had been able to get his hands on new red-sensitive photographic plates. Eastman Kodak had developed them for aerial detection of camouflaged enemy weapons and vehicles, but an ammonia bath made these emulsions suitable for long exposures in faint starlight.

---

[3]M. Capaccioli, *Red Moon*, cit., pp. 99–103.

[4]D.E. Osterbrock, *Walter Baade's Discovery of the Two Stellar Populations*, in P.C. van der Kruit, G. Gilmore (eds.), *Stellar Populations*, Proceedings of the IAU Symposium No. 164, Kluwer Academic Publishers, Dordrecht 1995, pp. 21–29.

[5]H.P. Robertson, *Edwin Powell Hubble, 1889–1953*, in "Publications of the Astronomical Society of the Pacific", 66, 390, 1954, pp. 120–125.

Being the right man in the right place at the right time, the peaceful Walter did not miss the opportunity to address a question that had been on his mind for some time. A few years earlier, while working with Hubble, he had resolved the dwarf galaxies Sculptor and Furnace and the outer regions of some galactic globular clusters into stars and found that the brightest objects were red. He wondered whether this behavior was shared by the diffuse central components of the disks of spiral and elliptical galaxies, or whether the brightest stars there were blue, as in open clusters such as the Pleiades.[6] To solve the problem, he used all his talents as an exquisite observer, focusing his efforts mainly on M31, the Great Nebula in Andromeda, and its satellites, M32 and NGC 205: nearby targets that pass high above the Mt. Wilson sky on autumn nights. He found what he expected. The most luminous stars in the diffuse regions of these systems were comparable in brightness and color to those in globular clusters. His conclusion was that blue and red stars form two distinct populations, both in their intrinsic properties and in the characteristics of the environments whre they live. He christened them Population I and II, respectively, without having the faintest idea of what this distinction might mean.

His work soon gained general acceptance, especially after the confirmation that the core of the Milky Way was also dominated by Population II stars. Thus, as the demobilization of the victors and defeated repopulated observatories around the world with revitalized astronomers, blood-tired and thirsty for celestial novelties, he became the leader in the fledgling field of extragalactic astrophysics, to the point where he could dictate the rules of the game. He did so explicitly in 1948, at the dedication of the 200 inch Palomar telescope, with a speech on the extragalactic research program to be conducted with this unique instrument.[7] It was an elegant way to assert his leadership, made easier by the fact that the other rooster in the henhouse, Edwin Hubble, was gradually reducing his involvement due to health problems. But, while Baade fully mastered the astronomical literature of his day, he was not a theorist and did not, or would not, immediately grasp what the division into two families meant for stellar evolution. He himself later confessed that already in 1944, while reading his reports on

---

[6] It is hardly worth mentioning that the color of stars is an indication of their photospheric temperature and that, according to Stefan-Boltzmann's law, temperature and intrinsic luminosity allow astronomers to estimate the size of objects.

[7] W. Baade, *A Program for Extragalactic Research for the 200 inch Hale Telescope*, in "Publications of the Astronomical Society of the Pacific", 60, 355, 1948, pp. 230–234.

M31 observations, George Gamow had sent him a postcard[8] on which he scribbled the conjecture that Population I was the tip of the iceberg of young stars and Population II of the old ones. Another of the Ukrainian physicist's many brilliant insights that Baade apparently did not heed.

Actually, the idea that the bluest objects represented the earliest stages of an evolution leading systems of coeval stars to be dominated by red giants was already clear by 1939 to theorists such as Martin Schwarzschild, Karl's brilliant son who had emigrated to America, and his friend Lyman Spitzer. The latter was a young "Yank" who had specialized with Eddington in Cambridge, the new Athens for the stars and stripes *feroces victores*.[9] But the time was not right, and Spitzer,[10] who had ventured this conjecture, was persuaded by two senior colleagues to remove it from his article. What a pity! Not infrequently, old age opposes the progress of science with a frightening viscosity, smuggled in as wisdom. It is not for nothing that the Nobel Prize in Physics, although often awarded at an advanced age, is given for results obtained when the laureate was young or very young.

In September 1952, astronomers from all over the world gathered to celebrate the 8th General Assembly of the International Astronomical Union, the organization founded after the Great War in the same utopian spirit of universal brotherhood that had led to the creation of the short-lived League of Nations in 1919. The meeting was to be held in Leningrad, but was moved to Rome,[11] following the escalation of tensions between the two blocs into which the world was divided. It sounded like a slap in the face to the Soviets. Orchestrated by the Americans, who were angered by the suffering and the results of the Korean War, it was made all the more stinging by the invitation to Pope Pacelli to give a welcoming address to the participants. The previous year, Pius XII had delivered an apologetic speech at the Pontifical Academy on Big Bang cosmology entitled *Proofs of God's Existence in the Light of Modern Natural Science*:[12] a kind of misappropriation of science by an unrepentant heir of Cardinal Bellarmine,

---

[8] Osterbrock, *Walter Baade's Discovery of the Two Stellar Populations*, cit., p. 25.

[9] See note at p. 46.

[10] B.G. Elmegreen, *Lyman Spitzer Jr. and the Physics of Star Formation*, in S. Röser (ed.), *Formation and Evolution of Cosmic Structures*, Wiley-VHC-GmbH & Co, Weinheim 2009, p. 159.

[11] H.H. Swope, *Meeting of the Eighth General Assembly of the International Astronomical Union*, in "Publications of the Astronomical Society of the Pacific", 65, 382, 1953, pp. 11–15.

[12] The full text of the Pope's speech in Italian can be found at http://satisf.org/pio-xii-prove-esistenza-dio.

disputed even by the obedient Lemaître, and a thesis totally unacceptable to Marxist-Leninist atheists behind the Iron Curtain.

No one doubted that the Pope would take the opportunity to reaffirm to the more than 400 astronomers from 35 countries gathered in the Swiss Hall of the Castel Gandolfo summer residence the cause-and-effect relationship between the divine act of creation, the biblical *"fiat lux"* (let there be light) and the Big Bang. And so it was.[13] As expected, the Soviet scientists left the hall in protest. Only the head of the delegation, academician Viktor Ambarzumjan, though a staunch supporter of dialectical materialism,[14] had to remain in the hall for institutional reasons; and in his capacity as Vice President of the IAU, he also had to respond to the Pontiff. The study of the cosmos, the great Armenian astronomer said mellifluously, would serve to bring different cultures closer together and promote the cause of world peace.[15] He didn't believe it at all, and he did well.

In the Eternal City and in a climate of open, if cold, warfare between nations, schools, economies, ideologies, and faiths, Baade finally decided to take a stand. Bolstered by the endorsement of his evolutionary thesis by a man as influential as Henry Norris Russell, he became the official spokesman for the idea that the two populations actually represented different stages of star life. He substantiated this belief by reporting his discovery, through the power of the Hale telescope, of the existence of two types of Cepheid variables, one for each stellar population.[16] Since then, the young variables of Population I have been called classical Cepheids and W Virginis the fainter and older pulsating stars belonging to Population II. It was a result of paramount importance for cosmology, since it led to a redefinition of the cosmic distance scale and thus the age of the universe, which doubled from 1.8 billion years, as estimated by Hubble, to 3.6 billion years. Still

---

[13]The full Italian text of the speech is available at https://www.vatican.va/content/pius-xii/it/speeches/1952/documents/hf_p-xii_spe_19520907_la-presence.html.

[14]A position well illustrated by the Soviet scientist Alexandr Bogdanov: «History shows that every system of ideas, be it religious, philosophical, juridical, or political, no matter how revolutionary it was at the time of its birth and when it began its struggle for supremacy, sooner or later becomes an obstacle and an impediment to further development, that is, it becomes a socially reactionary force. Only the theory that consciously rose above it, that was able to explain it and point out its causes, was able to escape this fatal degeneration. That theory was Marxism»; https://www.lasinistraquotidiana.it/materialismo-storico-e-filosofia-frattale/ (in Italian).

[15]H. Kragh. *The Universe, the Cold War, and Dialectical Materialism*, 2012, arXiv: 1204.1625.

[16]M. Catelan, H.A. Smith, *Pulsating Stars*, Wiley-VCH-GmbH & Co, Weinheim 2015, pp. 196–200.

too short a time, however, to solve the conundrum created by geologists who, by attributing to the Earth an age greater than that of the cosmos, made the son older than the father! A ridiculous paradox that had to be understood.

Upon his return from Europe, Baade and his students Sandage and Arp embarked on an extensive observing program with the 48 inch and 200 inch Palomar telescopes: the former to survey the vast sky, the latter to study a few well-chosen targets in depth. The primary goal was to test whether the faint stars in the globular clusters were the remaining faint tail of the Main Sequence[17] in a Hertzsprung-Russell diagram truncated by the loss of its brightest stars, which had turned into red giants due to faster evolution. Observing this tail would have meant finding the key for understanding sidereal evolution. It was an ambitious plan, requiring the study of individual faint stars in crowded clusters and the measurement of their brightness and color; but it was within the reach of the Hale telescope.

Around this time, Hubble's heart, now compromised by a first heart attack, gave out completely. The true cause of the death of the Major is actually unclear. He may have died of a cerebral thrombosis. His wife Grace and his friend Milt Humason took care of the funeral, keeping even the place of burial a secret. Apparently, the halo of mystery that surrounded the legendary explorer of outer space in life was to continue after his death. Like a good lone wolf, he left few heirs: the humble and faithful collaborator "Milt", and the young Sandage,[18] who by now had begun to work with Walter Baade as his Ph.D. student. Picking up the baton from his late hero, Sandage, while feeling like «a hick who fell off the turnip truck»[19] compared to the Major, devoted his scientific life entirely to continuing the project started by the Master: quantifying the expansion of the universe. At Caltech, he had received a first-rate education, learning stellar astrophysics from Jesse Greenstein, relativity from Howard P. Robertson, and quantum mechanics from Richard Feynman. Recalling these early years, he told an interviewer from "The Sydney Morning Herald": «I came at exactly

---

[17]The Main Sequence is the line of the H-R diagram populated by stars that live with burning hydrogen in their cores. The upper blue end is made up of bright, young stars that expend a lot of energy and live a short time. The lower red end contains dwarf and cold stars whose lives are longer than the current age of the universe.

[18]D. Overbye, *Lonely Hearts of the Cosmos: The Scientific Quest for the Secret of the Universe*, Little, Brown and Co., Boston (MA) 1991.

[19]K.R. Lang, *A Brief History of Astronomy and Astrophysics*, World Scientific Publishing Co., Singapore 2018, p. 183.

the right time. I was 26 years old and all the monks and priests down here were ready to retire. So I overlapped enough that I got to know them all».[20]

The problem on the table was not the validity of Hubble's law, which almost no one questioned at the time, but its isotropy and the trend at high redshifts, which is diagnostic of the metric and thus of the fate of the cosmos.[21] The goal was to establish the independence of the expansion rate from the direction of the line of sight, both to confirm the theoretical models based on it and to detect any possible peculiar motion of the observer relative to the cosmological substrate.[22] Once again, the big eye on Mount Palomar proved to be the perfect weapon for the enterprise. The soldiers were Humason and Nicholas Mayall, an astronomer at Lick Observatory with a bright future as its director; the strategist was young Allan.

Sandage also undertook to refine and document the morphological classification scheme of galaxies devised by Hubble with a series of magnificent images (still in black and white) collected at the 200 inch. «I have acted mainly as an editor», he proclaimed with genuine modesty in the preface to his *Atlas*, «of a set of ideas and conclusions that were implicit in the notes».[23] A concept he wanted to reinforce in an interview with a great metaphor that honors both the student and the Maestro: «If you were an assistant of Dante's — he said, looking into the interviewer's eyes — and Dante died and you had in your possession the entire Divine Comedy, what would you do? In fact, what would you do?».[24] Why this extraordinary devotion? Perhaps these other words of his about the revered Maestro will help us to understand: «Like Moses, Edwin Hubble had come down from the mountain with new laws of nature. But also like Moses, Hubble had to wander in the desert for decades, only to die within sight of the Promised Land».[25]

---

[20] *His reply to cosmic question was 55*; at https://www.smh.com.au/national/his-reply-to-cosmic-question-was-55-20101122-18434.html.

[21] Friedmann models show that an open, light universe is destined to live forever, while a closed, dense universe will sooner or later reverse the expansion into a contraction and end up in a Big Crunch.

[22] T. Ferris, *The Red Limit: The Search for the Edge of the Universe*, HarperCollins, New York 2002.

[23] A.R. Sandage, *The Hubble Atlas of Galaxies*, Carnegie Institution of Washington, Washington (DC) 1961, p. VII.

[24] G.E. Christianson, *Edwin Hubble: Mariner of the Nebulae*, Farrar, Straus & Giroux, New York 1995, p. 363.

[25] R. Panek, *The Loneliness of the Long-Distance Cosmologist*, in "The New York Times", 25 July 1999.

In the second half of the 1950s, Allan was completely absorbed by his gigantic project. Thus he did not feel like accepting the proposal to contribute an essay on galaxies to the volumes of the prestigious *Handbuch der Physik* — an encyclopedic handbook conceived in Germany to present the state of all experimental and theoretical physics — that the curator, the German theoretical physicist Siegfried Flügge, intended to devote to astrophysics. He chose to pass it on to a rising star, Gérard-Henri de Vaucouleurs.

A Frenchman by birth and education, de Vaucouleurs[26] had just landed in the United States with his wife Antoinette, who was also an astronomer. They came from a long stay in Australia, at the Mount Stromlo Observatory, where Gerard had earned a reputation as an authentic *connoisseur* of the heterogeneous extragalactic world. His intuitions, such as the identified galaxy clusters as a higher level of aggregation of matter, together with an unsurpassed ability to bring out the essence of phenomena in the *mare magnum* (big sea) of experimental data, had reported him to the postwar American market as one of the possible minds to enlist. The assignment at Lowell Observatory in Arizona had to begin in 1957, just as Sandage was looking for a viable replacement for the article on the *Handbook*. Allan and Gérard were very different in appearance: the American imposing and jovial, and the French small, snobbish, and reserved. But besides sharing a taste for a certain Western style of dress, the two seemed to understand each other's work perfectly. De Vaucouleurs, who, with the exception of a characteristic French accent, had a perfect control of the English language, having also worked for the BBC after the war, agreed to write the essay for Flügge. Sandage favored him by opening Hubble's archive in Pasadena: a treasure trove of unpublished images and spectra. This friendly collaboration produced two milestones in the literature on the shapes and optical properties of galaxies.[27] This brought de Vaucouleurs a certain notoriety and, shortly thereafter, an offer of a professorship at the University of Texas at Austin, where he stayed for the rest of his life.[28] The success of the French maverick made Sandage terribly jealous (sometimes total dedication

---

[26]M. Capaccioli, H.G. Corwin (eds.), *Gerard and Antoinette De Vaucouleurs: A Life for Astronomy*, World Scientific Publishing Co., Singapore 1989.

[27]G. de Vaucouleurs, *Classification and Morphology of External Galaxies*, and Id., *General Physical Properties of External Galaxies*, in "Handbook of Physics", 11, 53, 1959, pp. 275–310 and 311–372 respectively.

[28]R. Buta, *Obituary: Gerard Henri de Vaucouleurs, 1918–1995*, in "Bulletin of the American Astronomical Society", 28, 4, 1996, pp. 1449–1450.

leads to the development of a sense of ownership, as understandable as it is unjustified, which includes the world of intellectual property). He became convinced that de Vaucouleurs had gone far beyond the mandate he had received. This caused the first crack in the relationship between the two giants, which would later become a deep rift in the 1980s.

Meanwhile, with the help of Greenstein, Minkowski, and others, Baade continued his pioneering work on the properties of the various stellar populations, extending the investigation to the chemical composition derived from spectral analysis. Once again, it was the power of the Hale telescope that made this possible on a large number of stars. In 1952, Ira Bowen, the first director of the Mount Wilson and Palomar Observatories, had completed the construction of an innovative spectrograph placed in a special focus called *coudé* ("elbow" in French), extracted by transporting the Cassegrain beam to a fixed point by means of one or two auxiliary mirrors. The optical design of this very long focus — suitable for high-resolution spectrographs which, because of their weight and size, cannot be accommodated on board the moving structure of the telescope — had originally been designed in 1891 by Maurice Loewy for the *Grand Équatorial* of the Paris Observatory. Using the sophisticated machine built by Bowen, Greenstein, who was a tireless observer,[29] discovered that some stars had the same chemistry as the Sun. But he also found others with different percentages of elements such as carbon, oxygen, nitrogen, sodium, and calcium, either in excess or deficient. His work attracted the attention of an American nuclear physicist, William Fowler, also at Caltech. It was he, along with Fred Hoyle and Geoffrey and Margaret Burbidge, all three British, who in 1957 solved the puzzle of how stars synthesize the complex elements in their nuclei from lighter ones (hydrogen and helium), for which he was awarded the Nobel Prize in Physics twenty-six years later.

However, the hypothesis of a purely stellar genesis of chemical species, put forward by Eddington in the 1920s and supported by the fusion mechanisms discovered by Bethe, was not shared by George Gamow. The brilliant Ukrainian physicist had been living in the United States since 1934. He had reached the New World after an adventurous flight from the Soviet Union[30] to escape, together with his wife, the Stalinist purges that sowed

---

[29]V. Trimble, *Jesse Leonard Greenstein (1909–2002)*, in "Publications of the Astronomical Society of the Pacific", 115, 809, 2003, pp. 890–896.

[30]K. Hufbauer, *George Gamow, 1904–1968*, National Academy of Sciences Biographical Memoir, Washington (DC) 2009, p. 14.

terror, especially among the intelligentsia, the most exposed to jealousies and denunciations. In the quiet refuge of a private university in Washington D.C., in the company of his friend Edward Teller, a Hungarian refugee whom he himself had called from England to have someone to discuss with, George "the Joker" returned to his first love: that cosmology of which he had gotten a first taste in Leningrad by listening to Friedmann's lectures.

Until then, the dynamical models of Friedmann and Lemaître had been studied mostly in terms of the geometry of space and the density of matter. Going back in time, Gamow said to himself, the universe must become more and more compact and dense, and also hotter and hotter. At this point in his reasoning, his profound knowledge of nuclear physics came to his aid, suggesting a way to gradually produce the nuclei of all the elements in Mendeleev's table by successive neutron captures. Already in 1946 he wrote:

> Thus if free neutrons were present in large quantum ties in the beginning of the expansion, the mean density and temperature of expanding matter must have dropped to comparatively low values before these neutrons had time to turn into protons. We can anticipate that neutrons forming this comparatively cold cloud were gradually coagulating into larger and larger neutral complexes which later turned into various atomic species by subsequent processes of β-emission.[31]

To refine his thinking and make it quantitative, he had to proceed backwards in time, reconstructing the cosmic expansion and calculating step by step the physical conditions of the universe to find those suitable for nuclear accretion.[32] Not an easy task, even for a mind as sharp as his. It would have been useful to discuss it with his colleagues, but the best of them were busy with the super-bomb that would end the war in the Pacific. He had been excluded from the Manhattan Project despite his reputation as a brilliant and creative scientist. The intelligence services considered him unreliable because of his origin (the U.S. and the USSR were already suspicious of each other during the war) and especially because of the exuberance of his personality. He loved jokes and puns. For example, when asked why his many

---

[31] G. Gamow, *Expanding Universe and the Origin of Elements*, in "Physical Review", 70, 572, 1946, p. 3.

[32] P.J.E. Peebles, *Discovery of the Hot Big Bang: What happened in 1948*, in "The European Physical Journal H", 39, 2014, pp. 205–223.

popular books were so successful, he would answer with irony: «Well, it is a deep secret, so deep that I do not know the answer myself!».[33] A type of person willing to do anything to make a joke work, and therefore unsuitable for sharing classified information vital to national security.

Bubbly in all his manifestations, George overflowed with ideas in many fields of knowledge, but he had some difficulty with calculations, as he readily admitted. To solve the complicated equations involved in his cosmological exercise, he hired a graduate student. The son of Russian Jews who emigrated to the New World in the early 20th century, Ralph Alpher instead had a special turn for mathematics and the good fortune to live at the dawn of the electronic computing.[34] He was then able to get his hands on the first computers made available by wartime developments, and to make use of the new data on the probability of interaction between nuclear particles resulting from studies for the atomic bomb. This was a fundamental piece of information for Gamow's project, which until a few months earlier had been highly classified. The two scientists began their research by assuming that the cosmos, in its first moments, was filled with an extremely dense neutron gas. In this primordial broth, which they named *ylem*, a philosophical term derived from medieval English to indicate a primordial substance, neutrons should have decayed into protons and electrons as the expansion progressed. Then protons would have captured some surviving neutrons to form deuterium, the isotope of hydrogen with which oxygen readily combines to form heavy water (deuterium oxide). The iteration of the neutron capture mechanism would then have produced more and more complex nuclei, until the moment when the collapse of temperature would have inexorably stopped the process, rendering the *ylem* sterile.

The model seemed to work well, since it gave an estimate of the abundance of helium that was in good agreement with observations: one atom for every 10 atoms of hydrogen. Satisfied with the result, Alpher and Gamow sent a short note to the prestigious journal "Physical Review" with the explicit title: *The Origin of Chemical Elements*.[35] Not forgetting his reputation as an unrepentant prankster, George thought to include his

---

[33] K. Croswell, *The Universe at Midnight: Observations Illuminating the Cosmos*, The Free Press, New York 2001, p. 47.

[34] H.H. Goldstine, *The Computer from Pascal to Von Neumann*, Princeton University Press, Princeton (NJ) 1980.

[35] R.A. Alpher, H. Bethe, G. Gamow, *The Origin of Chemical Elements*, in "Physical Review", 73, 7, 1948, pp. 803–804.

friend Hans Bethe as a co-author, placing his name in an intermediate position to play on the assonance with the first three letters of the Greek alphabet: alpha (Alpher), beta (Bethe), and gamma (Gamow). Perhaps it wasn't just a joke, but a way to co-opt a very brilliant colleague into the group. In any case, Alpher took the move rather badly. Being at the beginning of his career, he did not like having to share the pie with a third party. But Gamow was adamant. He sent the paper to Bethe, who had no objection to appearing as the second author[36] of a possible masterpiece. The article appeared in the April 1 issue, to the great delight of Gamow, who adored the April Fish.

In fact, the mechanism of subsequent neutron capture collided with the lack of stable nuclei with 5 and even 8 nucleons. It was like pretending to climb, step by step and without jumping, a ladder missing the fifth and eighth rungs. We would be forced to stop at the fourth rung. This corresponds exactly to helium, which has four nucleons: two protons and two neutrons. The serial process envisioned by Alpher and Gamow failed when it came to producing more complex nuclei than the so-called "alpha particles".[37] Thus, in order to fill Mendeleev's table, the young cosmos had to pass the buck and leave the fabrication of the heavy elements to the factory of stars devised by Fowler & Co. Faced with these considerations, Bethe feared that he had invested his good name in a bankruptcy speculation and tried to clear himself by revealing that he had been brought in by a joke. But he was too hasty in his judgment and in his less than honorable flight. The Alpher-Bethe-Gamow theory correctly explained the cosmic abundance of hydrogen and helium, the two elements that together constitute more than 99% of the baryonic matter of the universe (the reference is to the protons and neutrons, baryons *par excellence*, which are the fundamental constituents of atomic nuclei and thus of ordinary matter). It also marked a turning point in cosmology, which fully acquired the connotations of a physical science after a long

---

[36] In alphabetical order; Gamow was third, which today, in a season obsessed with quantitative bibliometric analysis, would have been a career mistake.

[37] The instability of some atoms is a cause of radioactivity or radioactive decay. An unstable nucleus emits particles of different types in order to reach a higher level of stability. At the end of the 19th century, Rutherford and the Frenchman Paul Villard divided the radiation from nuclear decay into three main classes, which Rutherford called α, β, and γ, according to the degree to which they penetrate objects and are deflected by a magnetic field; α rays were the least penetrating. Later, neutron and proton emission and spontaneous fission were added.

period dominated mainly by mathematical physics (or at most mechanics). The insiders and even the media noticed this, as the following story shows.

At the time the paper was published in the "Physical Review", Alpher was still struggling with his doctoral dissertation, which he defended in the spring of 1948. Intrigued by the advances reported in the daily press, a few hundred curious onlookers came to the lecture hall at George Washington University. Echoing the young scientist's claim that the creation of hydrogen and helium in the hot soup of the primordial cosmos had taken only 300 seconds — «less time than you could cook a dish of duck and roasted potatoes»[38] —, the authoritative "Washington Post" headlined: *The World Began in Five Minutes.* A sensational statement, ridiculous to some, but not too far from the truth. Thirty years later, Steven Weinberg, winner of the 1979 Nobel Prize in Physics, would publish a best-selling popular science book about *The First Three Minutes.*[39]

Gamow's work on primordial nucleosynthesis made him famous as a cosmologist. But his fervent and restless mind was already moving on to new and different challenges. He continued to work in astrophysics for a while, then turned to molecular biology, attracted by the discovery of the structure of DNA. His ideas and conjectures would later be recognized by the English chemist Francis Crick, one of the fathers of the double helix, as fundamental to the development of this field of science.[40] Brilliant intuitions rather than real results, as his friend Bethe argued with a touch of resentment, adding that almost none of his alleged theories had ever managed to survive the sifting of time. Things went from bad to worse for Alpher. After a brief moment of fame, he was forgotten by the public and the academy, and had to settle for a research position at General Electric to make his living. But in 1948, in collaboration with Robert Herman, another *enfant prodige* from the New York enclave of Russian immigrants, he had formulated a second hypothesis of extraordinary scientific value: the existence of a homogeneous and isotropic pervasive cosmic radiation, heir to that primordial fire from which, according to the models of Friedmann and Lemaître, everything had

---

[38] G. Winstein, *George Gamow and Albert Einstein: Did Einstein Say the Cosmological Constant Was the "Biggest Blunder" He ever Made in His Life?*, 1913, https://arxiv.org/ftp/arxiv/papers/1310/1310.1033.pdf, p. 4.

[39] S. Weinberg, *The First Three Minutes: A Modern View of the Origin of the Universe*, Basic Books, New York 1993.

[40] G. Segrè, *The Big Bang and the genetic code*, in "Nature", 404, 2000, p. 427.

come out.[41] Incidentally, Herman successfully resisted Gamow's pressure to change his name to Delter, so that in the future he could be added to the chain of letters of the Greek alphabet that had characterized work on the chemistry of the cosmos.

The idea originated with Gamow, who imagined a primordial universe dominated by radiation that gradually cooled as it expanded. At some point, this ocean of photons had broken all relations with matter, taking a "sunset boulevard" that, after massive dilution due to the expansion of the cosmic substrate, gives it a very low temperature here and now. The story as we tell it today is this. About 380,000 years after the Big Bang, the matter-energy temperature of the universe had dropped to 3,000 K, a value low enough for electrons to combine stably with protons and form neutral hydrogen atoms. In the jargon of experts, this cosmic era is called recombination, although it is actually the first marriage of the two constituents. The disappearance of free electrons, favored partners in epic collisions with energy quanta, made the universe transparent to photons, which began to walk through it unimpeded. A freedom, however, paid dearly with a progressive loss of energy *pro capite* (for each head) to compensate the increasing volume of the space, with the same rate of dilution of the density of matter. But this is not enough! The lengthening of the wavelengths resulting from the expansion is a further source of weakening of the photons (being a redshift), so that matter, which is not affected by that, has gradually taken the place of energy, reduced to an increasingly cold fossil residue.

Alpher and Herman estimated the temperature of this cold remnant to be only 5 degrees from absolute zero (which is the lowest value a macroscopic system can approach without ever reaching it);[42] a prediction not far from the 2.73 K measured in 2013 by the European Space Agency's (ESA) Planck Surveyor satellite, the last of the sophisticated explorers of the fossil remnant of primordial fire. The conjecture aroused moderate interest in the community of physicists and cosmologists. Gamow was not one of the players. He may not have even thought it made sense at the time. Five years later, however, in an article[43] in the influential journal "Nature", in which

---

[41] R.A. Alpher, R. Herman, *Big Bang Cosmology and the Cosmic Black-Body Radiation*, in "Proceedings of the American Philosophical Society", 119, 5, 1975, pp. 325–348.

[42] Absolute zero corresponds to a temperature of −273.15 degrees Celsius; see note at p. 275.

[43] G. Gamow, *The Evolution of the Universe*, in "Nature", 162, 1948, pp. 680–682.

he summarized the path toward a theory of the hot Big Bang (i.e., based on light particles traveling faster than more massive ones for the same energy budget), he unscrupulously appropriated the result by recalculating the temperature of the cosmic background upward: 7 degrees Kelvin instead of 5. No doubt, this volcano of ideas did not really know how to handle calculations!

The proposal of a rather uniform cosmic background consisting of radiation in the microwave range, where the emission of a (black) body at a temperature of the order of a few degrees Kelvin reaches its maximum, provided a formidable testing ground for anyone who wanted to falsify the evolutionary paradigm of the cosmos. In fact, an alternative model to that of the Big Bang was already circulating, called "Steady State".[44] It contemplated a universe not only identical to itself in every place, i.e., for every observer (as in Friedmann's cosmology), but also at every time:[45] a homogeneous and isotropic distribution of matter-energy, invariant in geometries, morphologies, and in the average density of matter, which has neither a beginning nor an end. We must underline here that the idea of a single cosmic time does not contradict the principle of relativity. It materializes in the measurement of any physical parameter that has a monotonous trend over time and is justified by the postulates of homogeneity and isotropy.

In 1931, Einstein had also analyzed this type of assumption, which James Jeans had first considered in the 1920s but quickly abandoned. The concept was rescued 17 years later by two Austrian-born mathematicians and physicists, Hermann Bondi and Thomas Gold, who had moved to England to escape the anti-Semitism of the Third Reich. An anecdote tells us that the idea was conceived during a screening of the horror movie *Dead of Night*, in which the end is identical to the beginning (with the protagonist driving the car to a friend's country cottage after experiencing a nightmare that had begun with exactly the same action). The intention to introduce a "Perfect Cosmological Principle", not only in space but also in time, had both an aesthetic and an ideological motivation. It resolved at the root the paradox of a cosmos younger than its components. It also avoided the act of creation implicit in the models of Friedmann and Lemaître, so close to the

---

[44]This is different from Einstein's "static" model, where nothing moves because the active forces are in equilibrium. A "steady" universe can expand (or contract), but it must always remain equal to itself.

[45]F. Hoyle, *The Steady-State Universe*, in "Scientific American", 195, 3, 1956, pp. 157–169.

creed of the Church and therefore indigestible to those who sympathized with the ideas of Marx and Engels.

The proposal also convinced Fred Hoyle,[46] an exuberant young physicist and militant materialist who, towards the end of the war, had shared with Bondi and Gold an accommodation and a research program on radars for the Royal Air Force. The three heatedly discussed how to avoid the incredible assistance to the "priests" offered by a cosmology that, with the Big Bang, winked at a creative intervention of God. Pope Leo XIII Pecci had called for a new alliance between science and faith, with the proviso that, in the event of irreconcilable conflict, the former would remain subordinate to the latter. In the encyclical letter *Providentissimus Deus* (The God of All Providence) of November 18, 1893, aimed at promoting Catholic education, the Pope took up a statement of Saint Augustine on the subject of *Holy Scripture*:

> All that physicists, concerning the nature of things, will be able
> to prove with certain documents, it is our task to prove not to
> be even contrary to our Scriptures; that which they then present
> in their writings contrary to our Scripture, i.e. contrary to the
> Catholic faith, either we demonstrate with some argument to
> be false what they assert or let us believe it false without the
> smallest hesitation.[47]

A concept strongly reaffirmed by Pius XII in the encyclical *Divino afflante spiritu* (Inspired by the Divine Spirit) of September 30, 1943, a few days after the fall of Fascism: «This, then, is the doctrine which Our Predecessor Leo XIII so earnestly set forth, and which We also propose with Our authority and inculcate so that it may be scrupulously maintained by all».[48] A kind of appropriation, repeated in a speech (in Italian) to the members of the Pontifical Academy of Sciences in November of 1951:

> A mind enlightened and enriched by modern scientific knowl-
> edge, which serenely evaluates this problem, is led to break the
> circle of a completely independent and autochthonous matter,
> or because it is uncreated, or because it is created by itself, and

---

[46]S. Mitton, *Fred Hoyle: A Life in Science*, CUP, Cambridge 2011.

[47]Cf. https://www.vatican.va/content/leo-xiii/en/encyclicals/documents/hf_l-xiii_enc_18111893_providentissimus-deus.html.

[48]Cf. https://www.vatican.va/content/pius-xii/en/encyclicals/documents/hf_p-xii_enc_30091943_divino-afflante-spiritu.html.

to go back to a creator Spirit. With the same clear and critical gaze, with which he examines and judges the facts, he glimpses and recognizes the work of creative omnipotence, whose virtue, stirred by the powerful «fiat» pronounced billions of years ago by the creator Spirit, unfolded in the universe, calling into existence matter exuberant of energy with a generous gesture of love. It really seems that today's science, suddenly going back millions of centuries, has managed to bear witness to that primordial «fiat lux», when a sea of light and radiation burst out of nowhere with matter, while the particles of chemical elements split up and gathered in millions of galaxies.[49]

Hoyle, who combined an unstoppable drive and a strong communicative ability with an unquestionable scientific creativity, took it upon himself to convince public opinion of the absurdity of evolutionary cosmologies, which made everything come from nothing and demanded that the father be younger than the son. In fact, according to the figures then established by astronomers, the stars remained paradoxically older than the universe itself. It was in 1949, on the occasion of a radio broadcast on this subject produced by one of the BBC networks, that the sparkling Englishman coined the term "Big Bang" with a derogatory intent: «This instantaneous creation of the Universe is like a party girl jumping out of a birthday cake, it's ridiculous».[50] The colorful metaphor was not without consequences. Later, Hoyle would have to justify himself for a long time for this politically incorrect expression — «Words are like harpoons. Once they go in, they are very hard to pull out»,[51] he told a journalst in 1995 — and then watch helplessly as the theory of the "primeval explosion", universally renamed with its irreverent epithet, triumphed.

At first, the Steady State paradigm found favor with part of the community of physicists. With the Perfect Cosmological Principle it put space and time on the same level, removed the embarrassing problem of the initial singularity — Eddington, for example, had found «the notion of a beginning of the present order of Nature [...] philosophically [...] repugnant to

---

[49]Cf.   http://www.vatican.va/content/plus-xii/it/speeches/1951/documents/hf_p-xii_spe_19511122_di-serena.html.

[50]S.E. Vigdor, *Signatures of the Artist: The Vital Imperfections That Make Our Universe Habitable*, CUP, Cambridge 2018, p. 133.

[51]Cited by H. Kragh, *What's in a Name: History and Meanings of the Term "Big Bang"*, https://arxiv.org/ftp/arxiv/papers/1301/1301.0219.pdf.

him»[52] — and rendered useless the tricky question of what was there before the Big Bang occurred. However, the model could not ignore the empirical data on the expansion of the universe, as highlighted by Hubble's observations. There were at least two ways out. The first considered the spontaneous creation of matter from nothing in order to maintain the constancy of the average density of matter-energy despite the expansion of space. The explicit violation of the principle of conservation of mass, sanctified by the Russian polymath Mikhail Vasilyevich Lomonosov and by Lavoisier in the second half of the 18th century, made the skin crawl.[53] But, as Hoyle cheerfully pointed out, the requirement was minimal: it was enough for a single atom of hydrogen per cubic meter to pop out of nowhere every billion years. So little that the phenomenon would have escaped any experimental verification. It was like saying that if a sin is very small, it is not a sin. "Do you prefer this", he told the BBC in 1949, "or creating everything from a Big Bang?".[54] Christ or Barabbas? A substantial part of the scientific community replied that buying in installments seemed preferable to paying all at once. It would soon become clear that this was not the right answer.

Also (substantially) wrong were those who argued that the redshift had nothing to do with distance and was due to some mysterious gravitational effect associated with the sources. An extravagant conjecture that nevertheless offered the second (partial) way out of the problem posed by Hubble's law at the Steady State. In the 1960s, Halton Arp, a student of Baade and Hubble along with Sandage, would provide the example of some small groups of galaxies in which objects seemed to be physically connected by matter bridges, yet exhibited strikingly different redshifts.[55] According to this elegant New Yorker transplanted to California, these were indications of a pathology related to the purely Doppler interpretation of spectroscopic observations, a close cousin of the anomalies manifested by quasars. «It is with reluctance that I come to the conclusion that the redshifts of some extragalactic objects are not due entirely to velocity causes», he argued.[56]

---

[52] G. Lemaître, *The Beginning of the World from the Point of View of Quantum Theory*, in "Nature", 127, 1931, p. 127.

[53] N. von Hofsten, *Ideas of Creation and Spontaneous Generation prior to Darwin*, in "Isis", 25, 1, 1936, pp. 80–94.

[54] H. Kragh, *Naming the Big Bang*, in "Historical Studies in the Natural Sciences", 44, 1, 2014, pp. 3–36.

[55] G. Kirby, *Wacky and Wonderful Misconceptions about Our Universe*, Springer, Dordrecht 2018, pp. 214–217.

[56] J. Gregory, *Fred Hoyle's Universe*, Oxford University Press, Oxford 2005, p. 217.

But the luminous bridges do not prove the existence of any physical connection, since they can be caused by projection effects (as it turned out to be the case with Arp's galaxies). For these heresies, stubbornly defended even in the face of the strongest evidence to the contrary, "Chip" was later expelled from the American scientific community and exiled to Europe, proving that in every corner of the world the Inquisition changes its clothes but never dies.

On the other side of the Iron Curtain, in Stalin's Russia, neither cosmology met with the approval of the dictator and his sycophants, and as such, they couldn't be «either taught or defense or treated in any way», in the words of Bellarmine's admonition to Galilei in 1616,[57] at the cost of very severe penalties. The thought of Marx and Engels, uncritically understood as a secular religion, rejected the appearance out of nothing of that matter on which everything is based, both in the form of a single initial event, objectively embarrassing and too close to the biblical narrative, and in that of the "well-tempered" creative drip imposed by the Steady State. So, while in the United States Joseph McCarthy hypocritically donned the clothes of the Grand Inquisitor in order to expose and eradicate in his country the "communist cancer" ravaging officials, scientists, artists, and even ordinary people, in the Soviet Union the KGB, the Committee for State Security, was in power and controlled ideological integrity with the weapon of deportation to the deadly labor camps in Siberia (gulags and sharashkas, according to the rank and usefulness of the prisoner, and the severity of his guilt): a fate worse than death. For both opposing sides, it was an authentic return to the darkest periods of history, with new witches and new stakes. The sudden demise of "tovarisch Stalin", on March 5, 1953, produced little changes in the USSR, despite the condemnation of the dictator's crimes by a committee of the so-called successors and pretenders. It would take more than a decade before the Belarusian Jew Yakov Borisovich Zeldovich, having left nuclear physics where he had earned the recognition of power, a handful of honors, and a minimum of independence of thought, refounded a modern cosmology at the Moscow State University school.[58]

---

[57] Admonition of Cardinal Roberto Bellarmino to Galileo Galilei, February 26, 1616, in *The Vatican documents of the trial of Galileo Galilei (1611–1741)*, S. Pagano (ed.), Vatican Secret Archives, Vatican City 2009; http://www.pas.va/content/dam/accademia/pdf/sv112pas.pdf,p.\46.

[58] J.B. Zeldovich, M.V. Sazhin (eds.), *My Universe: Selected Reviews by Ja.B. Zeldovich*, Harwood Academic Publishers, Chur 1992.

Meanwhile, the white dome on the humped peak of Mount Palomar had continued to open on every clear night to collect and decipher the celestial messages to which only the giant glass eye of the Hale telescope had access. Like a prince's war-horse, the imposing instrument was constantly tended by teams of grooms charged with preventing any anomalies and, if necessary, curing them quickly. At a time of fierce confrontation between the two blocs into which the world was divided, it was essential to guarantee the perfect efficiency of the instrument, for the benefit of knowledge, but also for the image of the American system: both a five-star telescope and a five-star hospitality. Visiting astronomers, no matter where they came from, were housed in a cottage and could enjoy the privilege of dictating the rules of the table for lunch and dinner, according to the custom begun by Hubble at Mount Wilson. A British style of soft voices and affected manners that emphasized the honor of being admitted to the temple of astronomy. The rules did not change at night, despite the rigors of observing and the constant dangers, such as climbing in the dark to the cage that housed the direct-focus camera, suspended 50 feet (15 meters) above the large mirror.

One of the most frequent users of the telescope was Allan Sandage, who was fully immersed in the project of defining the fundamental parameters of the Friedmann cosmological model. First, the expansion rate here and now, measured by the Hubble constant H_o, which is essential for setting the scale length and an upper limit on the epoch of the Big Bang; then the acceleration parameter q_o, which, by qualifying the geometry of space, sheds some light on the future of the cosmos. As sometimes happens to men of fervent faith, more than a scientific research, this program became a mission and a purpose of Sandage's life.[59] He himself admitted it, using different words, in a 1961 paper with an emblematic title,[60] made it crystal clear that no one else in the world could enter the race. The project was as fundamental as it was ambitious. The estimation of q_o involved first the identification and then the observation of very distant and therefore very faint galaxies, whose luminosities and redshifts had to be measured. The project about H_o was even more difficult and uncertain, since it required combining two very different distance regimes: one where parallax measurements are still

---

[59] K. Wegter-McNelly, P. Clayton, R.J. Russell, *Science and the Spiritual Quest: New Essays by Leading Scientists*, Routledge, London 2002, pp. 52–63.

[60] A. Sandage, *The Ability of the 200 inch Telescope to Discriminate between Selected World Models*, in "Astrophysical Journal", 133, 1961, pp. 355–392.

possible, and another where the Hubble law holds. Let's see where the problem resides.

Parallax, as we already know (p. 3), is the only method we have to directly measure the distances of celestial bodies. Today, thanks to the specialized Hipparcos and Gaia space observatories launched by the European Space Agency, the limits of application of this technology have been greatly extended,[61] but until the 1980s it was not possible to go beyond 100 light-years (0.4% the distance of the Sun from the Galactic center). Hubble's law is also, in its own way, a direct, easy to use, and effective ruler, as long as one knows the value of the constant of proportionality between redshift and distance, which is H_o. But this parameter can be derived only empirically, through sufficiently distant celestial objects whose distance has been measured by other means. This is where things get complicated, since Hubble's law does not apply to sources closer than a few million light-years (which is why we said "sufficiently distant celestial objects"). The reason is that the cosmological velocity, i.e. the velocity due to cosmic expansion alone, is contaminated by the (peculiar) motions associated with local dynamics. For example, the orbital motion of the Earth around the Sun has nothing to do with cosmic expansion, nor does the motion of the Sun around the Galactic Center, nor that of the Milky Way toward the Andromeda Nebula, nor the motion by which the Local Group of Galaxies, the family of extragalactic nebulae to which the Milky Way belongs, falls toward the center of the Virgo Cluster, the cluster of 2,000 nebulae 65 million light-years away from us. These are motions caused by local inhomogeneities, in which excess density dominates over cosmic expansion. All that causes a kind of cumulative systematic error in measurements that can be removed once the set of motions we have mentioned is known. But there is yet another, more troublesome, source of uncertainty. Within the clusters to which they belong, galaxies defend their individuality by dynamically responding to the gravitational pull of other members of the association. Being limited (the specific velocities of galaxy groups and clusters relative to the cosmological substrate typically do not exceed one thousand kilometers per second), this

---

[61]The Gaia satellite (Global Astrometric Interferometer for Astrophysics), a continuation of the Hipparcos mission and part of ESA's Horizon 2000 science program, was launched in 2013. Its heliocentric orbit keeps it wandering around the second of the 5 stationary (Lagrangian) points of the Earth-Sun system. The main goal is to make astrometric measurements of a billion stars, with decreasing precision with distance, up to the limit of the Galactic Center (about 26,000 light-years from us).

(thermal) convulsion gradually loses importance with increasing redshift, that is, with increasing distance from us.

Let us try to illustrate this with an ideal example. Imagine an elastic sheet that expands uniformly over time. We sit in the middle and want to measure the (radial) velocity of the substrate, using some fireflies scattered on the flat surface as a reference. This will be easy to do via the Doppler effect if the insects are not moving on their own account. If instead they move randomly, each in its own way, at speeds that do not exceed a certain upper limit, the fireflies closest to us will give an unreliable result until their average peculiar velocities become negligible relative to the expansion velocity of the substrate, which is proportional to distance. Applying these considerations to the astronomical case, we realize that there is a large interval of distances in the sky, ranging from a few hundred[62] to 10 million light years, within which neither the parallaxes (too short) nor Hubble's law (too uncertain) are applicable. To measure distances there, we have to be content with grasping at straws. The situation is similar to trying to connect the standard meter kept in Sèvres, France, with the same sample kept overseas at the National Bureau of Standards in the United States, without being able to compare them directly. The only way out is to fill the gap by estimating cosmic distances using indirect methods, mainly empirical, based on the so-called "indicators": families of objects of known intrinsic size or brightness, easily recognizable and measurable, such as the Cepheids.

From this brief description, it is possible to appreciate how uncertain and subjective the coupling of the domain of trigonometric parallaxes to that of Hubble's law can be. For more than a decade, Sandage played the role of undisputed leader in the field. By 1960, the value of H_o had decreased by a factor of 5 from Hubble's original estimate (500 km/s/Mpc, or about 160 km/sec per million light-years) and the age of the universe had increased to 10 billion years. In the mid-1970s, Sandage halved it even further, doubling the lifetime of the cosmos, which eventually became longer than that of the oldest stars.[63] But not everyone was willing to accept the result provided by the Major and marketed by the most brilliant of his collaborators, Swiss astronomer Gustav Tammann. For example, Sidney van den Berg, a former student of Oort's in the Netherlands who was then working in Canada,

---

[62]Gaia has improved this lower limit by a factor of almost three; cf. http://ned.ipac.caltech.edu/level5/March19/Mignard/Mignard3.html.

[63]G.A. Tammann, *The Ups and Downs of the Hubble Constant*, in "Reviews in Modern Astronomy", 19, 2006, pp. 1–30.

questioned both the method and the value. «H_o is unbelievably bad now», complained in 1977 another Dutch astronomer, Bart Bok, who ran the Steward Observatory in Arizona. «We have Sandage's H_o of 50 km/s/Mpc, against van den Berg's H_o of 90. Now, the difference between 50 and 90 is almost a factor of two, which is a factor of eight in the volume and hence a factor of eight uncertainty in the densities in intergalactic space».[64]

At this point in the story, de Vaucouleurs came in with his vast experience and astronomical culture. He was convinced that his former friend Allan had underestimated some selection effects when polishing the data. A set of errors that, according to the French astronomer, would have caused an excessive contraction of H_o. The *querelle* between the two leaders took on the tones of a religious war, with hints of authentic folklore, as when Sandage, in transmitting his results to de Vaucouleurs, closed the covering letter by emphasizing that every morning before going to work he entered a church to implore the Lord's grace that his colleague might be enlightened and take the right path. The personal confrontation led to the creation of two opposing and strongly antagonistic parties (one centered in Pasadena and the other in Austin, Texas), with battles fought by the two champions and their squires in the most prestigious journals and in the meeting rooms. The conflict was effectively ended in 2001 by the results of a key project carried out with the Hubble Space Telescope. With small subsequent tweaks, Wendy Freedman and co-workers set the value of H_o just halfway between those of de Vaucouleurs and Sandage (see also the chapter 19), and the age of the universe at 13.72 billion years, with an uncertainty of one hundred million years. Meanwhile, the Steady State model of Hoyle, Bondi, and Gold had been largely falsified by the serendipitous discovery of the Cosmic Microwave Background Radiation (CMBR) and by the observation of some phenomena that incontrovertibly point to cosmic evolution. Let's see how.

Alpher and Herman's hypothesis that the cosmos was permeated by uniform radiation at a temperature close to absolute zero had found few adherents, especially after the two young proponents left the Johns Hopkins University Applied Physics Laboratory in Baltimore to earn a living in industry. In the early 1960s the theory was taken up by Zeldovich and, independently, by Robert Dicke and Jim Peebles on opposite sides of the Iron Curtain. In Moscow, in an academy now free from the ideological

---

[64] *Bart Bok's inteview (1978)*;   https://www.aip.org/history-programs/niels-bohr-library/oral-histories/4518-4.

idiosyncrasies of the regime, two brilliant students of Zeldovich, Andrei Doroshkevich and Igor Novikov, even went so far as to predict its measurability and propose the instrument for doing so.[65] At the same time Robert Dicke, a professor at Princeton, was building a radiometer for experimental verification. But he did not finish it in time. Once again, it was serendipity that led the dance and painted the story in the colors of legend.[66]

Arno Penzias and Robert Wilson, two physicists at Bell Laboratories,[67] were refurbishing a microwave antenna located about fifty kilometers from Princeton, in Holmdel, New Jersey, with which they hoped to observe the neutral hydrogen clouds of the galaxy. The imposing horn-shaped antenna had been built in 1960 by the large private U.S. telephone company for a very different purpose. It should pick up radio signals bouncing back from the Echo satellite, a 30-meter diameter aluminum-coated Mylar balloon placed in low Earth orbit to test a method of passive long-distance communication. Two years later, the experiment was shelved following the successful launch of the first active telecommunications satellite, the groundbreaking Telstar, capable of carrying 1,200 telephone channels or, alternatively, one television channel for worldwide communications. The Holmdel antenna was then reassigned to the two young scientists, experts in a new generation of cryogenic microwave receivers.

Penzias was born in Munich in 1933 to a Jewish family and narrowly escaped the Holocaust.[68] Transported to England along with tens of thousands of other children a few months after the war as part of a sporadic humanitarian initiative of those difficult times known as the *Kindertransport*, he had later reached the eastern United States where he had grown

---

[65] A.G. Doroshkevich, I.D. Novikov, *Mean Density of Radiation in the Metagalaxy and Certain Problems in Relativistic Cosmology*, in "Soviet Physics Doklady", Vol. 9, 1964, p. 111.

[66] P.J.E. Peebles, R.B. Partridge, *Finding The Radiation from the Big Bang*, 2007, http://staff.ustc.edu.cn/~wzhao7/c_index_files/main.files/CMBbook.pdf.

[67] Bell Laboratories (also known as Bell Labs) is a subsidiary of Western Electric Research Laboratories, a research division of AT&T Inc., a powerful U.S. telephone company based in Texas, which in turn was the daughter of the Bell Telephone System, founded in 1877 by Alexander Graham Bell after the controversial invention of the telephone. Bell Labs was created and made independent in 1925 and has since been one of the privileged places of experimentation and technological advancement of the 20th century. They can boast the assignment of nine Nobel Prizes, from 1937 to 2018, and four Turing awards.

[68] *Arno Penzias Nobel Lecture*, 1978, https://www.nobelprize.org/prizes/physics/1978/penzias/lecture/.

and formed. Wilson[69] was instead a quiet Texan, three years younger than his colleague and friend. The two had a big problem. Despite all their efforts to fine-tune the equipment, the signal coming back from the antenna showed a faint but persistent background noise, of constant amplitude and apparently equal intensity in all directions. They had tried everything to identify it. Interference from radio stations in the nearby metropolis had to be ruled out, since nothing changed when the antenna was pointed directly at New York City. Nor could the noise be attributed to the Milky Way, which had a much more stentorian but much less isotropic voice, or to the thermal emission from the lukewarm droppings of birds that had mistaken the big ear for a toilet. To be sure, Arno and Robert spent several days with shovels, brushes, and lots of water removing the guano. The mysterious sound did not disappear with the seasons, ruling out both the Solar System and atmospheric nuclear tests, since the radioactive pollution responsible for a possible spurious signal should have faded within a year. What could it be? The mystery was solved thanks to the intervention of the blindfolded goddess.

Dicke's theoretical and experimental work had been illustrated by Peebles to a fellow radio astronomer. The latter, returning by plane from Puerto Rico, where he had been observing with the large Arecibo dish (the one that collapsed in December 2020), had happened to run into another colleague, to whom he in turn had told the story, mainly to pass the time on the long flight. A few days later, the third link in the chain contacted Penzias for an entirely different reason. And so the unlikely circle closed. Although very intrigued by what he had heard on the phone, Arno was reluctant to bother the great Princeton guru. But finally, he made up his mind and deared to call. Dicke was having a frugal lunch at his office with his colleagues. He listened to what Penzias had to say, thanked him politely, and ended the call. Then he exclaimed laconically: «Well, boys, we've been scooped!».[70] The noise that had so disturbed the two Bell Labs employees was nothing more than the cosmic background postulated by Alpher and Herman. Penzias and Wilson agreed with Dicke on the strategy for communicating the news. Two separate articles were prepared — one for the experimental data and a second to illustrate the theoretical apparatus

---

[69] *Robert Woodrow Wilson Nobel Lecture, 1978*, https://www.nobelprize.org/prizes/physics/1978/wilson/lecture/.
[70] P.J.E. Peebles, L.A. Page Jr., R.B. Partridge, *Finding the Big Bang*, CUP, Cambridge 2009, p. 204.

supporting their interpretation — which were then submitted to the editor of the "Astrophysical Journal" with the request that they were published together; and so they were! The papers appeared on parade in the July 1965 issue of the journal. The least satisfied of the brigade was Robert Wilson, who, trained to believe in the Steady State model, distrusted the evolutionary interpretation of the phenomenon. «Just the fact»,[71] he repeated like a mantra. A conservatism that would not have prevented him from accepting a Nobel Prize in Physics 13 years later, shared with his friend Arno. Gamow was now dead, but Alpher and Herman, who were alive and well, felt bad about being ignored. They were absolutely right — we believe — but that's the way the world works, even the world of scientific genius awards! Dicke and Peebles were also disappointed, but in 2019 the latter finally saw his merits recognized by the Royal Swedish Academy of Sciences (Dicke could not, having died in 1997).

It must be said that the existence of a cosmic microwave background was only one of the tombstones under which Hoyle, Bondi, and Gold's theory was buried. Two other discoveries condemned it to rapid oblivion, which was resisted only by a few diehards, including Arp and Jayant Vishnu Narlikar, an Indian student of Fred Hoyle. The first test concerns the number of radio sources: an exercise that Hubble himself had already performed in the optical band. To understand what this is all about, let us first consider a static, Euclidean model of the universe, uniformly filled with a hypothetical population of time-invariant "standard candles". Using simple mathematics, it is easy to determine the total number of sources that an observer would see brighter than a given flux threshold. On a logarithmic scale, this number increases in proportion to the flux itself. If we repeat the exercise for cosmological models that require expansion, both Friedmann and Steady State, we still find a linear trend, but with a smaller slope, because the expansion of cosmic space causes the number of observed sources to grow less rapidly than in the static case, while pushing the counts to lower flux levels.

Well, in the mid-1950s, this theoretical prediction seemed sensationally contradicted by the detection of strong radio sources and radio-emitting quasi-stellar objects, whose density even increased with distance. «This discrepancy», was the comment of Martin Ryle, guru of radio astronomy

[71] V. Ginzburg, *Prime Elements of Ordinary Matter, Dark Matter & Dark Energy: Beyond Standard Model & String Theory*, Universal Publisher, Boca Raton (FL) 2007, p. 251.

in England, «seems too great to be explained by errors in the observations or their interpretation, and suggests that the Steady-State theory in its original form does not correspond to the actual Universe».[72] Not everyone agreed, however. Australians, for example, perpetually plagued by an inferiority complex toward their more celebrated British cousins, criticized the observations, questioned the interpretation, and thus rejected the conclusions. In the early 1960s, the picture became clearer. Jan Oort and his student Marteen Schmidt in Holland, and, independently, the Scottish Malcolm Longair, concluded that the results of the radio telescope surveys indicated an evolution of the brightness (or density) of cosmic sources with distance, and hence with look-back time. A property incompatible with the perfect cosmological principle. Amen!

At this point in history, a digression, albeit brief, on the birth and early steps of radio astronomy is in order.

---

[72]M. Ryle, *Radio astronomy and cosmology*, in "American Scientist", 50, 1, 1962, p. 7.

Chapter 18

Observing with closed eyes: The exploration of the universe at radio frequencies

> The total amount of energy
> from outside the Solar System
> ever received by all the radio telescopes
> on the planet Earth is less than
> the energy of a single snowflake
> striking the ground.
>
> Carl Sagan, *Cosmos*

> Once you have started,
> the only way back is to go forward.
>
> Imre Kertész, *Detective story*

It all began in 1873 with the *Treatise on Electricity and Magnetism*, a masterpiece of mathematical synthesis that unified two seemingly unrelated physical phenomena.[1] «The American Civil War will pale into provincial insignificance in comparison with this important scientific event of the same decade»,[2] Richard Feynman later commented. In this context, the author, James Clerk Maxwell, suggested that light was the sensitive fraction of the energy spectrum provided by his theory. The conjecture was motivated by the analogy between the characteristics of the two phenomena and by the concordance between the speed measured for light and that calculated by the Scottish scientist for the propagation of his electromagnetic waves. The price of Maxwell's synthesis was the introduction of a luminiferous (light-bearing) ether which *pro tempore* (for the time being) rescued the failure of the Galilean principle of equivalence among all observers in relative linear

---

[1] J.C. Maxwell, *A Treatise on Electricity and Magnetism*, 2 vols, Clarendon Press, Oxford 1873.
[2] R.P. Feynman, *The Character of Physical Laws*, British Broadcasting Company, London 1965, p. 12.

and uniform motion. Falsified by Michelson and Morley experiment,[3] at the end of the 19th century, the ether disintegrated and classical electromagnetism was reconstructed as a field theory.[4] It left open a problem that, after the partial attempts of the Frenchman Henri Poincaré and the Dutchman Hendrik Antoon Lorenz, would have been definitively solved by Einstein's new treatment of space-time.[5]

Only fifteen years later, Maxwell's conjecture was confirmed by an experiment designed by a professor of theoretical physics at the Technical University of Karlsruhe who had studied with Kirchhoff and von Helmholtz, apparently with excellent results. Heinrich Hertz's apparatus consisted of a transmitter, a copper ring open enough to produce a spark under the effect of a high-voltage discharge, and a receiver, also an open copper ring. The challenge was to observe a small spark in the receiver as a sign of the passage of radio waves produced by the electrical disturbance of the transmitter. And so it happened. Strangely enough, neither Hertz nor Maxwell before him could grasp the practical significance of their findings. It is said that when Hertz was asked about the possible applications of his discoveries, he replied frankly: «Nothing, I guess». It was the Italian Guglielmo Marconi who understood it and built his economic and scientific fortune on this insight.[6] In any case, Hertz continued his experiments out of pure scientific curiosity, realizing that radio waves were reflected by conducting materials and could thus be focused by metal mirrors, just like light.[7]

This being the case, astronomers had only to add one plus one to be convinced that celestial bodies, being hot, should themselves emit radio waves, and that these waves could be collected and studied with the technological

---

[3]R.S. Shankland, *The Michelson-Morley Experiment*, in "Scientific American", 211, 5, 1964, pp. 107–115.

[4]In physics, a field is a region of space (or space-time) to each point of which one can assign the value of some physical quantity: just a number or something more complicated endowed with directionality. Introduced by the Swiss Leonhard Euler in the eighteenth century to describe the behavior of fluids, this concept reached its maturity with Michael Faraday and Maxwell.

[5]For a historical excursus, see G.R.M. Garratt, *The Early History of Radio: From Faraday to Marconi*, The Institution of Engineering and Technology, London 2006.

[6]The Serbian genius Nikola Tesla, recognized as the inventor of alternating current and one of the fathers of modern robotics, had achieved the same results as Marconi. He managed to cover even greater distances, and much earlier than the Italian scientist. But not even this invention made him rich.

[7]Cf. G. Weightman, *Signor Marconi's Magic Box: The Most Remarkable Invention of the 19th Century & the Amateur Inventor Whose Genius Sparked a Revolution*, Harper-Collins Publishers, New York 2012.

recipe anticipated by Hertz. First and foremost target was the Sun, the most powerful of all astronomical sources.[8] However, early research in this direction at the turn of the century came to nothing. The reason was the inefficiency of the rudimentary receivers. Instead, the failure was attributed to a barrier surrounding the Earth's atmosphere that obstructed the passage of radio waves and reflected them back. This barrier really exists, but not without some important windows. Today we call it the Kennelly-Heaviside layer, after the two engineers — the American Arthur Kennelly and the Englishman Oliver Heaviside — who independently discovered it in the early 20th century. The corresponding layer is one of the components of the complex structure of our planet's atmosphere between 60 and 1,000 km altitude, called the ionosphere because the gas is kept ionized there[9] by solar radiation. It took 20 years and a Nobel Prize winner to verify the existence of the Kennelly-Heaviside layer. Meanwhile, Marconi's success in circumventing the Earth's curvature with his radio messages had confirmed the misconception that the ability to reflect high-frequency signals was not limited to medium waves (between 300 kHz and 3 MHz), but covered the entire electromagnetic spectrum. Were this true, the radio sky would be completely inaccessible from the ground. With this belief, any plans to explore the new world of sky messengers became purely unrealistic. In the absence of proactive action, progress had to rely on the intervention of fate.

It happened in 1930. The Bell Telephone Laboratories (yes, always them!), while experimenting with transatlantic voice transmissions through shortwave, between 10 and 20 m, had detected the presence of an unexpected source of noise whose nature had to be discovered (first of all, to get rid of it). The task of solving this problem was assigned to Karl Guthe Jansky,[10] a 25-year-old communications engineer from the wild Oklahoma Territory. He had worked for 3 years at Bell Labs in Holmdel, New Jersey, the same place that 35 years later would witness the discovery of the cosmic

---

[8] It is perhaps unnecessary to mention that the observed power is not just an intrinsic property. It depends (quadratically) on the distance of the object from the observer and is systematically eroded by the sources of absorption, scattering, etc. that the light encounters on its way to the observer.

[9] An ion is an atom made electrically charged by the addition (negative charge) or subtraction (positive charge) of one or more electrons. A fluid of ionized atoms (partially or completely) is called plasma.

[10] J. Kraus, *Jansky, Karl Guthe*, in "Biographical Encyclopedia of Astronomers", Springer, New York 2014, pp. 1109–1112. Cf. also the contributions in K. Kellermann, B. Sheets (eds.), *Serendipitous Discoveries in Radio Astronomy*, Proceedings of the NRAO Workshop, National Radio Astronomy Observatory, Green Bank (WV) 1984.

microwave background (see p. 477). The Bell Labs had just moved there to escape the electromagnetic pollution of New York, where they were originally located, taking advantage of the crash in land prices caused by the Great Depression. A fortunate strategic decision. That same year, Louis Bamberger, a wealthy businessman from Newark, and his sister Caroline Bamberger Fuld founded the Institute for Advanced Study in Princeton, not far from the Bell Telephone site, dedicated to theoretical research, which would soon host for the rest of his life a luxury refugee, Albert Einstein: an enclave with a very high density of brains, complemented by the University of Princeton, a prestigious academic institution.

Jansky was a son of art. From his father Cyril Methodius, who was born in America to Czech immigrants and at the time was serving as dean of the engineering department at the University of Oklahoma in Norman, Karl Guthe inherited his passion for mathematical science: a vocation also reflected in his name, which came from that of a German immigrant, a professor of physics at the University of Michigan and a former mentor of father Cyril. According to his older brother, Cyril Jr., Karl was anything but a bookworm, excelling at tennis, hockey, and ping-pong. Despite his academic and athletic achievements, however, he struggled to get hired at the Bell Labs because of chronic kidney disease. He eventually succeeded only because of Cyril's recommendation, which, as he himself would later recall, had unwittingly facilitated the realization of the last indispensable prerequisite for what actually happened: «All these factors, namely — basic elements of character, inherent abilities, family environment, scholastic environment, etc., together with the decisions of people — the choice of a course of study in college, the choice of a life work, the choice of an employer, the decision to employ, the assignment of a laboratory branch and to a specific project, had a vital influence upon what followed. The stage must be set before the play can begin».[11] In short, it is chance that prepares the background and provides the opportunities: it is then up to the individual to seize them.

Jansky rapidly succeeded in setting up a directional antenna to scan the horizon, which he coupled to a sensitive receiver tuned to the frequency of 20.5 MHz, equal to the wavelength of 14.6 m. The 30 m long and 6 m high frame of wooden slats and copper wire, which someone renamed "Jansky's Merry-go-round" although it was more like the Wright Brothers'

---

[11] C.M. Jansky Jr., *The Beginnings of Radio Astronomy*, in "American Scientist", 45, 1, 1957, p. 6.

biplane, was installed «on the flat, open expanse of a fallow, southern New Jersey potato field».[12] A circular track and the four wheels of a Model T Ford automobile (the "Tin Lizzie") driven by an electric motor allowed the structure to rotate completely about the vertical. The device had an angular resolution that would have been ridiculous for an optical telescope: 25 degrees in azimuth and no discrimination in elevation. The collected signal produced a current that was amplified in real time and read by an ammeter connected to a paper tape recorder. Dry "electrocardiograms" instead of the beautiful images produced in color on the retina of the eye by optical telescopes, or rather, but in black and white, on photographic emulsions, then in common use by astronomers.

«I wish to dwell for a moment on the subject of data taking», Cyril Jr., also a transmission engineer, would later say in remembrance of his illustrious brother.

> It is one thing to sit at a desk and on the basis of information and data gathered by others hypothesize with respect to the laws of nature. It is another to design, develop and build apparatus and equipment and then to embark upon a long, tedious program of taking accurate, dependable data on what actually takes place. There are some who look upon data taking as beneath the dignity of true scientists preferring to leave this detailed work to assistants. I do not wish to disparage the value of theory and speculation as necessary elements in any scientific study but rather to emphasize that Karl Jansky's work demonstrates how essential is the accumulation of adequate, accurate data. This is as true with respect to studies of radio propagation phenomena as it is in astronomy».[13]

After nearly a year of measurements, Jansky was able to isolate three components of the noise. The first two were related to the electrical activity of thunderstorms: a stronger, sporadic signal from local phenomena and another, more constant and weaker signal as a combined effect of distant meteorological perturbations. The third consisted of a weak continuous

---

[12] J. Kraus, *The First Five Years of Radio Astronomy – Part One – Jansky, Karl and his Discovery of Radio Waves from Our Galaxy*, in "Cosmic Search", Vol. 3, No. 12, 1981, p. 9.

[13] C.M. Jansky Jr., *My Brother Karl Jansky and His Discovery of Radio Waves from Beyond the Earth*, in "Cosmic Search", 1, 4, 1979, p. 12.

hissing sound. At first Jansky attributed it to some kind of industrial pollution, but later he realized that the maximum intensity of this interference seemed to follow the daytime path of the Sun. Was there a connection between the two? Instead of trumpeting the possible discovery from the rooftops and risking an embarrassment, Karl rolled up his sleeves and doggedly continued the measurements throughout 1932. It was the right decision, because as time went on he had to change his mind. The peak of noise emission moved farther and farther away from the position of the Sun. The phenomenon — he discovered with infinite patience — had the same periodicity as the sky of fixed stars, which, as we know, completes its revolution around the Earth four minutes earlier than the Sun each day. It took him another year to determine that the radio emission came from a mysterious astronomical source projected into Sagittarius, the same region of the sky where Shapley had recently placed the center of the Milky Way (p. 364). Jansky, who was not an astronomer, much less one of the yuppies of modern science in search of easy fame, merely whispered it. «In conclusion, data have been presented which show the existence of electromagnetic waves in the earth's atmosphere which apparently come from a direction fixed in space», he wrote in his report[14] at the meeting of the International Scientific Radio Union held in Washington in April of 1933. The discovery found space in "Nature" and earned a moment of celebrity enough to be featured on the May 5, 1933 pages of "The New York Times".[15] However, there was no further follow-up. It seems plausible that, had Jansky lived long enough, he might have been awarded the Nobel Prize. But by the time radio astronomy became a mature science, in the late 1950s, Karl had died, and as we know, the prize is not given in memory. Unaware of even the rudiments of radio transmission technology, astronomers preferred to continue along the old, well-known, and well-resourced path.

Jansky's discovery was promptly confirmed by an experiment conducted at Caltech in 1936 by the Russian-born astronomer Gennady Potapenko, whose work, however, was not publicized and was even cut for lack of funds: «You designed a large antenna?», «Yes, and I did it for Dr. [Robert Andrews] Millikan», Potapenko recalled in a December 1974 interview,[16]

---

[14]K.G. Jansky, *Electrical Phenomena that Apparently Are of Interstellar Origin*, in "Popular Astronomy", Vol. 41, 1933, p. 555.

[15]"*May 5, 1933: The New York Times Covers Discovery of Cosmic Radio Waves*, in "APS News", 24, 5, 2015.

[16]*Interview with Gennady W. Potapenko on 26 December 1974*, in "NRAO Archives", https://www.nrao.edu/archives/items/show/15124.

«and explained everything to him and the next day he told me, "You know it would cost $1,000? Then [...] my work was done». Thus, in the absence of strong scientific demands and justified commercial motivations, the Bell Labs management dropped Jansky's proposal to build a 30-bond antenna with which to further study the phenomenon. Reassigned to another task, Karl died in the shadows in 1950, struck down by a heart attack. He was only 44 years old. However, the discovery of the existence of radio signals from space, reported by Jansky himself in a journal for radio technicians in 1933, managed to capture the imagination of Grote Reber,[17] a radio amateur born in 1911 in a village on the outskirts of Chicago, where he still lived.

What this enthusiastic 25-year-old dreamer liked best was tinkering with radios, both transmitters and receivers. But he had also developed a certain curiosity about astronomy, thanks to a book on *The Realm of the Nebulae*, written by Hubble[18] and given to him by his mother, Harriet Grote, a schoolteacher who had the famous astronomer among her students. As soon as he received his master's degree in electronics, Reber asked Bell Labs to hire him to "listen" to the sky with Jansky, his hero, but received a flat refusal. He did not fare any better at the Yerkes Observatory, near his home, where only one graduate student paid attention to his proposals. Jesse Greenstein was working with Fred Whipple — a great comet expert — on the (erroneous) hypothesis that cosmic dust was the cause of the radiation detected by Jansky. But the opinion of this promising young researcher, future opinion leader of American astronomy,[19] still mattered little. The numerous stars of the assault group assembled at Williams Bay by director Otto Struve, himself a scion of a noble line of Russian astronomers — people of the caliber of Chandrasekhar, Bengt Strömgren, Gerhard Herzberg, William Morgan — were too absorbed in their classical research to seize the fleeting moment. Struve, for his part, «had never heard of Jansky and didn't want to».[20] Shapley's reception of the young dreamer at Harvard was equally cold, proving that age often dulls the courage of even the strongest.

---

[17] K.I. Kellermann, *Grote Reber (1911–2002): Obituary*, in "Publications of the Astronomical Society of the Pacific", 116, 822, 2004, pp. 703–711.

[18] Hubble, *The Realm of the Nebulae*, cit.

[19] J. Greenstein, *The Future of Astronomy*, http://calteches.library.caltech.edu/338/1/greenstein.pdf.

[20] K.I. Kellermann, *Grote Reber's Observations on Cosmic Static*, in "Astrophysical Journal", 525C, 1999, p. 371.

But Reber did not give up and decided to cultivate his dream on his own, with personal resources and the time left over from his day job at a radio equipment factory. «I consulted with myself — he said later — and decided to build a dish»,[21] a large metal parabola to collect the radio waves coming from the cosmos and focus them on a dipole. The instrument was conceptually no different from an optical mirror, but with much less stringent shape requirements, since the wavelengths involved (on which the tolerance of the imperfections of the reflecting surface is judged) were hundreds of thousands of times greater than those of visible light. This is an enormous advantage of radio astronomy over optical astronomy, which is counterbalanced by a severe handicap on the resolution front. A parabola tuned to the frequency of 30 GHz, i.e., at the wavelength of 1 cm, would have to have a diameter roughly equal to the distance between downtown Paris and Charles de Gaulle airport (20 km) to obtain the same visual acuity as a 1 m mirror observing in the green (at the center of the optical window)!

Grote asked a specialized company for a quote on the complete turnkey instrument, but the answer he received was out of his budget. "I'll do it myself", he decided, and within four months, with the help of two willing friends, he built an anodized metal dish with a focal length of 20 feet and a diameter of 31, as tall as a three-story building. He had invested $1,300 of his own money in it, a fair sum equal to the cost of the top-of-the-line version of the Ford Model A automobile: quite a sacrifice for a young man! The supporting structure of the radio telescope, made of wooden beams, could tilt, but the azimuth remained fixed to the south, in the manner of ancient wall quadrants. Taking advantage of the diurnal motion of the celestial sphere, the instrument waited for the transit of the sources like the tip of a gramophone for the groove of a record. Grote started his observations in 1937. Since he worked for a living during the day, he used his antenna at night, when, among other things, the electromagnetic pollution caused by candle sparks from internal combustion engines — an unbearable source of low-frequency noise, falling precisely where the astronomical signal was strongest — was minimal. His hobby consumed him. He slept only a few hours after dinner and used weekends and holidays to analyze the data.

Based on current knowledge, Reber was apparently convinced that the radio emission had a thermal origin.[22] Therefore, in an attempt to

---

[21] Kellermann, *Grote Reber (1911–2002): Obituary*, cit.
[22] J.D. Kraus, *Grote Reber: Founder of Radio Astronomy*, in "Journal of the Royal Astronomical Society of Canada", 82, 3, 1988, pp. 107–114.

intercept a stronger signal and at the same time improve the resolution, he had decided to significantly increase the frequency of observations compared to that used by Jansky, i.e. to shift the bandwidth of the detector in the direction of the optical window. In fact, the emissivity of a hot, blackbody-like source rises with frequency to a maximum that for stars is usually in the optical range. Furthermore, the resolution improves as the observing frequency increases. Initially, he chose to tune the receiver to the wavelength of 9 cm, 150 times shorter than that of his idol, but he was unable to detect anything. After a year and a major technical upgrade to the receiver, he reversed course a bit and moved the center of the observation band back to 33 cm, still without result. Stubborn as explorers must be, he moved further to 1.87 m and ... bingo! His homemade telescope fully verified the presence of an alien radio signal from the direction of Sagittarius. Just as Jansky had suspected. It was not a simple confirmation: the whole of Reber's experience included an important physical information that was not intended at the time. The radio signal from the sky was more intense at low frequencies than at high frequencies. An absurdity for a thermal blackbody spectrum, whose intensity should instead increase rapidly with frequency (on the radio side), as Lord Rayleigh and James Jeans, two "big shots" of the University of Cambridge, had demonstrated in 1905 using the principles of classical physics and electromagnetism.

When Reber decided to publicize his work, he realized that «the astronomers of the time did not know anything about electronics or radio, and the radio engineers did not know anything about astronomy. They thought the whole affair was at best a mistake, and at worst a hoax».[23] His short article, submitted to the "Astrophysical Journal" under the title *Cosmic Static*, remained on the editor's desk because there was no referee able to evaluate it.[24] The situation was resolved[25] when Otto Struve, then the editor of the "Astrophysical Journal", after inspecting the instrument, decided to take upon himself the responsibility of breaking the protocol and publishing the exotic research anyway, even in the absence of a third

---

[23]K.I. Kellermann, *Grote Reber (1911–2002)*, in "Nature", 421, 2003, p. 596.

[24]G. Reber, *Cosmic Static*, in "Astrophysical Journal", 91, 1940, pp. 621–624; note that in this case "static" means electrical noise and interference.

[25]In the meantime Reber had also sent the manuscript to the technical journal of the Institute of Radio Engineers, which thus beat the "Astrophysical Journal", by publishing the fundamental work first.

judgment.[26] All this happened in June 1940, in a nation that was still at peace. On the Old Continent, on the contrary, France was experiencing the hour of a humiliating defeat and Fascist Italy was shamelessly caressing the chimera of an easy victory in the wake of its Nazi comrades.

Unsatisfied with the results he had achieved, Reber continued to work on his own. He planned to build the radio analog of the optical image of the Milky Way: a grandiose project, comparable in its aims to the star map of another lone wolf, William Herschel (see ch. 9). The radio map, drawn by the contours of equal intensity of the signal of the galactic source, was published in 1944[27] while the war was still raging in Europe and Asia, causing many millions of deaths. The article, titled like the previous one, contained three fundamental discoveries: 1) the coincidence between the diffuse radio emission and the distribution of matter in the Galactic disk; 2) the first evidence of isolated radiation peaks in the constellations of Sagittarius, Cassiopeiae [sic!], Canis Major, and Puppis (discrete sources, in today's language); 3) the existence of an intense solar emission of a non-thermal nature. A masterpiece that did not have the effect it deserved. The time was not yet ripe; a not uncommon condition in astronomy, as we have already had occasion to say.

The war, into which even the United States had now fallen, had stifled any nascent interest in radio astronomy, multiplying instead that for radio antennas used by the combatants to monitor air and sea space.[28] It was the radar network for the defense of the British island that in 1942 detected a loud noise at the frequency of 60 MHz. At first, it was interpreted as an enemy jamming to prevent interception. The Germans had already successfully used the tactic of camouflaging some of their warships in the English Channel by polluting the ether with powerful radio emissions. But in a stroke of genius, a young physicist, James Hey,[29] realized that the origin of the phenomenon should be searched elsewhere, in the transit of a particularly active spot on the Sun's photosphere. Though purely scientific,

---

[26] It is the practice of scientific journals to subordinate the acceptance of a paper to the anonymous evaluation of one or more referees chosen on a case-by-case basis by the editor from among the experts in the field (peer reviewers).

[27] Id., *Cosmic Static*, in "Astrophysical Journal", 100, 1944, pp. 279–287.

[28] R. Buderi, *The Invention That Changed the World: How a Small Group of Radar Pioneers Won the Second World War and Launched a Technical Revolution*, Touchstone, New York 1998.

[29] A. Hewish, *James Stanley Hey, MBE 3 May 1909–27 February 2000*, in "Biographical Memoirs of Fellows of the Royal Society", 48, 2002, pp. 167–178.

the discovery, comparable in scope to Reber's findings, had been classified so as not to benefit the enemy in any way. Eventually the war ended, and though their pockets were empty, the scientists who survived the massacre were able to get their hands on some of the technological discoveries made for the war, as well as some of the equipment that peace had rendered useless. Bernard Lovell,[30] later knighted for his scientific achievements, was one of the first to convert the military surplus for the purpose of radar observations of meteors.

It all began with a wrong intuition. Lovell, a pioneer in the study of cosmic rays, was convinced that some sporadic echoes collected by His Majesty's Air Force radar umbrella came from the transit of charged particles through the atmosphere. To investigate the matter, he needed an isolated location away from the sparks of internal combustion engines. He then arranged with the Botany Department of Manchester University to use a plot of land on the banks of a stream in the village of Jodrell Bank, 30 km south of the city, for his radar observations. It did not take him long to discover that the occasional radar echoes were not caused by cosmic rays, but by meteors: real kamikazes, which tickle the atmosphere in their last dives to produce all kinds of photons. They are insignificant vanguards of a very large family of celestial radio sources with different fingerprints. Thus was born, with very few resources and no poetry, one of the first and most famous outposts of a new science, radio astronomy.[31] Its realm seemed to be limited to the low frequency region of the thermal spectrum: a Cinderella to the far more powerful optical signals. Instead, it was a small, if not marginal, province of a much larger kingdom. In fact, physicists had already identified new energy generation mechanisms alternative to that associated with the (thermal) motion of the matter particles and measured by temperature, which are very efficient and capable of producing spectra with an intensity increasing with frequency. Here are some of the most common examples relevant to astrophysics.[32]

First there is synchrotron emission, produced by hyper-fast (relativistic) electrons forced by an intense magnetic field to spiral around its lines of

[30]D. Saward, *Bernard Lovell: A Biography*, Robert Hale, Farringdon 1984; B. Lovell, *Astronomer by Chance*, Basic Books, New York 1990.

[31]Id., *Story of Jodrell Bank*, Oxford University Press, Oxford 1968.

[32]For an in-depth study of non-thermal emission processes and physical processes in astrophysics, see H. Bradt, *Astrophysical Processes: The Physics of Astrophysical Phenomena*, CUP, Cambridge 2014.

force; the (centripetal) acceleration causes electric charges to emit electromagnetic radiation. Then there is Compton scattering, from the name of the American physicist who theorized the phenomenon that won him the Nobel Prize in 1927; it originates from the collision between very energetic photons (X or γ) and slow electrons, with consequent transfer of energy from the richest to the poorest (of course, the reverse phenomenon also exists, in which photons of low frequency, such as those of the cosmic microwave background, acquire energy at the expense of relativistic electrons). Finally, there is the stimulated emission of a photon by an atom in an excited state with a very long mean lifetime. Let's see why. In general, an excited atom spontaneously discharges after a hundred millionth of a second, emitting a "well-tempered" quantum of light, unless it first interacts with another particle or photon with which it can dynamically exchange the excess energy. However, there are some excited states, called "metastable", whose average lifetime is millions of years. The spontaneous excitation of these long-stay parking can only occur under conditions of absolute hermitage for the atoms, because they must avoid any possibility of contact with the rest of the world. This means extremely low densities, unattainable on Earth even in the most advanced laboratories, but present in the boundless spaces among the stars. It should be emphasized that in all these phenomena, as well as in others, the spectrum of the source is no longer related to that of a blackbody (typical of a thermal emission) and can manifest its peak either at very high frequencies, in the range of X and γ rays, or in the domain of radio waves, without primarily affecting the optical band where mankind has traditionally built its astronomical knowledge. It is such a wealth of possibilities that one is tempted to say, to paraphrase Hamlet's remark to Horatio, that there are more things in heaven and Earth than our physical understanding can dream of.

But there is something else that makes observing the sky at radio frequencies particularly advantageous. While light is scattered and absorbed by the dust and gas met on its way to the observer, including the passage through the Earth's atmosphere, radio waves penetrate diffuse matter more easily and with fewer consequences. This makes it possible to see beyond the obstacles that instead are real barriers to visible radiation; for example, within and beyond the dark clouds that stand out in the Milky Way's disk, potential incubators of new stars. This property has a greedy consequence: the observation of the sky at radio frequencies is less conditioned by terrestrial meteorology than the optical one. Nobody today would even dream of setting up a big optical telescope in foggy Albion or on the coast of the Netherlands, to name two particularly striking cases. But even the most unsuitable locations for traditional astronomy become plausible when it comes to listening to radio transmissions from the firmament.

Jan Oort[33] had begun to ponder these possibilities immediately after learning of the news of Reber's findings about the Milky Way, brought to him by a postman whom not even the wrath of Mars could stop. In 1944, Holland, isolated from the rest of the world, was still suffering under the yoke of the Nazi occupation, which had imposed a total embargo. Fortunately, in Zurich, in neutral Switzerland, there was still a service of Astronomical Telegrams thanks to which scientific information could cross borders and trenches (as the three papers on general relativity had done in 1916). The now famous professor was no longer in Leiden.[34] Two years earlier, shocked by the brutality of the German troops, so far away from the people he had always considered as an example of civilization, and firmly determined not to collaborate with them, he had resigned from all public positions, including those of university teacher and deputy director of the observatory. A decision that was not without consequences and dangers, so much so that he was advised to go into hiding for a while. So he moved with his family to the countryside, to a cottage about 60 kilometers east of Amsterdam. However, he maintained contact with his school: a few sporadic visits to Utrecht and Leiden on his bicycle, some seminars, a sack of potatoes sent whenever possible to friends in the starving university citadel, and special attention to those students whose tutors had fallen into the Gestapo's net.

Among them was a graduate student at the University of Utrecht, the 26-year-old Hendrik van de Hulst.[35] Oort, who had already evaluated this young fellow for a prize, held him in the palm of his hand the point of selecting him for a task of responsibility: to investigate the possible presence and observability of emission lines in the radio spectra of celestial bodies. It was a kind of senior project. "Henk" was to report on the work at the annual meeting of the *Nederlandse Astronomenclub*, the Astronomical Society of Holland and Belgium, to be held in Leiden in mid-April 1944. To understand what the general climate was like, it is enough to remember

[33]P.C. van der Kruit, *Jan Hendrik Oort: Master of the Galactic System*, Springer, Berlin 2019.
[34]A. Blaauw, M. Schmidt, *Jan Hendrik Oort (1900–1992)*, in "Publications of the Astronomical Society of the Pacific", 105, 1993, pp. 681–685.
[35]H.J. Habing, *Obituaries: Hendrik Christoffel van de Hulst, 1918–2000*, in "Astronomy & Geophysics", 42, 1, 2001, pp. A33–A37.

that within three months the SS would have sunk their claws into Anne Frank and her family: an emblematic story that will never cease to arouse horror in every good soul, just like all the atrocities perpetrated today on the heads of innocent populations.

Van de Hulst made a masterpiece by showing that neutral hydrogen, which he thought was probably present in massive quantities in the Milky Way despite the very low temperature, could be undermined by a spectral line at the centimeter wavelength. This is the radiation emitted by the most elementary of all atoms when its single electron has already reached the fundamental level, on the ground floor of its energy staircase. Although reduced to misery by the lack of invigorating collisions with other material or luminous particles, in the coldest cold, the hydrogen atom has still an asset to play with. Its fundamental state — van der Hulst noted — is actually divided into two different but very close energy sublevels called hyperfine. It works as if one half of the bottom rung of the metaphorical ladder were slightly higher than the other half. Jumping down from this last marginal height difference produces the emission of a weak photon with a frequency of 1,420 MHz, which corresponds to a wavelength of 21.1 cm in vacuum. This is technically called a "spin transition" and corresponds to a configuration change in the coupling between the proton and the electron. The atom goes from a state of higher energy, in which the spins of the two particles are parallel,[36] to the one in which they become antiparallel, which is the true bottom of the barrel. It must be said that hydrogen is in no hurry to get rid of this small excess of energy. But after thinking about it for an average of 10 million years, it does so spontaneously, provided that no one has come to disturb its isolation in the meantime. As we said above, this state of extreme loneliness is only possible at extremely low densities and temperatures.

The study, presented by van der Hulst with a brilliant discussion of the pros and cons, convinced Oort that it was time to take the plunge. Misty Holland could become an open window on a new face of the cosmos, hitherto overlooked and unknown: vast prairies of rarefied atomic hydrogen, tracked by a monochromatic emission that could also provide a measure of the radial velocities of the sources through the Doppler effect. Without even waiting for the end of hostilities, the former professor at Leiden set out to

---

[36]They can be visualized as the directions around which each particle rotates about itself, although the spin is not associated with an actual rotation of the particle, according to the common concept applied to macroscopic objects.

explore with Philips, a company that managed to survive by maintaining fair relations with the Nazis, the feasibility of a receiver to be associated with a hypothetical radio antenna of 20 m aperture. It was just a matter of waiting for the right moment to make the dream come true.

In May 1945, Holland was finally liberated, but there was no return to normality. With the population exhausted by hunger and a particularly cold winter, and with the economy on its knees, there was clearly little money for pure research. The Leiden Observatory had survived the night thanks to its historic and glorious director, Ejnar Hertzsprung, the man who in 1911 had discovered the diagram that today bears Henry Norris Russell's name along with his own (cf. chapter 14). His motto, "astronomy first", was deliberately intended by the occupiers as an implicit sign of acquiescence; and perhaps it was (another of the many *"über alles"*, over everything, that have poisoned and continue to poison the world). To get off on the right foot, a new commander was needed, younger, less compromised, but equally authoritative and visionary. Oort, in the middle of his life, was recalled from his voluntary exile and placed at the head of a prestigious institution with the task of revitalizing it. It was a winning bet.

Among the various initiatives to be implemented immediately, the new director included his radio astronomy project. But, as we know, no money, no honey. After consulting with Reber about the cost of an instrument similar to his, Oort realized that he would not be able to raise the necessary funds quickly enough. Then, like Lovell, he thought of using the radar antennas abandoned by the Germans after the surrender. So he installed one in Kootwijk, a town in the heart of the country, near a research center of the Dutch Post and Telegraph Company. He also managed to find what he needed to make a receiver tuned to 21 cm. At the finish line, however, he was beaten by two American researchers at Harvard, Harold Ewen and Edward Purcell,[37] who in 1951 used a horn antenna (now on display outside the Jansky Lab at the National Radio Astronomy Observatory in Green Bank, WV) to reveal the line emission from neutral hydrogen clouds scattered among the stars.[38] We will see what extraordinary consequences this observation would have in consolidating the idea that in addition to the matter we know, which is mainly composed of atoms, there is another ingredient, much more abundant but invisible because it is indifferent to

---

[37] Parcell shared the 1952 Nobel Prize in Physics for his research in nuclear physics.
[38] *"Doc" Ewen and the Discovery of Radio Emission from Hydrogen*, sd, National Radio Astronomy Observatory, https://www.nrao.edu/whatisra/hist_ewenpurcell.shtml.

photons. Instead, for the sake of brevity, we will not speak of the pioneering radio observations of the atmosphere and the ionosphere, of the Sun and its principal son, Jupiter, nor of the discovery, potentially vital for mankind, of the molecules scattered in the cosmos, starting with that of the OH radical, seen for the first time in 1963. These elementary chemical structures, unraveled by comparing the lines observed in the sky with those tabulated in the laboratory, are the possible link between the immense inanimate nature and us, "lumps of organic matter", to paraphrase Bertrand Russell[39] and Macbeth,[40] that stir for brief moments on the crust of a tiny planet in the vastness of the cosmos. Instead, we will focus on emissions from outside the Solar System.

The new science began to take shape in the Commonwealth countries, which, along with Germany, had focused more than most on radar technologies for defense and offense. The first to scan the sky at radio frequencies and catalog its sources were in fact the English radio astronomers and their Australian cousins. The angular resolution of the radio antennas available just after the war was so modest that the solar disk itself was indistinguishable from a point source. To get an idea of the problem, it is enough to know that a 100 m diameter dish, such as the one at the Max Planck Institute for Radio Astronomy in Bad Münstereifel, Germany, at a wavelength of 10 cm has the same resolving power as the human eye or, which is the same thing, an optical telescope with an aperture of 5 mm. The embarrassing size of the so-called error box, the area of the sky within which observations identify the position of a source, prevented the association of radio phenomena with optically known objects, making attempts to understand their nature difficult, if not all in vain.

To overcome this obstacle without resorting to gigantic parables, which would have been impossible because of mechanical difficulties and expense,

---

[39] «How much there may be beyond what our telescopes show, we cannot tell; but what we can know is of unimaginable immensity. In the visible world, the Milky Way is a tiny fragment; within this fragment, the solar system is an infinitesimal speck, and of this speck, our planet is a microscopic dot. On this dot, tiny lumps of impure carbon and water, of complicated structure, with somewhat unusual physical and chemical properties, crawl about for a few years, until they are dissolved again into the elements of which they are compounded»; B. Russell, *Dreams and Facts*, 1919, https://users.drew.edu/~jlenz/br-dreams.html.

[40] «Life's but a walking shadow, a poor player, / That struts and frets his hour upon the stage, / And then is heard no more. It is a tale / Told by an idiot, full of sound and fury, / Signifying nothing.»; W. Shakespeare, *Four Tragedies: Hamlet, Othello, King Lear, Macbeth*, Bantam Doubleday Dell Publishing Group Inc, New York 1988.

Martin Ryle[41] and Derek Vonberg built the first radio interferometer at Cambridge: an instrument consisting of four fixed antennas spaced a few hundred meters apart and working together.[42] In principle, this technique is analogous to the optical Michelson interferometer that was used at the end of the 19th century to falsify the existence of the ether; a discovery that had important consequences because it initiated Einstein's thinking that led him to special relativity. It enables measurements with the angular resolution of a single radio telescope with an aperture equal to the distance between the elements of the interferometer. In other words, the alliance of several individual antennas only slightly increases the depth of the overall instrument, which depends on the actual collecting area, but produces much clearer "images. A truly fundamental feature for classifying the nature of radio sources, as we will soon understand, starting with the distinction between stars and galaxies.

This consideration brings us to Dover Heights, an eastern suburb of Sidney, Australia, where in the immediate postwar period a handful of young researchers worked to advance our understanding of the Sun's activity at the radio frequencies discovered by Hey. The most enterprising and creative of these pioneers was a 25-year-old Englishman who had returned from four years' service in the Royal Navy as a communications engineer. A long and hard wartime experience that he himself would have recognized as far more formative than any university course. By day John Bolton[43] observed the Sun. But since 1947 he had begun to use his equipment at night to detect particularly bright "radio stars". They had to exist if, as Reber had already pointed out, the radio power of the Milky Way exceeded by 10 billion times what all its stars could produced if they behaved identically to the Sun. It was a reasonable congecture, but it was falsified by the facts, as would soon become clear.

To begin with, Bolton had been looking for the discreet source that Reber had found in the constellation Cygnus in 1939. The mysterious signal behaved strangely. Unlike everything else in the radio sky, its intensity seemed to change very quickly, on the order of a minute. Proof, it was

[41]F. Graham-Smith, *Sir Martin Ryle (September 27, 1918–October 14, 1984): A Biographical Memoir*, in "Biographical Memoirs of Fellows of The Royal Society", 32, 1986, pp. 497–524.

[42]See B.F. Burke, F. Graham-Smith, *An introduction to Radio Astronomy*, CUP, Cambridge 1999.

[43]J.P. Wild, V. Radhakrishnan, *John Gatenby Bolton (June 5, 1992–6 July 1993)*, in "Biographical Memoirs of Fellows of the Royal Society", 41, 1995, pp. 72–86.

concluded, of the stellar nature of the object that produced it, which could hardly have been larger than the distance that light travels in the same amount of time (to get a sense of this dimension, remember that light from the Sun takes 8.5 minutes to reach the Earth). Otherwise, the necessary causal connection between its various parts would have failed.[44] Even this reasoning, though flawless, was marred by a misinterpretation of the observations and a preconceived notion that the source was composed of one or more stars. An accumulation of misunderstandings that led to the correct conclusion.

During the war, Bolton had become very familiar with an interferometric technique developed to estimate the altitude of aircraft approaching the British coast by combining the direct radar signal with that reflected from the sea surface. By recycling this method to observe signals from the sky, he was able to reduce the uncertainty about the size of the source in Cygnus to 8 arc minutes. This was a second clue in favor of the "point-like" hypothesis, i.e., a stellar origin, in addition to the one provided by the variability. By the end of the year, John felt ready to reveal in an article in "Nature" the exact location of the object and its properties, true or guessed:[45] the existence of two components, one with constant intensity and the other variable with time, and the non-thermal nature of the spectrum (which, unlike just "hot things", causes a source to emit more energy at low frequencies).

The first to be informed were astronomers at the Commonwealth Solar Observatory near the capital, Canberra, who tried to scan the sky within the error box given by Bolton in the hope of identifying the optical counterpart of the radio phenomenon. They looked for a monster, but found only a few insignificant star images. Then the news reached Mount Wilson, where it aroused lively interest but yielded no new results. The same thing happened at Yerkes, where the Dutch planetary expert Gerard Peter Kuiper suggested that the variability was not an intrinsic phenomenon, but the consequence of density oscillations in the ionospheric plasma; something like seeing in optical astronomy, but with a lower frequency. He was right, but due to a strange conspiracy of adverse circumstances, this was discovered only later.

Meanwhile, the search for more radio stars continued. In late 1947, after some substantial improvements in the detection system, Bolton found three

---

[44] Causal connection implies communication, which cannot be faster than the speed of light.
[45] J.G. Bolton, G.J. Stanley, *Variable Source of Radio Frequency Radiation in the Constellation of Cygnus*, in "Nature", 161, 1948, pp. 312–313.

more sources in Taurus, Coma Berenices (later moved to nearby Virgo), and Centaurus, all weaker than the prototype in Cygnus; and then two more. It was necessary to establish a rule for naming them. The final agreement was to take the name of the constellation followed by a letter of the English alphabet to mark the chronological order of the discovery. At that time, no one thought that radio sources were very numerous. Instead, their number doubled in just one year and promised to multiply rapidly. Meanwhile, Bolton, obsessed with finding the optical counterpart of Cygnus A, had reduced the error bar on the position to a single minute of arc, still without consequence. Finally, in 1951, Francis Graham-Smith, later knighted and promoted Astronomer Royal, used the Cambridge interferometer to measure the positions of the four brightest sources in the northern sky to within 1 minute of arc. This refinement allowed Baade and Minkowski, who had access to the Californian 100 and 200 inch telescopes, to identify the counterparts of Cassiopeia A and Gygnus A respectively with a remnant of a young supernova in the Milky Way and with a very distant galaxy at a redshift equal to 6% of the speed of light. A record! "The deck is rich, so I dive in" is a poker player's rule that also applies to the most enterprising scientists. The discovery made Ryle realize that radio astronomy could be a powerful tool for studying the deep universe.

Gygnus A highlighted the fact that the emission from the Milky Way is non-thermal in nature. It is caused by hyper-fast (relativistic) electrons forced to spiral along the lines of force of a magnetic field: a mechanism we have already encountered and called synchrotron because it also occurs in these particular particle accelerators. Thanks to the sharper eyes of the new British interferometer, it was also seen that the energy emitted by the distant radio galaxy, a hundred million times more powerful than the Milky Way, was confined to two lobes of plasma outside the optical image of the object; two jets arranged diametrically and much farther apart than the size of the galaxy as painted by visible light. Baade and Minkowski[46] described the source as «an extragalactic object, two galaxies in actual collision».[47] Today we know that this phenomenology is associated with the presence, at the core of the nebula, of a supermassive black hole with

---

[46] G. Burbidge, *Baade & Minkowski's Identification of Radio Sources*, in "Astrophysical Journal", 525C, 1999, pp. 569–570.

[47] W. Baade, R. Minkowski, *Identification of the Radio Sources in Cassiopeia, Cygnus A, and Puppis A*, in "Astrophysical Journal", 119, 1954, p. 206.

a mass equivalent to 2.5 billion Suns: a starting point for another branch of modern astronomy, high-energy astrophysics.

Beginning in 1950, researchers at the Cavendish Laboratory published a series of nine radio source catalogs. The objects are listed by an order number preceded by the letter C. The most important catalog for us is the third, 3C, which appeared in 1959. It was produced using the Cambridge interferometer at a frequency of 159 MHz (188.5 cm), rebuilt on a disused military site southeast of Cambridge by the usual Ryle and by Antony Hewish,[48] a young man destined for a bright career. Both would share the 1974 Nobel Prize in Physics «for their pioneering research in radio astrophysics: Ryle for his observations and inventions, in particular of the aperture synthesis technique, and Hewish for his decisive role in the discovery of pulsars».[49] A record! Never before had the most coveted scientific prizes been awarded to scientists for their astronomical work.

The 3C contained 471 radio sources. The brightest were reobserved at a higher frequency to improve the resolution beyond one arcminute. This work resulted in a revised catalog, the 3CR, whose 328 objects became the standard reference in the northern sky. But what kind of objects? The search for optical counterparts, the only way to answer this question, was particularly easy and convincing when an extended object such as a galaxy or a gaseous nebula appeared in the field around the position of the radio source defined by the instrumental resolution. It became much more difficult and uncertain when only point-like images were available instead. Therefore, while the identification of extended extragalactic radio sources was growing, that of stellar-like objects was still an empty box. But few doubted that sooner or later some prey would fall into the net. So much so that Greenstein had personally offered a case of bourbon as a prize to the lucky discoverer. What a difference in style between the solemn atmosphere of the European observatories and the easygoing attitude of the Americans!

Finally, in 1962, using the positions provided by the twin 27.5 m antennas of the new Owens Valley Radio Interferometer — sought by Caltech

---

[48] F.P. Miller, A.F. Vandome, J. McBrewster (eds.), *Antony Hewish: Radio Astronomy, Nobel Prize in Physics, Martin Ryle, Aperture Synthesis, Pulsar, Jocelyn Bell Burnell, Eddington Medal*, Alphascript Publishing, Riga 2010.

[49] *The Nobel Prize in Physics 1974*, https://www.nobelprize.org/prizes/physics/1974/summary/%E2%80%9D.

astronomers and built 400 km north of Los Angeles with the direct collaboration of Bolton, who had temporarily moved to California for this purpose — Sandage and Thomas Matthew were able to identify three radio sources (3C 48, 3C 196, and 3C 286) with as many stellar-like optical sources.[50] Shortly thereafter, Sandage himself, Greenstein, and Mexican astronomer Guido Munch obtained spectra using the Palomar 200 inch telescope showing strong emission lines that the three brilliant scientists were unable to interpret. Bolton suspected that an unusually high redshift might be the key, but he could not convince his collaborators. The strongest objection concerned the inconsistency between the rapid variability and the requirement of a size sufficient large to produce the observed brightness if the object were indeed very distant. Distance implied greater intrinsic luminosity and thus larger physical size, incompatible with the particularly short coherence lengths imposed by rapid variability. Correct reasoning, but based on the erroneous assumption that the production of a torrent of light requires very large volumes.

Meanwhile, the British Cyril Hazard had specialized in the application of lunar occultations to radio domain.[51] This classical technique of determining the position of a point-like source by measuring the times of disappearance and reappearance of a star behind the dark limb of the Moon (during a kind of micro-eclipse) takes advantage of the perfect knowledge of the orbit of our satellite and of the fact that it is described in the same frame of reference as the optical stars. It should be noted that the celestial coordinate system for optical sources is set up independently of that for radio sources (since the observations in both cases contemplate very different instruments). In order to fit one to the other, one matches the coordinates of as many common objects as possible by a minimization procedure that is not without systematic error. The problem disappears when a radio source is placed directly over the optical reference system, as in the case of lunar occultations.

A skill, Hazard's, that came just in time. In fact, Her Britannic Majesty's "Nautical Almanac" reported that in 1962 there would be five occultations of 3C 273, the sixth most powerful radio source in the northern

---

[50]T.A. Matthews, A.R. Sandage, *Optical Identification of 3C 48, 3C 196, and 3C 286 with Stellar Objects*, in "Astrophysical Journal", 138, 1963, pp. 30–56.
[51]R.W. Clarke, *Locating Radio Sources with the Moon*, in "Scientific American", 214, 6, 1966, pp. 30–41.

sky.[52] Three of them would have been visible from Parkes, New South Wales, where the Commonwealth Scientific and Industrial Research Organization, an independent federal agency of the Australian government, had just inaugurated a giant 64 m diameter parabola, jokingly called "the dish". The opportunity was too good and too rare to miss. It would not come around again for another 20 years or so, and perhaps it would be the one to finally flush out a *bona fide* radio star. Hazard was brought in to coordinate the operation.

It was a heroic undertaking. The first of the three occultations highlighted an unfavorable circumstance. The moment of emergence from the lunar disk, as indispensable as that of immersion to carry out the measurement, caught the Moon and the source beyond the limits of the horizon of the instrument. Bolton, who had become director of the Australian Observatory after the American parenthesis, took the risk of removing the protective systems that prevented further tilting of the "dish". He even dug a trench to prevent the parabola from rubbing against the ground when it reached an almost vertical position. With an unprecedented investment of energy, precaution, and daring, the position of 3C 273 was determined with an accuracy of less than one second of arc. It was a fantastic achievement. Here is what happened then, in the words of the young Marteen Schmidt, who, on the retirement of Rudolph Minkowski, had inherited the project of studying spectroscopically the optical counterparts of the radio sources, real or supposed, using the Palomar facilities:

> The next object among quasars-to-be was 3C 273. Cyril Hazard *et al.* had been observing lunar occultations of this strong source in April, August and October 1962, and found that it was a double. John Bolton sent us the first accurate positions obtained in August 1962. Tom Matthews found that the two sources coincided to within a few arcseconds with a thirteenth magnitude star and a nebular wisp or jet. I suspected that the jet was a peculiar nebula associated with the radio source and that the 13th mag. star was a foreground object. Since the jet was exceedingly faint and would require long exposures, I decided to take a spectrum of the bright stars so that it could be

---

[52]More specifically, 3C 273 was in the top ten 3CR sources located within 75 degrees of the North Pole.

eliminated from consideration. The first spectrum of the star taken at the end of the night of December 27/28, 1962 was badly overexposed — I was not used to observing such bright objects [. . .] Two nights later I obtained a spectrum with the correct exposure and found several more emission lines. It was clear that 3C 273 belonged to the class of 3C 48.[53]

Surprise! The star showed an unusual excess of light in the ultraviolet and a seemingly indecipherable small set of emission lines. Faced with this problem, even the gurus of spectroscopy threw in the towel. 3C 273 was temporarily forgotten until, as is often the case, fate provided the conditions for a revival. Hazard intended to publish in "Nature" the summary of the measurements and observations made on the object, including the spectra collected at Palomar. So he wrote to Schmidt, inviting him to contribute. Urged to take back the material on 3C 273, the Dutchman had a stroke of genius. He tried to compare four lines in the spectrum of the stellar-like object with those of the so-called hydrogen Balmer series,[54] and he realized that everything was in order if one assumed that the source had a redshift equal to 16% of the speed of light. Interpreted in a cosmological key, this very high radial velocity placed 3C 273 at an enormous distance (over 2 billion light years) and gave it a luminosity a thousand times greater than normal galaxies. To maintain such a standard of living, the object would have to send to "smoke", that is to convert into energy, every month, the equivalent of one solar mass. Its ability to vary its brightness over the years implied that the region where all this was happening could not exceed that in which the Sun reigns supreme: a handful of parsecs. A supermassive black hole?

---

[53]M. Schmidt, *The Discovery of Quasars*, in Kellermann, Sheets (eds.), *Serendipitous Discoveries in Radio Astronomy*, cit., p. 172. About this story, see also C. Hazard *et al.*, *The Sequence of Events that Led to the 1963 Publications in Nature of 3C 273, the First Quasar and the First Extragalactic Radio Jet*, in "Publications of the Astronomical Society of Australia", 35, E006, 2018, and K.I. Kellermann, *The Discovery of Quasars*, in "Bulletin of the Astronomical Society of India", 41, 2013, pp. 1–17.

[54]The Balmer series is a set of hydrogen lines that lie entirely in the optical region, between 6562.8Å (Hα line) and 3645.6Å. Named after the Swiss mathematician Johann Jakob Balmer, who discovered the empirical formula for calculating the wavelengths of the line series in 1885, it corresponds to the electronic transitions from any quantum level to the first excited level (principal quantum number $n = 2$). From red to blue, the lines are denoted by the letter H, followed by a letter of the Greek alphabet, starting with the first. So we have Hα, Hβ, Hγ, etc.

That's right! But in the beginning, no one thought about it, so the energy incontinence of 3C 273 became a mystery. Anyway, within a short time, other objects with the same properties and characteristics were discovered. They had to be named. It was decided to call them quasi-stellar radio sources, a long and deliberately ambiguous name, then shortened to "quasars" by Hong-Yee Chiu in a popular article in 1964. No one remembers this Chinese-American astrophysicist, but today everyone uses the lucky nickname he coined: «Because the nature of these objects is completely unknown, it is hard to prepare a short, appropriate nomenclature for them so that their essential properties are obvious from their name. For convenience, the abbreviated form "quasar" will be used throughout this paper».[55]

In the years following the discovery of the extragalactic nature of 3C 273, it became clear that all or most of the radio sources seen outside the plane of the Milky Way were distant galaxies or quasars — a sufficient sample to build a test of the stationarity of the cosmos. Despite Hoyle's desperate defense, his theory began to fall apart. High redshift, i.e. distant quasars exhibited both a spatial density (number of objects per unit volume) and some intrinsic properties at variance with those of the nearest ones: a scenario incompatible with the model of an ever-changing universe. It did not take long to understand the reason for these differences. They depend on the time evolution of the supply of matter that fuels the giant black holes nested at the centers of quasars. «Feeding the monster»,[56] was the happy image coined by the American cosmologist James Gunn. We will then see how and why of this when we speak of "active galactic nuclei".[57]

A touch of color and costume. As a result of his exploit, Schmidt quickly achieved great popularity, including media attention. In 1966, "Time" magazine dedicated the cover of its March 11 issue to him; an honor reserved for only a few scientists, including Nikola Tesla, Hubble, and Einstein (who was even named "Man of the Century"). All this fame made Sandage jealous. In his heart, he believed that he and Tom Matthews were the true discoverers of quasars. «The quasars were stolen from both

---

[55] Hong-Yee Chiu, *Gravitational Collapse*, in "Physics Today", 17, 1964, p. 21.

[56] J.E. Gunn, *Feeding the Monster: Gas Discs in Elliptical Galaxies*, in C. Hazard, S. Mitton (eds.), *Active Galactic Nuclei*", CUP, Cambridge 1979, pp. 213–225.

[57] For a brief history of active galactic nuclei (AGN), see G.A. Shields, *A Brief History of AGN*, in "Publications of the Astronomical Society of the Pacific", 111, 760, 1999, pp. 661–678.

of us [...] It's not oh-we're-all-in-this-together-just-seeking-the-secrets-of-nature. The competition between colleagues is really much stronger than anybody on the outside understands».[58] An explicit confession of the belligerent attitude of scientists, which, at least in the Anglo-Saxon world, is an integral part of the business. Borrowing the words of Michiavelli's *Principe*, we can conclude that, like all other human beings, researchers «armati vissono ed 'e disarmati ruinorono» (lived if armed and fell to ruin if unarmed).

Before concluding this excursion into the rich universe of radio waves, let us go back in time to recall another formidable discovery of the fledgling radio astronomy, that of "pulsars". In 1948, Martin Ryle had joined the Cavendish team with a young man he had met during the war at the Royal Air Force Telecommunications Laboratory, where he himself had been posted. Antony (Tony) Hewish had theoretical and experimental skills and a natural managerial ability. A particularly interesting mix in an area of research where there was still little certainty and much to be done. Among the problems on the table was the propagation of radio waves in an inhomogeneous transparent medium. Hewish succeeded in modeling the scintillation produced on the signal by the small, rapid, and random deviations in the direction of motion of a wavefront passing through the ionospheric plasma. In other words, he succeeded in mathematically describing the phenomenon that produces the fluctuations in the position and intensity of radio sources, which in the case of Cygnus A had been mistaken for intrinsic variability. The idea was to apply the result to the measurement of the electron density fluctuations responsible for the amplitude of scintillation. To do this, he had to build an instrument capable of observing the phenomenon on particularly short time scales, on the order of fractions of a second. The study would also serve to distinguish point sources from extended sources, which are insensitive to scintillation (the same happens in optical vision, which is why planets twinkle less than stars).

In 1965, thanks to a successful fund-raising campaign, Hewish began construction of a large low-frequency interferometer covering more than one hectare. In August 1967, a month after the instrument went into service, it produced an unexpected signal. Jocelyn Bell, a 24-year-old postdoc who had joined the project a few years earlier, noticed that a particular

---

[58]Panek, *The Loneliness of the Long-Distance Cosmologist*, cit.

radio source produced a strong scintillation despite being in opposition to the Sun: a situation that theoretical predictions suggested should produce a minimum of the phenomenon because of the reduced exposure of starlight to the wind. But that was not enough. The source was also extremely variable, to the point of disappearing from time to time. A truly exciting mystery. Some hasty astronomers and the usual smart alecks, always ready to speculate on people's dabbling, began to talk about the "little green men" and their signals. The mysterious object was observed again in November of the following year, when it was finally possible to separate its pulses. In the blink of an eye, the sensational discovery was publicized by an article on "a rapidly pulsating radio source", which appeared in the February 1968 issue of "Nature".[59] It was the first "pulsar" to be discovered — a name coined, as with quasars, by combining the words "pulsation" and "star". Within a short time, three more pulsars were found, one of them with 4 hiccups per second: a timing sufficient to rule out that the phenomenon, immediately associated with stellar rotation, had to do with a white dwarf. The required rotation speed would have been sufficient to disintegrate the star. Something more robust than that was in order. Within months, the solution to the problem was found by the Austrian astrophysicist Thomas Gold and the Italian Franco Pacini, both then researchers at Cornell University. They showed that a magnetized, rotating neutron star is capable of producing the observed phenomena, provided that the rotational axis of the object is misaligned with respect to the direction of the magnetic poles.

The discovery, as we already know, earned Hewish a Nobel Prize which he shared with Ryle. Jocelyn Bell gasped, but did not want to express any disappointment at this sexist-flavored exclusion. Among those who reacted there was the irreverent Fred Hoyle. An outspoken man, he first accused Hewish of stealing Bell's data and then, with an article in the "Times", called the prize selection committee to account for a blatant injustice. There is good reason to believe that someone in Stockholm smeared Hoyle's name with black ink for this generous fight, as evidenced by his absence from the 1983 Nobel Prize, which was awarded to Fowler for joint research (cf. p. 462). Perhaps George Bernard Shaw was right when he refused the check accompanying the 1926 prize: «I can forgive Alfred Nobel for having

---

[59] A. Hewish, S.J. Bell, J.D.H. Pilkington, P.F. Scott, R.A. Collins, *Observation of a Rapidly Pulsating Radio Source*, in "Nature", 217, 1968, pp. 709–713.

invented dynamite, but only a fiend in human form could have invented the Nobel Prize».[60] Sometimes, however, time is a gentleman. In 2018, Jocelyn Bell was awarded the Breakthrough Prize for Fundamental Physics, the richest of all scientific research awards. Although belated, justice was done!

---

[60]M. Holroyd, *Introduction*, in *G.B. Shaw, Pygmalion and Major Barbara*, Bantam Dell, New York 1992, p. X.

Chapter 19

From here to eternity: New generation instruments, amazing discoveries, and open problems

> I am a slow walker,
> but I never walk back.
>
> Abraham Lincoln,
> newspaper interview (1864)

> Have faith in progress:
> it is always right
> even when it is wrong,
> because it is the movement,
> the life, the struggle, the hope.
>
> Filippo Tommaso Marinetti,
> *Teoria e invenzione futurista*

The 1960s opened with the astonishing success of the Soviets in the race to conquer space: the pioneering flight of Yuri Gagarin in April 1961, the "equal opportunity" mission of Valentina Tereshkova in 1963, and the extravehicular walk of Aleksey Leonov in 1965, which followed the launch of Sputnik in 1957 and the non-return flight of the dog Laika, the first living creature to orbit the Earth. It took little more than a decade for the greatest of human adventures to be accomplished with a change of baton. The arrival of NASA's Apollo 11 mission on the Moon, the landing of its lunar module, and the first «small step» of Neil Armstrong on the ground of our lunar satellite, on July 20, 1969, marked the victory of the United States in the last round of this epic match.[1]

The bloodless competition had been fostered by a continuous war without direct confrontation (hence the adjective "cold"), but with frequent and bloody diversions, low blows, and a relentless pursuit of strategic

---

[1]Cf. M. Capaccioli, *Luna rossa*, cit.

superiority.[2] The Soviets wanted aircraft carriers capable of flying their nuclear warheads across the ocean to strike American territory. This was a strategic need the Yanks considered less pressing, since they could rely on their bases in Europe and the Middle East, which were close enough to the USSR's borders for B-52 flights. It was the skill of Ukrainian engineer Sergei Korolev and the flair of Nikita Khrushchev that enabled the transformation of military ballistic missiles into space launchers. It was the charisma of John Kennedy, the boundless economic power of the United States, and the know-how of Nazi scientists bred at the Peenemunde Research Center and handed over to the Allies at the fall of the Third Reich that led the Americans to victory after the resounding knockouts in all the early rounds. «We choose to go to the Moon in this decade and do the other things, not because they are easy, but because they are hard», Kennedy told students at Rice University in Houston on September 12, 1962; «because that goal will serve to organize and measure the best of our energies and skills, because that challenge is one that we are willing to accept, one we are unwilling to postpone, and one we intend to win, and the others, too».[3] Americans were able to keep this bold prediction despite the deadly and still unclear ambush in Dallas on November 22, 1963.

The frantic and costly struggle to conquer space was only one of the many manifestations of the state of permanent tension between the two blocs into which the world was divided. It was only the tip of the iceberg of the widespread malaise felt by a humanity teeming with billions of beings increasingly aware of their rights and less and less willing to accept their systematic abuse. The sound of church bells after six years of senseless slaughter had lulled the survivors of the massacres of the Second World War into hoping for a lasting peace. Instead, almost immediately, the arms race resumed, now thundering in Asia, the Middle East, Africa, and Latin America, reaching Europe and nipping at the heels of the United States: both the revolts of oppressed peoples and severely discriminated ethnic groups, and the cynical military actions explicitly aimed at maintaining or acquiring power and raw materials in strategic areas of the world. Wars fought with conventional weapons, now enhanced by technology, or with the more subtle and devious tools of economics, propaganda, political

---

[2] J.L. Harper, *The Cold War*, Oxford University Press, Oxford 2011.
[3] Cf. *Text of President John Kennedy's Rice Stadium Moon Speech*, https://er.jsc.nasa.gov/SEH/ricetalk.htm.

and religious ideology, and more recently, information technology, with the manipulation of minds through fake news.[4] Obviously, these turmoils have also affected science, which had been elevated from wellspring of human curiosity to a crutch of power, thanks to the impact of its sensational discoveries and the prestige that comes with knowledge (but also because of the need for researchers to come to terms with those who control the purse strings and careers). In the trenches of Ypres, invaded by mustard gas, and in the martyred cities of Hiroshima and Nagasaki, the noble "natural philosophy" had experienced sin. «This is a knowledge which [physicists] cannot lose»,[5] Robert Oppenheimer, the father of the first atomic bomb, would later confess.

Despite its low strategic value, astronomy also entered the competition between the two blocs, and grew to the point of becoming, together with physics, the leading science of the 20th century.[6] Simplifying the historical analysis, one can probably argue that the explosion of this ancient discipline, which had remained elitistically under track for thousands of years, was the result of three favorable conjunctures: the application of the revolutionary technological developments to the study of the sky, an ever-closer synergistic alliance with physics (also growing exponentially), and the use of information technology to manage the rising tide of data and solve complex problems concerning computation and analysis. The full-scale confrontation between the two superpowers also favored and even motivated the development of newer and more powerful astronomical instruments,[7] either for reasons of prestige or as by-products of research activities with military purposes (for instance, in the field radio astronomy, as we saw in the previous chapter).

An example of this peculiar growth process is indeed the *Bolshoi Teleskop Alt-azimutalnyi*[8] or BTA-6, the giant telescope built by the Soviets

---

[4]The modern version of «the slander [which] is a little breeze/a very nice little breeze/that subtly, imperceptibly/begins to murmur,» from Rossini's memory.

[5]J.R. Oppenheimer, *Physics in the Contemporary World*, in "Bulletin of the Atomic Scientists ", 4, 1948, p. 66.

[6]M. Longair, *The Cosmic Century*, CUP, Cambridge 2006.

[7]C.R. Benn, S.F. Sánchez, *Scientific Impact of Large Telescopes*, in "Publications of the Astronomical Society of the Pacific", 113, 781, 2001, pp. 385–396.

[8]B.K. Ioannisiani *et al.*, *The Zelenchuk 6 m Telescope (BTA) of the USSR Academy of Sciences*, in C.M. Humphries (ed.), *Instrumentation for Astronomy with Large Optical Telescopes*, Proceedings of the Sixty-seventh Colloquium (Zelenchukskaya, USSR), Reidel, Dordrecht 1982, pp. 3–10.

on Mount Pastukhov, in the northern slopes of the Caucasus — the imposing chain of high peaks between the Black and Caspian Seas — to respond
with a high C to the decades-long supremacy of the Americans in the optical observation of the sky. The idea was born in the 1950s as a reaction
to the commissioning of the Hale telescope on Mount Palomar. It was a
matter of making up for the delay accumulated since the end of the 19th
century, when the 76 cm refractor, which for a short time had been the
largest telescope in the world, arrived at the Pulkovo Observatory. But
after this last flash of imperial grandeur, the Russians had far more to
worry about, and they gradually lost step with the West. At the end of the
1950s, in the context of the general exaltation of the role of superpower that
the USSR had gained through the victorious Great Patriotic War against
the Nazis and the spread of communist ideology throughout the world, the
Moscow Academy of Sciences believed that the time had come to catch up
with the "capitalists" in this field of scientific knowledge as well. A winning shot was needed. For these reasons it was decided to build a telescope
with an aperture of 6 m, considerably larger than the 5 m mirror of the
Palomar giant and still within the feasibility limits of a monolithic primary.
The task of producing the blanks for the primary and secondary optics
and polishing them was entrusted to the Lytkarino Optical Glass Plant
(LZOS), a factory for special glass for military purposes founded in 1934
in the Moscow region. In 1946 it had been converted to the production
of opto-mechanical components for civilian use, also thanks to the equipment, technologies, and skills of Schott of Jena, acquired in the USSR as
war booty.

Now known for its high quality products, LZOS had to pay the price
for its lack of experience with extra-large optics. A *dejà vu*, it comes to
say. As a result, the casting of the main mirror blank turned out to be
rather poor. The Pyrex millstone of 6 × 1 m, weighing 42 tons, showed
countless bubbles and cracks. The production was repeated two years later
to learn from the mistakes. This time the blank was better than the first,
but still not completely satisfactory. Thus, despite of the skill with which
the optical surface was polished with a tool designed for the purpose and
built with Soviet exuberance of iron and poverty of frills, the eye of the
BTA remained far from the theoretical limit of resolution of 0.02 seconds
of arc expected by its record size. However, this was not the only problem
of the instrument: others were the thermal inertia of the primary, unable
to adjust in a reasonable time to temperature changes greater than 2°C,
as well as the mediocre meteorological characteristics of the Zelenčukskaja

site, aggravated by the enormous mass of an imposing 53 m high dome, beautiful to look at but deadly for image quality.

The fact is that the design and logistical decisions were guided not only by scientific requirements. There was also the intention to astonish the world, both at home and abroad (much as the Roman Baroque had done for the Catholic Church), and the desire to please the Caucasus, a strategic region for the Kremlin, yet not perfectly integrated into the diverse archipelago of the USSR. For the same reason the Akademija Nauk SSSR, the very powerful Academy of Sciences of the Soviet Union, had decided since 1974 to place in the same geographical area also the radio telescope RATAN-600, a ring of iron panels with a diameter of 600 m and the appearance of a scrapped spaceship, the second largest collector of radio waves in the world.

Majestic yet sleek, with a primary focal length of 24 m, and unusually bare, with all of its cabling taped underneath, the BTA was completed in 1975 and remained the world's largest telescope until the 1990s. However, despite its record aperture, it never managed to be the best. In fact, it was a half-hearted scientific and propaganda failure, largely mitigated by a winning technological innovation: the adoption of a mount that revisited, in a modern key, the classical altazimuth design abandoned by the great telescopes of the 19th and 20th centuries in favor of the equatorial mount.[9] The choice of orienting one of the axes in the direction of gravity made it easier and more precise to control passive bending and optical alignment, thus stiffening the supporting structure while reducing weight (a possibility not yet fully exploited by the Russian telescope, however). Strongly recommended because of the large total mass of the instrument, the revival was made possible by the use of electronic calculators for the guidance. Indeed, the tracking of the celestial sphere with an altazimuth mount involves accelerated movements in both axes, extremely difficult to manage for a human being, but within the reach of the lightning reflexes of even a modest artificial intelligence.

The BTA maintained its sterile record for three decades, then was surpassed in all respects by the first of the two Kecks: giant telescopes[10] atop Mauna Kea, one of the extinct volcanoes of Hawaii's Big Island: a site at

---

[9] D. Leverington, *Observatories and Telescopes of Modern Times*, CUP, Cambridge 2017.

[10] S.J. Medwadowski, *Structure of the Keck Telescope: An Overview*, in "Astrophysics and Space Science", 160, 1–2, 1989, pp. 33–43.

4,150 m above the level of the Pacific Ocean that provides ideal weather conditions for astronomical observations. The most striking feature of the Keck is its primary optics, which consists of 36 hexagonal segments assembled into a mosaic 10 m in diameter. This was an ingenious strategy to overcome the technical and cost limitations of monolithic mirrors, which had been conceived and pioneered half a century earlier by Guido Horn d'Arturo, director of the Astronomical Observatory of the University of Bologna.

Guido[11] was born in 1879 into a family, the Horns, belonging to the good Jewish bourgeoisie of the Austro-Hungarian city of Trieste, on the northeastern corner of the Italian peninsula. After graduating in Vienna with a thesis on the orbits of comets, he had chosen to work in the Italian observatories. A choice that became definitive at the outbreak of the Great War when, at the age of thirty-six, he volunteered as an artilleryman in the Royal Savoy Army. Fearing that he might be mistaken for a deserter if he ever fell into the hands of the Austro-Germans, he assumed a new identity, choosing to call himself d'Arturo in memory of his father, who had died prematurely;[12] a surname he retained at the end of the conflict by adding it to his own family name. In 1921 he was appointed director of the University Observatory of Bologna with the responsibility for teaching astronomy. This was the start of a brilliant career. A man of considerable ingenuity, he set himself the problem of creating large light collectors by replacing the expensive monoliths, beyond the reach of a small institute, with mosaics of light and small mirrors, which he called "tasselli" (tiles). In 1932 he wrote in the popular magazine "Coelum" (which he had founded for the dissemination of astronomy in the Italian society):

> The reader may wonder why we want to assemble the mirror
> in pieces if it can be made in one piece: I answer that above
> a certain diameter, casting and polishing are beyond human
> strength, while the assembly of tiles [*tasselli*] with a thickness
> of one centimeter and an area of about one square decimeter
> is a very easy task [...]. It is reasonable to think that, having
> succeeded in the experiment with a diameter of one meter and

---

[11]M. Zuccoli, F. Bònoli (eds.), *Guido Horn d'Arturo and the dowel mirror*, CLUEB, Bologna 1999.

[12]Perhaps he was also playing on the fact that Arturo, the "guardian of the bear", is the brightest star in the constellation of Boote.

ten [centimeters], it will be possible without undue difficulty to obtain considerably larger apertures.[13]

The idea was put into practice with a 110 cm zenithal mirror prototype composed of 80 independent segments. Each of them was supported on three adjustable points by means of screws and slides, attached to a single horizontal marble surface, and adjusted so that its focus matched that of all the other "tasselli". The mosaic was installed in the highest room of the Specola tower, which had been modified to open to the sky. It looked only in the vertical direction and thus functioned as an instrument of transit. Horn spent the rest of his life perfecting this project, producing multimirrors with apertures of up to 180 cm, 1.5 times larger than the classical telescope built between 1940 and 1942 by Officine Galileo in Florence for the University Observatory of Padua on the Asiago plateau, for a while the largest reflector in Europe.[14] However, this activity had to be suspended for seven years due to the implementation of racial laws. After the fall of Fascism, the elderly astronomer was reinstated in his various roles and died in Bologna at the age of 88, surrounded by the affection of his family, students and friends, including the painter Giorgio Morandi, with whom he played endless games of chess. A fine example of all-round intellectual vivacity.

Guido Horn d'Arturo's brilliant but premature insight remained buried until it was dusted off by engineers at Caltech and Lawrence Berkeley Labs in the late 1970s. The time was ripe to harness the postwar technological impulse and make the segmented mirror the eye of a modern, versatile Cyclops. After a first attempt in 1979 in Arizona with the Multi-Mirror

---

[13]G. Horn d'Arturo, *Telescopes of the future and speckled mirrors*, in "Coelum", 2, 1932, pp. 25–27 (in Italian). See also V. Picazzi, *Il progetto di telescopio a tasselli di Guido Horn d'Arturo: forefather of the new generation multi-mirror telescopes*, thesis, University of Bologna, 2015–2016, p. 127, http://www.lelucidihorn.it/hex/wp-content/uploads/2018/02/Tesi-Picazzi.pdf (in Italian, with abstract in English). It is interesting to note that even in 1932, a mirror of 1.1 m in diameter still appeared to Horn d'Arturo as a true giant; a sign of the profound gap between the standard of observational astronomy in Europe and in the United States. This explains the emigration of so many excellent European astronomers to the New World.

[14]It is said that, at the behest of Mussolini, the size of the mirror was chosen to exceed, albeit slightly, the aperture of the reflector of the Berlin-Babelsberg Observatory, then the largest in Europe, completed by Carl Zeiss in 1924 after an interruption due to the First World War.

Telescope (MMT)[15] — an instrument with a primary made by assembling into a single frame six traditional 1.8 m-diameter mirrors, equivalent to a 4.5 m-aperture monolith, housed in a rotating, compact, and well-ventilated building that, reversing previous ideas, greatly reduced the impact on seeing quality — in 1985 also came the resources to finance the entire project: seventy million dollars given to the Caltech by the wealthy oilman Howard Keck for the first truly multi-mirror telescope.[16] Five years later, thanks to the genius of Jerry Nelson,[17] Kavli Prize for Astrophysics in 2010, the innovative instrument saw its first light on Mauna Kea, albeit with a very incomplete mosaic. Only 1/4 of the 36 hexagonal, 1.8 m wide and 7.5 cm thick menisci were in place, each made from half a ton of Zerodur[18] and housed in a common structure designed to provide maximum rigidity.

Although unfinished, the telescope, named Keck in honor of its patron, performed so well that it attracted the interest of NASA. Substantial donations were made to Caltech to complete the instrument as early as 1993 and to build a twin unit to create an optical interferometer with a base of 85 m and a resolving power of 5 thousandths of an arc second in the near infrared. In addition to the computer-controlled USSR-style altazimuth mount, the Keck mosaic of mirrors had its alignment readjusted twice per second by a system of active supports. Each *"tassello"* differed from conventional mirrors in that its thickness was reduced from the traditional ratio of one to six to four percent of its diameter, making it more responsive to forces applied to its back surface. This allowed the use of force actuators for the control of aberrations due to changes in the telescope inclination[19] and the external temperature. The innovative concept was applied in Europe to the 3.6 m-aperture altazimuthal New Technology Telescope (NTT), a monolithic

---

[15]S.C. West *et al.*, *Toward First Light for the 6.5 m MMT Telescope*, in "Optical Telescopes of Today and Tomorrow", Proceedings SPIE, 2871, 1997, pp. 38–48.

[16]M.M. Waldrop, *Telescope Gets Largest Private Gift Ever*, in "Science", Vol. 227, No. 4684, 1985, p. 275.

[17]J. E. Nelson, T.S. Mast, *Construction of the Keck Observatory, in Very Large Telescopes and their Instrumentation*, ESO Conference and Workshop Proceedings, European Southern Observatory, ESO, Garching 1988, pp. 7–15.

[18]Zerodur is a glass-ceramic material patented by the glassmaking company Schott of Mainz, Germany, which is owned by the Carl Zeiss Foundation. Its molecular structure, both amorphous and crystalline, offers an excellent response to mechanical machining, polishing, and coating with other materials, and above all is almost insensitive to changes in temperature.

[19]Here we see the superiority of the altazimuth mount over the equatorial mount. Bending of the structure is more easily compensated when it depends on only one angle.

active-mirror instrument designed and built in the 1980s for the observatory at Cerro La Silla, Chile, using the contributions paid by Italy in 1982 to join the European Southern Observatory. This state-of-the-art organization, which combines the forces of several nations to form a critical mass of expertise and resources, has enabled the European optical astronomy community to catch up with its overseas counterpart after nearly a century of American leadership.

The idea of an astronomical facility common to the countries of the Old World, singularly outnumbered by the two superpowers that emerged from World War II, blossomed in the 1950s through the vision and lobbying of Baade and Oort, and the exemplary model of the Geneva-based CERN. To get started, there were a number of options on the table for what to do and where to do it. After a decade of heated debate, it was agreed to focus on the southern hemisphere[20] because it is rich in interesting objects, including the center of the Milky Way, and also because it had not yet been explored as extensively as its northern counterpart (where the high concentration of telescopes of all sizes made the competition tougher). Finally, in 1962, the five founding nations — Belgium, France, Germany, the Netherlands, and Sweden — signed the agreement to formally establish ESO. Its headquarters were first located at the CERN laboratories in Geneva. Then, in 1980, they moved to a beautiful, purpose-built building on the outskirts of Munich, in the vast research area of Garching: a jewel just a few kilometers from the infamous Dachau concentration camp. In the meantime, other countries had joined the European organization: Denmark, Switzerland, and also Italy.[21] Today the number of member states has grown to 16.

A very extensive test campaign was carried out to select the site for the telescopes that ESO planned to build. The exceptional weather conditions of the Chilean Andes won out,[22] wisely judged more important than the advantages offered by a pre-existing colonization of South Africa by European astronomers (especially British, who at the time were not yet members

---

[20]In chapter 21 of his *The Voyage of the Beagle*, Charles Darwin writes: «Among the other most remarkable spectacles which we have beheld, may be ranked, the Southern Cross, the cloud of Magellan, and the other constellations of the southern hemisphere»: see https://www.literature.org/authors/darwin-charles/the-voyage-of-the-beagle/chapter-21.html.

[21]For a detailed history of ESO, see C. Madsen, *The Jewel on the Mountaintop: The European Southern Observatory through Fifty Years*, Wiley-VCH, Weinheim 2012.

[22]See the beautiful collection of images in G. Schilling, L.L. Christensen, *Europe to the Stars: ESO's First 50 Years of Exploring the Southern Sky*, Wiley-VCH, Weinheim 2012.

of ESO). The final decision favored a 2,400 m high mountain called Cerro La Silla in the southern extension of the Atacama Desert, 600 km north of the capital Santiago de Chile. The first relevant astronomical facility, commissioned in 1977, was a traditional telescope with a 3.6 m aperture: one of the most powerful instruments of its time, but of an old design that could not take full advantage of the unique features of the Andean site.[23] For this reason, ESO, which had rapidly grown in importance and in economic and human resources, decided to focus its efforts on the new technology of thin monolithic mirrors that could be deformed by the application of appropriate forces: a concept developed by a prince of opticians, the Englishman Raymond (Ray) Wilson.[24] NTT was designed around this idea. Its primary mirror, a light and very thin Zerodur meniscus with a diameter of 3.6 m and a thickness of only 24 cm, and with a short focal length ($F/2.2$) to facilitate the rigidity of the supporting structure, rested on an active system of 75 axial actuators with 3 fixed points and 24 lateral supports. In all, a set of pistons and levers capable of correcting aberrations by responding to the commands of an image analyzer pointed at a bright star used as a reference. It was this "activation" of the primary optics, traditionally passive, which, among other things, made the quality of the reflecting surface less critical; a significant step forward in the unrelenting struggle of astronomers against defects in optical systems. Active positioning was also applied to the secondary and tertiary mirrors of the two Nasmyth foci, using robots capable of both translation and rotation, i.e. operating in all five degrees of freedom.[25] Following the path traced by MMT, the altazimuth mount was protected by a compact dome that rotated with the telescope, designed to minimize the thermal effects that cause dome seeing.

Commissioned in 1989, NTT was a great success[26] and quickly set a new standard. It was a short-lived supremacy, however. The following year, NASA launched into low Earth orbit an optical/IR telescope designed

---

[23]The telescope was completely upgraded in 1999 and since 2008 equipped with only one focal plane instrument, the High Accuracy Radial velocity Planet Searcher (HARPS) dedicated to the search for extrasolar planets.

[24]R.N. Wilson, *The History and Development of the ESO Active Optics System*, in "ESO Messenger", 113, 2003, pp. 2–9.

[25]The movement of the secondaries was then entrusted to hexapods, six-legged instruments used, for example, in the TNG, LBT, and VST telescopes, and which also finds applications in fields quite different from astronomy, such as neurosurgery and flight simulators.

[26]B. Goss Levi, *New Technology Telescope Actively Corrects for Misalignments*, in "Physics Today", 43, 5, 1990, pp. 17–18.

primarily to accurately calibrate the cosmic distance scale, and thus named in memory of Edwin Hubble. The flagship of the American space agency,[27] the Hubble Space Telescope (HST) had an aperture of only 2.4 m and suffered from an initial manufacturing defect that was corrected by astronaut specialists in 1993 with a spectacular extravehicular activity,[28] the first such maneuver in space history. But despite these limitations, it had one feature that made it unique. It benefited from the reduced presence of the atmosphere at an altitude of 540 km above sea level. The astonishing resolution of its images — which became a scientific propaganda tool for ordinary people thanks to a clever public relations policy pursued by NASA and the contribution of the Hubble Heritage Project, an institute with a mission that straddles science and art — cornered ground-based telescopes, which had much larger collecting areas (at the same cost) but were hampered by air turbulence. To keep the terrestrial instruments in the game, a practical way had to be found to limit the devastating effects of seeing. But how?

Conceptually, the problem has a fairly simple solution. Consider, for example, the wavefront of an intrinsically point-like source such as a star. When it reaches the Earth, it is essentially flat, because the spherical surface over which the radiation is distributed has an almost infinite radius of curvature relative to the pupil of the telescope. Therefore, if the beam were intercepted and focused by an ideal instrument in the absence of an atmosphere, the image in the focal plane would have the resolution given by the aperture of the telescope.[29] But in reality, even if we ignore the unavoidable deficiencies of the optics and the receiver, things are much worse. The turbulence of the air layers just above the telescope corrugates the (almost) plane wave so that its shape changes at a frequency of 100 Hz or more. You can try to visualize this with a sheet hung out to dry and waving

---

[27]D.J. Shayler, D.M. Harland, *The Hubble Space Telescope: From Concept to Success*, Springer, Chichester 2016.

[28]The mission was possible because the Hubble Space Telescope was deliberately placed at an altitude accessible to the shuttle and thus to servicing teams.

[29]In general, when radiation emitted by a monochromatic point source at infinite distance is focused by a circular optical system, such as a mirror or lens, due to the effects of diffraction, the resulting image is a spot with an angular radius ϕ_o that is directly proportional to the wavelength λ of the incoming radiation and inversely proportional to the diameter D of the optical system: $\phi_o = 1.22 \times \lambda/D$ radiant. For example, a 10 m aperture telescope observing in the green ($\lambda = 0.5$ micron) will ideally produce images of about 0.01 arc seconds. To get the same result at $\lambda = 5$ cm in the radio domain, the diameter of the mirror would have to be 100,000 times larger.

in the wind. The challenge is to flatten it in real time and with continuity, following the whims of the air vortices. The winning strategy for stars is to reflect the distorted wave off a mirror deformed in the opposite direction, so that the sum of the two defects cancels them out. It is a very tricky operation, but not impossible. The principle is the same as in active noise-control headsets. All you need is an instrument that maps the distortions generated by the atmosphere onto a point-like source. The result, updated one hundreds of times per second, is then applied to create the "negative" of the distortion on a thin reflective plate. The military interest in such a device[30] has stimulated the practical implementation of terrestrial units capable of previously inconceivable resolutions.

Among the first examples of telescopes designed to operate with an active secondary is the Large Binocular Telescope (LBT).[31] This large reflector, armed with two twin mirrors, is located on Mount Graham, a 3,300 m high peak in the Pinaleno Mountains of southeastern Arizona. When the project was conceived in the 1980s, it was called "Columbus" because its completion was expected to coincide with the 500th anniversary of the discovery of America. Technical problems due to the extraordinary nature of its design as well as the opposition to its use of the site by environmentalists and members of the Apache tribe of the San Carlos Reservation, who consider the site sacred, delayed its first light until January 2008. Today, the imposing observatory is a joint venture of the United States (50%), Germany (25%), and Italy (25%). The main optics of the LBT — an instrument designed for observations in the visible and near infrared — consist of two 8.4 m monolithic mirrors made by Roger Angel of the Steward Observatory at the University of Arizona in Tucson using an innovative technique based on silicon carbide and carbon fiber. Each blank was lightened by a honeycomb structure and preformed by slowly rotating the tank during the long post-casting cooling of the glass mass. The two primary mirrors, supported by a system of active actuators, rest on a common altazimuth structure built by Ansaldo Camozzi in Milan, Italy, and can operate as a single instrument or as two independent telescopes sharing the same pointing. In combined mode, the collecting power becomes that of a single 11.8 m aperture pupil, and since the centers of the two mirrors

---

[30] J.W. Hardy, *Adaptive optics for astronomical telescopes*, Oxford University Press, Oxford 1998.

[31] J.M. Hill *et al.*, *The Large Binocular Telescope, in Ground-based and Airborne Telescopes II*, Proceedings SPIE, 7012, 2008, 701203.

are 14.4 m apart, in interferometric mode the resolution is equivalent to that of a 22.8 m monolith.

Beyond its remarkable dimensions and technological innovations, the real specialty of the LBT is its adaptive secondaries, which make this instrument competitive even with the Hubble Space Telescope in terms of resolution. These are Zerodur menisci, each nearly a meter in diameter and only 1.4 mm thick, truly ultra-fragile sheets of tissue paper aluminized on both sides. They can be deformed in real time under the control of a computer fed by a wavefront sensor. The mechanical operation is performed by a few hundred electromagnets coupled to as many magnets glued to the back of each meniscus. The forces exerted at different points on the thin glass plate shape the light-exposed surface in order to match the deformations of the wavefront. The instrument produces spectacular results, but it has some important drawbacks. It is critical, difficult to operate, and has a narrow field of view (only partly mitigated by the large primary collecting area compared to HST). An additional problem is the need for a bright star on which to perform the wavefront analysis. The template must be very close to the position of the target, since the characteristics of the seeing change with the direction of the line of sight. But bright stars are rare. The obstacle has been overcome with artificial light sources created by high-powered lasers that excite the sodium layers of the atmosphere at an altitudes of about 90 km above sea level. This is why today we can see intense green beams fired into the night sky from the dome of LBT or other large telescopes in the Chilean Andes or on the summit of Mauna Kea, to compensate for the lack of bright templates[32] and control the "wrinkles of light".

But how do we measure wavefront distortion? There are several ways. One widely used sensor is called Shack-Hartmann, from the names of the two scientists who contributed most to the design and construction of the instrument: Roland Shack, an American optical engineer with a Ph.D. from England and a long association with the Optical Sciences Center at the University of Arizona, and Johannes Franz Hartmann, a German astronomer and physicist, director of the La Plata Observatory in Argentina between the two world wars and discoverer of the existence of diffuse matter scattered among the stars (the so-called interstellar medium). The working principle of the Shack-Hartmann wavefront sensor had already been

---

[32] Cf. N. Ageorges, C. Dainty (eds.), *Laser Guide Star Adaptive Optics for Astronomy*, Springer, Berlin 2010.

outlined in the 1620 *Oculus, hoc est: Fundamentum opticum* (The eye, that is: The optical foundation) by the Austrian Jesuit Christoph Scheiner, a true genius of optics. It is based on the point-by-point sampling of the wavefront tilt by a matrix of identical microlenses. They focus the image of the pupil of the primary optics illuminated by a reference star (real or artificial). The idea is to simulate a telescope made up of as many apertures as there are lenses, so that each one looks at a different part of the wave; a bit like a team of traffic cops, each one busy watching the part of town assigned to them. Let's look at one of these microlenses. If the corresponding part of the wavefront is not flat, the image of the star will be focused at a different point than the projection of the optical axis of the microlens. The map of these deviations ("tilt" in the jargon) allows us to reconstruct the geometry of the entire wavefront and to derive, through appropriate mathematical models and look-up tables, the matrix of forces to be applied to the adaptive secondary. The same deviations can be measured by focusing the light of the telescope on the vertex of a glass pyramid which acts like an image splitter. This solution provides higher sensitivity because the sensor sees the focused light from the entire telescope rather than from individual sub-apertures.[33]

In its various technological forms, adaptive optics corrects for the effects of atmospheric turbulence. Active optics, on the other hand, as implemented for the first time at the ESO NTT, keeps under control the deformations of the thin primaries (menisci) due to weight, temperature variations, and manufacturing defects, by acting from time to time during an observing night on the actuators that support the mirror both axially and laterally. Of course, none of this would have been possible without the intervention of electronics in the form of fast processors, sophisticated optomechanical devices, and digital detectors to interpret the information coming from the wavefront sensors, calculate the corrections, and apply them in real time.

Modern detectors for astronomical use appeared at the University of Illinois in the early 1900s, following the variable star experiments of Joel Stebbins,[34] who, with American humor, claimed that his wife had motivated him to find a way to speed up the work and get home earlier.[35] In the beginning,

---

[33] See R. Ragazzoni, *Pupil Plane Wavefront Sensing with an Oscillating Prism*, in "Journal of Modern Optics", 43, 1996, pp. 289–293.

[34] K. Krisciunas, *A Brief History of Astronomical Brightness Determination Methods at Optical Wavelength*, 2001; https://arxiv.org/pdf/astro-ph/0106313.pdf.

[35] A.E. Whitford, *Stebbins Joel, 1878–1966: a Biographical Memoir*, National Academy of Sciences, Washington (DC) 1978, p. 297.

selenium photocells were used; special tubes inserted into an electric circuit and capable of regulating the intensity of the current according to the amount of light falling on a metal plate called the "cathode". They were replaced by the photoelectric cells of the Swiss Jakob Kunz, up to 200 times more sensitive. In any case, these were all relatively omnivorous analog photometers (which could be made monochromatic by special filters) with no resolving power. They measured all the incoming light through an aperture set by the observer, giving a photometric accuracy of better than 1%. At the end of World War II, this technology was greatly improved by the development of photomultiplication, a cascade amplification of the number of electrons released by the photoelectric effect at the cathode. The quantum efficiency, i.e., the relative number of photons detected, which does not exceed 1% in the best photographic emulsions, could reach 10%. Since there is always another side to every coin, the new technique generated the need to harmonize the photoelectric measurements of the stellar fluxes with those made photographically, both for the coexistence of the two types of detectors and for the conservation of the previous, vast database in the visual and photographic spectral bands:[36] a heritage that could not be alienated also because it bears witness to the sky of past epochs.

The real leap in quality came when optical astronomers got their hands on charge-coupled digital detectors (CCDs), which had been developed at Bell Labs in the late 1950s for possible military use. The two-dimensional version, which can return a matrix of numbers representing the signal on a grid of pixels, was tested in 1970, patented the following year, and improved ever since. It works like this.[37] When light strikes the photosensitive surface of the CCD, it releases pairs of particles of opposite charge, negative (electrons) and positive (called holes). The negative charges are accumulated in an underlying layer, within potential wells corresponding to the pixels, physically marked by suitable gate electrodes. This process continues additively throughout the exposure time on the CCD, at least until saturation occurs, i.e., until the "container" does not overflow. At the end of the exposure, the charge packets accumulated in each well, with a value

---

[36] A spectral band is a region of the electromagnetic spectrum bounded between two wavelengths (or frequencies). The visual band corresponds to the sensitivity range of the human eye and is centered around the yellow-green color. The photographic band, on the other hand, corresponds to the sensitivity range of the first emulsions created for astronomical use, called orthochromatic (blue-sensitive).

[37] Cf. M. Lasser, *A Summary of Charge-Coupled Devices for Astronomy*, in "Publications of the Astronomical Society of the Pacific", Vol. 127, No. 957, 2015, pp. 1097–1104.

proportional to the input signal, are read by gradually shifting the rows of the pixel matrix to bring them one by one into an auxiliary reading register. At the output of this register, an analog-to-digital converter turns the potential differences into codes that can be read by a computer. If that sounds complicated, you're damn right it is. But this is exactly what happens on your cell phone when you take a picture.

CCDs have revolutionized the field of optical detectors (and not only). They are panoramic like photographic plates, but have a quantum efficiency 100 times higher, a linear response,[38] and a digital output that can be readily interfaced to a computer. But they also have some problems with use (they need to be cooled to prevent hot electrons from contaminating the signal) and with their physical dimensions. While CCDs are now commercially available in remarkably large formats (thousands of pixels per side), they are inadequate for collecting the light from telescopes with fields of view on the order one square degree or more and resolutions below arcseconds. But be careful! One might be led to believe that the more "tiny" the pixel, the higher the resolution of the collected image, but this is not the case. Information theory warns us to avoid spatial sampling steps that are too large, so as not to lose detail (although in doing it there is a gain in signal-to-noise ratio), but also too small, so as not to introduce unwanted additional noise. The new monsters, such as the VLT Survey Telescope (VST)[39] — a wide-field reflector at the Cerro Paranal Observatory in northern Chile, built by the National Institute for Astrophysics (INAF, the Italian public agency for astronomy and astrophysics) in a joint venture with ESO and the international consortium OmegaCAM —, use CCD mosaics that bring the total number of pixels to over one billion.

Thanks to increasingly powerful and efficient telescopes, new focal-plane instruments (such as spectrographs, which, like octopuses, use their fiber-optic tentacles to collect light from many stars simultaneously and record their spectra in a single exposure), sophisticated computer techniques for data analysis and querying huge databases (both data and software), and the extension of observations to the entire electromagnetic spectrum, from

---

[38]Linearity, i.e., the direct proportionality between input and output, is an important characteristic of CCDs that links this class of detectors to photoelectric photometers. Photographic emulsions, instead, give a response that has a logarithmic trend in the interval between between underexposure and overexposure.

[39]M. Capaccioli, P. Schipani, *The VLT Survey Telescope Opens to the Sky: History of a Commissioning*, in "ESO Messenger", 146, 2011, pp. 2–7.

gamma rays to long radio waves, and to other "messengers" such as neutrinos and gravitational waves, using space observatories, interferometers,
and underground laboratories, astrophysics and cosmology have taken a
resounding step forward in the last half century, solving several puzzles and
proposing new ones. And even if the fundamental questions remain the
same, the answers we give today are far more compelling and clear than
the paradigms that seemed unbreakable only yesterday (cf. the quote at
p. 541).

For example, we have confirmed that the Sun is not the only star
crowned by a planetary system. In 1995, Michel Mayor and Didier Queloz[40]
have discovered the existence of a "hot Jupiter" in close orbit around the star
51 Pegasi, using the same meticulous technique that led Bessel to hypoth-
esize the presence of a companion in orbit around Sirius. Quite a coup
indeed, which earned the two Swiss astronomers a belated Nobel Prize in
Physics in 2019. Since then, through *ad hoc* observations from the ground
and mainly from space, and with all the tricks of the trade,[41] thousands of
other extrasolar planets invisible to direct observation have been found.
Indications of their presence have been, for example, the tiny periodic
eclipses produced by the transit of a cold body in front of the primary,
observable when the plane of the orbit is seen (almost) edge-on. The most
sought-after prey of this safari are the Earth-sized planets around solar-
type stars, located at distances from their primaries such that the water, if
any, is in a liquid state: the so-called "habitable zones". The hidden hope is
that these bodies may harbor the same life that flourished on Earth 3.9 billion years ago, perhaps intelligent and technologically advanced. The "little
green men" that were for a moment suspected to be the cause of the pulsar
phenomenon have not yet been found. On the other hand, the dense spinning tops made of neutrons have given us many surprises and have provided
a valuable tool for validating the general theory of relativity, together with
the black holes so often mentioned, about which, after this indigestion of
technologies, a digression is in order.

Singularities created in 1916 by Karl Schwarzschild's exact solution of
the field equations of general relativity outside of a spherical mass, black
holes are mathematical monsters to which observations have conferred

---

[40]M. Mayor, D. Queloz, *A Jupiter-Mass Companion to a Solar-Type Star*, in "Nature",
378, 1995, pp. 355–359.
[41]*How to find an extrasolar planet*; https://www.esa.int/Science_Exploration/
Space_Science/How_to_find_an_extrasolar_planet.

the dignity of physical entities.[42] They are massive entities contained within their own horizon, an ideal spherical surface that marks the boundary between real space-time and that non-world from which not even light can escape. The first convincing evidence for their existence at the stellar-mass level was provided by Tom Bolton in 1971. Here's how it happened.

In the early 1960s, the Geiger counters of some suborbital rockets launched from the White Sands Missile Range in New Mexico had detected the existence of an X-ray source, later named Cygnus X1. An astonishing fact, since no one had imagined that stars, as thermal machines, could produce a large flux of high-energy photons. In order to clarify the matter, it was necessary to carry out a systematic search to enlarge the sample. An Italian physicist, Riccardo Giacconi, took care of this.[43] In 1963, as head of a research group at a private company, the American Science and Engineering based near Boston, he conceived the idea of a satellite to capture from space those celestial X-rays that the atmosphere does not let pass through. The team already had some experience with the specific instrumentation, having worked from 1961 to 1962 for a U.S. Department of Defense program to detect high-altitude nuclear explosions in violation of international treaties. The Cold War continued to stimulate space ventures. One of the problems to be solved was the optics for focusing the high-energy photons which, instead of being reflected like the weaker photons (UV, optical, etc.), penetrate the surface of conventional mirrors. Fortunately, in the same way that flat stones thrown above water in a pond do not sink but bounce, the X-rays are also reflected when their incidence is grazing. The loophole required an unusual geometry of the mirrors, more like funnels than soup bowls as for ordinary photons.

After the technical problems were resolved, the satellite Uhuru ("freedom" in the Swahili language, *lingua franca* of the African Great Lakes) was launched from the Italian platform San Marco, off the coast of Kenya, to be placed in a low orbit at an altitude of 540 km. From December 1970 to March 1973, it detected more than 300 X-ray sources, opening a new window on an unknown class of celestial phenomena.[44] Something

---

[42]K. Thorne, *Black Holes and Time Warps: Einstein's Outrageous Legacy*, W.W. Norton & Co., New York 1994.

[43]*Riccardo Giacconi Biographical*, The Nobel Prize, https://www.nobelprize.org/prizes/physics/2002/giacconi/biographical/.

[44]Cf. H. Gursky, R. Ruffini, L. Stella (eds.), *Exploring the Universe: A Festschrift in Honor of Riccardo Giacconi*, World Scientific Publishing Co., Singapore 2000.

similar to what Galileo had done with his telescope 360 years earlier, for which Giacconi won the Nobel Prize in Physics in 2002. Since then, many more satellites with improved technology and different flags have followed Uhuru's example. Among their many discoveries are gamma-ray bursts: massive explosions of energy associated with a new class of stars called hypernovae,[45] of which the classic supernovae would be only the poor cousins.

Uhuru also revealed that the emission from Cygnus X1 oscillates with a periodicity of 0.3 seconds. Given the need to ensure a causal connection to the phenomenon,[46] these periodic hiccups reduced the extension of the region where the energy is produced to only 100,000 kilometers, thus putting normal stars out of the picture! It was necessary to find an explanation for a situation similar to that of a normal household fountain that produces as much water as the Iguazú Falls. But how?

In 1971, some Dutch and American radio astronomers succeeded in associating the X-ray source with a supergiant star, but it was too large and too cold to produce the large photon flux observed. The enigma was solved by Bolton and, at the same time, by two Britons from the Royal Greenwich Observatory, with the further discovery that the star had a compact companion. This was to be the mysterious source responsible for the shocking temperatures. As it could not be a neutron star, since its mass exceeded the gravitational limits dictated by Oppenheimer and Volkoff's rule (p. 440) for this class of objects, only one hypothesis remained: that it was a supermassive black hole.[47] A little scientific masterpiece that Einstein probably

---

[45]Cf. *It's Brighter than an Exploding Star, It's a Hypernova!*, 1999, https://imagine.gsfc.nasa.gov/news/20may99.html.

[46]The components of a physical system must be causally connected to coordinate themselves in order to realize, for example, a global change in brightness. In other words, if on July 14, 1789 we wanted all the Parisians *sans-culottes* to participate in the storming of the Bastille, we would have to wait for the time necessary for them to be duly informed, wherever they were in the city. Now, this time depends on the speed at which news travels and on the size of the region in which the information must be spread. Since nothing travels faster than light, if the phenomenon is contained in a time interval Δt, the region to be coordinated must have a size Δr not greater than that covered in the same time span by the luminous messenger, i.e., $\Delta r \leq c \times \Delta t$, where c is the speed of light.

[47]C.T. Bolton, *Identification of Cygnus X-1 with HDE 226868*, in "Nature", 235, 1972, pp. 271–273; K. Culp, *The Proof Is Out There: In the early 1970s, black holes were only the subject of scientific speculation. Then astrophysicist Tom Bolton began to ponder the matter*, in "University of Toronto Magazine", https://magazine.utoronto.ca/research-ideas/science/astrophysics-tom-bolton-black-holes-research/.

would not have digested well! No wonder: even the greatest geniuses sometimes run the risk of indigestion.

But how does this happen? To understand this, one must have at least
a rudimentary understanding of the evolutionary history of stars; a scenario built in the second half of the 20th century through the synergistic
collaboration of observers and theorists and the use of electronic computers to solve systems of equations that resist analytical solutions. The main
parameter that conditions and determines the fate of stars is their initial
mass. The most massive stars live splendidly but briefly, ending their existence with a bang. Those with low mass survive for a long time and die
of old age, slowly fading through the phases of white and gray dwarf, in
a kind of unpaid retirement. For all stellar sizes (from 0.08 to 150 solar
masses), however, life begins with the gravitational collapse of a gas cloud
that heats the protostar's core and culminates in the ignition of a central
nuclear furnace. The fusion of hydrogen into helium provides the energy to
replace the radiative losses and maintains the *status quo* for a long time. In
fact, after an initial settling phase, each newborn star reaches a perfect balance between gains and losses, and keeps this state for millions or billions
of years, as long as the core is depleted of the fuel (a longevity that depends
on the initial mass). After the hydrogen in the center is exhausted, the core
collapses in an attempt to reach a higher temperature suitable for burning
helium, a less abundant but more demanding fuel than hydrogen. At the
same time, the outer shell expands. This is because the star has now lost
the feedback mechanism that, during the hydrogen-burning phase, allowed
it to calibrate energy production as needed, thus keeping the antagonistic forces in balance. Therefore, when the envelope can no longer control
the flow of heat produced by the collapse of the core, it is forced to swell
in order to dissipate it. Since (at the same temperature) the luminosity
increases in proportion to the emitting surface, the star, although cooler,
becomes a very bright red supergiant due to its hypertrophic photosphere.
It is during this phase of overwhelming expansion that a star can transfer a
fraction of its mass to a nearby companion, if any, which often does not like
the gift and gets rid of it with periodic minor explosions. If the recipient
is a compact object, such as a neutron star or a black hole, the matter
released by the supergiant is arranged in a disk from which it plummets
toward its new master in an increasingly violent spiral. In the deadly vortex
inside the space curved by the degenerate object, the mutual interactions
of the kamikaze particles raise the temperature to the millions of degrees
necessary to produce X-rays. This is the phenomenology of Cygnus X1.

A similar mechanism, though on a completely different scale, has been invoked to decode quasars and all those cores of galaxies that astronomers have named Active Galactic Nuclei (AGN)[48] because their energy does not seem to come, as for the rest of the nebula, from the usual components, i.e. stars, dust, and interstellar gas. The phenomenology of AGNs is very rich, varied, and complex, and includes morphology as well as luminosity and spectral properties.[49] In all cases, the puppeter is a giant black hole with a mass of millions or even billions of suns, confined to a space smaller than the orbit of Pluto. This celestial Minotaur feeds on the matter that falls upon it, allowing only the light produced by the interaction of matter in free fall to escape (because the emission takes place above the horizon). If the black hole rotates and has a magnetic axis, it can also develop a jet that stretches to enormous distances before fading away, up to several percent of the radius of the host galaxy, where ionized matter travels at nearly the speed of light.

What happens to these Pantagruels in times of famine? When put on a forced diet, the monster goes quiet, which explains why the number of galaxies per unit volume with active nuclei changes with distance, i.e. with lookback time. Food becomes increasingly scarce because it has been consumed and not replaced. So no more fireworks! But as long as you get close enough to the silent monster, you can still appreciate its gravitational effects, so much so that in 1978 a team of Caltech astronomers led by the British astronomer Wallace Sargent[50] detected the signs of these tugs in the hectic traffic of the stars near the center of M87, the giant nebula that dominates the Virgo cluster; a galactic nucleus pierced, like an astral Saint Sebastian, by a magnificent jet discovered by Curtis at Lick in 1918. Who is "frightening" the stars — the authors of the paper wondered — forcing them into a paroxysmal agitation far greater than would be sufficient for them if visible matter were the only source of gravitational pull. The only plausible explanation led to a hypothetical black hole with a mass of 5 billion suns: a gamble, but also a prophetic intuition, later confirmed by

---

[48] H. Netzer, *Active Galactic Nuclei: Basic Physics and Main Components*, in D. Alloin, R. Johnson, P. Lira (eds.), *Physics of Active Galactic Nuclei at all Scales*, Springer, Berlin, 2006, pp. 1–38.

[49] For more information, see B.M. Peterson, *An Introduction to Active Galactic Nuclei*, CUP, Cambridge 1997.

[50] W.L.W. Sargent, P.J. Young, C.R. Lynds, A. Boksenberg, K. Shortridge, *Dynamical evidence for a central mass concentration in the galaxy M87*, in "Astrophysical Journal", 221, 1978, pp. 731–744.

observations of numerous other galaxies.[51] A touch of color: a year earlier, in 1977, the same interpretation, based only on photometry, had been proposed for the galaxy NGC 3379 by Gérard de Vaucouleurs, Massimo Capaccioli, and Peter Young in a paper submitted to the "Astrophysical Journal". Its acceptance was strongly opposed by a reviewer who later published in the same journal the discovery of the supermassive black hole in the very same galaxy. That's how the world works!

In summary, black holes can be detected by their victims, which excite themselves when they come close to the monster and thus get inflamed (by mutual interactions) before disappearing into the non-universe. But nobody really believed that it was possible to photograph this extreme hell. Instead, the extraordinary feat was achieved in 2019, thanks to the synergy of numerous radio telescopes around the Earth, which, virtually set up to form a worldwide interferometer, have reached the required degree of resolution.[52] The project, called the "New Horizon Telescope, was led by Shepard Doeleman of Harvard's Center for Astrophysics. It captured the shadow of the black hole in M87, confirming previous mass estimates and proving once again that Einstein was right in writing his equations, but wrong in rejecting with some disgust the solution found by his friend Schwarzschild. The same team has also set its sights on the black hole lurking at the center of the Milky Way. This elusive object has been pursued for years also by another team of astronomers led by a professor at the Ludwig Maximilian University in Munich, the German astrophysicist Reinhard Genzel (who will share the 2020 Nobel Prize with the American astronomer Andrea Ghez and the British theorist Roger Penrose). Today we know that the black hole of the Milky Way is a medium-sized object, only a thousandth of the mass of the M87 black hole and even less well fed, but still able to put a feather in the cap of the stars orbiting it. These unfortunates are forced to travel at nearly 10% of the speed of light as they pass near their pericenter: a fantastic speed that would take us to the Moon in a fistful of seconds.

Black holes are physical singularities, non-places inaccessible to experience, whose horizons are semi-permeable but imperfect membranes. Steven

---

[51]F. Melia, *The Galactic Supermassive Black Hole*, Princeton University Press, Princeton 2007.
[52]K. Akiyama *et al.*, *First M87 Event Horizon Telescope Results. I. The Shadow of the Supermassive Black Hole*, in "The Astrophysical Journal Letters", 875, L1, 2019, pp. 1–17.

Hawking — whose scientific genius few know how to assess, while most are instead aware of the severe handicap that plagued him throughout his life — has shown that such stingy and omnivorous objects have an Achilles' heel. They can evaporate, and thus lose weight, at first extremely slowly, then faster and faster, until they disappear with a violent sneer. The process could involve very small black holes created by the cosmos at its very birth.[53] Astronomers are still scanning the early universe for signs of the traumatic disappearance of the micro-monsters. Will they find them? Hard to say. But in a different and completely unexpected way, they have recently found black holes of intermediate size, a few tens of solar masses.

The discovery came when the herculean effort to measure Einstein's ripples in spacetime caused by a violent and asymmetric release of energy succeeded with the first detection of gravitational waves.[54] Scientists had chased the signal for decades, knowing full well that it was extremely difficult, if not impossible, to measure the marginal effects of its passage drowned in a cocktail of natural and anthropogenic noise: a real needle in a haystack. Repeated failures of the interferometers built for this purpose in the USA and Italy, with airless and kilometer-long arms, ultra-powerful lasers, and top-of-the-line optics suspended on thin dampers, seemed to confirm the position of those who considered the search utopian, if not a useless waste of time and money. Then, on September 14, 2015, a barely perceptible shudder tickled the twin instruments of the Laser Interferometer Gravitational-Wave Observatory (LIGO) in the United States to announce the merger of two black holes of 29 and 36 solar masses, respectively, that occurred 1.4 billion years ago somewhere in the deep space. It was the first detection of a series of dramatic and sensational events that culminated on August 14, 2017 with the spectacle of a deadly encounter between two neutron stars: magnificent combatants whose merger, unlike that of two very stingy black holes, also produces observable effects in the domain of electromagnetic waves. The search for their counterparts has ushered in an era of multiband collaboration that has been called, perhaps with a hint of exaggeration, multimessenger astronomy, in which the various

---

[53]For more information see: V. Frolov, I. Novikov, *Black Hole Physics: Basic Concepts and New Developments*, Springer, Berlin 1998.

[54]G. Schilling, *Ripples in Spacetime: Einstein, Gravitational Waves, and the Future of Astronomy*, Harvard University Press, Harvard, 2017; M. Coleman Miller, N. Yunes, *The new frontier of gravitational waves*, in "Nature", 568, 2019, pp. 469–476.

"messengers" are photons, gravitational waves, neutrinos, and cosmic rays, properly observed, compared, and interpreted.[55]

What comes next? It is reasonable to predict that the successes of LIGO and the Italian-French Virgo Observatory could convince the governments of the richest and most powerful nations to invest in a giant space-based interferometer: an equilateral triangle of three satellites in heliocentric orbit, one million kilometers apart, with the goal of listening to the space-time vibrations of the early universe, all the way back to the time of the Big Bang. This will happen one way or another, unless a foolishly suicidal humanity squanders resources on absurd and cruel wars. In the meantime, gravitational waves have given yet another satisfaction to Einstein, who was skeptical about the detectability and usefulness of the phenomenon, and yet another Nobel Prize in Physics, split in two parts: one half to the German Rainer Weiss and the other half to the Americans Barry Barish and Kip Thorne, all in their 80s, for their «crucial contribution to the LIGO detector and to the observation of gravitational waves»,[56] as the motivation for the award states. It just goes to show that it is never too late!

Medium-size black holes have further complicated the picture of the population of these off-limits entities: dark chasms from which there is no return, scattered across the fabric of a seemingly shining universe. It has indeed been authoritatively written: *"Fiat lux"*. But in fact, light seems to be a minor component of the universe. Today, few doubt that 85 percent of all cosmic matter is in a dark, unknown form, of which only its repulsion at any interaction with electromagnetic waves is clearly recognized. This is the same ingredient (or phenomenon) identified in 1937 by Fritz Zwicky as a conspicuous mass deficit[57] in the Coma cluster of galaxies (see also p. 442). The irascible genius reached this conclusion by estimating the amount of matter required to maintain the gravitational equilibrium of the cluster's "gas of galaxies", the temperature of which he had deduced from measurements of the radial velocities of the nebulae. Zwicky's result implied a percentage of dark matter that far exceeded direct observations of the mass of the Milky Way and those of individual galaxies derived from their internal dynamics. A premature discovery, weakened by the criticism

---

[55] Cf. A. Neronov, *Introduction to Multi-Messenger Astronomy*, 2019, https://arxiv.org/pdf/1907.07392.pdf.

[56] https://www.nobelprize.org/prizes/physics/2017/summary/.

[57] F. Zwicky, *On the Masses of Nebulae and of Clusters of Nebulae*, in "Astrophysical Journal", 86, 1937, pp. 217–246.

of the equilibrium hypothesis adopted in the analysis of the data and, unfortunately, by the antipathy of the scientific community towards the author. Even in science, reason sometimes gives way to passion. In short, the times were not ripe for dark matter.[58] Things changed in the early 1970s, when radio astronomers discovered, thanks largely to observations of the 21 cm line of neutral hydrogen, that galaxies exhibit a gravitational response at their outskirts far in excess of that implied by their visible masses. A little like seeing a huge weight lift itself into the air pulled up seemingly only by an infant. This fundamental and now topical discovery is associated with the name of a tenacious and intelligent woman, Vera Cooper Rubin,[59] and his lifelong collaborator Kent Ford.

Born in Philadelphia in 1928 to Jewish immigrants from northeastern Europe, Vera enjoyed an excellent education, but she soon experienced firsthand the persistence of stubborn gender discrimination even in the democratic scientific institutions of North America. Hateful obstacles, such as being denied admission to Princeton, and trivial vetoes, such as not being able to go to her advisor's office because that area of Cornell University was off-limits to the fairer sex. Thus, in parallel with her scientific research, she began her polite but determined fight for equal rights for women, which consisted of emblematic words and actions that have marked the path of feminism. An example for all.[60] When Vera arrived at Palomar headquarters to observe with the 200 inch telescope, she was told she could not stay on the mountain because there were no women's restrooms (an argument also used in other circumstances by the hardened male chauvinists; see p. 297). Undaunted, she chose one of the men's toilets and marked it with an unmistakable message written in her own handwriting in block letters on an A4 sheet of paper: WOMEN!

Since her graduate thesis, Vera had been involved in the study of the redshifts of galaxies, looking for deviations from the uniform expansion discovered by Hubble. She thus confirmed the existence of the Local

---

[58]A. Del Popolo, *Invisible Universe, The: Dark Matter, Dark Energy, and the Origin and End of the Universe*, World Scientific Publishing, Singapore 2021; S. van den Bergh, *A Short History of the Missing Mass and Dark Energy Paradigms*, in "Historical Development of Modern Cosmology", ASP Conference Proceedings, Vol. 252, 2001, pp. 75–83; G. Bertone, D. Hooper, *History of Dark Matter*, FERMILAB-PUB-16-157-A, https://arxiv.org/pdf/1605.04909.pdf.

[59]V.C. Rubin, *An Interesting Voyage*, in "Annual Review of Astronomy and Astrophysics", 49, 2011, pp. 1–28.

[60]M. McKinnon, *How colleagues remember astrophysics pioneer Vera Rubin*, https://astronomy.com/news/2017/01/vera-rubin-remembered.

Supercluster proposed by de Vaucouleurs in the 1950s: a megastructure of thousands of galaxies centered on the Virgo cluster, of which the Local Group and our Milky Way are a peripheral extension. But her main contribution to cosmology concerns the internal dynamics of individual galaxies. Her interested was shared by a couple of British astrophysicists, Margaret and Geoffrey Burbidge.[61] The trio, consisting of two female observers and a man devoted to theory, used the 82 inch telescope of the McDonald Observatory in West Texas, not far from El Paso, and obtained rather classic results: a valuable work but nothing revolutionary.

The turning point came in the mid-1960s, when Kent Ford, who was essentially an engineer, offered Vera the opportunity to use a spectrograph equipped with an innovative image intensifier tube to reach the faint peripheral regions of nebulae. The instrument, developed for U.S. defense purposes, converted photons to electrons so that they could be multiplied and then converted back to photons with a net gain in sensitivity. The new instrument enabled the two scientists to discover[62] that in the extragalactic *banlieues* the rotational velocities required for gravitational equilibrium remained constant with increasing distance from the center, instead of decreasing as in the planets of the Solar System (the so-called "Keplerian decrement"). At very beginning, few trusted these observations. Then, thanks to growing evidence and the emergence of other needs — not least that of cosmologists to have a glue that would allow galaxies to form and resist the disintegrating action of pressure — the idea of a physical entity that cannot be seen, but whose gravitational effects are nevertheless felt, gained more and more ground.

Dark matter acquired further fascination for the vast family of physicists when it became clear that this ingredient could not share the nature of ordinary matter; the one that today we call "baryonic", of which the atoms and all the things we perceive and "see" are made, both on Earth and in the sky. In the course of its rapid evolution, the early universe did not have enough time to form baryons in sufficient quantity to match observations.

---

[61] Margaret Burbidge, incidentally, was the first woman to become president of the prestigious American Astronomical Society and might have been the first woman to hold the title of Astronomer Royal, which for three centuries had been reserved for the director of the Greenwich Observatory. However, when she was appointed director of the observatory in 1972, the prestigious title was spun off and given to Martin Ryle. Another case of blatant gender discrimination, made all the more bizarre by the fact that it occurred in a country ruled by a queen.

[62] V. Rubin, *Bright Galaxies Dark Matter*, Copernicus Books, Göttingen 1997.

Dark matter had to be something else. So physicists began to search and fantasize about this new ingredient, but still in vain, because no one really knows what it could be. It is claimed that it must be relatively cold, move at modest speeds relative to light (which implies that "dark particles" have relatively large individual masses), and be non-collisional, i.e., not interacting either with itself or with ordinary matter (but gravitationally). The hunt for the culprit, conducted both in underground laboratories, such as that of the Italian National Institute for Nuclear Physics (INFN) dug into the bowels of Gran Sasso, and with specialized instruments launched into space, has so far yielded no results, despite the considerable number of candidate particles considered by theoretical physicists.

This absence of smoking guns has encouraged the search for solutions to satisfy appearances that are alternative to the existence of a pure material constituent. This has generated, for example, some proposals for a revision of Newtonian dynamics by the followers of the Israeli physicist Mordekai Milgrom. According to these new "heretics", in weak force fields acceleration is no longer proportional to the acting force.[63] Few believe it, but this lack of consensus is not a good reason to reject a theory, as the "*damnatio*" of heliocentrism in the classical and medieval world demonstrates. Rather, Modified Newtonian Dynamics, abbreviated as MOND, should ward off the blows of a certain lack of formal elegance and, in particular, explain the existence of phenomena such as the bullet cluster,[64] which seems to falsify it mercilessly. A look at this case introduces us to the use of another powerful tool of modern astrophysics, gravitational lensing, already experienced by Arthur Eddington & Co. in 1919 to validate Einstein's theory of relativity through a solar eclipse.

In fact, our star acts as a very modest but very useful lens. Far more effective are the giant galaxies at the center of large clusters, whose enormous masses bend space just enough to bring into focus, albeit poorly, objects far away in space and time: distant nebulae whose images are both amplified (i.e. made visible) and reduced to thin arcs that crown the gravitational lenses. Thanks to its keen eye, the Hubble Space Telescope has found many of them. What do we do with these features? One now common use is to reconstruct the gravitational field that produced them, thus providing

---

[63]Cf. R.H. Sanders, S.S. McGaugh, *Modified Newtonian Dynamics as an Alternative to Dark Matter*, in "Annual Review of Astronomy and Astrophysics", 40, 2002, pp. 263–317.
[64]D. Paraficz *et al.*, *The Bullet cluster at its best: weighing stars, gas, and Dark Matter*, in "Astronomy & Astrophysics", 594, A121, 2016, pp. 1–14.

an independent measure of the mass of the lens. Impressive halos of matter have been discovered in this way. Let's come to the case of the bullet cluster.[65] The visible and X-ray images collected by space telescopes reveal the consequences of a titanic collision between two clusters of galaxies. While the baryons, represented by the intracluster gas, are still fighting over the accident site, the dark matter components, identified by the gravitationally lensed background objects even in places where there is no starlight left, have passed through unscathed. A sign of their total apathy, as expected. Needless to say, MOND's followers still found a way to parry the blow, and so the evidence has been downgraded from suggestive to circumstantial.

In short, does dark matter really exist? Vera Rubin hoped not. «If I could have my pick, I would like to learn that Newton's laws must be modified in order to correctly describe gravitational interactions at large distances. That's more appealing than a universe filled with a new kind of sub-nuclear particle».[66] If this were really the case, extragalactic astrophysicists and cosmologists would be left with a lit match in their hands, because they would not know how, for example, spiral nebulae, which are fragile structures threatened by gravitational perturbations, are kept in shape. They would not be able to explain how giant galaxies and clusters manage to hold a restless gas (extremely hot, millions of degrees), and how matter that tries to collapse into galaxies in the early universe can avoid being dispersed by radiation pressure. Without dark matter, it seems very hard to explain the large-scale structure of baryonic matter that has been revealed by photometric and spectroscopic surveys and reproduced so well in numerical simulations. It is true that the inability to find an alternative explanation is not in itself a strong reason to accept the veracity of a hypothesis that provides satisfactory answers but lacks the positive detection of its main actor. However, in the absence of a better one, most astronomers act as if dark matter really does exist.

The narrative becomes fast-paced because there are so many recent astronomical discoveries worth telling about, involving men, instruments, techniques, successes, failures, and bluffs, ingenious tricks and brilliant theories. Sensational facts, such as the discovery of a large number of dwarf planets beyond the orbit of Uranus, which degraded Pluto and upgraded Ceres, to the posthumous delight of Father Piazzi; the detailed and plausible

---

[65] See the image at the address https://apod.nasa.gov/apod/ap170115.html.
[66] *History of Scientific Women. Vera Rubin*, https://scientificwomen.net/women/rubin-vera-86.

reconstructions of the history of the cosmos, rendered visually by the movies of the "Illustris Project", one of the many cosmological simulations of the processes of galaxy formation, using the most advanced numerical codes and the most advanced physical knowledge;[67] the interpretation of gamma-ray bursts capable of producing XXL photons, each carrying more energy than a trillion quanta of visible light;[68] visits to the planets of the Solar System and the landing of robotic probes on asteroids and comets;[69] the discovery of the Higgs boson, the God particle that gives mass to other particles;[70] the census of more than a billion stars in the Milky Way by the Global Astrometric Interferometer for Astrophysics (Gaia) satellite, ESA flagship;[71] the discovery of the secondary evolution of galaxies, showing that even in the firmament the adage *mors tua vita mea* (your death is my life) applies, with interactions ranging from simple harassment to crude cannibalism.[72] Examples of compelling advances in knowledge, neither exhaustive of the state of the art nor respectful of its merits, flanked by a prodigious development of the tools of observation, measurement, and interpretation. But perhaps the field that has benefited most from this impressive leap forward in knowledge of the physical world concerns cosmology itself, that is, the science that studies the whole with the aim of explaining its origins and transformations. We will limit ourselves to mentioning two recent revolutions in this field,[73] concerning the first minutes of the universe's life and the bizarre run-up that began 4.5 billion years ago.

Half a century ago, Penzias and Wilson's serendipitous discovery had confirmed a Friedman-type model of the cosmos characterized by a decelerated expansion of the primordial fluid starting from a singular state of extreme temperature and density. A triumph, considering how elementary cosmology was at the beginning of the 20th century. The excitement over this discovery was short-lived, however, because new observations of the

[67] See *The Illustris Simulation. Towards a predictive theory of galaxy formation*, https://www.illustris-project.org/.

[68] P. Schady, *That gamma-ray flashes and their use as cosmic probes*, in "R. Soc. Open sci.", 4, 2017.

[69] *NASA: Solar System Exploration*, https://solarsystem.nasa.gov/.

[70] L.M. Lederman, C.T. Hill, *Beyond the God Particle*, Prometheus Books, Amherts (NY) 2013.

[71] G. Gilmore, *Git Gaia: three-dimensional census of the Milky Way Galaxy*, in "Contemporary Physics", 59, 2, 2018, pp. 155–173.

[72] P. Schneider, *Extragalactic Astronomy and Cosmology: An Introduction*, Springer, Berlin 2014.

[73] Cf. D.H. Lyth, *The History of the Universe*, Springer, Berlin 2016.

ice-cold cosmic microwave background (CMB) confirmed the new paradigm of a universe in the making. The U.S. satellites COBE (short for Cosmic Background Explorer) and WMAP (Wilkinson Microwave Anisotropy Probe), and more recently the European satellite named after Max Planck, have revealed an unexpected uniformity and isotropy of the CMB temperature from one end of the sky to the other, with differences of only 0.001 percent. A paradoxical situation, like a perfect performance by an orchestra without a conductor. In fact, it turned out that two diametrically opposed regions of the sky could not communicate being further apart than the distance the fastest messenger, light, can travel in the lifetime of the cosmos. The mystery of the "violated incommunicability" was solved in the early 1980s by the Russian Alexei Starobinsky and the American Alan Guth. They appealed to an initial paroxysmal exponential expansion that, by inflating space to excess, would have causally separated cells of the cosmos that had already gotten to know each other anyway. In short, today the cosmic Riccardo Muti is no longer on the podium, but the concert players had the opportunity to practice with him at the very beginning, and they remember it. Guth christened the phenomenon "inflation" in analogy to the rise in prices following one of the many political-military crises in the Middle East.

A revolutionary idea that, in addition to solving the problem of an inexplicably coherent horizon, gave cosmologists a number of life-saving donuts. First of all, the reason why, among so many possible curvatures of space, the universe had chosen the only one corresponding to the familiar Euclidean space. One of those peculiar situations that scientists neither like nor dislike. Instead, inflation explained this suspicious behavior in a natural way, as a consequence of local flattening. To capture the essence of the idea, let us use a two-dimensional analogy. If you blow up a balloon, the surface becomes locally less and less curved as the radius increases, until it appears almost flat. Another bonus of Guth's expedient concerns the justification for the absence of magnetic monopoles, those strange particles with a single pole instead of the usual two (i.e., with a net magnetic charge). They are predicted by the Grand Unified Theory,[74]. which unifies the three fundamental non-gravitational interactions into a single force, and the physics

---

[74]For more information see S. Raby, *Supersymmetric Grand Unified Theories. From Quarks to Strings via SUSY GUTs*, Springer, Cham, 2017.

of superstrings,[75] which interprets the cosmos as a very complex stringed instrument. So monopoles are in hiding because they are diluted by the overwhelming inflation. In fact, it would be quite difficult to find a single raisin out of a handful scattered in a plum pudding that has grown so much that it has doubled in size a hundredfold. It sounds like nothing, but try doing it by folding a sheet of paper the same number of times!

Wonderful! Yet neither Guth nor Starobinsky have received the Nobel Prize for this sensational idea of a newborn cosmos literally exploding in space and time (though both have made up for it with other prestigious and rich awards, including the Kvali Prize in 2014). In fact, the smoking gun is still being sought in this case as well. The hope of cosmologists would be to find signs of the immense initial bang in the form of fossil gravitational waves, but the attempts made so far have only produced a resounding false alarm: the hasty announcement of a non-existent discovery, in 2015. A sign of the times, where the need to appear burns away the most basic prudence, and the sense of the ridiculous does not frighten. «I don't mind you thinking slowly, Doctor, but I do mind you publishing faster than you think»;[76] words that the prickly Wolfgang Pauli used to admonish his students, and which today, perhaps because of the large number of researchers, sometimes seem to have been forgotten.

The other fundamental cosmological revolution of our time concerns the way in which the expansion of the space-time substrate proceeds. Until 1998, no one imagined that the universe had the resources to start an engine capable of accelerating it. It was believed, based on observations, that space gained volume at the expense of an initial impulse that was gradually consumed without ever being cancelled out by the lack of a consistent gravitational pull. In short, matter did not seem dense enough to stop the race and force space into a ruinous collapse, a Big Crunch. However, the problem of accurately measuring the expansion parameters, in particular the Hubble constant H_o, remained open. Using the best available instruments and the flashes of selected supernovae treated as "standard candles", two teams led respectively by Saul Perlmutter in the USA and Brian Schmidt and Adam Riess in Australia showed that the universe has not slowed down,

---

[75] B. Greene, *The Elegant Universe: Superstrings, Hidden Dimensions, and the Quest for the Ultimate Theory*, W.W. Norton & Co., New York 2003.

[76] R. Peierls, *Bird of Passage: Recollections of a Physicist*, Princeton University Press, Princeton (NJ) 1985, p. 47.

but accelerated in the last third of its lifetime.[77] The discovery, which was awarded a triple Nobel Prize in 2011, poses the problem of the engine and the fuel, solved by postulating the existence of a dark energy that would take a slice equal to 3/4 of the current matter-energy budget. A *deux ex machina*[78] conjured out of the hat of that great magician who is the subatomic world, where the void is not empty at all and actually has resources that grow in proportion to the expansion of space. It sounds like the ravings of a drunkard. Instead, there are several different observations that confirm the paradigm of a universe where ordinary matter is now little more than a scattered trace in a sea of dark matter. Is there no other way out? Perhaps there is, for example by starting a revision of general relativity. An alternative solution to dark energy could be an intrinsically curved space, since it would appear to us in the same way as a flat metric shaped by a component that is unknown (for the simple reason that it does not exist).

Why this obsessive search for flaws in theories? Because scientific truth is not absolute, and the best opportunities for the growth of knowledge arise precisely when cracks in current paradigms, caused by the discrepancy between facts and models, are discovered. This could be the case with the Hubble constant, which has been measured with astonishing accuracy, but has different values depending on the method used. We normally encounter such a situation in experimental practice, with different determinations all falling within the global error bar. It becomes pathological in the case of H_o if, as it seems, the measurements referring to very different cosmic times are estimated to be very precise and yet differ more than their errors. If confirmed, this anomaly will require a drastic revision of the theory that predicts the constancy of this cosmological parameter.[79] We will also need more powerful and sharp eyes to peer even deeper and more detailed into the cosmos, using mediators of all kinds in multi-messenger mode to create win-win synergies. Novelties are already on the horizon, such as the 39 m aperture adaptive optics telescope that ESO will install on Cerro Armazones in Chile's Atacama Desert: a giant called the Extremely Large Telescope; or the James Webb Space Telescope, HST's big brother, with a mosaic of gold

---

[77]Adam Riess' *Nobel Lecture* in https://www.nobelprize.org/uploads/2018/06/riess_lecture.pdf.

[78]Latin phrase referring to the god who descends to earth and solves the situation. The phrase comes from the performance of Greek tragedies, where the actor playing the god was positioned on a suspended, moving platform called a *mechane*.

[79]W.L. Freedman, *Cosmology at a Crossroads: Tension with the Hubble Constant*, https://arxiv.org/ftp/arxiv/papers/1706/1706.02739.pdf.

mirrors for an aperture of 6. 5 m and record-breaking resolution, placed at a magical point in space opposite the Sun (named L_2 after the mathematician Lagrange) where it can be anchored to remain aligned with the Earth; or the grandiose Square Kilometer Array (SKA), a system of radio antennas so huge that. it must be housed simultaneously on two continents as far apart as Africa and Australia, and which will require the invention of new solutions to handle as much Internet traffic as to day the rest of the world. There is plenty to marvel at.

Our story, which ideally began with the first humans looking curiously at the sky, has come to an end for the moment. What we have seen in these and the preceding pages represents only a glimpse of the state of the art in astrophysics, cosmology, and related technologies. Most likely, in the short time between the completion of the manuscript and the printing of the book, new initiatives, new machines, and new discoveries will be added to those we have mentioned, in a process of growth that is now almost exponential. As the Belgian-born American chemist and historian George Sarton, considered the founder of the history of science as an independent discipline, wrote:

> The saints of today are not necessarily more saintly than those of a thousand years ago; our artists are not necessarily greater than those of early Greece; they are likely to be inferior; and, of course, our men of science are not necessarily more intelligent than those of old; yet one thing is certain, their knowledge is at once more extensive and more accurate. The acquisition and systematization of positive knowledge is the only human activity that is truly cumulative and progressive. [...] Progress has no definite and unquestionable meaning in other fields than the field of science.[80]

Although it is an increasingly arduous task for those who are called upon to metabolize new discoveries and their consequences, research is still a wonderful experience. «To me there has never been a higher source of earthly honor or distinction than that connected with advances in science», Newton seems to have declared.[81] So let us rejoice! But let us also be vigilant that

---

[80] G. Sarton, *The History of Science and the New Humanism*, Transaction Books, New Brunswick 1988. Cf. also M.J. Crowe, *Modern Theories of the Universe: from Herschel to Hubble*, Dover, New York 1994, p. 2.

[81] I. Ushakov, *Histories of Scientific Insights*, A. Mortensen (ed.), Lulu.com, 2007, p. 141.

this sublime activity of the human mind, a true umbilical cord with the cosmos, does not end up slipping out of our hands and surrendering itself, in the glory of its power, to a soulless oligarchy. Only in this way we will be able to continue the ascent towards a knowledge that is functional for happiness, without which everything loses its meaning.

To conclude, I can find no better words than those of Christiaan Huygens in his *Cosmotheoros: or, conjectures concerning the inhabitants of the planets*:

> How vast those Orbs must be, and how inconsiderable this Earth, the Theatre upon which all our mighty Designs, all our Navigations, and all our Wars are transacted, is when compared to them. A very fit consideration, and matter of Reflection, for those Kings and Princes who sacrifice the Lives of so many People, only to flatter their Ambition in being Masters of some pitiful corner of this small Spot.

Acknowledgments

While the other animals look prone to the earth,
the man was given a face turned upwards and
his gaze aims at the sky and rises to the stars

Publius Ovidius Naso, *Metamorphoses*

The eternal silence of these infinite spaces
terrifies me

Blaise Pascal. *Thoughts*

The drafting of the Italian version of this book began in the winter of 2011 — when I was visiting the European Southern Observatory in Garching near Munich — to fill the long lonely nights with stories of great men of the past, while I watched the snow fall from the large window of my apartment. It was still snowing when I wrote the last sentences as a guest at Karazin University in Kharkiv, Ukraine. It took more snow than I had hoped to complete the work: blame my ignorance, the vastness of the subject, and perhaps the unconscious desire not to abandon the seductive navigation in the sea of history. It is easier and less painful to learn what you do not know than to confront what you think you know.

I used the opportunity of the English translation to improve and enrich the text. The downside of this process is that in correcting errors and expanding topics, I have probably created new ones, to the point where I wonder if there is a law of conservation of errors.

«What grows old early? Gratitude», said the wise Aristotle. A bitter and often true remark, which does not apply here and now. In fact, it is a real pleasure for me to acknowledge the many people who have helped me, consciously or unconsciously, to complete this project. First of all, my late father Anzio, who, among the many teachings he passed on to me by word and example, instilled in me a love of knowledge and a passion for a panoramic vision of history; my beloved mother Lilian, in whose sweetest eyes I first saw the sky and to whose memory I dedicate this work; Leonida Rosino, my learned and unforgotten teacher in Padua; and my other master, Gérard de Vaucouleurs. I remember with nostalgia the vivid stories of the greats of the twentieth century whom he knew and visited, and about whom he told me in the rare moments when we paused from work, under the patio of his house in Austin, Texas, with a glass of chilled white wine, strictly French.

Then I have to thank Rosanna, who accepted my mental and physical absences to work on the book (and now on the translation) and listened to my frequent readings of fragments, especially those I liked least and for which I needed her opinion, politely critical; my daughter Barbara, who is the purpose of my life; Alessandro Bettini, the colleague and friend from Padua, always punctual and sharp in his remarks; Filippo Zerbi, who knows the past of astronomy and manages the future; Pietro Schipani and Roberto Regazzoni, who know telescopes because they design and build them; Elena Bannikova, for lively and endless discussions on the history of astronomy and the principles of physics; my friend Margaret Rasulo, a native English-speaking university lecturer, and my dear colleague Ginevra Trinchieri, who generously tried to rid my text of errors and inevitable "Italianisms"; and the many with whom I have been confronted on specific issues, whom I do not mention so as not to risk offending anyone. Perhaps I should also thank the great musicians whose masterpieces kept me company as I tried to line up the facts and characters of a long and complicated story and make it clearer, first of all, to myself.

Finally, and only for chronological reasons, a heartfelt thank you to Luisa Castellani, who sifted through the Italian manuscript, identifying with enviable acuity inaccuracies, errors, logical or linguistic lapses, suggesting improvements and mitigating with polite firmness my aversion to bibliographical rigor; and to Gianluca Mori, an expert helmsman on whom a traveler always willingly relies.

Index

Printed in the United States
by Baker & Taylor Publisher Services